PRECALCULUS
with
EARLY TRIGONOMETRY

3rd edition

BY

G. VIGLINO   and   M. BERGER

RAMAPO COLLEGE OF NEW JERSEY
April, 2017

CONTENTS

CHAPTER 1 BASIC CONCEPTS

1.1 Exponents 1

1.2 Algebraic Expressions 8

1.3 Unit Conversion 23

1.4 Lines, Linear Equations and Inequalities 29

1.5 Systems of Linear Equations 41

1.6 Quadratic Equations 52

Chapter Summary 62

Chapter Review Exercises 66

Cumulative Review Exercises 70

CHAPTER 2 AN INTRODUCTION TO FUNCTIONS AND GRAPHS

2.1 General Concepts 73

2.2 The Graph of a Function 88

2.3 One-To-One Functions and their Inverses 104

Chapter Summary 115

Chapter Review Exercises 119

Cumulative Review Exercises 122

CHAPTER 3 TRIGONOMETRIC FUNCTIONS OF ANGLES

3.1 Trigonometric Functions of Acute Angles 125

3.2 Right Triangle Applications 134

3.3 Radian Measure and Oriented Angles 144

3.4 Trigonometric Functions of Oriented Angles 154

Chapter Summary 164

Chapter Review Exercises 168

Cumulative Review Exercises 169

CHAPTER 4 TRIGONOMETRIC FUNCTIONS OF A REAL VARIABLE

4.1 The Trigonometric Functions 171

4.2 Trigonometric Identities 184

4.3 Inverse Trigonometric Functions 198

4.4 Trigonometric Equations 211

Chapter Summary 218

Chapter Review Exercises 221

Cumulative Review Exercises 223

ii   Contents

CHAPTER 5 SOLVING OBLIQUE TRIANGLES

5.1   Law of Sines and Law of Cosines 225
5.2   Oblique Triangle Applications 235
Chapter Summary 242
Chapter Review Exercises 243
Cumulative Review Exercises 244

CHAPTER 6 POLYNOMIAL FUNCTIONS

6.1   Quadratic Functions 245
6.2   Polynomial Functions and Equations 263
6.3   Polynomial Functions and Inequalities 276
6.4   Graphing Polynomial Functions 287
Chapter Summary 294
Chapter Review Exercises 297
Cumulative Review Exercises 301

CHAPTER 7 RATIONAL FUNCTIONS

7.1   Rational Functions, Equations, and Inequalities 303
7.2   Graphing Rational Functions 320
7.3   Partial Fractions 331
Chapter Summary 339
Chapter Review Exercises 342
Cumulative Review Exercises 345

CHAPTER 8 ABSOLUTE VALUE AND RADICAL FUNCTIONS

8.1   Absolute Value Functions, Equations, and Inequalities 347
8.2   Radical Equations and Inequalities 358
Chapter Summary 370
Chapter Review Exercises 372
Cumulative Review Exercises 373

CHAPTER 9 EXPONENTIAL AND LOGARITHMIC FUNCTIONS

9.1 Exponential Functions 375
9.2 Logarithmic Functions 387
9.3 Additional Exponential and Logarithmic Equations and Inequalities 401
9.4 Exponential Growth and Decay 411
9.5 Additional Applications 422
Chapter Summary 434
Chapter Review Exercises 439
Cumulative Review Exercises 441

APPENDIX A CONIC SECTIONS A-1

SOLUTIONS TO CHECK YOUR UNDERSTANDING BOXES B-1

ANSWERS TO ODD-NUMBERED EXERCISES C-1

INDEX I-1

iv

PREFACE

This book is designed to provide a sound foundation for subsequent mathematics and math-related courses.

Chapter 1 presents a comprehensive review of basic algebraic concepts, and Chapter 2 offers a general introduction to functions and graphs. Chapters 3 through 5 are devoted to a study of trigonometry. With its early introduction, we are able to reinforce trigonometric concepts throughout the remainder of the text.

Included in the text are numerous ***Check Your Understanding*** boxes with problems that challenge the student's understanding of newly introduced concepts. Detailed solutions of those problems appear in an Appendix. Exercises at the end of each section, covering the gamut of difficulty, provide ample opportunities to hone mathematical skills. Moreover, at the end of each chapter, in addition to ***Chapter Review Exercises*** there are ***Cumulative Review Exercises*** addressing topics of the current and previous chapters. A Student Solutions Manual is available.

Graphing calculator glimpses primarily designed to illuminate concepts appear throughout the text. In the final analysis, however, one can not escape the fact that

MATHEMATICS DOES NOT RUN ON BATTERIES
A case in point:
While graphing calculators can certainly graph most functions better and faster than any of us, learning to sketch them by hand requires an understanding of important concepts, and serves to reinforce those concepts.

Dear Student: May we suggest that after you feel you understand a concept, you revisit a related example in the text, write down the problem, close the book, and try to solve it on our own. You can then compare your effort with the solution in the book. When attempting a Check Your Understanding problem, only turn to its solution in the appendix after making a valiant effort to solve it by yourself or with others. In the words of Descartes:

We never understand a thing so well, and make it our own, when we learn it from another, as when we have discovered it for ourselves.

Finally, we wish to thank our colleagues, Professors Sara Kuplinski and Anthony Russo for their invaluable input.

CHAPTER 1
Basic Concepts

A robust calculus student is not likely to grow from poor algebraic soil. A sound algebraic foundation provides necessary manipulative skills, and in the process helps you to develop the ability to think and reason in logical steps. We therefore suggest that you give this first chapter your serious consideration.

We begin with a review of the properties of exponents in Section 1, and then move on in Section 2 to a consideration of polynomial, rational, and radical expressions. This is certainly not going to be your first encounter with these topics, but you really need to learn them well, for they have been the undoing of many an unwary calculus student. Conversions between different units is essential in the sciences, and is discussed in Section 3. Sections 4 and 5 deal with linear and quadratic equations, respectively.

§1. EXPONENTS

Mathematics is, in part, a visual art, and much of its power stems from the fact that a great deal of information can be represented compactly. An example of this is the familiar exponent form:

DEFINITION 1.1

INTEGER EXPONENTS

For any positive integer n and any number a:

$$a^n = \underbrace{a \cdot a \cdots \cdot a}_{n\text{-times}}$$

a raised to the nth power

In addition, if $a \neq 0$:

$$a^0 = 1 \quad \text{and} \quad a^{-n} = \frac{1}{a^n}$$

For example:

$$2^{-4} = \frac{1}{2^4} = \frac{1}{2 \cdot 2 \cdot 2 \cdot 2} = \frac{1}{16}, \text{ and } 2^0 = 1$$

If a and b are not zero, then: $\left(\dfrac{a}{b}\right)^{-1} = \dfrac{1}{\frac{a}{b}} = \dfrac{b}{a}$

Also: $\dfrac{2a^{-3}}{3b^{-4}} = \dfrac{2b^4}{3a^3}$ since: $\dfrac{2a^{-3}}{3b^{-4}} = \dfrac{2 \cdot \frac{1}{a^3}}{3 \cdot \frac{1}{b^4}} = \dfrac{\frac{2}{a^3}}{\frac{3}{b^4}} = \dfrac{2}{a^3} \cdot \dfrac{b^4}{3} = \dfrac{2b^4}{3a^3}$

2   Chapter 1 Basic Concepts

While a is a **factor** of ab, it is not a factor of $ab + c$. It is, however, a factor of $ab + ac$, since:

$$ab + ac = a(b + c)$$

When you evaluate $3 \cdot 2^2$ you do not square the 3. Likewise, don't square the -1 in:

$$-2^2 = (-1) \cdot 2^2$$

Answers: (a) $\frac{3}{8}$ (b) 16

As you can see, a **factor** of the numerator was moved to the denominator and vice versa by simply changing the sign of the exponent. But please don't try these maneuvers with non-factors. For example:

$$\frac{2 + a^{-3}}{3 + b^{-4}} \neq \frac{2 + b^4}{3 + a^3} \qquad \text{(just substitute } a = -1 \text{ and } b = 1\text{)}$$

We also caution you that while $(-2)^2 = (-2)(-2) = 4$, -2^2 is **not** equal to 4:

$$-2^2 = -(2 \cdot 2) = -4$$

CHECK YOUR UNDERSTANDING 1.1

Evaluate:

(a) $\dfrac{(2 - \frac{1}{3})^0 2^{-3}}{3^{-1}}$

(b) $\left[\dfrac{-2^{-2} + 2^{-1}}{(-2)^2} \right]^{-1}$

ROOTS AND RATIONAL EXPONENTS

To find the nth root of a number a, is to find that number b whose n^{th} power equals a. To be more precise:

DEFINITION 1.2

n^{th} ROOTS

The nonnegative restriction, is imposed because no real number raised to an even power can be negative.

If n is an **odd** positive integer and a is any number, then the nth **root of a** is defined to be that number, denoted by $a^{\frac{1}{n}}$, such that:

$$\left(a^{\frac{1}{n}} \right)^n = a$$

If n is an **even** positive integer, then a must be nonnegative, and the above notation is reserved for the nonnegative n^{th} root of a (called the **principal** n^{th} root of a).

In the case that $n = 2$, $a^{\frac{1}{2}}$ can also be denoted by $\sqrt{a}$, and is called the (principal) **square root of a.**

For example:

$$(-32)^{\frac{1}{5}} = -2 \quad \text{since: } (-2)^5 = -32$$

$$\left(\frac{1}{8}\right)^{\frac{1}{3}} = \frac{1}{2} \quad \text{since: } \left(\frac{1}{2}\right)^3 = \frac{1}{8}$$

$$16^{1/4} = 2 \quad (\textbf{NOT} \pm 2) \quad \text{and} \quad \sqrt{25} = 5 \quad (\textbf{NOT} \pm 5)$$

Having defined expressions of the form a^m and $a^{\frac{1}{n}}$ for integers m and n, we are now able to extend the exponent notation to all rational numbers:

> A **rational number** is an expression of the form $\frac{m}{n}$ where m and n are integers, with $n \neq 0$.

DEFINITION 1.3

RATIONAL EXPONENTS

For $\dfrac{m}{n}$ a positive rational number in lowest terms (no common factors), and any a if n is odd, or any nonnegative a if n is even:

$$a^{\frac{m}{n}} = \left(a^{\frac{1}{n}}\right)^m = (a^m)^{\frac{1}{n}}$$

In addition, if $a \neq 0$, then:

$$a^{-\frac{m}{n}} = \frac{1}{a^{\frac{m}{n}}}$$

> The denominator of the fractional exponent indicates the root, while the numerator is the power.

For example:

$$8^{\frac{2}{3}} = \left(8^{\frac{1}{3}}\right)^2 = 2^2 = 4 \quad \text{or, if you prefer: } 8^{\frac{2}{3}} = (8^2)^{\frac{1}{3}} = 64^{\frac{1}{3}} = 4$$

$$25^{-\frac{3}{2}} = \frac{1}{25^{\frac{3}{2}}} = \frac{1}{(\sqrt{25})^3} = \frac{1}{5^3} = \frac{1}{125}$$

We again underline the fact that, in the real number system, expressions of the following form are **NOT** defined:

$$(-8)^{\frac{3}{2}}$$

negative   even root

CHECK YOUR UNDERSTANDING 1.2

Evaluate:

> Answers: (a) 1   (b) $\dfrac{71}{8}$

(a) $\sqrt{9} - 16^{\frac{1}{4}}$

(b) $(-27)^{\frac{2}{3}} - 4^{-\frac{3}{2}}$

4   Chapter 1 Basic Concepts

EXPONENT RULES

The following results can be shown to hold for all **rational exponents** r and s:

THEOREM 1.1
EXPONENT RULES

In each of the following, when all expressions are defined, we have:

(i) $\quad a^r a^s = a^{r+s}$ — When multiplying add the exponents

(ii) $\quad \dfrac{a^r}{a^s} = a^{r-s}$ — When dividing subtract the exponents

(iii) $\quad (a^r)^s = (a^r)^s = a^{rs}$ — A power of a power: multiply the exponents

(iv) $\quad (ab)^r = a^r b^r$ — A power of a product equals the product of the powers

(v) $\quad \left(\dfrac{a}{b}\right)^r = \dfrac{a^r}{b^r}$ — A power of a quotient equals the quotient of the powers

In particular, from (iv) we see that:

$$\sqrt{8} = \sqrt{4 \cdot 2} = \sqrt{4}\sqrt{2} = 2\sqrt{2}$$

CANCELLATION LAW
For $c \neq 0$

$$\frac{a\cancel{c}}{b\cancel{c}} = \frac{a}{b}$$

Because equations are two-way streets, you can read (iv) from right to left, and see, for example, that:

$$16^3 \cdot \left(\frac{1}{8}\right)^3 = \left(16 \cdot \frac{1}{8}\right)^3 = \left(\frac{2 \cdot \cancel{8}}{\cancel{8}}\right)^3 = 2^3 = 8$$

Please remember that the rules of exponents deal with products and **NOT** with sums. For example:

$$\sqrt{9 \cdot 16} = \sqrt{9}\sqrt{16} = 3 \cdot 4 = 12 \quad \text{BUT: } \sqrt{9+16} \neq \sqrt{9} + \sqrt{16}$$
$$5 \quad\ \neq 3 + 4$$

$$(2 \cdot 3)^2 = 2^2 \cdot 3^2 = 4 \cdot 9 = 36 \quad \text{BUT: } (2+3)^2 \neq 2^2 + 3^2$$
$$25 \quad\ \neq 4 + 9$$

Note also that the exponent rules of Theorem 1.1 hold only when all expressions are defined. In particular, rule (iii) $(a^r)^s = a^{rs}$ does not hold if $a = -5$, $r = \frac{1}{2}$, and $s = 2$

$$(\sqrt{-5})^2 \neq -5$$

$$\left[(-5)^{\frac{1}{2}}\right]^2 \text{ is NOT equal to } (-5)^{\frac{1}{2} \cdot 2} = (-5)^1 = -5$$
$$\underset{\text{undefined}}{\uparrow}$$

Henceforth: When instructed to "simplify" an algebraic expression, it is assumed that the variables in the expression are restricted to those values for which the expression is meaningful, that expressions in the denominator are not zero, and that the answer is to contain only positive exponents.

EXAMPLE 1.1 Simplify:

$$\frac{-3a^5\sqrt{a}}{\left(2a^{\frac{4}{3}}\right)^3}$$

SOLUTION: The first step is to write the radical in exponent form, and then apply the appropriate exponent rules:

$$\frac{-3a^5\sqrt{a}}{(2a^{\frac{4}{3}})^3} = \frac{-3a^5a^{\frac{1}{2}}}{2^3a^{\frac{4}{3}\cdot 3}} = \frac{-3a^{5+\frac{1}{2}}}{8a^4} = \frac{-3a^{5+\frac{1}{2}-4}}{8} = -\frac{3}{8}a^{\frac{3}{2}}$$

EXAMPLE 1.2 Simplify:

$$\left[\frac{(4b-2a)^2}{3a-6b}\right]^3$$

SOLUTION: First factor and then apply the exponent rules:

Since $x^2 = (-x)^2$, you can multiply any expression that is being squared (or raised to any even power) by -1. In particular:
$$(4b-2a)^2 = [-(4b-2a)]^2$$
$$= (2a-4b)^2$$

$$\left[\frac{(4b-2a)^2}{3a-6b}\right]^3 \overset{\text{see margin}}{=} \left[\frac{(2a-4b)^2}{3(a-2b)}\right]^3 = \left[\frac{[2(a-2b)]^2}{3(a-2b)}\right]^3$$

$$= \left[\frac{2^2(a-2b)^2}{3(a-2b)}\right]^3 = \left[\frac{4(a-2b)}{3}\right]^3$$

$$= \frac{4^3(a-2b)^3}{3^3} = \frac{64}{27}(a-2b)^3$$

CHECK YOUR UNDERSTANDING 1.3

Simplify:

Answers: (a) $\dfrac{1}{a^{\frac{17}{6}}}$

(b) $\dfrac{1}{512}(a-b)^{12}$

(a) $\dfrac{\sqrt{a}}{\left(-a^{-2}a^{\frac{1}{3}}\right)^{-2}}$

(b) $\left[\dfrac{(b-a)^2(a-b)^5}{(2a-2b)^3}\right]^3$

6   Chapter 1 Basic Concepts

E X E R C I S E S

Exercises 1–6: Evaluate each of the following.

1. $\dfrac{2^{-1}+2}{9\cdot 3^{-1}}$

2. $\dfrac{\left(3-\frac{1}{2}\right)^2 3^{-1}}{4^0\cdot\frac{1}{3^2}}$

3. $\dfrac{3^{-2}-2^{-3}}{2^4(-3^{-2})}$

4. $\dfrac{2^{-1}-3^{-1}}{3^{-1}+2^{-1}}$

5. $-[-(2^{-3}-3^{-2})]^{-1}$

6. $(-2)^2-2^2+\dfrac{-16}{\left(1-\frac{1}{2}\right)^{-2}}$

Exercises 7–16: Evaluate the expression, if it is defined. If it is not defined, explain why.

7. $-9^{\frac{1}{2}}$

8. $(-9)^{\frac{1}{2}}$

9. $-27^{-\frac{2}{3}}$

10. $\left(-16^{\frac{1}{2}}-64^{\frac{1}{3}}\right)^{\frac{1}{3}}$

11. $32^{-\frac{2}{5}}-(-27)^{\frac{1}{3}}$

12. $\sqrt{18}-\sqrt{8}$

13. $(-16)^{\frac{3}{4}}-8^{\frac{1}{3}}$

14. $-8^{\frac{2}{3}}+(-8)^{\frac{2}{3}}-\left(\frac{1}{16}\right)^{-\frac{3}{4}}$

15. $\sqrt{8}+\sqrt{50}+\sqrt{32}$

16. $\left(\frac{1}{3}-\frac{2}{9}\right)^{\frac{1}{2}}$

Exercises 17–41: Simplify.

17. a^3a^5

18. a^3a^{-5}

19. $\dfrac{a^{-3}}{a^5}$

20. $b\left(\dfrac{a}{b}\right)^5$

21. $(ab)^3$

22. $-(-2a)^2-(-a)^4$

23. $\dfrac{(ab)^2}{a^3(-b)^3}$

24. $\dfrac{(-ab)^2a^3}{(ab)^3}$

25. $\dfrac{-b^2a^3}{(ba)^2}$

26. $\dfrac{(ab)^2+a^2}{(ab)^3}$

27. $\dfrac{(a^2b^3)^4}{a^8b^4-a^9b^5}$

28. $\dfrac{a^{-2}b^{-3}(-b)^2}{(ab)^{-2}b^{-2}}$

29. $\left[\dfrac{(-a)^2(-b)^3}{(-a^2b^3)^2}\right]^4$

30. $\left[\dfrac{(ab)^3(ab^2)^3}{b^{-2}a^2}\right]^{-2}$

31. $\dfrac{\sqrt{a}}{a^3}$

32. $a^{\frac{1}{3}}a^{\frac{3}{2}}$

33. $\dfrac{a^{\frac{2}{3}}a^{-\frac{1}{2}}}{a^{\frac{5}{6}}}$

34. $\dfrac{b^2}{a}\left(\dfrac{a}{b}\right)^{\frac{2}{3}}$

35. $\dfrac{(2a-2b)^2(b-a)^3}{(-a+b)^{-\frac{1}{3}}(-b+a)^4}$

36. $\dfrac{(-a-b)^3}{(2a+2b)^2}\cdot\dfrac{\sqrt{b+a}}{\left(\frac{1}{2}b+\frac{1}{2}a\right)^{-1}}$

37. $\dfrac{b^{\frac{1}{2}}a}{c^{-\frac{1}{4}}}\cdot\dfrac{c^{-2}}{(2ab)^3}$

38. $\dfrac{a^{\frac{3}{2}}b^{-2}c^{-\frac{1}{3}}}{c^{-2}(ab)^3}$

39. $\sqrt{\dfrac{b^2c^{\frac{1}{2}}}{c^{-\frac{1}{2}}}}$

40. $\left(\sqrt{\dfrac{ab^4}{9c^4}}\right)\left(\dfrac{3ab^{\frac{1}{2}}}{c^2}\right)$

41. $\left(\dfrac{a^{\frac{1}{4}}b^{-2}}{c^{\frac{1}{2}}ab}\right)^{\frac{1}{2}}\left[\dfrac{(ba^{\frac{1}{4}})^2}{ca^{-3}}\right]^{-2}$

Exercises 42–45: Evaluate the expression $a^2b^{-2}-c^3$ if

42. $a=3, b=2, c=-2$

43. $a=-2, b=-1, c=2$

44. $a=-2, b=a^{-1}, c=ab^{-2}$

45. $a=3^0, b=a^2, c=a^{-2}b^2$

1.1 Exponents 7

Exercise 46:   If $a = 2^{87}$, express 2^{90} and 2^{86} as multiples of a.

Exercises 47–52: (**Rationalizing the Denominator**) Express each number as an equivalent number having a *rational* denominator. For example $\dfrac{3}{\sqrt{7}} = \dfrac{3}{\sqrt{7}} \cdot \dfrac{\sqrt{7}}{\sqrt{7}} = \dfrac{3\sqrt{7}}{7}$, and from $(a-b)(a+b) = a^2 - b^2$, we have, $\dfrac{2}{4-\sqrt{2}} = \dfrac{2}{4-\sqrt{2}} \cdot \dfrac{4+\sqrt{2}}{4+\sqrt{2}} = \dfrac{8+2\sqrt{2}}{16-2} = \dfrac{8+2\sqrt{2}}{14} = \dfrac{4+\sqrt{2}}{7}$.

47. $\dfrac{1}{\sqrt{5}}$ 48. $\dfrac{\sqrt{3}}{\sqrt{2}}$ 49. $\dfrac{9}{5-\sqrt{7}}$ 50. $\dfrac{5}{\sqrt{3}+1}$ 51. $\dfrac{1}{\sqrt{5}+2\sqrt{3}}$ 52. $\dfrac{\sqrt{2}}{3\sqrt{2}-2\sqrt{3}}$

Exercises 53–58: Combine the terms and express the answer with a rational denominator. (See Exercises 47–52.)

53. $\dfrac{1}{\sqrt{3}} - \dfrac{4}{\sqrt{2}} - \sqrt{\dfrac{9}{2}}$

54. $\left(\dfrac{1}{\sqrt{2}} - \dfrac{1}{\sqrt{3}} \right)^2$

55. $\dfrac{2}{\sqrt{2}-\sqrt{7}} - \dfrac{1}{\sqrt{2}+\sqrt{7}}$

56. $\dfrac{2}{1+\sqrt{3}} - \dfrac{1}{\sqrt{3}-1}$

57. $\dfrac{\sqrt{2}}{1 - \dfrac{1}{1-\sqrt{2}}}$

58. $\dfrac{2\sqrt{3}+1}{\sqrt{3}-\sqrt{2}} \cdot \dfrac{4\sqrt{3}}{\sqrt{2}+\sqrt{3}}$

Exercises 59–62: (**Scientific Notation**) Any number can be represented in the form $a \times 10^n$, where $1 \le a < 10, n$ is an integer, and $\times$ denotes multiplication. This is called scientific notation, and it is particularly useful for expressing very large and very small quantities. For example, $0.0000023 = 2.3 \times 10^{-6}$, and $40,213,000,000 = 4.0213 \times 10^{10}$.

In the following exercises, rewrite each number in scientific notation, and then perform the indicated operations. Express your answer in scientific notation.

59. $\dfrac{(40,000)(0.0006)}{0.00000008}$

60. $\dfrac{(4,200)(0.00008)(16)}{(0.02)(280,000)(0.12)}$

61. $(0.0000002)^5 (0.003)^{-1} (0.0004)^3 (0.008)^{-2}$

62. $\dfrac{(200,000)^3 (4,000,000)^{-2}}{(30,000)^2 (6,000)^{-8}}$

Exercises 63–68: Rewrite the given number in scientific notation. (See Exercises 59-62.)

63. The speed of light in a vacuum: 29,979,250,000 centimeters per second.

64. The average distance of Pluto from the sun: 3,600,000,000 miles.

65. The mass of an electron: 0.00000000000000000000000000091 grams.

66. The radius of the earth: 6,371,000 m.

67. The mass of the sun: 1,987,000,000,000,000,000,000,000,000,000 kg.

68. Avogadro's number (the number of molecules in a mole):
        602,400,000,000,000,000,000,000.

8   Chapter 1 Basic Concepts

---

§2. ALGEBRAIC EXPRESSIONS

Any arithmetic combination of rational powers and roots of x, such as:

$$2x^2 - \frac{1}{2}x + 3, \quad \frac{4x-1}{3x^2+2}, \quad \text{and} \quad \sqrt{2x-5} + \frac{1}{x}$$

is called an **algebraic expression** in x. The first expression on the left is a polynomial expression, the middle one is a rational expression, and the last one is a radical expression.

POLYNOMIAL EXPRESSIONS

Real numbers, with the exception of 0, are said to be polynomials of degree 0. The number 0 is also a polynomial, but has no degree.

DEFINITION 1.4

POLYNOMIAL

A **polynomial of degree n** (in the variable x) is an algebraic expression of the form:

$$a_n x^n + a_{n-1} x^{n-1} + \ldots + a_1 x + a_0$$

where n is a positive integer and a_0 through a_n are numbers, with $a_n \neq 0$.

The arithmetic operations on the real numbers extend naturally to polynomials. In particular:

$$(3x^2 - 2x + 1) + (x^3 + 4x^2 - 2) = \mathbf{3x^2} - 2x + 1 + x^3 + \mathbf{4x^2} - 2$$
$$= x^3 + \mathbf{7x^2} - 2x - 1$$

$$(3x^2 - 2x + 1) - (x^3 + 4x^2 - 2) = \mathbf{3x^2} - 2x + 1 - x^3 - \mathbf{4x^2} + 2$$
$$= -x^3 - \mathbf{x^2} - 2x + 3$$

and:

$$(3x^2 - 5)(4x^2 + 2x + 7) = 3x^2(4x^2 + 2x + 7) - 5(4x^2 + 2x + 7)$$
$$= 12x^4 + 6x^3 + 21x^2 - 20x^2 - 10x - 35$$
$$= 12x^4 + 6x^3 + x^2 - 10x - 35$$

(see margin for long multiplication method)

$$\begin{array}{r} 4x^2 + 2x + 7 \\ 3x^2 - 5 \\ \hline 12x^4 + 6x^3 + 21x^2 \\ -20x^2 - 10x - 35 \\ \hline 12x^4 + 6x^3 + x^2 - 10x - 35 \end{array}$$

Dividing one polynomial by another is also analogous to dividing one integer by another. While the integer division terminates when the remainder is less that the divisor, polynomial division ends when the degree of the remainder is less than the degree of the divisor. Consider the following example.

1.2 Algebraic Expressions   **9**

Analogous to
Divide 145 by 6

$$\begin{array}{r} 2 \\ 6\overline{)145} \end{array}$$

$$\begin{array}{r} 2 \\ 6\overline{)14\,5} \\ \underline{12} \\ 25 \end{array}$$

$$\begin{array}{r} 24 \leftarrow \text{quotient} \\ 6\overline{)14\,5} \\ \underline{12} \\ 25 \\ \underline{24} \\ 1 \leftarrow \text{remainder} \end{array}$$

We have shown that $\frac{145}{6} = 24 + \frac{1}{6}$ and can check our result by multiplying the quotient 24 by the divisor 6, and adding the remainder 1 to see if we obtain the dividend 145:
$24 \cdot 6 + 1 = 145$ Check!

EXAMPLE 1.3   Divide $3x^2 + x - 5$ by $x + 1$.

SOLUTION:

Consider the <u>leading</u> <u>terms</u>, x and $3x^2$, and determine the "number of times" that x divides $3x^2$; that is: $\frac{3x^2}{x} = 3x$. Position the $3x$ as indicated:

$$\begin{array}{r} 3x \\ x+1\overline{)3x^2 + x - 5} \end{array}$$

Next, multiply $3x$ by $x + 1$ and position the product $3x^2 + 3x$ as indicated:

$$\begin{array}{r} 3x \\ x+1\overline{)3x^2 + x - 5} \\ \underline{3x^2 + 3x } \end{array}$$

And **subtract:** $-2x - 5$

Since the degree of $-2x - 5$ (one) is not **less** than that of $x + 1$, repeat the process and divide x into $-2x$, positioning $\frac{-2x}{x} = -2$ as indicated: Then multiply the -2 by $(x + 1)$: and subtract as before:

$$\begin{array}{r} 3x - 2 \leftarrow \boxed{\text{quotient}} \\ x+1\overline{)3x^2 + x - 5} \\ \underline{3x^2 + 3x } \\ -2x - 5 \\ \underline{-2x - 2} \\ -3 \leftarrow \boxed{\text{remainder}} \end{array}$$

We have shown that $\dfrac{3x^2 + x - 5}{x + 1} = 3x - 2 - \dfrac{3}{x + 1}$, and can check our result by multiplying the quotient $3x - 2$ by the divisor $x + 1$, and adding the remainder -3 to see if we obtain the dividend $3x^2 + x - 5$:

$$(3x - 2)(x + 1) - 3 = 3x^2 + x - 5 \quad \text{Check!}$$

Actually, the main reason for performing the above division process or algorithm is featured in the "Check." Specifically, for the given polynomial $p(x) = 3x^2 + x - 5$ and the given divisor $d(x) = x + 1$, the algorithm enabled us to find a (quotient) polynomial $q(x) = 3x - 2$ and a (remainder) polynomial $r(x) = -3$ such that:

$$\underbrace{3x^2 + x - 5}_{p(x)} = \underbrace{(3x - 2)}_{d(x)}\underbrace{(x + 1)}_{q(x)} + \underbrace{(-3)}_{r(x)}$$

In general:

10   Chapter 1 Basic Concepts

Also:

$$\frac{p(x)}{d(x)} = q(x) + \frac{r(x)}{d(x)}$$

THEOREM 1.2

DIVISION ALGORITHM

For any polynomial $p(x)$ and any nonzero polynomial $d(x)$ there exist (unique) polynomials $q(x)$ and $r(x)$, with $r(x) = 0$ or the degree of $r(x)$ **less** than that of $d(x)$, such that:

$$p(x) = d(x)q(x) + r(x)$$

EXAMPLE 1.4   Find the polynomials $q(x)$ and $r(x)$ of Theorem 1.2 such that:

$$3x^4 + 10x^3 - 15x^2 + 3x - 2 = (3x^2 - 2x - 1)q(x) + r(x)$$

SOLUTION: We divide $3x^2 - 2x - 1$ into $3x^4 + 10x^3 - 15x^2 + 3x - 2$:

$$\frac{3x^4}{3x^2}$$

$$\begin{array}{r} x^2 \\ 3x^2 - 2x - 1 \overline{\smash{\big)}\, 3x^4 + 10x^3 - 15x^2 + 3x - 2} \\ \underline{3x^4 - 2x^3 - x^2 } \\ 12x^3 - 14x^2 + 3x - 2 \end{array}$$

$$\frac{12x^3}{3x^2}$$

$$\begin{array}{r} x^2 + 4x \\ 3x^2 - 2x - 1 \overline{\smash{\big)}\, 3x^4 + 10x^3 - 15x^2 + 3x - 2} \\ \underline{3x^4 - 2x^3 - x^2 } \\ 12x^3 - 14x^2 + 3x - 2 \\ \underline{12x^3 - 8x^2 - 4x } \\ - 6x^2 + 7x - 2 \end{array}$$

$$\begin{array}{r} x^2 + 4x - 2 \longleftarrow \boxed{\text{quotient}} \\ 3x^2 - 2x - 1 \overline{\smash{\big)}\, 3x^4 + 10x^3 - 15x^2 + 3x - 2} \\ \underline{3x^4 - 2x^3 - x^2 } \\ 12x^3 - 14x^2 + 3x - 2 \\ \underline{12x^3 - 8x^2 - 4x } \\ - 6x^2 + 7x - 2 \\ \underline{- 6x^2 + 4x + 2} \\ 3x - 4 \longleftarrow \boxed{\text{remainder}} \end{array}$$

Leading us to:

$$\underbrace{3x^4 + 10x^3 - 15x^2 + 3x - 2}_{p(x)} = \underbrace{(3x^2 - 2x - 1)}_{d(x)}\underbrace{(x^2 + 4x - 2)}_{q(x)} + \underbrace{(3x - 4)}_{r(x)}$$

Answers:

(a) $2x^6 - 2x^5 - 21x^3 + 5x + 32$

(b) quotient: $2x - 3$
     remainder: 7

(c) $q(x) = 4x^2 - 11x - 10$
     $r(x) = 22x + 19$

CHECK YOUR UNDERSTANDING 1.4

(a) Combine and simplify:

$$(2x^4 + 3x - 5)(x^2 - x) - 8(3x^3 - x^2 - 4)$$

(b) Divide $8x^2 - 14x + 10$ by $4x - 1$

(c) Find polynomials $q(x)$ and $r(x)$ such that

$$4x^4 - 11x^3 - 2x^2 - 1 = (x^2 + 2)q(x) + r(x)$$

with $r(x) = 0$ or the degree of $r(x)$ less than that of $x^2 + 2$.

FACTORING

Factoring is the reverse process of multiplication. By reading the following equation from right to left:

$$(a+b)(a-b) = a^2 - b^2$$

we obtain the formula for factoring a difference of two squares:

THEOREM 1.3
DIFFERENCE OF TWO SQUARES

$$a^2 - b^2 = (a+b)(a-b)$$

For example: $\quad 4x^2 - 9 = (2x+3)(2x-3)$

And: $\quad 4x^2 - 5 = (2x+\sqrt{5})(2x-\sqrt{5})$

Some trinomials (three terms), such as:

$$5x^2 + 27x - 18$$

can be factored by trial and error. We look to the "template":

$$(5x \quad)(x \quad)$$

which at least gives us the term $5x^2$. We then envision pairs of integers in the template whose product is -18, in the hope of finding one for which the middle term also turns out to be $27x$. In particular, one quickly sees that the first combination below is wrong, while the second is correct:

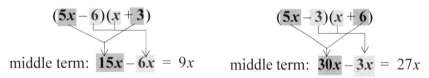

We have obtained the factorization: $5x^2 + 27x - 18 = (5x-3)(x+6)$

EXAMPLE 1.5 $\quad$ Factor $\ 4x^3 + 6x^2 - 18x$

SOLUTION: The first step is to pull out the common factor $2x$, and then proceed as above to factor the resulting trinomial:

$$4x^3 + 6x^2 - 18x = 2x(2x^2 + 3x - 9) = 2x(x+3)(2x-3)$$

EXAMPLE 1.6 $\quad$ Factor $\ 3x^3 + 5x^2 - 6x - 10$

FACTORING BY GROUPING

SOLUTION: The given polynomial is certainly not the difference of two squares, nor is it a trinomial. If you look at it closely however you can see that by grouping the first two terms together:

12   Chapter 1 Basic Concepts

$3x^3 + 5x^2 = x^2(3x + 5)$, and the last two terms together: $-6x - 10 = -2(3x + 5)$, the common factor $(3x + 5)$ emerges:

$$3x^3 + 5x^2 - 6x - 10 = (3x^3 + 5x^2) + (-6x - 10)$$
$$= x^2(3x + 5) - 2(3x + 5)$$
$$= (3x + 5)(x^2 - 2) = (3x + 5)(x + \sqrt{2})(x - \sqrt{2})$$

Answers:
(a) $(5x + 1)(5x - 1)$
(b) $(\sqrt{3}x + \sqrt{2})(\sqrt{3}x - \sqrt{2})$
(c) $(2x - 1)(x + 4)$
(d) $(x + 1)(x + 4)(x - 4)$

CHECK YOUR UNDERSTANDING 1.5

Factor: (a) $25x^2 - 1$ (b) $3x^2 - 2$

(c) $2x^2 + 7x - 4$ (d) $x^3 + x^2 - 16x - 16$

For a given polynomial $p(x)$, $p(c)$ denotes the value obtained when the number c is substituted for the variable x in $p(x)$. For example:

If $p(x) = 2x^2 + x - 1$, then $p(3) = 2 \cdot 3^2 + 3 - 1 = 20$

A **zero of a polynomial** $p(x)$ is a number which when substituted for the variable x yields zero. For example, -1 is a zero of the polynomial $p(x) = x^3 - 3x - 2$, since:

$$p(-1) = (-1)^3 - 3(-1) - 2 = -1 + 3 - 2 = 0$$

If
$$p(x) = (x - c)q(x)$$
then:
$$p(c) = (c - c)q(x) = 0$$

It is easy to see that if $(x - c)$ is a factor of the polynomial $p(x)$, then c is a zero of that polynomial (margin). The converse is also true:

THEOREM 1.4
ZEROS AND FACTORS

If c is a zero of a polynomial then $(x - c)$ is a factor of the polynomial.

PROOF: Let c be a zero of $p(x)$. Applying Theorem 1.2, we know that:
$$p(x) = (x - c)q(x) + r(x)$$

for some polynomials $q(x)$ and $r(x)$, where $r(x) = 0$ or the degree of $r(x)$ is less than that of $(x - c)$. Since $(x - c)$ is of degree 1, $r(x)$ must be a constant, say d, and we have:
$$p(x) = (x - c)q(x) + d \qquad (*)$$

Substituting c for x in the above equation we obtain:
$$p(c) = (c - c)q(x) + d$$
$$p(c) = d$$
since c is a zero of $p(x)$: $0 = d$

Hence, (*) reduces to $p(x) = (x - c)q(x)$, and $(x - c)$ is seen to be a factor of $p(x)$.

COROLLARY If $p(x)$ is a polynomial of degree n with leading coefficient A and zeros $a_1, a_2, ..., a_n$, then:

$$p(x) = A(x - a_1)(x - a_2) \cdots (x - a_n)$$

PROOF: Employing Theorem 1.4, we have:

$$p(x) = (x - a_1)(x - a_2) \cdots (x - a_n)q(x)$$

for some polynomial $q(x)$. Since the degree of $p(x)$ is n and the degree of $(x - a_1)(x - a_2)...(x - a_n)$ is also n, the polynomial $q(x)$ must be a constant polynomial, say $q(x) = C$. Thus:

$$p(x) = (x - a_1)(x - a_2) \cdots (x - a_n)C$$
$$= C(x - a_1)(x - a_2)...(x - a_n)$$

So, C is the leading coefficient of $p(x)$, and it is given to be A. Thus:

$$p(x) = A(x - a_1)(x - a_2) \cdots (x - a_n)$$

As is illustrated in the following example, Theorem 1.4 can help you factor certain polynomials:

EXAMPLE 1.7 Factor:

$$p(x) = 2x^3 - 3x^2 - 8x - 3$$

SOLUTION: Noting that -1 is a zero of $p(x)$:

$$p(-1) = 2(-1)^3 - 3(-1)^2 - 8(-1) - 3 = -2 - 3 + 8 - 3 = 0$$

Theorem 1.4 assures us that $x - (-1) = x + 1$ is a factor of $2x^3 - 3x^2 - 8x - 3$, and as you can easily check:

$$\begin{array}{r} 2x^2 - 5x - 3 \\ x + 1 \overline{\smash{\big)}\, 2x^3 - 3x^2 - 8x - 3} \end{array}$$

From its factored form, we see that in addition to -1, $-\dfrac{1}{2}$ and 3 are also zeros of

$$p(x) = 2x^3 - 3x^2 - 8x - 3$$

At this point we have: $2x^3 - 3x^2 - 8x - 3 = (x + 1)(2x^2 - 5x - 3)$

which we can factor further: $= (x + 1)(2x + 1)(x - 3)$

CHECK YOUR UNDERSTANDING 1.6

Verify that 2 is a zero of the polynomial

$$3x^3 - 13x^2 + 16x - 4$$

then use it to completely factor the polynomial

Answer: $(x - 2)^2(3x - 1)$

14   Chapter 1 Basic Concepts

Although there is no factorization formula for the **sum** of two squares (when restricted to real numbers), there are formulas for both the sum and difference of two cubes:

> Later, we will show that neither $x^2 - ax + a^2$ nor $x^2 + ax + a^2$ can be factored further (in the real number system).

$$\begin{array}{r} x^2 - ax + a^2 \\ x + a \overline{\smash{)}\, x^3 + 0x^2 + 0x + a^3} \\ \underline{x^3 + ax^2} \\ -ax^2 + 0x + a^3 \\ \underline{-ax^2 - a^2x} \\ a^2x + a^3 \\ \underline{a^2x + a^3} \\ 0 \end{array}$$

THEOREM 1.5

SUM AND DIFFERENCE OF TWO CUBES

$$x^3 + a^3 = (x + a)(x^2 - ax + a^2)$$
$$x^3 - a^3 = (x - a)(x^2 + ax + a^2)$$

If you forget these formulas, you can derive them. For instance, since $-a$ is a zero of the polynomial $x^3 + a^3$, $x + a$ is a factor, and you can divide to find the other factor:

$$\begin{array}{r} x^2 - ax + a^2 \\ x + a \overline{\smash{)}\, x^3 + 0x^2 + 0x + a^3} \end{array} \quad \text{(see margin)}$$

Thus: $x^3 + a^3 = (x + a)(x^2 - ax + a^2)$

(You are asked to derive the difference formula in the exercises.)

EXAMPLE 1.8   Factor:

(a) $x^3 - 8$ (b) $27x^3 + 1$

SOLUTION:

(a) $x^3 - 8 = x^3 - 2^3 = (x - 2)(x^2 + 2x + 4)$

(b) $27x^3 + 1 = (3x)^3 + 1^3 = (3x + 1)[(3x)^2 - 1 \cdot (3x) + 1^2]$
$$= (3x + 1)(9x^2 - 3x + 1)$$

EXAMPLE 1.9   Factor:

$$x^6 - 64$$

SOLUTION: The polynomial $x^6 - 64$ can be looked at as a difference of two cubes or as a difference of two squares:

$$x^6 - 64 = (x^2)^3 - 4^3 \quad \text{and} \quad x^6 - 64 = (x^3)^2 - 8^2$$

As the difference of two cubes, we arrive at the factorization:

$$x^6 - 64 = (x^2)^3 - 4^3 = (x^2 - 4)(x^4 + 4x^2 + 16)$$
$$= (x + 2)(x - 2)(x^4 + 4x^2 + 16)$$

But it isn't clear how (if at all) $x^4 + 4x^2 + 16$ can be factored.

We will have better luck looking at $x^6 - 64$ as a difference of two squares:

$$x^6 - 64 = (x^3)^2 - 8^2 = (x^3 + 8)\,(x^3 - 8)$$

From Theorem 1.5: $(x + 2)(x^2 - 2x + 4)\,(x - 2)(x^2 + 2x + 4)$

Since the two factorizations of the sixth degree polynomial must be equal, we can conclude that:

$$\cancel{(x+2)(x-2)}(x^4 + 4x^2 + 16) = \cancel{(x+2)}(x^2 - 2x + 4)\cancel{(x-2)}(x^2 + 2x + 4)$$

$$x^4 + 4x^2 + 16 = (x^2 - 2x + 4)(x^2 + 2x + 4)$$

a fact which can easily be verified by performing the multiplication.

Answers:

(a) $(2x+3)(4x^2 - 6x + 9)$

(b) $(x-1)(x+1)$
$(x^2 + x + 1)(x^2 - x + 1)$

CHECK YOUR UNDERSTANDING 1.7

Factor:

(a) $8x^3 + 27$ (b) $x^6 - 1$

RATIONAL EXPRESSIONS

Just as a rational number is an expression of the form $\dfrac{p}{q}$, where p and q are integers with $q \neq 0$, so then we have:.

DEFINITION 1.5

RATIONAL EXPRESSION

A **rational expression** (in the variable x) is an algebraic expression of the form $\dfrac{p(x)}{q(x)}$, where $p(x)$ and $q(x)$ are polynomials, with $q(x) \neq 0$.

Each of the following is a rational expression:

$$\frac{3x^3 + x^2}{x^2 - x + 2}, \qquad \frac{x-1}{x^5}, \qquad 3x^2 + x - 1, \quad \text{and} \quad 7$$

EXAMPLE 1.10 Simplify:

$$\frac{30x(x+2)^3 - 45x(2-x)(x+2)^2}{10(x+2)^5}$$

SOLUTION: Factor out the largest common factor of the terms in the numerator, $15x(x+2)^2$, and then cancel as much of it as you can with corresponding factors in the denominator:

$$\frac{30x(x+2)^3 - 45x(2-x)(x+2)^2}{10(x+2)^5} = \frac{15x(x+2)^2[2(x+2) - 3(2-x)]}{10(x+2)^5}$$

$$= \frac{\cancel{5(x+2)^2} \cdot 3x[2(x+2) - 3(2-x)]}{\cancel{5(x+2)^2} \cdot 2(x+2)^3}$$

$$= \frac{3x(2x + 4 - 6 + 3x)}{2(x+2)^3}$$

$$= \frac{3x(5x - 2)}{2(x+2)^3} = \frac{15x^2 - 6x}{2(x+2)^3}$$

16   Chapter 1 Basic Concepts

EXAMPLE 1.11 Simplify and express answer in factored form:

$$\frac{2x^2 - 5x - 3}{x^2 - 9} \cdot \frac{x^2 + 3x}{4x^2 + 4x + 1} - \frac{2x + 1}{x}$$

SOLUTION: The first step is to factor each expression and then cancel wherever possible:

$$\frac{2x^2 - 5x - 3}{x^2 - 9} \cdot \frac{x^2 + 3x}{4x^2 + 4x + 1} - \frac{2x + 1}{x}$$

$$= \frac{(x - 3)(2x + 1)}{(x - 3)(x + 3)} \cdot \frac{x(x + 3)}{(2x + 1)(2x + 1)} - \frac{2x + 1}{x}$$

$$= \frac{x}{2x + 1} - \frac{2x + 1}{x} = \frac{x^2}{x(2x + 1)} - \frac{(2x + 1)^2}{x(2x + 1)}$$

$$= \frac{x^2 - (4x^2 + 4x + 1)}{x(2x + 1)} = \frac{-3x^2 - 4x - 1}{x(2x + 1)} = -\frac{(3x + 1)(x + 1)}{x(2x + 1)}$$

> It is important to remember that you can only cancel a factor that is common to the numerator and the denominator of an expression.

EXAMPLE 1.12 Simplify:

$$2(4 - 3x)^{-1}(x + 1) - x(4 - 3x)^{-2}$$

SOLUTION: We offer two solutions, side by side, for your consideration:

Fraction approach:

$$2(4 - 3x)^{-1}(x + 1) - x(4 - 3x)^{-2}$$

$$= \frac{2(x + 1)}{4 - 3x} - \frac{x}{(4 - 3x)^2}$$

$$= \frac{2(x + 1)(4 - 3x) - x}{(4 - 3x)^2}$$

$$= \frac{-6x^2 + x + 8}{(4 - 3x)^2}$$

Factoring approach (take out any common factor to the **smallest** exponent):

smallest exponent →

$$2(4 - 3x)^{-1}(x + 1) - x(4 - 3x)^{-2}$$

$$= (4 - 3x)^{-2}[2(4 - 3x)(x + 1) - x]$$

$$= (4 - 3x)^{-2}[-6x^2 + x + 8]$$

$$= \frac{-6x^2 + x + 8}{(4 - 3x)^2}$$

Answers: (a) $\dfrac{x^2(9x + 4)}{2(5x + 2)^3}$

(b) $\dfrac{18x^3 - 8x^2 + 2x}{(3x - 1)^3}$

CHECK YOUR UNDERSTANDING 1.8

(a) Simplify and express the answer in factored form:

$$\frac{6x^2(5x + 2)^4 - 3x^3(5x + 2)^3}{6(5x + 2)^6}$$

(b) Simplify: $2x(3x - 1)^{-1} + 4x^2(3x - 1)^{-3}$

1.2 Algebraic Expressions   17

RADICAL EXPRESSIONS

In the calculus, you will often need to simplify expressions involving radicals (Example 1.13) or fractional exponents (Example 1.14).

EXAMPLE 1.13   Simplify:

$$\frac{\dfrac{\sqrt{x+1}}{3} - \dfrac{2}{\sqrt{x+1}}}{\sqrt{x+1}}$$

SOLUTION: We begin by multiplying the numerator and denominator of the given expression by the lowest common denominator of the fractions appearing in the numerator or denominator, namely $3\sqrt{x+1}$:

$$\frac{\dfrac{\sqrt{x+1}}{3} - \dfrac{2}{\sqrt{x+1}}}{\sqrt{x+1}} = \frac{\left(\dfrac{\sqrt{x+1}}{3} - \dfrac{2}{\sqrt{x+1}}\right)(3\sqrt{x+1})}{(\sqrt{x+1})(3\sqrt{x+1})}$$

$$= \frac{\dfrac{\sqrt{x+1}}{3} \cdot 3\sqrt{x+1} - \dfrac{2}{\sqrt{x+1}} \cdot 3\sqrt{x+1}}{3(x+1)}$$

$$= \frac{(x+1)-6}{3(x+1)} = \frac{x-5}{3x+3}$$

EXAMPLE 1.14   Simplify:

$$x(x+1)^{-\frac{1}{2}} + (x+1)^{\frac{3}{2}}$$

SOLUTION: We again offer two solutions: the "fractional approach" on the left, and the "factoring approach" on the right:

$x(x+1)^{-\frac{1}{2}} + (x+1)^{\frac{3}{2}}$

$\displaystyle = \frac{x}{(x+1)^{\frac{1}{2}}} + (x+1)^{\frac{3}{2}}$

$\displaystyle = \frac{x}{(x+1)^{\frac{1}{2}}} + \frac{(x+1)^{\frac{3}{2}}(x+1)^{\frac{1}{2}}}{(x+1)^{\frac{1}{2}}}$

$\displaystyle = \frac{x+(x+1)^2}{(x+1)^{\frac{1}{2}}} = \frac{x^2+3x+1}{\sqrt{x+1}}$

---

smallest exponent

$x(x+1)^{-\frac{1}{2}} + (x+1)^{\frac{3}{2}}$ $\boxed{\dfrac{3}{2} - \left(-\dfrac{1}{2}\right) = 2}$

$= (x+1)^{-\frac{1}{2}}[x + (x+1)^2]$

$\displaystyle = \frac{x + x^2 + 2x + 1}{(x+1)^{\frac{1}{2}}} = \frac{x^2+3x+1}{\sqrt{x+1}}$

18   **Chapter 1 Basic Concepts**

Answers: (a) $-\dfrac{1}{2\sqrt{x+2}}$

(b) $\dfrac{1-x}{(2x+1)^{\frac{1}{3}}}$

CHECK YOUR UNDERSTANDING 1.9

Simplify:

(a) $\dfrac{\dfrac{x}{2\sqrt{x+2}}-\sqrt{x+2}}{x+4}$

(b) $(2x+1)^{\frac{2}{3}}-3x(2x+1)^{-\frac{1}{3}}$

1.2 Algebraic Expressions   19

EXERCISES

Exercises 1–8: Combine and simplify.

1. $(6x^2 - 3x + 2) + (-7x^2 + x - 4)$

2. $(-2x^3 + 4x^2 - 5x) + (3x^4 - 5x^3 - 2x^2 + 3)$

3. $(4x^5 - 3x^3 + 2x) - (6x^5 + x^4 - x^3 + x + 1)$

4. $(3x^4 + 4x^2 - 6x + 1) - (4x^4 + 2x^2 - 4x + 3)$

5. $(3 - x - x^2)(4x^2 + x - 2)$

6. $(7x^3 + 2x)(2x^2 - x + 1)$

7. $2(5x - 2)(6x + 1) - 3(x^4 - 3x + 3)$

8. $4x^3(2x^4 - 3x^2 - 1) - 2x^2(-2x^3 + 4x - 5)$

Exercises 9–16: Carry out the division and then check your result, as in Example 1.3.

9. Divide $6x^2 - 11x + 3$ by $x - 1$

10. Divide $-4x^2 - 17x - 7$ by $x + 2$

11. Divide $3x^2 - 7$ by $x - 3$

12. Divide $x^3 - 8x^2 + 14x + 10$ by $x - 4$

13. Divide $-3x^3 - x^2 + 19x - 19$ by $x + 3$

14. Divide $-4x^3 + 3x^2 + 6$ by $x + 1$

15. Divide $-x^4 + 2x$ by $x - 1$

16. Divide $x^4 - 1$ by $x - 2$

Exercises 17–24: Carry out the division, determining the quotient $q(x)$ and remainder $r(x)$, and then write the result in the form given by the Division Algorithm (Theorem 1.2) as in Example 1.4.

17. Divide $x^4 + 6x^3 - x^2 - 7x - 9$ by $x + 6$

18. Divide $2x^4 - 10x^3 + 11x^2 + 6x + 2$ by $x - 3$

19. Divide $x^3 - 3x^2 - 9x + 2$ by $x - 2$

20. Divide $2x^4 - 5x^3 + 2x - 3$ by $x + 1$

21. Divide x^3 by $x^2 - 3x - 1$

22. Divide $10x^3 + x^2 - 9x - 6$ by $2x^2 - x - 2$

23. Divide $x^5 + x^3 - 2x^2 - 3x - 2$ by $x^3 - 2$

24. Divide $6x^5 - 19x^4 + 19x^3 - 10x^2 - 10x - 6$ by $2x^2 - 3x + 4$

Exercises 25–28: Carry out the division, determining the quotient $q(x)$ and remainder $r(x)$, and then write the result in the alternate form given by the Division Algorithm (Theorem 1.2), as $\dfrac{p(x)}{d(x)} = q(x) + \dfrac{r(x)}{d(x)}$.

25. Divide $5x^3 - 2x^2 + x$ by $x^2 - 2$

26. Divide $-6x^4 - 3x^3 + 8x^2$ by $2x^2 + x - 2$

20   Chapter 1 Basic Concepts

27. Divide $16x^5 - 12x^3 + 12x^2 - 20x + 14$ by $4x^3 - 6x + 3$

28. Divide $2x^7 - x^4 + 2x^3 + 2x^2 - 6$ by $2x^4 - 3x + 2$

Exercises 29–36: Factor the polynomial as a difference of squares or by trial and error.

29. $6x^2 + 7x - 20$ 30. $3x^2 + 2x - 1$ 31. $4x^2 - 16x + 16$ 32. $2x^4 - 3x^2$

33. $\dfrac{x^2}{5} - 5$ 34. $8x^2c - 32b^2c$ 35. $81x^4 - a^4$ 36. $2x^2 + \dfrac{4}{3}x - \dfrac{5}{2}$

Exercises 37–42: Factor the polynomial by the method of grouping.

37. $x^3 - x^2 - x + 1$ 38. $3a^2 + ba - 6a - 2b$ 39. $x^6 + x^4 + 3x^2 + 3$

40. $2x^{10} + x^6 + 4x^4 + 2$ 41. $2x^3 + x^2 - 8x - 4$ 42. $4x^3 + 6x^2 - 4x - 6$

Exercises 43–48: Factor the polynomial, given that one zero (or two zeros, in the case of a polynomial of degree 4) is $1, -1, 2,$ or -2, as in Example 1.7.

43. $6x^3 - 5x^2 - 17x + 6$ 44. $9x^3 - 30x^2 - 81x + 30$

45. $x^4 + x^3 - 7x^2 - x + 6$ 46. $6x^4 - x^3 - 7x^2 + x + 1$

47. $x^4 - 2x^3 - 3x^2 + 4x + 4$ 48. $12x^4 + 29x^3 + 4x^2 - 11x + 2$

Exercises 49–57: Factor the polynomial, involving a sum or difference of cubes.

49. $8x^3 - y^3$ 50. $a^3 - 8x^3$ 51. $a^3 + 27b^3$

52. $27x^3 + 64a^3$ 53. $x^3(x+1) - (x+1)$ 54. $(3a+b)^3 - (2a-b)^3$

55. $(2a-b)^3 + (5a+2b)^3$ 56. $64x^6 - 1$ 57. $27^2x^6 - 64$

Exercises 58–68: Factor the polynomial.

58. $20x^2 + 48x + 27$ 59. $9x^2 + 30x + 25$ 60. $-xa^2 - 25x$

61. $4x^4 - x^2$ 62. $36x^3 - 15x^2 - 6x$ 63. $2a^2 - 3b^2$

64. $x^3 + 1$ 65. $3x^4 - 48$ 66. $x^3 - 3x^2 - x + 3$

67. $6x^4 + 5x^2 - 4$ 68. $2x^4 - 3x^2 - 2$

Exercises 69–72: Combine and simplify.

69. $\dfrac{3}{x} - \dfrac{2}{3x-1} + \dfrac{x}{6x-2}$ 70. $\dfrac{x^2+x}{x^2-4x+3} \cdot \dfrac{x-3}{x^2+2x+1} + x - 1$

71. $\dfrac{3x-1}{4x^2+4x-8} + \dfrac{x-5}{6x^2-12x+6}$ 72. $\dfrac{x^2+3}{x^2+2x-3} + \dfrac{2x+3}{x^2+5x+6}$

1.2 Algebraic Expressions   **21**

Exercises 73–76: Simplify.

73. $\dfrac{5-x}{x-2} \cdot \dfrac{x^3-2x^2-x+2}{x^2-6x+5}$

74. $\dfrac{15x(2x^2+3)^4 - 12(2x^2+3)^3(2x^2)}{x(3+2x^2)^6}$

75. $\dfrac{(3x+1)^3(-3)-(1-3x)3(3x+1)^2 3}{(3x+1)^6}$

76. $\dfrac{(x^2+1)^2(-4x)-(2x^2+2)2(x^2+1)2x}{(x^2+1)^3}$

Exercises 77–81: Simplify, and express the answer in factored form.

77. $\dfrac{3x^2-x-2}{3x-2} \cdot \dfrac{5x^2+x-4}{x^2-1}$

78. $\dfrac{(1-x^2)(-6x)-(5-3x^2)(-2x)}{(1-x^2)^2}$

79. $\dfrac{(x+3)^2(2x+6)-(x-2)(x+8)2(x+3)}{(x+3)^4}$

80. $\dfrac{2x^3-25x^2+84x-36}{2x-1}$

81. $\dfrac{18x^2(6x+2)^2 - 2x(6x+2)^3}{64(3x+1)^6}$

Exercises 82–89: Simplify each of the following.

82. $\sqrt{x+1} - \dfrac{x}{\sqrt{x+1}} - \dfrac{x-1}{x^2-1}$

83. $\dfrac{\dfrac{\sqrt{3x-1}}{4} - \dfrac{2}{\sqrt{3x-1}}}{6\sqrt{3x-1}}$

84. $-x(4x+3)^{-2} - 2x^{-1}$

85. $\dfrac{1}{4}(x-4)^{-3} - \dfrac{1}{2}(x-4)^{-2}$

86. $\dfrac{1}{2}(x+4)^{-\frac{1}{2}} - \dfrac{1}{2}x^{-\frac{1}{2}}$

87. $\dfrac{1}{2}\left(\dfrac{x}{x+4}\right)^{-\frac{1}{2}}\left[\dfrac{4}{(x+4)^2}\right]$

88. $\dfrac{(1-x^2)^{\frac{3}{2}} - x \cdot \frac{3}{2}(1-x^2)^{\frac{1}{2}}(-2x)}{(1-x^2)^3}$

89. $3(2x-5)^{-\frac{3}{2}} - (2x-5)^{-\frac{1}{2}} + 2x(2x-5)^{-\frac{5}{2}}$

Exercises 90–95: Simplify, and express the answer in factored form.

90. $\dfrac{2}{3}x^{-\frac{1}{3}} - \dfrac{2}{3}x^{-\frac{4}{3}}$

91. $\dfrac{3}{2}x^{\frac{1}{2}} - 3x^{-\frac{1}{2}}$

92. $\dfrac{3\sqrt{4x^2+1} - \dfrac{12x}{\sqrt{4x^2+1}}}{\sqrt{4x^2+1}}$

93. $\sqrt{1-x} - \dfrac{1}{2}\left(\dfrac{x}{\sqrt{1-x}}\right)$

94. $2x^7(x^2+1)^{-\frac{1}{4}} + 4x^6(x^2+1)^{\frac{3}{4}}$

95. $\dfrac{(x^{\frac{2}{3}}-4)(x^{-\frac{2}{3}}) - x^{\frac{1}{3}}(x^{-\frac{1}{3}})}{(x^{\frac{2}{3}}-4)^2}$

22   **Chapter 1 Basic Concepts**

Exercise 96:   Construct a polynomial whose only zeros are 1, -2, and 3.

Exercise 97:   Construct a polynomial of degree three whose zeros are 1, -2, and 3, and whose leading coefficient is 4.

Exercise 98:   Show that $x^4 + a^4$ can be factored even though it has no linear factors. [Suggestion: Write $x^4 + a^4 = (x^4 + 2a^2x^2 + a^4) - (2a^2x^2)$ and factor as the difference of two squares.]

Exercise 99:   Derive the factorization formula for the difference of two cubes:
$$x^3 - a^3 = (x - a)(x^2 + ax + a^2).$$

§3. UNIT CONVERSION

More than a number is required to represent a measurement. One cannot, for example, say that a board is two long. Units must also be specified: Is it two feet long? Two inches? Two yards? Two meters? The length of the board is certainly constant, but the numbers describing that length differ, depending on the unit of measurement used. For example, a board that is 2 feet long is also 24 inches long, and we write:

$$2 \text{ feet } = 24 \text{ inches} \quad \text{ or } \quad 2 \text{ ft } = 24 \text{ in.}$$

The units in the above equation are critical; removing them would result in the ridiculous assertion that $2 = 24$.

A unit of measure, such as a "foot," is a quantity that can be cancelled when it appears as a common factor of the numerator and denominator of a fraction. For example,

$$\frac{12 \text{ ft}}{3 \text{ ft}} = \frac{12}{3} = 4$$

In short, units can be treated like nonzero constants; for indeed, they are:

$7a + 2a = 9a$	$7\,ft + 2\,ft = 9\,ft$
$(360\,a)\left(3 \cdot \dfrac{b}{a}\right) = 1080\,b$	$(360\,yd)\left(3\dfrac{ft}{yd}\right) = 1080\,ft$
$\left(9 \cdot \dfrac{a}{b}\right)\left(40\,\dfrac{b}{c}\right)\left(52\,\dfrac{c}{d}\right) = 18{,}720\dfrac{a}{d}$	$\left(9\,\dfrac{\$}{hr}\right)\left(40\dfrac{hr}{wk}\right)\left(52\dfrac{wk}{yr}\right) = 18{,}720\dfrac{\$}{yr}$
$\dfrac{3a}{16 \cdot \dfrac{a}{b}} = \dfrac{3a}{16} \cdot \dfrac{b}{a} = \dfrac{3}{16} \cdot b$	$\dfrac{3\,oz}{16\dfrac{oz}{lb}} = \dfrac{3\,oz}{16} \cdot \dfrac{lb}{oz} = \dfrac{3}{16}lb$
invert and multiply	invert and multiply

24   Chapter 1 Basic Concepts

Suppose you want to convert 3 pounds to ounces. Knowing that 1 lb = 16 oz allows you to write:

$$\frac{1\,\text{lb}}{16\,\text{oz}} = 1 \quad \text{and} \quad \frac{16\,\text{oz}}{1\,\text{lb}} = 1$$

and you can, of course, multiply an expression by 1 without changing its value:

$$3\,\text{lb} = (3\,\text{lb})(1) = (3\,\cancel{\text{lb}})\left(\frac{16\,\text{oz}}{1\,\cancel{\text{lb}}}\right) = 48\,\text{oz}$$

Let us agree to call a relation of the type 1 *lb* = 16 *oz* a conversion bridge. Here are some of the more familiar conversion bridges:

1 lb = 16 oz	1 ft = 12 in.	1 *yd* = 3 *ft*
1 mi = 5280 ft	1 gal. = 4 qt	1 qt = 2 pt
1 hr = 60 min	1 min = 60 sec	1\$ = 100 ¢

EXAMPLE 1.15   Convert 2.3 gallons to pints.

SOLUTION: To convert 2.3 gallons to pints, we take the path:

$$\text{gal.} \to \cancel{\text{gal.}}\left(\frac{\text{qt}}{\cancel{\text{gal.}}}\right)\left(\frac{\text{pt}}{\cancel{\text{qt}}}\right) \to \text{pt}$$

Along with the bridges 1 gal. = 4 qt and 1 qt = 2 pt:

$$2.3\,\text{gal.} = (2.3\,\text{gal.})\left(\frac{4\,qt}{1\,\text{gal.}}\right)\left(\frac{2\,\text{pt}}{1\,qt}\right) = (2.3)(4)(2)\text{pt} = 18.4\,\text{pt}$$

EXAMPLE 1.16   Convert 50 feet per minute to miles per hour.

SOLUTION: To convert 50 feet per minute to miles per hour we take the path:

$$\frac{\text{ft}}{\text{min}} \to \frac{\cancel{\text{ft}}}{\cancel{\text{min}}} \cdot \frac{\text{mi}}{\cancel{\text{ft}}} \cdot \frac{\cancel{\text{min}}}{\text{hr}} \to \frac{\text{mi}}{\text{hr}}$$

Along with the bridges 1 min = 60 sec and 1 mi = 5280 ft:

$$50\frac{\text{ft}}{\text{min}} = 50\frac{\text{ft}}{\text{min}} \cdot \frac{1\,\text{mi}}{5280\,\text{ft}} \cdot \frac{60\,\text{min}}{1\,\text{hr}} = \frac{50 \cdot 60}{5280}\frac{\text{mi}}{\text{hr}} \approx 0.57\frac{\text{mi}}{\text{hr}}$$

approximately equal

Answer: $45\frac{\text{lb}}{\text{hr}}$

CHECK YOUR UNDERSTANDING 1.10

Convert 12 ounces per minute to pounds per hour.

As is illustrated in the following example, one has to be careful when converting units that are raised to a power other than one.

EXAMPLE 1.17 Convert 7 pounds per cubic foot to ounces per cubic inch.

SOLUTION: The following path will take you from pounds per cubic foot to ounces per cubic inch:

$$\frac{\text{lb}}{\text{ft}^3} \to \frac{\cancel{\text{lb}}}{\cancel{\text{ft}^3}} \cdot \frac{\cancel{\text{ft}^3}}{\text{in.}^3} \cdot \frac{\text{oz}}{\cancel{\text{lb}}} \to \frac{\text{oz}}{\text{in.}^3}$$

Noting that, $(1 \text{ ft})^3 = (12 \text{ in.})^3 = 12^3 \text{in.}^3$, we have:

$$7\frac{\text{lb}}{\text{ft}^3} = 7\frac{\text{lb}}{\text{ft}^3} \cdot \frac{1^3 \text{ft}^3}{12^3 \text{ in.}^3} \cdot \frac{16 \text{ oz}}{1 \text{ lb}} = \frac{7 \cdot 16}{12^3}\frac{\text{oz}}{\text{in.}^3} \approx 0.065 \frac{\text{oz}}{\text{in.}^3}$$

Answer: $\approx 0.13 \frac{\text{qt}}{\text{ft}^2}$

CHECK YOUR UNDERSTANDING 1.11

Convert 0.3 gallons per square yard to quarts per square foot.

THE METRIC SYSTEM

One meter was originally intended to be, and is very nearly equal to, one ten-millionth of the distance between the equator and the north pole, measured along a meridian. Presently, it is defined as 1,650,733.73 wavelengths of the orange-red radiation of krypton86 under specified conditions.

The three basic units of measure in the metric system are the meter (m), gram (g), and liter (l). By combining these basic units with the following prefixes, multiple and fractional units are obtained:

$$kilo = one\ thousand$$
$$deci = one\ tenth$$
$$centi = one\ hundredth$$
$$milli = one\ thousandth$$

For example:

one kilogram $= 1000$ grams, and we write: $1\,kg = 1000\,g$ and $\frac{1}{1000}\,kg = 1g$

one decimeter $= \frac{1}{10}$ meter, and we write: $1\,dm = \frac{1}{10}m$ and $10\,dm = 1\,m$

one centimeter $= \frac{1}{100}$ meter, and we write: $1\,cm = \frac{1}{100}m$ and $100\,cm = 1\,m$

one milliliter $= \frac{1}{1000}$ liter, and we write: $1\,ml = \frac{1}{1000}l$ and $1000\,ml = 1\,l$

It should be noted that the convergence bridge $1\,g = 0.0353\,oz$ only holds near the surface of the earth, as gram is a mass-unit while ounce is a force-unit.

Here are some (approximate) conversion bridges between the English system and the metric system:

$$1\,m = 3.28 \text{ ft} \qquad 1\,g = 0.0353\,oz \qquad 1\,l = 0.264 \text{ gal.}$$

26   Chapter 1 Basic Concepts

EXAMPLE 1.18   Convert 13 pounds to kilograms.

SOLUTION: To convert 13 pounds to kilograms without a direct bridge from pounds to grams, first convert from pounds to ounces, and then from ounces to grams. Here is the path:

$$lb \to \cancel{lb} \cdot \frac{\cancel{oz}}{\cancel{lb}} \cdot \frac{\cancel{g}}{\cancel{oz}} \cdot \frac{kg}{\cancel{g}} \to kg$$

And here is the completed journey:

$$13\,lb = (13\,lb)\left(\frac{16\,oz}{1\,lb}\right)\left(\frac{1\,g}{0.0353\,oz}\right)\left(\frac{1\,kg}{1000\,g}\right)$$

$$= \frac{(13)(16)}{(0.0353)(1000)}\,kg \approx 5.89\,kg$$

EXAMPLE 1.19   The currency in Venezuela is the Bolivar (Bs), and in 1974 the exchange rate was 4.23 Bs per dollar. At that time, a liter of gasoline sold for 0.15 Bs. What was the cost in dollars of a gallon of gasoline?

SOLUTION: Taking the path:

$$\frac{Bs}{l} \to \frac{\cancel{Bs}}{\cancel{l}} \cdot \frac{\$}{\cancel{Bs}} \cdot \frac{\cancel{l}}{gal.} \to \frac{\$}{gal.}$$

we have:

$$0.15\frac{Bs}{l} = \left(0.15\frac{Bs}{l}\right)\left(\frac{1\,\$}{4.23\,Bs}\right)\left(\frac{1\,l}{0.264\,gal.}\right)$$

$$= \frac{0.15}{(4.23)(0.26)}\frac{\$}{gal.} \approx 0.13\frac{\$}{gal.}$$

Conclusion: A gallon of gasoline cost $0.13.

CHECK YOUR UNDERSTANDING 1.12

Convert 4 grams per cubic centimeter to ounces per cubic inch.

Answer: $\approx 2.32\frac{oz}{in.^3}$

1.3 Unit Conversion   27

EXERCISES

Exercises 1–18: Convert

1. 3.2 feet to miles   2. 3.2 feet to millimeters   3. 4 pt to ml

4. 20 grams to kilograms   5. 50 lb to kg   6. 3.2 feet per hour to yards per minute

7. 55 mph to ft/sec   8. 55 miles per hour to kilometers per hour

9. 25 in./sec to mph   10. 35 ft/sec to cm/min

11. 5 cubic centimeters to cubic inches   12. 5 liters per second to gallons per minute

13. 8 square meters to square feet   14. 4 pounds per cubic foot to ounces per cubic inch

15. $0.2450 \dfrac{\text{lb}}{\text{in.}^3}$ to $\dfrac{\text{kg}}{\text{m}^3}$ 16. $4 \dfrac{\text{ft}^2 \text{ gal.}}{\text{lb}^3}$ to $\dfrac{\text{cm}^2 \text{ ml}}{\text{mg}^3}$

17. 3 kilograms per cubic meter to pounds per cubic yard

18. 2 pounds per square foot to kilograms per square centimeter

Exercise 19:   The speed of light in a vacuum is 2.9979×10^8 m/sec. Convert this to

(a) kilometers per second   (b) miles per second   (c) miles per hour

Exercise 20:   How many seconds are there in a week? a month? a year?

Exercise 21:   A quire of paper is 25 sheets of paper, and a ream is 500 sheets. How many quires are there in a ream?

Exercise 22:   A fathom is 6 feet. How many meters are there in 10 fathoms? How many fathoms in a mile?

Exercise 23:   One nautical mile is 1.852 kilometers. How many miles are there in 150 nautical miles?

Exercise 24:   A furlong is $\frac{1}{8}$ of a mile, and a league is 3 miles. How many furlongs to a league?

Exercise 25:   Air pressure at sea level is defined as 1 Atmosphere and is 2,116.102 lb/ft^2.

(a) Convert this to lb/in.^{2} (b) Convert this to kg/cm^2

(c) What is a pressure of 75 grams per square meter in Atmospheres?

Exercise 26:   Determine the cost in dollars of a British automobile valued at £52,890 (pounds sterling) if, at the time of transaction, £1 equaled $1.63.

Exercise 27:   The radius of a wire is 0.058 inches. What is it in millimeters?

Exercise 28:   How many pounds of clay can you buy for $10, if a kilo (kilogram) sells for £0.4 (pounds sterling) and £1 = $1.63?

28   Chapter 1 Basic Concepts

Exercise 29:   Oil flows through a pipe at the rate of 50 gal./min. Find the rate in liters/sec.

Exercise 30:   What is the cost of a cubic foot of gravel that sells for $1.50 per cubic yard?

Exercise 31:   A factory assembly line moves at the rate of 20 ft/min. Find the rate in inches per second.

Exercise 32:   If you are driving at 50 mph, how many meters do you travel in 2 minutes?

Exercise 33:   How many square feet of a certain material can you buy for $5, if a square yard costs $18.20?

Exercise 34:   How many grams of dried mushrooms can you buy for 2500 lira, if a pound costs $60 and $1 = 1718 lira?

Exercise 35:   When the planet Mars is closest to the Earth, it is about 35 million miles away. Since light travels at 3×10^8 meters per second, how many minutes does it take light from Mars to reach the Earth when Mars is closest to the Earth?

Exercise 36:   How many gallons of water are necessary to uniformly coat an acre of land with one inch of rain? (An acre is 43,560 square feet, 1 cubic foot of pure water weighs about 62.4 pounds, and 1 gallon weighs 8.345 pounds.)

§4. LINES, LINEAR EQUATIONS AND INEQUALITIES

It is common to associate numerical values with geometrical objects: the area of a rectangle, the circumference of a circle, and so on. The following definition attributes a measure of "steepness" to any nonvertical line in the plane.

> The slope of a line does not depend on the particular points chosen. To see this, consider the two pairs of points (a_1, b_1), (a_2, b_2), and (c_1, d_1), (c_2, d_2) on the line below. Since the two triangles are similar, the ratios of corresponding sides are equal:
> $$\frac{b_2 - b_1}{a_2 - a_1} = \frac{d_2 - d_1}{c_2 - c_1}$$
>
> Note also that:
> $$\frac{b_2 - b_1}{a_2 - a_1} = \frac{b_1 - b_2}{a_1 - a_2}$$

DEFINITION 1.6

SLOPE OF A LINE

For any nonvertical line L and any two distinct points (x_1, y_1) and (x_2, y_2) on L, we define the **slope** of L to be the number m given by:

$$m = \frac{y_2 - y_1}{x_2 - x_1} = \frac{\text{change in } y}{\text{change in } x}$$

The slopes of the lines labeled L_1, L_2, and L_3 in Figure 1.1 are easily determined:

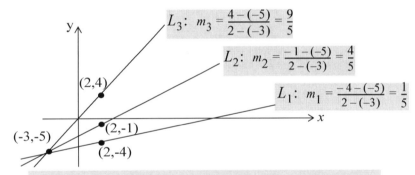

The steeper the climb, the more positive the slope

Figure 1.1

While lines of positive slope climb as you move to the right, those of negative slope fall:

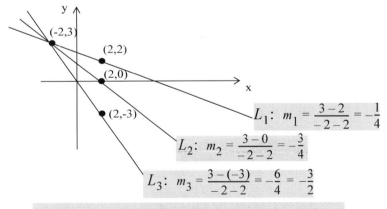

The steeper the fall, the more negative the slope

Figure 1.2

If you calculate the slope of any horizontal line, you will find that it is zero. Consider, for example, the line L of Figure 1.3(a). Since $(2, 3)$ and $(4, 3)$ are on L:

$$m = \frac{3-3}{4-2} = \frac{0}{2} = 0$$

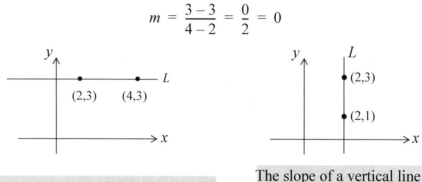

Horizontal lines have a slope of 0
(a)

The slope of a vertical line is not defined.
(b)

Figure 1.3

If you try to calculate the slope of a vertical line, you will encounter a problem. Consider, for example, the vertical line L of Figure 1.3(b) that contains the points $(2, 1)$ and $(2, 3)$. An attempt to use these two points (or any other two points on the line) to calculate its "slope" will lead to an undefined expression:

$$m = \frac{3-1}{2-2} = \frac{2}{0} \leftarrow \textbf{undefined}$$

CHECK YOUR UNDERSTANDING 1.13

(a) Determine the slope of the line L that passes through the points $(-1, 6)$ and $(3, -2)$.

(b) If you move 8 units to the right of the point $(-1, 6)$ and then vertically to the line L of part (a), at what point on L do you arrive?

Answers: (a) -2
(b) $(7, -10)$

PARALLEL LINES

PERPENDICUAR LINES

Lines are said to be **parallel** when they have the same slope, or when they are vertical. Lines that intersect at right angles are said to be **perpendicular**. It follows that every horizontal line is perpendicular to every vertical line. The following theorem, which you are invited to verify in the exercises, asserts that a necessary and sufficient condition for two nonvertical lines to be perpendicular is that the slope of one is the negative reciprocal of the slope of the other:

THEOREM 1.6 Two lines with nonzero slopes m_1 and m_2 are perpendicular if and only if:

$$m_1 = -\frac{1}{m_2}$$

1.4 Lines, Linear Equations and Inequalities   31

> **CHECK YOUR UNDERSTANDING 1.14**
>
> Determine the slope of a line that:
> (a) is parallel to the line through $(2, 7)$ and $(-6, 1)$.
> (b) is perpendicular to the line through $(2, 7)$ and $(-6, 1)$.

Answers: (a) $\frac{3}{4}$ (b) $-\frac{4}{3}$

SLOPE-INTERCEPT EQUATION OF A LINE

Consider the line L of slope m in the adjacent figure. Being nonvertical, it must intersect the y-axis at some point $(0, b)$, where b is the **y-intercept** of the line. Since any two distinct points on a line determine its slope, for any point (x, y) on L other than $(0, b)$:

$$m = \frac{y-b}{x-0}$$

$$m = \frac{y-b}{x}$$

$$y - b = mx$$

$$y = mx + b$$

Direct substitution shows that the above equation also holds at the point $(0, b)$; thus:

THEOREM 1.7
Slope-Intercept Equation of a line

If the point (x, y) is on the line of slope m and y-intercept b then:.

$$y = mx + b$$

EXAMPLE 1.20 Find the slope-intercept equation of the line that contains the points $(3, 2)$ and $(-4, 6)$.

SOLUTION: Using the given points, we find that:

$$m = \frac{6-2}{-4-3} = -\frac{4}{7}$$

We now know that the equation is of the form:

$$y = -\frac{4}{7}x + b$$

Since the point $(3, 2)$ is on the line, the above equation must hold when 3 is substituted for x and 2 for y, and this enables us to solve for b:

$$2 = -\frac{4}{7}(3) + b$$

$$b = 2 + \frac{12}{7} = \frac{26}{7}$$

It is a good idea to sketch the line in question:

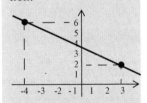

We see that the line has a negative slope, and that its y-intercept is about 4.

The other point, $(-4, 6)$, could be used instead of $(3, 2)$, and it would lead to the same result (try it).

Leading us to the equation: $y = -\frac{4}{7}x + \frac{26}{7}$

EXAMPLE 1.21 Find the slope-intercept equation of the line through the point (3, 1) that is perpendicular to the line $3x + 5y = 4$.

SOLUTION: Since the given line: $3x + 5y = 4$
$$5y = -3x + 4$$
$$y = -\frac{3}{5}x + \frac{4}{5}$$

has slope $-\frac{3}{5}$, the line we want has slope $m = -\frac{1}{-\frac{3}{5}} = \frac{5}{3}$.

We now know that our line is of the form: $y = \frac{5}{3}x + b$

Since it is to pass through the point (3,1): $1 = \frac{5}{3}(3) + b$

Or: $b = -4$

Equation: $y = \frac{5}{3}x - 4$

CHECK YOUR UNDERSTANDING 1.15

Find the slope-intercept equation, of the line containing the point (–2, 4) that is parallel to the line $x - 5y = 3$.

Answer: $y = \frac{1}{5}x + \frac{22}{5}$

POINT-SLOPE EQUATIONS OF A LINE

A **point-slope** equation of a (nonvertical) line is obtained by fixing an arbitrary point (x_0, y_0) on L [not necessarily the point $(0, b)$]. With reference to this point and **any other point** (x, y) on L, we have:

$$m = \frac{y - y_0}{x - x_0}$$

or: $y - y_0 = m(x - x_0)$

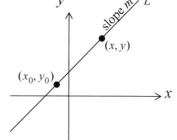

Noting that the equation is also satisfied when $(x, y) = (x_0, y_0)$, we have:

THEOREM 1.8
POINT-SLOPE EQUATION OF A LINE

A point (x, y) is on the line of slope m passing through the point (x_0, y_0) if and only if:
$$y - y_0 = m(x - x_0)$$

EXAMPLE 1.22 Find a point-slope equation of the line in Example 1.20.

SOLUTION: Unlike the slope-intercept form, which is unique, there are infinitely many point-slope equations of a line (because there are infinitely many points on the line from which to choose). In particular, using the point $(3, 2)$ and the slope $m = -\frac{4}{7}$ that was determined in Example 1.20, we arrive at the point-slope equation:
$$y - y_0 = m(x - x_0)$$
$$y - 2 = -\frac{4}{7}(x - 3)$$

If you choose the point $(-4, 6)$, you will get:
$$y - 6 = -\frac{4}{7}(x + 4)$$
Either way, you can rewrite the resulting point-slope equation in its slope-intercept form:
$$y = -\frac{4}{7}x + \frac{26}{7}$$

CHECK YOUR UNDERSTANDING 1.16

Find a point-slope equation of the line containing the point $(-2, 4)$ that is perpendicular to the line $x - 5y = 3$.

Answer: $y - 4 = -5(x + 2)$

EQUATIONS OF HORIZONTAL AND VERTICAL LINES

The horizontal line L in Figure 1.4(a) has y-intercept 3, and slope 0. Consequently, its equation is $y = 0 \cdot x + 3$, or simply, $y = 3$. The equation of the x-axis is $y = 0$.

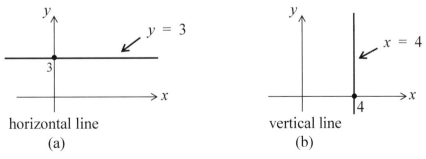

horizontal line
(a)

vertical line
(b)

Figure 1.4

While slopes are not defined for vertical lines, such lines can still be described in terms of an equation. In particular, since every point on the line in Figure 1.4(b) has x-coordinate 4, the equation of that line is simply $x = 4$. The equation of the y-axis is $x = 0$.

34   Chapter 1 Basic Concepts

Answer: $y = 2x - 6$

CHECK YOUR UNDERSTANDING 1.17

Find the equation of the line of slope 2 passing through the point of intersection of the vertical line $x = 5$ and the horizontal line $y = 4$.

LINEAR EQUATIONS

The **solution set** of an equation is simply the set of numbers that satisfy the equation (left side of equation equals right side of equation). When solving equations, the following result plays a dominant role:

The fact that you can add the same quantity to both sides of an equation without altering its solution set allows you to move terms from one side of the equation to the other as long as you change the sign of those terms.   For   example:

$$2x + 3 = 5$$
$$(2x + 3) + (-3) = 5 + (-3)$$
$$2x = 5 - 3$$

As you can see that "+3" on the left side of the equation ended up being a "-3" on the right side of the equation.

THEOREM 1.9 Adding (or subtracting) the same quantity to (or from) both sides of an equation, or multiplying (or dividing) both sides of an equation by the same nonzero quantity will not alter the solution set of the equation.

EXAMPLE 1.23 Solve the equation:

$$\frac{3x}{5} - \frac{2-x}{3} + 1 = \frac{x-1}{15}$$

SOLUTION: We first eliminate all denominators by multiplying both sides of the equation by 15:

$$\frac{3x}{5} - \frac{2-x}{3} + 1 = \frac{x-1}{15}$$

$$15\left(\frac{3x}{5}\right) - 15\left(\frac{2-x}{3}\right) + 15 = 15\left(\frac{x-1}{15}\right)$$

$$3(3x) - 5(2-x) + 15 = x - 1$$

$$9x - 10 + 5x + 15 = x - 1$$

$$14x + 5 = x - 1$$

$$14x - x = -1 - 5$$

$$13x = -6$$

$$x = -\frac{6}{13}$$

CHECK YOUR UNDERSTANDING 1.18

Answer: $\frac{5}{9}$

Solve:

$$-\frac{2x}{3} - \frac{x}{2} + 1 = \frac{2x+1}{6}$$

1.4 Lines, Linear Equations and Inequalities   35

EXPRESSING ONE VARIABLE IN TERMS OF ANOTHER

Sometimes, an equation will contain more than one variable; or variables and some unknown constants, as in:

$$2x + \frac{3}{2}y = 4 + \frac{5}{3}x + y \text{ and } 2(x - 2a) = \frac{5x + a}{3}$$

In such a situation, you may need to express one variable in terms of other variables or constants. Consider the following example.

EXAMPLE 1.24 Solve for y in terms of x, if:

$$2x + \frac{3}{2}y = 4 + \frac{5}{3}x + y$$

Solution: We begin by multiplying both sides of the equation by 6 to clear the denominators, and go on from there:

$$6 \cdot 2x + 6 \cdot \frac{3}{2}y = 6 \cdot 4 + 6 \cdot \frac{5}{3}x + 6 \cdot y$$

$$12x + 9y = 24 + 10x + 6y$$

$$9y - 6y = 24 + 10x - 12x$$

$$3y = 24 - 2x$$

$$y = 8 - \frac{2}{3}x$$

Answer:

$y = \frac{x-1}{3}, x = 3y + 1$

CHECK YOUR UNDERSTANDING 1.19

Express y in terms of x, and x in terms of y if:

$$3x + 2(x - y) = y + 4x + 1$$

36 Chapter 1 Basic Concepts

LINEAR INEQUALITIES

One solves linear inequalities in exactly the same fashion as one solves linear equations, with one notable exception:

> WHEN MULTIPLYING OR DIVIDING BOTH SIDES OF AN INEQUALITY BY A **NEGATIVE** QUANTITY, **REVERSE** THE DIRECTION OF THE INEQUALITY SIGN.

If you multiply both sides of the inequality $-2 < 3$ by the positive number 2, then the inequality sign remains as before:
$$-2 < 3$$
multiply by 2: $-4 < 6$

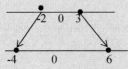

But if you multiply both sides by a negative quantity, then the sense of the inequality is reversed:
$$-2 < 3$$
multiply by -2: $4 > -6$

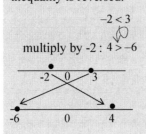

To illustrate:

Equation	Inequality
$3x - 5 = 5x - 7$	$3x - 5 < 5x - 7$
$3x - 5x = -7 + 5$	$3x - 5x < -7 + 5$
$-2x = -2$	$-2x < -2$ ←$reverse$ dividing by a negative number
$x = 1$	$x > 1$

EXAMPLE 1.25
Solve:
$$-3 \leq 4x + 2 < 9$$

SOLUTION: The above is shorthand for writing:
$$-3 \leq 4x + 2 \quad \text{and} \quad 4x + 2 < 9$$

You can solve the two inequalities separately and then take the intersection of their solution sets (below left), or you can solve the two inequalities simultaneously (below right).

$-3 \leq 4x + 2$ and $4x + 2 < 9$	$-3 \leq 4x + 2 < 9$
$-4x \leq 2 + 3$ and $4x < 9 - 2$	Subtract 2: $-3 - 2 \leq (4x + 2) - 2 < 9 - 2$
$-4x \leq 5$ and $4x < 7$	$-5 \leq 4x < 7$
$x \geq -\dfrac{5}{4}$ and $x < \dfrac{7}{4}$	Divide by 4: $-\dfrac{5}{4} \leq x < \dfrac{7}{4}$
$-\dfrac{5}{4} \leq x < \dfrac{7}{4}$	

CHECK YOUR UNDERSTANDING 1.20

Solve:
$$\frac{3x}{5} - \frac{2-x}{3} + 1 < \frac{x-1}{15}$$

Suggestion: Begin by multiplying both sides of the inequality by 15 so as to eliminate all denominators.

Answer: $x < -\dfrac{6}{13}$

1.4 Lines, Linear Equations and Inequalities   37

E X E R C I S E S

Exercises 1–8: Determine the slope of the line containing the given points.

1. $(2,5), (3,-4)$

2. $(4,-4), (5,-5)$

3. $(-2,3), (5,3)$

4. $(-3,-2), (-1,0)$

5. $(4,15), (15,4)$

6. $(0,0), (-3,-9)$

7. $(a,b), (2a,3b),$ where $a \neq 0$

8. $(a,b), (c,d),$ where $a \neq c$

Exercise 9:   On the same set of axes, sketch the four lines that pass through the origin and the additional point:

$$L_1 : (1,1); \quad L_2 : (1,5); \quad L_3 : (1,-1); \quad L_4 : (1,-5)$$

Determine the slope of each line.

Exercise 10:   On the same set of axes, sketch four distinct lines of slope 4.

Exercises 11–13: Determine another point on the line L, if

11. the slope of L is 2 and L contains the point (1,1).

12. the slope of L is $-\frac{1}{2}$ and L contains the point $(-2,0)$.

13. L is parallel to the line that contains the points (1,1) and $(-2,0)$, and L contains the point (2,7).

Exercise 14:   Determine a point (x,y) such that the line passing through the points $(1,-1)$ and (x,y) is perpendicular to the line that contains the points (3,7) and $(-1,2)$.

Exercise 15–18:   Determine the slope-intercept equation of the line that satisfies the given conditions. The line

15. contains the points $(-3,10)$ and $(2,0)$.

16. contains the points $(0,1)$ and $(-1,0)$.

17. contains the point (3,0) and is perpendicular to the line $y = 2x + 4$.

18. contains the point $(-2,0)$ and is parallel to the line $x + 3y = 6$.

Exercises 19–22: Determine a point-slope equation of the line that

19. contains the point $(3,-2)$ and has a slope of -1.

20. contains the points $(1,-2)$ and $(4,2)$.

21. contains the point $(-1,8)$ and is parallel to the line $y = -x - 3$.

22. contains the point $(4,5)$ and is perpendicular to the line $2x + 5y = 4$.

38   Chapter 1   Basic Concepts

Exercises 23–28:   Determine the equation of the horizontal line that satisfies the given conditions. The line

23. intersects the y-axis at 1.

24. has a y-intercept of 5.

25. contains the point $(3, -2)$.

26. passes through the origin.

27. passes through the y-intercept of the line containing $(6, -2)$ and $(-3, 4)$.

28. intersects $y = 3x - 1$ when $x = 1$.

Exercises 29–34:   Determine the equation of the vertical line that satisfies the given conditions. The line

29. intersects the x-axis at 1.

30. has an x-intercept of 5.

31. passes through the origin.

32. intersects $y = 3x - 1$ when $y = 1$.

33. passes through the x-intercept of the line containing $(6, -2)$ and $(-3, 4)$.

34. contains the point $(3, -2)$.

Exercises 35–44:   Determine the equation of the line that satisfies the given conditions, then sketch the line. The line

35. contains the points $(3,7)$ and $(1, -2)$.

36. contains the point $(3,7)$ and has a slope of -4.

37. passes through the origin and the point $(3,7)$.

38. has x-intercept 7 and y-intercept -4.

39. has x-intercept -2 and y-intercept 3.

40. contains the point $(2, -4)$ and is perpendicular to the line $x = -2$.

41. contains the point $(2, -4)$ and is perpendicular to the line $y = -2$.

42. contains the point $(3,6)$ and is parallel to the line $y - 2x + 4 = 0$.

43. is perpendicular to the line $2y - 6x = -7$ and contains the point $(1,4)$.

44. contains the point $(3,6)$ and passes through the point of intersection of the lines $y = 4$ and $x = -1$.

Exercises 45–50:   Solve the linear equation.

45. $-3x + 7 = 4x - 5 + x$

46. $-(-2x - 4) + (-x + 3) = 0$

47. $\dfrac{x - \frac{1}{2}}{\frac{1}{4}} = 0$

48. $\dfrac{2x}{3} - \dfrac{3}{2} + \sqrt{2}\,x = x - 1$

49. $-\dfrac{-2x + 5}{7} + x - 1 = 2x + 1$

50. $7x + \dfrac{2x - 4 - 3x}{2} = \dfrac{7x - 5 + 2x}{4}$

1.4 Lines, Linear Equations and Inequalities   **39**

Exercises 51–56: Solve for x in terms of y, and y in terms of x.

51. $2x - 3y = 5x - 6$

52. $2(x - 3y) = 5x - y\left(1 - \frac{3}{5}\right)$

53. $\dfrac{x}{2} - \dfrac{2x}{3} = \dfrac{5y}{2} + y - x$

54. $3yx - 2x + 4 = 3x - 4y - 1$

55. $\dfrac{2x - 3}{y} = \dfrac{x + 2}{y + 1}$

56. $\dfrac{x}{y} - \dfrac{3}{2} + \dfrac{2x}{3y} = 0$

Exercise 57:   Solve for a in terms of b, and then b in terms of a, if $2a - 3b = b + a - 4ab$.

Exercises 58–59: Determine a value of c for which the equation $4x - 5 + c = 3x - 2 - 2c - x$
has the given solution.

58. $x = -\frac{1}{2}$

59. $x = 0$

Exercises 60–82: Solve the linear inequality.

60. $4x - 5 > 9x + 10$

61. $6x + 5 \geq 2x + 3$

62. $4 - 6x \leq -(3x + 6) + 2$

63. $-(-2x - 3) \leq -(5x - 7) - \frac{1}{2}$

64. $2x + 1 \leq 2x - 1$

65. $2x - 1 < 2x + 1$

66. $x - \frac{1}{2} + \frac{1}{4}(2x - 5) \geq 0$

67. $\dfrac{x}{-2} + 3 - \dfrac{3x}{4} \leq 0$

68. $\dfrac{x - 3}{3} \geq \dfrac{4x - 1}{2} - 1$

69. $\dfrac{-3x - 1}{2} \leq \dfrac{2x + 4}{-4} + 5$

70. $-2(x - 3) + \dfrac{x + 5}{3} \leq -(-\frac{1}{2}x - \frac{1}{4})$

71. $\frac{2}{5}x + 3 > -\frac{1}{2} + x$

72. $-3 < -2x - 1 < 9$

73. $-5 < 1 - 3x \leq 7$

74. $\frac{2}{3} \geq \frac{1}{6}x + \frac{1}{2} > -\frac{1}{3}$

75. $3 \leq 5x - \frac{1}{2} < 9$

76. $0 > -5x - 1 > -1$

77. $-x + 2 < -2x - 1 < x + 1$

78. $\dfrac{3x - 1}{-2} \leq \dfrac{2x + 1}{4} < x - 5$

79. $4x - 5 \leq 3x + 1 \leq x - 7$

80. $2x - 3 < 3x + 5 < x$

81. $-2 < 2x - 2 < 2$ *AND* $2x - 3 > -2x + 1$

82. $-2 < 2x - 2 < 2$ *OR* $2x - 3 > -2x + 1$

Exercise 83: Prove Theorem 1.7; that two nonvertical lines with slopes m_1 and m_2 are perpendicular if and only if $m_1 = -\dfrac{1}{m_2}$.

[Hint: In the figure below, either show that triangles OAC and OCB are similar so that corresponding sides are proportional; or use the Pythagorean Theorem on the right triangle OAB. First, note that the lengths of AC and BC, resp., are m_1 and $-m_2$ (why?).]

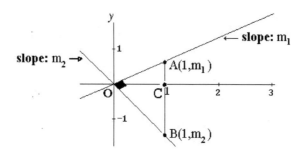

Exercise GC84: Use a graphing calculator to view several lines

$$\frac{x}{a} + \frac{y}{b} = 1$$

for different values of a and b. From those graphs it appears that the meaning of a is _____ and the meaning of b is _____.

Exercise 85: Show that (in the plane) a line perpendicular to a line perpendicular to a line L is parallel to L.

Exercise 86: Show that the line containing the points (a, b) and (c, d) is perpendicular to the line containing $(b, -a)$ and $(d, -c)$. What if $a = c$? What if $b = d$?

Exercise 87: Discuss why the solution set of the inequality $ax + b < cx + d$ can be expressed in one of the following forms:

(i) $\emptyset$, (ii) $(-\infty, \infty)$, (iii) $(-\infty, C)$, (iv) (C, ∞)

What conditions determine the form of the solution set? If the solution set is of type (iii) or (iv), what is the value of C in terms of a, b, c, and d?

§5. SYSTEMS OF LINEAR EQUATIONS

The following example illustrates two methods that can be used to solve a system of two linear equations in two unknowns.

To solve a system of several equations in several variables is to determine values of the variables which simultaneously satisfy each equation in the system. For example, $x = 1, y = 2$ is a solution of the following system of two equations in two unknowns:

$$\left. \begin{array}{r} 2x + y = 4 \\ x - 3y = -5 \end{array} \right\}$$

since both equations are satisfied when the indicated values are substituted for the variables:

$$2 \cdot 1 + 2 = 4 \quad \text{Check!}$$
$$1 - 3 \cdot 2 = -5 \quad \text{Check!}$$

EXAMPLE 1.26 Solve the following system of two equations in two unknowns.

$$\left. \begin{array}{rl} (1): & -3x + y = 2 \\ (2): & 2x + 2y = 5 \end{array} \right\}$$

SOLUTION:

ELIMINATION METHOD: Add (or subtract) a multiple of one equation to a multiple of the other, so as to eliminate one of the variables and arrive at one equation in one unknown:

$$\begin{array}{rl} \text{multiply equation 1 by 2:} \quad 2 \times (1): & -6x + 2y = 4 \\ (2): & \underline{2x + 2y = 5} \\ \text{subtract:} \quad -8x = -1 \end{array}$$

$$\boxed{x = \frac{1}{8}}$$

Substituting $x = \frac{1}{8}$ in (1) [we could have chosen (2)], we have:

$$-3\left(\frac{1}{8}\right) + y = 2$$

$$\boxed{y} = 2 + \frac{3}{8} = \boxed{\frac{19}{8}}$$

SUBSTITUTION METHOD:

$$\left. \begin{array}{rl} (1): & -3x + y = 2 \\ (2): & 2x + 2y = 5 \end{array} \right\}$$

Solving for y in (1), we have:

$$y = 3x + 2 \tag{*}$$

Substituting this value in (2) yields:

$$2x + 2(3x + 2) = 5$$
$$2x + 6x + 4 = 5$$
$$8x = 1$$

$$\boxed{x = \frac{1}{8}}$$

Returning to (*), we find the corresponding y-value:

$$\boxed{y} = 3 \cdot \frac{1}{8} + 2 = \boxed{\frac{19}{8}}. \quad \text{Answer:} \left(x = \frac{1}{8}, y = \frac{19}{8} \right).$$

Our glimpses feature the TI-84+ graphing calculator.

The determination of a window to adequately display the graph of a given function is an important part of the graphing process. In this chapter, a suitable window size will be provided. Hints and suggestions that will help you find your own window size will surface in subsequent chapters.

GRAPHING CALCULATOR GLIMPSE 1.1

The graphing capability of calculators can be used to find an approximate solution to the system of equations in Example 1.26 [actually, in this case, it yields the exact solution (in decimal form)]. We first rewrote the given equations in function form (solved for y in terms of x):

(1): $y = f(x) = 3x + 2$ and (2): $y = g(x) = -x + \frac{5}{2}$ (see [A])

The point at which the graphs of those two functions intersect, $x = 0.125, y = 2.375$, satisfies both of the given equations, and is therefore the solution of the system. (see [B]).

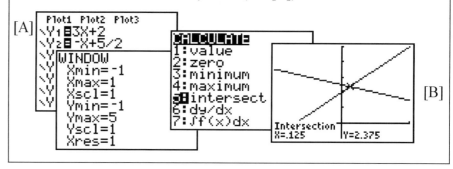

Answer: $\left(x = -1, y = \frac{1}{2}\right)$

CHECK YOUR UNDERSTANDING 1.21

Solve:
$$\left. \begin{array}{r} 3x + 4y = -1 \\ x + 2y = 0 \end{array} \right\}$$

LARGER SYSTEMS OF LINEAR EQUATIONS

The Elimination Method is generally used when solving ("by hand") a system of three or more linear equations. Basically, one applies the elimination method repeatedly, each time lowering the number of equations and unknowns by one, until ultimately arriving at a single equation in one unknown. The following examples illustrate the process.

EXAMPLE 1.27 Solve:
$$\left. \begin{array}{rl} (1): & x + 3y + 2z = 9 \\ (2): & 2x + y - z = -2 \\ (3): & -x - y + 2z = 1 \end{array} \right\}$$

SOLUTION: We begin by "extracting" a 2×2 system in the variables y and z [see equations (4) and (5) below]:

Add equation (1) to equation (3) to eliminate x:

$$(1): \quad x + 3y + 2z = 9$$
$$(3): \quad -x - y + 2z = 1$$
$$\text{add:} \quad 2y + 4z = 10$$
$$(4): \quad y + 2z = 5$$

Using the remaining equation (2), and either of the other two original equations, we again eliminate the variable x:

$$(2): \quad 2x + y - z = -2$$

multiply equation 3 by 2: $2 \times (3): \quad -2x - 2y + 4z = 2$

$$\text{add:} \quad (5): \quad -y + 3z = 0$$

At this point, we have a system of two linear equations in two unknowns [equations (4) and (5)], which can easily be solved:

$$(4): \quad y + 2z = 5$$
$$(5): \quad -y + 3z = 0$$
$$\text{add:} \quad 5z = 5$$
$$\boxed{z = 1}$$

We eliminated the variable x, and ended up with a 2×2 (read: two-by-two) system involving the variables y and z. We could, just as well, have eliminated the variable z, (or the variable y), to end up with a 2×2 system involving the variables x and y (or the variables x and z).

Substituting $z = 1$ in equation (4), we have:
$$y + 2(1) = 5$$
$$\boxed{y = 3}$$

Substituting $z = 1$ and $y = 3$ in (1):
$$x + 3(3) + 2(1) = 9$$
$$\boxed{x = -2}$$

A TI-92 teaser:

We have shown that $(x = -2, y = 3, z = 1)$ is the solution of the given system of equations; a claim that is easily checked by substituting directly in each of the three equations:

(1): $x + 3y + 2z = 9$ (1): $-2 + 3(3) + 2(1) = 9$ Yes
(2): $2x + y - z = -2$ $\begin{array}{c} x = -2 \\ y = 3 \\ z = 1 \end{array}$ (2): $2(-2) + 3 - 1 = -2$ Yes
(3): $-x - y + 2z = 1$ (3): $-(-2) - 3 + 2(1) = 1$ Yes

CHECK YOUR UNDERSTANDING 1.22

Solve:
$$\left.\begin{array}{r} x + y - 2z = 2 \\ 2x + y + z = 0 \\ -3x + 2y - z = -4 \end{array}\right\}$$

Answer: $(x = 1, y = -1, z = -1)$

44   Chapter 1 Basic Concepts

EXAMPLE 1.28 Solve:

$$\left.\begin{array}{rl} (1): & 2x + 3y = 7 \\ (2): & y + 2z = -1 \\ (3): & x + 2y + 3z = 1 \end{array}\right\}$$

SOLUTION: There are only two variables involved in the first two equations, but they are not the same two variables. We chose to eliminate x from (1) and (3). This will give us two equations in the two variables y and z: equation (2), and the resulting equation (4) below.

$$\begin{array}{rl} (1): & 2x + 3y \qquad\quad = 7 \\ \text{multiply equation 3 by 2:}\quad 2 \times (3): & 2x + 4y + 6z = 2 \\ \hline \text{subtract:}\quad (4): & \qquad -y - 6z = 5 \end{array}$$

Using equations (2) and (4), we solve for y and z:

$$\begin{array}{rl} (2): & y + 2z = -1 \\ (4): & -y - 6z = 5 \\ \hline \text{add:} & -4z = 4 \\ & \boxed{z = -1} \end{array}$$

Substituting -1 for z in (2):

$$y + 2(-1) = -1$$
$$\boxed{y = 1}$$

Substituting 1 for y in (1):

$$2x + 3(1) = 7$$
$$\boxed{x = 2}$$

We have shown that $(x = 2, y = 1, z = -1)$ is the only solution of the given system of equations, and leave it for you to check the result.

CHECK YOUR UNDERSTANDING 1.23

Solve:

$$\left.\begin{array}{rl} x + y - 2z & = -3 \\ 2x + y & = 4 \\ y + z & = 5 \end{array}\right\}$$

Answer:
$(x = 1, y = 2, z = 3)$

INDEPENDENT, INCONSISTENT, AND DEPENDENT SYSTEMS OF EQUATIONS

When a system of equations has exactly one solution, it is said to be **independent**, and when it has more than one solution, it is said to be **dependent**. A system that has no solution is called **inconsistent**. These mutually exclusive situations are depicted in Figure 1.5 for systems of two linear equations in two unknowns. In the independent case, the two lines representing the two equations intersect at exactly one point (exactly one solution) [Figure 1.5(a), and Example 1.26]. In the dependent case the "two" lines coincide and the system has infinitely many solutions [Figure 1.5(b) with solutions: $x = t, y = t + 1$, where t is any number]. In the inconsistent case, the two lines are distinct and parallel and have no point of intersection (no solution) [Figure 1.5(c)].

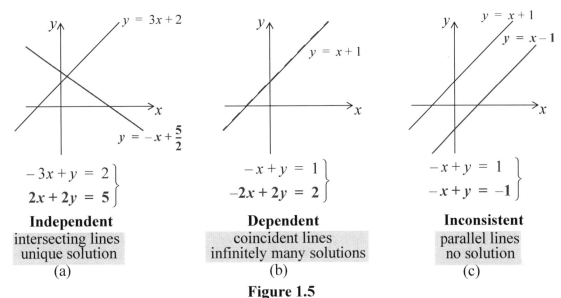

Figure 1.5

It can be shown that just as the graph of a linear function of one variable $y = f(x) = ax + b$ is a line in two-dimensional space, the graph of a linear function of two variables, $z = g(x, y) = ax + by + c$, is a plane in three-dimensional space. Figure 1.6 displays the geometrical distinction between independent, dependent, and inconsistent systems of three linear equations in three unknowns.

There are many other ways that the three planes can lead to the three cases depicted in Figure 1.6.

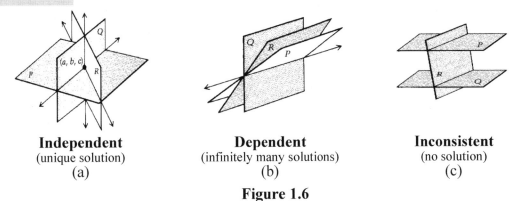

Figure 1.6

46   Chapter 1 Basic Concepts

> ### APPLICATIONS

We now return to the challenge of "translating" word problems into equations.

EXAMPLE 1.29

INVESTMENT OPTIONS

You have $20,000 to invest in high-risk, medium-risk, and low-risk stocks, with estimated yearly returns of 15%, 11%, and 7%, respectively. To play it "safe," you decide to put four times as much money in the low-risk stock as in the high risk stock. How much should be invested in each commodity, if your investment goal is to realize an (estimated) rate of return of 9.5% on your investment?

SOLUTION:

SEE THE PROBLEM		
RISK	**$**	**RETURN**
High	h	$.15h$
Medium	m	$.11m$
Low	l	$.07l$
Sum	20,000	$.15h + .11m + .07l$

Requirement: $l = 4h$ To equal 9.5% of 20,000

Let h, m, and l denote the dollar amounts to be invested in the high-, medium-, and low-risk stocks, respectively. We have to find three equations in those three unknowns. We SEE:

$$(1): \quad h + m + l = 20,000$$

$$\frac{15}{100}h + \frac{11}{100}m + \frac{7}{100}l = \frac{9.5}{100} \cdot 20,000$$

multiply both sides by 100

$$(2): \quad l = 4h$$

$$\rightarrow (3): 15h + 11m + 7l = 190,000$$

The rest is routine. Substituting $4h$ for l in (1) and (3), we have:

$$h + m + 4h = 20,000: \quad (4): \quad 5h + m = 20,000$$

$$15h + 11m + 28h = 190,000: \qquad 43h + 11m = 190,000$$

$$11 \times (4): \quad \underline{55h + 11m = 220,000}$$

$$\text{subtract:} \quad -12h \qquad = -30,000$$

$$h = 2,500$$

From (2): $l = 4h = 4(2,500) = 10,000$

From (4): $m = 20,000 - 5(2,500) = 7,500$

Conclusion: Invest $2,500 in the high-risk stock, $7,500 in the medium-risk stock, and $10,000 in the low-risk stock.

1.5 Systems of Linear Equations 47

CHECK YOUR UNDERSTANDING 1.24

A health-conscious individual wants to get her daily requirements of three minerals (A, B, and C) by eating the three food items (X, Y, and Z) shown in the table. How many ounces of each food item should she eat to exactly attain her daily requirements?

Mineral	Requirement (in mg)	X (mg/oz)	Y (mg/oz)	Z (mg/oz)
A	24	5	3	3
B	11	2	1	3
C	15	1	5	2

Answer: X: 3 oz; Y: 2 oz; Z: 1 oz

48   Chapter 1   Basic Concepts

EXERCISES

Exercises 1–8:   Solve the system of equations.

1. $\left.\begin{aligned} 3x + y &= -1 \\ 5x + 3y &= 5 \end{aligned}\right\}$

2. $\left.\begin{aligned} 4x - 7y &= 3 \\ 3x - 2y &= 12 \end{aligned}\right\}$

3. $\left.\begin{aligned} 3x - y &= 2 \\ \frac{2x}{3} + 1 &= \frac{y}{7} + 2 \end{aligned}\right\}$

4. $\left.\begin{aligned} \frac{5x + y}{6} &= \frac{-x}{6} + \frac{1}{3} \\ \frac{2x}{3} &= \frac{1}{9} - \frac{y}{3} \end{aligned}\right\}$

5. $\left.\begin{aligned} x + y &= 23 \\ y + z &= 25 \\ z + x &= 24 \end{aligned}\right\}$

6. $\left.\begin{aligned} 3x + 2y - z &= 4 \\ 3x - 2y + z &= 5 \\ 4x - 5y - z &= -1 \end{aligned}\right\}$

7. $\left.\begin{aligned} x - 4y + z &= -2 \\ 2x + y - z &= -1 \\ 3x + 2y - z &= 0 \end{aligned}\right\}$

8. $\left.\begin{aligned} 2x + 3y &= 4 \\ -x + 4z &= 1 \\ y + 3z &= 4 \end{aligned}\right\}$

Exercises GC9–GC12:   For each system, use a graphing calculator to simultaneously view the graphs of the two equations, and answer (a)–(c) below.

9. $\left.\begin{aligned} 6x - y &= 2 \\ 4x + \frac{2y}{3} &= 4 \end{aligned}\right\}$

10. $\left.\begin{aligned} x + 5y &= \frac{1}{2} \\ 2x + \frac{y}{2} &= \frac{3}{2} \end{aligned}\right\}$

11. $\left.\begin{aligned} 6x + 5y &= -6 \\ 21x + 14y &= -17 \end{aligned}\right\}$

12. $\left.\begin{aligned} 3x + 6y &= 1 \\ -5x + 4y &= -4 \end{aligned}\right\}$

(a) From the graphs, determine, approximately, the points of intersection, as in Graphing Calculator Glimpse 1.1.

(b) Solve the system by hand (algebraically) to determine the exact solutions.

(c) Compare your answers in (a) and (b).

Exercise 13:   Find the points of intersection of the two lines

$$7(x - y) = 0 \quad \text{and} \quad 2x - y = 10$$

Exercise 14:   Determine the constant c such that the following system is dependent.

$$\left.\begin{aligned} 6x - 9y &= -3 \\ -4x + 6y &= c \end{aligned}\right\}$$

1.5 Systems of Linear Equations   49

Exercise 15:   Determine any specific values of c and d such that the system

$$\left.\begin{array}{r} 3x - 4y = 1 \\ cx - dy = 2 \end{array}\right\}$$

(a) is dependent.   (b) is inconsistent.

Exercise 16:   Find a, b, and c such that $(x = 3, y = -1, z = 1)$ is a solution of

$$\left.\begin{array}{r} 6x + ay + bz = c \\ cx - by - az = 42 \\ bx + cy + az = 10 \end{array}\right\}$$

Exercise 17: (**Finding Two Numbers**) Find two numbers whose sum is -1 and whose difference is 10.

Exercise 18: (**Mixture**) A certain mixture consists of acid and water. If 5 gallons of acid are added, it will produce a mixture that is one-half acid and one-half water. If, on the other hand, 5 gallons of water are added, the resulting mixture is one-third acid. How many gallons of acid does the original mixture contain?

Exercise 19: (**River Current**) A boat can travel 50 miles downstream in 3 hours, and 40 miles upstream in 5 hours. Find the speed of the current.

Exercise 20: (**Baseball**) Batting averages indicate the percentage of times a player hit safely in the season. A player who hit safely 32% of the time, for example, has a batting average of .320. When a rookie came up to bat early in the season, the commentator noted that if the rookie gets a hit, his batting average would be .300, but his average would drop to .250 if he fails to get a hit. How many hits out of how many at bats did the rookie have?

Exercise 21: (**Determining a Parabola**) Just as two points determine a line in the plane, three points not lying on a line, determine a parabola. Find the quadratic function, $f(x) = ax^2 + bx + c$, whose parabolic graph passes through the three points $(1, 2)$, $(-1, 8)$, and $(2, 11)$.

Exercise 22: (**Health Foods**) A health-conscious individual wants to get her daily requirements of three minerals by eating the three food items shown in the table. How many ounces of each food should she eat?

Mineral	Req. (in mg)	Item #1 (mg/oz)	Item #2 (mg/oz)	Item #3 (mg/oz)
1	23	2	2	3
2	35	3	4	1
3	30	4	1	2

50   Chapter 1 Basic Concepts

Exercise 23: (**Mixture**) A pharmacist requires 10 ounces of a solution which is 35% alcohol. Solutions of 10%, 40%, and 50% are readily available. How much of each available solution should be used so that the mixture has the desired percentage of alcohol, if the pharmacist uses twice as much of the 50% solution as the 40 % solution?

Exercise 24: (**Investment**) A wealthy VIP has $1,000,000 to put into three investments paying, respectively, 8%, 10%, and 15%. The VIP wants an average return of 12% on the total investment, and 30% of the total must be invested at 10%. How much should be invested at each rate?

Exercise 25: (**Investment**) An investor wants an annual 7% dividend return on a $14,500 investment. He has narrowed consideration to two stocks – a speculative stock selling for $75 a share which pays an annual dividend of $6.50 a share, and a "blue-chip" that sells for $125 a share with an annual dividend of $6 a share. How many shares of each stock should he purchase?

Exercise 26: (**Mixture**) How many milliliters of a 25% acid solution must be mixed with a 60% acid solution to obtain seventy milliliters of a 50% solution?

Exercise 27: (**Basketball**) A basketball team scored 113 points with a combination of foul shots (1 point), field goals (2 points), and some 3-pointers. How many 1-point, 2-point, and 3-point shots were made, if a total of 58 shots hit their mark, with the number of successful foul shots exceeding the combined total of 2- and 3-pointers by 2?

Exercise 28: (**Loans**) A bank loaned $50,000 to a company for the development of three products. How much was loaned for each product, if the loan for product B was $6,000 more than the loan for product A, and the loan for product C was $10,000 less than the sum of the other two loans?

Exercise 29: (**Investment**) Mr. Smith invests $5,000 in two securities with anticipated annual returns of 8% and 15%, and another $5,000 in two securities with anticipated annual returns of 9% and 12% (the larger the anticipated return, the greater the risk). If the anticipated returns materialize, Mr. Smith will realize a return of $1,135 on his investment at year's end. Taking all of the money invested in the 15% option and dividing it equally in the remaining three options, would yield a return of $975 (assuming anticipated percentages). How much did Mr. Smith invest in the 15% security?

Exercises GC30–GC33: Use a calculator to solve the system approximately.

30.
$$\left.\begin{array}{r} 4x + 6y - 6z = -10 \\ 3x - 5y + 2z = 34 \\ -4x + 3y + 10z = -9 \end{array}\right\}$$

31.
$$\left.\begin{array}{r} -2x - 5y + z = -5 \\ x + 3y + 3z = -7 \\ x - 4y - 5z = 3 \end{array}\right\}$$

32.
$$\left.\begin{array}{r} 5x - 2y + 8z = 143 \\ 4x + 11y + 3z = 310 \\ 7x + 15y + 5z = -26 \end{array}\right\}$$

33.
$$\left.\begin{array}{r} 3x + 3y - 2z = 8.6 \\ -2x - 4y + 3z = -23.4 \\ x + 6y + 7z = 2.4 \end{array}\right\}$$

1.5 Systems of Linear Equations 51

Exercise 34: Show that the system

$$ax + by = E \, \Big\} \\ cx + dy = F \, \Big\}$$

is independent if and only if $ad - bc \neq 0$.

Exercise 35: (**Sales**) A university's book store stocks TI-83, TI-85, TI-89, and TI-92 graphing calculators. The TI-83 sells for \$85; the TI-85 for \$110; the TI-89 for \$125; and TI-92 for \$155. In the first week of the semester, the revenue from the sales of those calculators amounted to \$11,600. How many of each type of calculator were sold that week, if as many TI-83s were sold as the combined sales of the TI-85s and TI-89s, as many TI-85s were sold as the combined sales of the TI-89s and TI-92s, and twice as many TI-89s as TI-92s were sold?

Exercise 36: (**Mixture**) A pharmacist requires 10 ounces of a solution which is 23% alcohol. Solutions of 10%, 20%, 30%, and 40% are available. How much of each available solution should be used so that the mixture has the desired percentage of alcohol, if the pharmacist uses twice as much of the 30% solution as of the 20% solution, and one-third as much of the 40% solution as the 10% solution?

52   Chapter 1 Basic Concepts

§6. QUADRATIC EQUATIONS

An equation that can be expressed in the form $ax^2 + bx + c = 0$, with $a \neq 0$ is said to be a **quadratic equation**. As is illustrated in Example 1.30, such an equation can easily be solved when the quadratic polynomial is expressed in factored form. All one need do is apply the following result:

THEOREM 1.10
ZEROS AND FACTORS
A product is zero if and only if one of its factors is zero.

EXAMPLE 1.30   Solve:

$$2x^2 + x = -4x + 3$$

SOLUTION:

$$2x^2 + x = -4x + 3$$

Set equal to zero: $2x^2 + 5x - 3 = 0$

Factor: $(2x - 1)(x + 3) = 0$

Apply Theorem 1.10: $2x - 1 = 0$ or $x + 3 = 0$

$$x = \frac{1}{2} \text{ or } \quad x = -3$$

CHECK YOUR UNDERSTANDING 1.25

Solve the quadratic equation:

$$3x^2 + 3x = x^2 - 2x - 3$$

Answer: $-\frac{3}{2}, -1$

The quadratic equation $x^2 = a$, for $a \geq 0$, is easily solved:

$$x^2 = a$$
$$x^2 - a = 0$$
$$(x + \sqrt{a})(x - \sqrt{a}) = 0$$

$$x = -\sqrt{a} \text{ or } x = \sqrt{a} \quad \leftarrow \quad \begin{array}{c} \text{also written:} \\ x = \pm\sqrt{a} \end{array}$$

To summarize:

The equation $x^2 = a$ has no (real) solution if $a < 0$.

THEOREM 1.11   For $a \geq 0$, if $x^2 = a$, then $x = \pm\sqrt{a}$

1.6 Quadratic Equations 53

For example, if $x^2 = 9$, then $x = \pm\sqrt{9} = \pm 3$. But do not miss out on the pattern:

For $a \geq 0$: if $\boxed{}^2 = a$ then $\boxed{} = \pm\sqrt{a}$

For example: if $(x-3)^2 = 5$ then $x - 3 = \pm\sqrt{5}$

In the following example, we use this pattern to solve a fourth degree polynomial equation.

EXAMPLE 1.31 Solve:
$$(x^2 - 3x - 2)^2 = 4$$

SOLUTION: From the pattern: $x^2 - 3x - 2 = \pm 2$

$$
\begin{array}{lll}
x^2 - 3x - 2 = 2 & \textbf{OR} & x^2 - 3x - 2 = -2 \\
x^2 - 3x - 4 = 0 & & x^2 - 3x = 0 \\
(x - 4)(x + 1) = 0 & & x(x - 3) = 0 \\
x = 4 \text{ or } x = -1 & & x = 0 \text{ or } x = 3
\end{array}
$$

We see that $-1, 0, 3, 4$ are the solutions of the given equation.

CHECK YOUR UNDERSTANDING 1.26

Solve:
$$(x^2 - 6x - 8)^2 = 64$$

Answer: $-2, 0, 6, 8$

COMPLETING THE SQUARE

PERFECT SQUARES

When polynomials of the form:
$$(x + a)^2 \quad \text{and} \quad (x - a)^2$$

(called **perfect squares**) are expanded, they take on the following form:

$$
\begin{array}{l}
(x + a)^2 = x^2 + 2ax + \boxed{a^2} \leftarrow \\
(x - a)^2 = x^2 - 2ax + \boxed{a^2} \leftarrow
\end{array}
$$

The square of $\frac{1}{2}$ the coefficient of x

In particular, to be a perfect square, the question mark in the expression:
$$x^2 + 8x + \boxed{?}$$

must be replaced by the square of one-half the coefficient of x:
$$x^2 + 8x + \mathbf{16} = (x + 4)^2$$
$$\left(\frac{8}{2}\right)^2 = \mathbf{16}$$

54   **Chapter 1 Basic Concepts**

This technique is called **completing the square**, and it can be used to solve any quadratic equation. In particular, on the left we use it to solve the equation $2x^2 - x - 2 = 0$, and mimic the procedure, step by step, to find a formula for the solution of the general quadratic equation $ax^2 + bx + c = 0$:

<u>**To solve the quadratic equation:**</u>

$$2x^2 - x - 2 = 0 \qquad\qquad\qquad ax^2 + bx + c = 0$$

Move the constant term to the right of the equal sign:

$$2x^2 - x = 2 \qquad\qquad\qquad ax^2 + bx = -c$$

Divide both sides of the equation by the coefficient of x^2:

$$x^2 - \frac{1}{2}x = 1 \qquad\qquad\qquad x^2 + \frac{b}{a}x = -\frac{c}{a}$$

Add to both sides of the equation
the square of one-half of the coefficient of x:

$$x^2 - \frac{1}{2}x + \left(\frac{1}{4}\right)^2 = 1 + \left(\frac{1}{4}\right)^2 \qquad\qquad x^2 + \frac{b}{a}x + \left(\frac{b}{2a}\right)^2 = -\frac{c}{a} + \left(\frac{b}{2a}\right)^2$$

Express the left side as a perfect square:

$$\left(x - \frac{1}{4}\right)^2 = \frac{17}{16} \qquad\qquad\qquad \left(x + \frac{b}{2a}\right)^2 = \frac{b^2 - 4ac}{4a^2}$$

Use Theorem 1.11: If $x^2 = a$ then $x = \pm\sqrt{a}$:

$$x - \frac{1}{4} = \pm\sqrt{\frac{17}{16}} \qquad\qquad\qquad x + \frac{b}{2a} = \pm\sqrt{\frac{b^2 - 4ac}{4a^2}}$$

Solve for x:

$$x = \frac{1}{4} \pm \frac{\sqrt{17}}{4} \qquad\qquad\qquad x = -\frac{b}{2a} \pm \frac{\sqrt{b^2 - 4ac}}{2a}$$

$$x = \frac{1 \pm \sqrt{17}}{4} \qquad\qquad\qquad x = \frac{-b \pm \sqrt{b^2 - 4ac}}{2a}$$

We have established:

The solution of a quadratic equation may have been known by the Hindus as early as 500 B.C. Sridhara (c. 1025) was the first, as far as we know, to give a formula which in modern day symbols is the Quadratic Formula. Similar formulas, expressing the solutions of general third and fourth degree polynomial equations in terms of radicals, were developed by Del Ferro (1465-1526, Italian) in 1515, and by Ferrari (1522-1565, Italian), respectively. One of the cornerstones of modern mathematics, leading to the development of the branch of mathematics called Galois Theory, was the work of Abel (1802-1829, Norwegian) and Galois (1811-1832, French) which established the fact that no algebraic formula is possible for the general solution of fifth and higher degree polynomial equations.

THEOREM 1.12

QUADRATIC FORMULA

The solutions of the quadratic equation:

$$ax^2 + bx + c = 0$$

are given by:

$$x = \frac{-b \pm \sqrt{b^2 - 4ac}}{2a}$$

The term $b^2 - 4ac$ under the square root symbol in the quadratic formula is called the **discriminant** of the quadratic equation $ax^2 + bx + c = 0$. There are three possibilities:

Positive discriminant: Since $b^2 - 4ac > 0$, the quadratic equation has two distinct solutions:

$$x = \frac{-b \pm \sqrt{b^2 - 4ac}}{2a}$$

Zero discriminant: Since $b^2 - 4ac = 0$, the quadratic equation has only one solution:

$$x = \frac{-b \pm \sqrt{b^2 - 4ac}}{2a} = -\frac{b}{2a}$$

Negative discriminant: Since $b^2 - 4ac < 0$, the quadratic equation has no (real) solutions, since no real number is associated with the expression $\sqrt{b^2 - 4ac}$ when $b^2 - 4ac$ is negative.

Note: When the discriminant is a perfect square the solutions are rational, and the factorization process is the easiest way to solve the equation. For example:

$$2x^2 - 5x - 3 = 0$$

Quadratic Formula:

$$x = \frac{-b \pm \sqrt{b^2 - 4ac}}{2a}$$

$$= \frac{-(-5) \pm \sqrt{(-5)^2 - 4(2)(-3)}}{2(2)}$$

$$= \frac{5 \pm \sqrt{49}}{4} = \frac{5 \pm 7}{4} = \frac{12}{4} \text{ or } \frac{-2}{4}$$

$$= 3 \text{ or } \frac{-1}{2}$$

Factorization Method:

$$2x^2 - 5x - 3 = 0$$

$$(2x + 1)(x - 3) = 0$$

$$x = -\frac{1}{2}, x = 3$$

56   Chapter 1 Basic Concepts

EXAMPLE 1.32   Solve the quadratic equation:

$$3x^2 - 4x - 3 = 0$$

SOLUTION: Since the discriminant is positive:

$$b^2 - 4ac = (-4)^2 - 4(3)(-3) = 52$$

the   equation   has   two   distinct   solutions:

$$x = \frac{-b \pm \sqrt{b^2 - 4ac}}{2a} = \frac{-(-4) \pm \sqrt{52}}{2 \cdot 3} = \frac{4 \pm \sqrt{4 \cdot 13}}{2 \cdot 3}$$

$$= \frac{2 \cdot 2 \pm 2\sqrt{13}}{2 \cdot 3} = \frac{2 \pm \sqrt{13}}{3}$$

That is: $x = \dfrac{2 - \sqrt{13}}{3}$ and $x = \dfrac{2 + \sqrt{13}}{3}$

Incidentally, by the Corollary to Theorem 1.4, page 13, we can now factor $3x^2 - 4x - 3$:

Since $\dfrac{2 - \sqrt{13}}{3}$ is a zero, $\left(x - \dfrac{2 - \sqrt{13}}{3}\right)$ is a factor

$$3x^2 - 4x - 3 = 3\left(x - \frac{2 - \sqrt{13}}{3}\right)\left(x - \frac{2 + \sqrt{13}}{3}\right)$$

Since $\dfrac{2 + \sqrt{13}}{3}$ is a zero, $\left(x - \dfrac{2 + \sqrt{13}}{3}\right)$ is a factor

Required to make the leading coefficient 3

EXAMPLE 1.33   Solve the quadratic equation:

$$4x^2 + 2x + 1 = 0$$

SOLUTION: Since the discriminant is negative:

$$b^2 - 4ac = 2^2 - 4(4)(1) = -12$$

the equation has no (real) solution.

CHECK YOUR UNDERSTANDING 1.27

Solve the quadratic equation $x^2 - 5 = -2x^2 + 5x - 7$ by:
(a) Completing the square
(b) Factoring
(c) Using the Quadratic Formula.

Answer: $\frac{2}{3}, 1$

EXAMPLE 1.34 Show that for any $d \neq 0$, the quadratic polynomials in the factorization of a sum and difference of two cubes:
$$x^3 + d^3 = (x + d)(x^2 - dx + d^2)$$
$$x^3 - d^3 = (x - d)(x^2 + dx + d^2)$$
can not be factored further.

We changed the "a" in Theorem 1.6, page 14, to a "d" so as to avoid confusion with the "a" in the quadratic formula.

SOLUTION: The discriminant of both $x^2 - dx + d^2$ and $x^2 + dx + d^2$ are negative:
$$b^2 - 4ac = d^2 - 4(1)(d^2) = -3d^2$$

It follows that neither of those quadratic polynomials has a (real) zero, and consequently, neither can be further factored. Such a polynomial is said to be **irreducible**.

EXAMPLE 1.35

CONSTRUCTING A CARTON

From a rectangular piece of cardboard, with one side twice the length of the other, a 5-inch-deep carton is to be constructed by cutting the same size square from each corner of the rectangle, and then folding up the resulting cross-like configuration. What must be the dimensions of the cardboard, if the carton is to have a volume of 500 cubic inches?

SOLUTION:

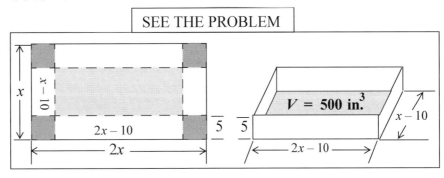

58   Chapter 1 Basic Concepts

> The volume of the carton is the area of its base times its height.

Let x denote the length (in inches) of the shorter side. Focusing on the 500 cubic inch volume leads us to the equation:

> Note that the units match up, and we can cancel the common unit, $in.^3$:

$$[(2x-10)\text{ in.}][(x-10)\text{ in.}][5\text{ in.}] = 500\text{ in.}^3$$
$$(2x-10)(x-10) = 100$$
$$(x-5)(x-10) = 50$$
$$x^2 - 15x + 50 = 50$$
$$x(x-15) = 0$$
$$x = 0 \quad \text{or} \quad x = 15$$

Of the two roots, only 15 is a candidate for the solution of the problem, since length must be a positive quantity. Let's check our result in the problem itself:

If $x = 15$, then the dimensions of the cardboard are 15 in. by 30 in. Cutting a 5 inch square from each corner, and folding the cross-like configuration as instructed, we end up with a box of height 5 in. and a base of area:

$$[(15-2\cdot5)\text{ in.}][(30-2\cdot5)\text{ in.}] = (5\text{ in.})(20\text{ in.}) = 100\text{ in.}^2$$

For a volume of: $(100\text{ in.}^2)(5\text{ in.}) = 500\text{ in.}^3$ —Check!

CHECK YOUR UNDERSTANDING 1.28

A farmer wishes to enclose a 160,000 square foot rectangular field using 1200 ft of fencing. By taking advantage of the straight bank of a river, only three sides of the region have to be enclosed. Find the dimensions of all such possible fields.

> Answer: 400 ft by 400 ft; or 200 ft by 800 ft, with the 800 ft parallel to the river.

1.6 Quadratic Equations   59

EXERCISES

Exercises 1–6: Solve the quadratic equation by factoring.

1. $x^2 - 2x - 15 = 0$

2. $7x^2 - 10x - 8 = 0$

3. $2x^2 + 3x - 5 = x^2 - 1$

4. $2x^2 - \frac{1}{2}x - \frac{1}{4} = 0$

5. $5x^2 - 5 = x^2 - 3x + 11x$

6. $12x(3x + 5) = -25$

Exercises 7-14: Use the pattern that if $x^2 = a$ then $x = \pm\sqrt{a}$ to solve the quadratic equation.

7. $49x^2 = 4$

8. $(x - 6)^2 = 3$

9. $(3x + 5)^2 = 9$

10. $\left(\dfrac{x-3}{2}\right)^2 - \dfrac{1}{2} = 0$

11. $(4x - 5)^2 = x^2$

12. $(x - 5)^2 = (2x + 3)^2$

13. $(x^2 + 3x - 5)^2 = 25$

14. $(x^2 - 3x - 1)^2 = 9$

Exercises 15–18: Solve by completing the square.

15. $x^2 + 8x - 2 = 0$

16. $x^2 + 5x + 3 = 0$

17. $4x^2 + 8x - 1 = 0$

18. $-2x^2 + 3x + 3 = 0$

Exercises 19–26: (a) Use the quadratic formula to solve the equation.
(b) Factor the quadratic polynomial using the zeros obtained in (a).

19. $-3x^2 + 2x + 7 = 0$

20. $5x^2 - 3x + 2 = 0$

21. $\dfrac{(x + 2)^2}{3} = x$

22. $5x^2 - 2x - 6 = -4x^2 + x$

23. $-x^2 - 4x + 3 = 0$

24. $-2x^2 + 3x = \dfrac{1}{4}$

25. $(3x - 2)(x + 3) = 5x^2 - 2$

26. $(4x - 1)(x + 2) = (x - 3)(x + 1)$

Exercises 27–36: Solve the quadratic equation.

27. $8x^2 + 2x = 15$

28. $\frac{1}{2}(3x + 2)^2 - 8 = 0$

29. $x^2 - 5x - \dfrac{3}{4} = 0$

30. $x(x + 3) = -3$

31. $\left(\dfrac{2x}{3} - 1\right)^2 = 0$

32. $x^2 + 3x - 8 = 0$

33. $-x^2 - 4x - 1 = 0$

34. $-\frac{1}{2}x^2 + \frac{1}{4}x + \frac{1}{16} = 0$

35. $3x^2 + 3x = 0$

36. $3(x - 1)^2 = (x - 1) + 2$

60   Chapter 1 Basic Concepts

Exercise 37:   (**Area**) A letter is to be typed on an 8 inch by 10 inch sheet of paper. The side margins of the letter are to be twice that of the common dimension of the top and bottom margins. What is the dimension of the top margin, if the writing area is to be 54 square inches?

Exercise 38:   (**Area**) What is the width of the frame of a 10 inch by 12 inch painting, if the framed painting covers an area of 224 square inches?

Exercise 39:   (**Volume**) A 4-inch deep carton is constructed by cutting the same size square from each corner of a rectangular piece of cardboard whose length is 3 times its width, and then bending up the sides. What are the dimensions of the cardboard, if the carton has a volume of 448 in.^{3}?

Exercise 40:   (**Integers**) It can be shown that the sum of the first n positive integers is $\frac{n^2 + n}{2}$. (For example, the sum of the first three positive integers is $1 + 2 + 3 = 6$, and substituting 3 for n in the formula also gives 6.) Determine how many consecutive positive integers must be added together to obtain a sum of 171.

Exercise 41:   (**Stopping Distance**)The distance (in feet) required to stop a car traveling at m miles per hour is given approximately by $x = \frac{1}{20} m^2 + m$.

  (a) Determine the distance required to stop, if the car is traveling at 50 mph?

  (b) What is the fastest a car can be moving and still stop in 100 ft?

Exercise 42:   (**Volume**) A cylinder has a diameter, d, and a height of the same length. By what amount can the diameter be increased and the height decreased, so as to leave the volume unchanged?
($V = \pi r^2 h$,   where r is the radius and h is the height of the cylinder.)

Exercises 43–46:   (**Free Fall**) Near the surface of the earth, the height $s(t)$ above the ground (in feet) of an object subject only to the force of gravity is given by

$$s(t) = -16t^2 + v_0 t + s_0$$

where t is time in seconds, v_0 is initial velocity and s_0 is the initial height of the object. Velocity is positive when the object is going up, and negative when it is coming down.

  43. A ball is thrown downward from the top of a 200 foot building with a speed of 40 ft/sec. How long will it take the ball to hit the ground?

  44. An object is thrown downward with a velocity of -10 ft/sec from a height of 134 ft. When will it be 20 ft above the ground?

  45. An object is dropped from a height of 93 ft. How long will it take to hit the ground?

  46. A ball is tossed directly upward from the edge of the roof of a building at 62 ft/sec and hits the ground 8 seconds later. How high is the building?

Exercise 47: (**Area**) A window has the shape of a rectangle 3 feet high capped by a semicircle (see adjacent figure).

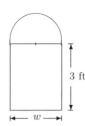

(a) Express the area of the window as a function of the width w of the rectangular part.

(b) Find w, if the area is 18 ft^2.

Exercise 48: (**Working Backwards**) Determine values of a and b so that $x = 1$ and $x = 5$ are solutions of the quadratic equation.

(a) $(x-a)(x+b) = 0$. Can there be more than one solution? Justify your answer.

(b) $\left(x + \frac{a}{2}\right)(x + 2b) = 0$

Exercise 49: (**Construct Equation**) Construct a quadratic equation with integer coefficients whose solutions are $\frac{1}{2}$ and $\frac{2}{3}$.

CHAPTER SUMMARY

INTEGER EXPONENTS	For any positive integer n and any number a: $$a^n = \underbrace{a \cdot a \cdots a}_{\text{n-times}}$$ If $a \neq 0$: $\qquad a^0 = 1$ and $a^{-n} = \dfrac{1}{a^n}$
ROOTS	If n is an **odd** positive integer and a is any number, then the $\boldsymbol{n^{\text{th}}}$ **root of a** is defined to be that number, denoted by $a^{\frac{1}{n}}$, such that: $$\left(a^{\frac{1}{n}}\right)^n = a$$ If n is an **even** positive integer, then a must be nonnegative, and the above notation is reserved for the nonnegative n^{th} root of a (called the **principal** n^{th} root of a).
RATIONAL EXPONENTS	For $\dfrac{m}{n}$ a positive rational number in lowest terms (no common factors), and any a if n is odd, or any nonnegative a if n is even: $$a^{\frac{m}{n}} = \left(a^{\frac{1}{n}}\right)^m = (a^m)^{\frac{1}{n}}$$ If $a \neq 0$: $\qquad a^{-\frac{m}{n}} = \dfrac{1}{a^{\frac{m}{n}}}$
In each of the following, when all expressions are defined:	For r and s rational numbers: (i) $\quad a^r a^s = a^{r+s}$ — When multiplying add the exponents (ii) $\quad \dfrac{a^r}{a^s} = a^{r-s}$ — When dividing subtract the exponents (iii) $\quad (a^r)^s = (a^r)^s = a^{rs}$ — A power of a power: multiply the exponents (iv) $\quad (ab)^r = a^r b^r$ — A power of a product equals the product of the powers (v) $\quad \left(\dfrac{a}{b}\right)^r = \dfrac{a^r}{b^r}$ — A power of a quotient equals the quotient of the powers

Cancellation Law	$$\dfrac{a\overset{\shortmid}{c}}{b\underset{\llcorner\;-\;-\;-\;}{c}} = \dfrac{a}{b}\quad \textbf{If } c \neq 0$$
POLYNOMIAL	A **polynomial of degree n** (in the variable x) is an algebraic expression of the form: $$a_n x^n + a_{n-1} x^{n-1} + \ldots + a_1 x + a_0$$ where n is a positive integer and a_0 through a_n are numbers, with $a_n \neq 0$.
Difference of two squares	$$a^2 - b^2 = (a+b)(a-b)$$
Sum and difference of two cubes	$$x^3 + a^3 = (x+a)(x^2 - ax + a^2)$$ $$x^3 - a^3 = (x-a)(x^2 + ax + a^2)$$
ZERO OF A POLYNOMIAL	c is a **zero of the polynomial** $p(x)$ if $p(c) = 0$.
Zeros and Factors	c is a zero of a polynomial if and only if $(x - c)$ is a factor of the polynomial.
RATIONAL EXPRESSION	A **rational expression** is an algebraic expression of the form $\dfrac{p(x)}{q(x)}$, where $p(x)$ and $q(x)$ are polynomials, with $q(x) \neq 0$.
Unit Conversion	Just use the appropriate conversion bridges.
SLOPE OF A LINE	For any nonvertical line L and any two distinct points (x_1, y_1) and (x_2, y_2) on L, the **slope** of L is the number m given by: $$m = \dfrac{y_2 - y_1}{x_2 - x_1} = \dfrac{\text{change in } y}{\text{change in } x}$$ The steeper the climb, the more positive the slope. The steeper the fall, the more negative the slope
PARALLEL AND PERPENDICULAR LINES	Lines are said to be **parallel** when they have the same slope, or when they are vertical. Lines that intersect at right angles are said to be **perpendicular**.
Two nonvertical lines are perpendicular if and only if the slope of one is the negative reciprocal of the slope of the other.	Two lines with nonzero slopes m_1 and m_2 are perpendicular if and only if: $$m_1 = -\dfrac{1}{m_2}$$

EQUIVALENT EQUATIONS	Equations that have the same solution sets are said to be equivalent. Adding (or subtracting) the same quantity to (or from) both sides of an equation, or multiplying (or dividing) both sides of an equation by the same nonzero quantity will not alter the solution set of the equation.
EQUATIONS OF LINES	**Slope-Intercept:** A point (x, y) is on the line of slope m and y-intercept b if and only if: $$y = mx + b$$ **Point-Slope:** A point (x, y) is on the line of slope m passing through the point (x_0, y_0) if and only if: $$y - y_0 = m(x - x_0)$$ **Vertical Line** (crossing the x-axis at a): $x = a$.
LINEAR EQUATION	An equation that can be written in the form $ax + b = 0$, for $a \neq 0$.
LINEAR INEQUALITY	One solves a linear inequality in exactly the same fashion as one solves a linear equation, with one notable exception: When multiplying or dividing both sides of an inequality by a **NEGATIVE** quantity, **REVERSE** the direction of the inequality sign.
SOLUTION OF A SYSTEM OF EQUATIONS	To solve a system of equations in several variables is to determine values of the variables which simultaneously satisfy each equation in the system.
Solution Process	To solve a system of equations, one repeatedly lowers the number of equations and unknowns by one, until ultimately arriving at one equation in one unknown.
INDEPENDENT, DEPENDENT, AND INCONSISTENT SYSTEMS OF EQUATIONS	When a system of equations has exactly one solution, it is said to be **independent**, and when it has more than one solution, it is said to be **dependent**. A system that has no solution is called **inconsistent**.
QUADRATIC EQUATION	An equation that can be written in the form $ax^2 + bx + c = 0$, for $a \neq 0$.
Products Equal to Zero	A product is zero if and only if one of its factors is zero.
Solutions of $x^2 = a$	$$x = \pm\sqrt{a}, \text{ if } a \geq 0$$ No (real) solutions, if $a < 0$.

Completing The Square	$$x^2 + ax + \boxed{?} \Rightarrow x^2 + ax + \left(\frac{a}{2}\right)^2 = \left(x + \frac{a}{2}\right)^2$$ leading coefficient must be 1

The Quadratic Formula	The solutions of the quadratic equation $ax^2 + bx + c = 0$ are given by: $$x = \frac{-b \pm \sqrt{b^2 - 4ac}}{2a}$$ The term $b^2 - 4ac$ is called the **discriminant** of the equation.
	     Positive discriminant: two distinct solutions.
	     Zero discriminant: one solution.
	     Negative discriminant: no (real) solution.

SOME COMMON PITFALLS

WRONG:	$\dfrac{3^{-2}}{a^{-1} + 2^{-2}} = \dfrac{a + 2^2}{3^2}$	Because:	The a^{-1} and 2^{-2} are **not factors** of the denominator. This is okay: $\dfrac{3^{-2}}{a^{-1}2^{-2}} = \dfrac{a \cdot 2^2}{3^2}$
WRONG:	$-2^2 = 4$	Because:	$-2^2 = -(2 \cdot 2) = -4$ This is okay: $(-2)^2 = 4$
WRONG:	$\sqrt{4} = \pm 2$	Because:	$\sqrt{4} = 2$ **principal** square root
WRONG:	If $x^2 = 4$, then $x = 2$	Because:	If $x^2 = 4$, then $x = \pm 2$
WRONG:	For all a and b: $\sqrt{a + b} = \sqrt{a} + \sqrt{b}$	Because:	$\sqrt{4 + 16} \neq \sqrt{4} + \sqrt{16}$ This is okay: $\sqrt{4 \cdot 16} = \sqrt{4}\sqrt{16}$

66   Chapter 1 Basic Concepts

C H A P T E R   R E V I E W   E X E R C I S E S

Exercises 1–16: Simplify.

1. $\dfrac{\frac{2}{3} - \frac{5}{4}}{1 - \frac{1}{2+4}}$

2. $\dfrac{2 + \frac{1}{1+3}}{5 + \frac{1}{2}}$

3. $(-3a^3 b)^2$

4. $\left(\dfrac{a^3}{2b^2}\right)^3$

5. $\left(-\dfrac{3a^{-3}}{b^{-2}}\right)^2$

6. $(2a^{-1}b^{-3})^2$

7. $\dfrac{(2a)^3 b^2 - b^3 a^2}{(ab)^2}$

8. $(2a^{\frac{2}{3}})^2$

9. $(a^{\frac{4}{3}} b^{\frac{2}{5}})^5$

10. $\dfrac{a^{-\frac{1}{3}} b^{-2}}{a^2 \sqrt{b}}$

11. $\dfrac{a^{\frac{3}{2}} b^2}{\sqrt{ab}}$

12. $\left(\dfrac{b^2 c}{c^{-3} b^4}\right)^{-\frac{1}{4}}$

13. $\dfrac{(-2a)^2(-\sqrt{b})^3}{(a^2 b)^{-\frac{1}{2}}}$

14. $\dfrac{(8a)^{\frac{2}{3}} b^{-\frac{1}{2}}}{\left(\dfrac{3b^{-1}}{2a}\right)^3}$

15. $\dfrac{-(9a)^{\frac{1}{2}} b^{\frac{1}{4}} c^{\frac{1}{2}}}{b^3 \left(\dfrac{-c}{3}\right)^2}$

16. $\dfrac{\sqrt{a}\, b^{-\frac{2}{3}} (a^2 b)^{-1}}{(a^2 b^{-3})^{\frac{1}{2}}}$

Exercises 17–20: Carry out the division, and then check your result by multiplying the quotient by the divisor and adding the remainder.

17. $(3x^3 - 16x^2 + 21x - 26) \div (x - 4)$

18. $(x^4 + 2x^3 - x + 2) \div (x + 3)$

19. $(x^7 + 2x^4 - 3x^3 - 4x - 1) \div (x^4 - 3)$

20. $(18x^4 + 18x^3 + 6x^2 - 3x - 5) \div (6x^3 + 2x^2 - 1)$

Exercises 21–26: Factor the polynomial.

21. $3x^2 + 7x + 2$

22. $3x^3 - 6x^2 + 5x - 10$

23. $2x^5 + 54x^2$

24. $\dfrac{y^2}{6} - \dfrac{5y}{6} + 1$

25. $(x + 5)^2 - (y + 3)^2$

26. $\dfrac{1}{4}x^4 - x^3 - x^2 + 4x$

Exercises 27–41: Combine and simplify.

27. $\left(\dfrac{2a - 3}{4 - 2a}\right)\left(\dfrac{2 - a}{3 - 2a}\right) - \dfrac{10 - 2a}{a - 5}$

28. $\dfrac{\left(\dfrac{a - b}{a - 2b}\right)\left(\dfrac{4b - 2a}{2a + 2b}\right)}{a - b}$

29. $\dfrac{-(3c + 1)(a - b) - (b - a)}{(-1 - 3c)(b - a)}$

30. $\dfrac{3x}{x - 1} + \dfrac{2}{x^2 + 2x - 3} - \dfrac{x}{x + 3}$

31. $\dfrac{4}{3x + 2} - \dfrac{x - 1}{9x^2 - 4} + \dfrac{1}{3x - 2}$

32. $4\left(\dfrac{2x - 1}{4 + 3x}\right)^3 \left[\dfrac{(4 + 3x)2 - (2x - 1)3}{(4 + 3x)^2}\right]$

33. $\dfrac{-4(3x - 1)^3 - 5x(1 - 3x)^2}{(6x - 2)^5}$

34. $\dfrac{-(2 - x)}{x^3 + 3x^2 - x - 3} - \dfrac{2x - 5}{x^3 + 7x^2 + 15x + 9}$

Chapter Review Exercises   **67**

35. $\left(\dfrac{x^2 + x - 6}{x^2 + 4}\right)\left(\dfrac{x^2 + x - 2}{x^3 + 4x^2 + x - 6}\right)$ 36. $\left(\dfrac{x^3 - 8}{x + 2}\right)\left(\dfrac{2x^2 + 8}{x^3 - 4x}\right)\left(\dfrac{x^3 + 2x^2}{x^3 + 2x^2 + 4x}\right)$

37. $\dfrac{2x^2 - 2x - 4}{3x^2 + 2x - 5} - \left(\dfrac{x^2 - 6x - 7}{x^2 - 8x + 7}\right)\left(\dfrac{2x + 1}{3x + 5}\right)$ 38. $-x(1 - x^2)^{-\frac{1}{2}} + \frac{1}{2}$

39. $\frac{10}{3}\, x^{-\frac{1}{3}} - \frac{4}{3}\, x^{-\frac{2}{3}}$ 40. $x^4(x^3 - 2)^{-\frac{2}{3}} + 2x(x^3 - 2)^{\frac{1}{3}}$ 41. $\frac{1}{2}(1 - x)^{-\frac{3}{2}} - (1 - x)^{-\frac{1}{2}}$

Exercises 42–46: Given $1\text{m} = 3.28$ ft, 1 mi $= 5280$ ft, $1\,l = 0.264$ gal.

42. Convert 255 mph to kilometers per hour.

43. Convert 40 cents per foot to dollars per meter.

44. Convert 15 l/sec^2 to gal./hr^2.

45. A German bolt has 10 threads per centimeter of length. How many threads per inch does it have?

46. The speed of light is approximately 3×10^8 meters per second. What is the speed of light in feet per second?

Exercises 47–49: Determine another point on the line L, if

47. the slope of L is 5 and L contains the point (3,7).

48. it is parallel to the line containing the points (2,3) and $(-1, 5)$ and contains the point (3,7).

49. it is perpendicular to the line containing the points (2,3) and $(-1, 5)$ and contains the point (3,7).

Exercises 50–54: Determine the slope-intercept equation of the line which

50. contains the points (3,5) and (2,7).

51. contains the point $(-3, 4)$ and has a y-intercept of -1

52. contains the point $(-4, -4)$ and has an x-intercept of -8.

53. contains the point (1,4) and is parallel to the line $y = 2x - 3$.

54. contains the point (1,4) and is perpendicular to the line $y = 2x - 3$.

68   Chapter 1 Basic Concepts

Exercises 55-58: Determine a point-slope equation of the line which

55. contains the point (4,2) and has a slope of -3.

56. contains the points (1,3) and (2,4).

57. contains the point $(-2,5)$ and is parallel to the line $y = 2x - 375$.

58. contains the point (1,2) and is perpendicular to the line $y - 4 = 3(x + 1)$.

Exercises 59–66: Determine the equation of the line that

59. passes through the points $(6, -1)$ and $(-7, 3)$.

60. is parallel to the line $2y + 3x = 4$ and passes through the origin.

61. is perpendicular to the line $x = 2$ and passes through (4,1).

62. is parallel to the line $x = 2$ and contains the point (4,1).

63. is perpendicular to the line $y = 1$ and contains the point $(4, -2)$.

64. is parallel to the line $y = 1$ and passes through $(4, -2)$.

65. contains the x-intercept of the line $3x - 2y = 6$ and is perpendicular to the line $2x + 5y = 1$.

66. is vertical and passes through an x-intercept of the graph of $f(x) = (x - 4)^2 - 1$.

Exercise 67: Show that the line passing through the points (1,1) and (2,6) is parallel to the line that contains the points $(-2, 3)$ and $(-1, 8)$.

Exercises 68–76: Solve the linear equation or inequality.

68. $2(x - 5) + 3x - 4 = 0$

69. $3x + 4\left(-x - \frac{1}{2}\right) = 1 - x$

70. $\sqrt{2}\,x + 2x = 5$

71. $\left(\frac{1}{8}\right)^{-\frac{2}{3}} x + \sqrt{3^2 + 4^2}\, x = (-8)^{\frac{4}{3}}$

72. $-(2x + 5) \leq -(3x + 4)$

73. $-5x + 5 \geq 2x - 15$

74. $2x - 7 > 6 + 2x$

75. $-\frac{1}{2}x + \frac{1}{4} > \frac{1}{2} + \frac{1}{4}x$

76. $\frac{7}{2}x - 5 < \frac{2}{3}x + \frac{1}{2}$

Chapter Review Exercises   69

Exercises 77–81: Solve the linear system of equations.

77. $\left.\begin{array}{l} \dfrac{y}{4} - \dfrac{1}{2} = x - 6 \\[2mm] \dfrac{x-5}{3} - 1 = y + \dfrac{x+1}{6} \end{array}\right\}$ 78. $\left.\begin{array}{l} 5x = 3y + 3 \\[1mm] 4y = -x - 4 \end{array}\right\}$ 79. $\left.\begin{array}{r} 2x + 3y - 4z = 4 \\ 3x - 4y + 20z = -15 \\ 4x + 5y - 8z = 4 \end{array}\right\}$

80. $\left.\begin{array}{r} 2x - 4y + z = 8 \\ x - 3y - 2z = 0 \\ x + y + 3z = 6 \end{array}\right\}$ 81. $\left.\begin{array}{r} 2x + y - z = -1 \\ 3x + 2y - z = 0 \\ x - 4y + z = -2 \end{array}\right\}$

Exercise GC82:   Use a graphing calculator to simultaneously view the graphs of the equations in the system, and determine, approximately, the solutions of the system of equations.

$$\left.\begin{array}{l} \dfrac{3x}{2} = \dfrac{y}{4} \\[2mm] \dfrac{x-y}{4} = 1 \end{array}\right\}$$

Exercises 83–86: Solve the quadratic equation.

83. $(2x - 5)^2 = 4$

84. $6x^2 = (x + 1)^2$

85. $x^2 + 8 = -2x^2 + 10x$

86. $x^2 - 3x - 3 = 0$

Exercises 87–90: Solve the quadratic equation by completing the square.

87. $\frac{1}{4}x^2 + x - 1 = 0$

88. $x^2 + 6x + 7 = 0$

89. $9x^2 - 18x + 8 = 0$

90. $3x^2 - 5x - 5 = 0$

Exercise 91:   (**Fencing**) A farmer used 1600 ft of fencing to enclose a rectangular field. What are the dimensions of the field, if one side is 37 ft longer than three-fourths the length of the other side?

Exercise 92:   (**Integers**) Twice the square of an integer is three less than seven times the integer. Find the integer.

Exercise 93:   (**Area**) The area of a triangle is given by the formula $A = \frac{1}{2}bh$, where b is the length of the base and h is the height of the triangle. Determine the height of a triangle, if the height is 5 inches more than the base, and the area is 33 square inches.

Exercise 94:   (**Radioactive Decay**) The rate of decay of a radioactive substance is proportional to the amount of substance present at that time.[1] A certain radioactive substance is found to decay at the rate of 2 grams per year when 90 grams of the substance are present. What is the rate of decay when 80 grams are present?

---

[1] A quantity y is said to be proportional to a quantity x if there is a nonzero constant k such that $y = kx$.

70   Chapter 1 Basic Concepts

Exercise 95:   (**Gasoline Consumption**) The number G of gallons of gasoline required per hour to fuel a car traveling at a constant speed v between 50 mi/hr and 70 mi/hr varies directly with the square of the speed ($G = kv^2$). How many gallons are required to maintain a speed of 65 mi/hr for 2 hours, if 6 gallons are needed to maintain a speed of 55 mi/hr for 3 hours?

Exercise 96:   (**Mixture**) A 10-gallon can contains 5 gallons of a solution that is 35 percent alcohol and 65 percent water. How much water must be added to the can if the alcohol concentration is to be reduced to 26 percent ?

Exercise 97:   (**Mixture**) A puppy mix is to contain 4 parts protein to each part carbohydrate. The mix is to be a blend of stock-A and stock-B mixes. Stock-A is 10% protein and 7% carbohydrate, while Stock-B is 15% protein and 3% carbohydrate. How many pounds of Stock-A and Stock-B are to be included in a 50-pound bag of the puppy mix?

Exercise 98:   (**Area**) A gravel border of uniform width is to frame a 25 feet by 20 feet flower garden. How wide should the border be, if there is enough gravel to cover 364 square feet?

CUMULATIVE   REVIEW   EXERCISES

Exercise 1–8:   Simplify.

1. $\dfrac{2\sqrt{a^2+1} - \dfrac{3}{\sqrt{a^2+1}}}{\left(a^2+1\right)^2}$

2. $\dfrac{(2a-4)^{-2}(a+b)^3}{(4-2a)^{-3}(-2a-2b)^2}$

3. $\dfrac{4(3x-4)(2x+1) - 3(2x+1)^2}{(3x-4)(6x-19)}$

4. $\dfrac{2x^2+6x}{x^2+2x} + \dfrac{2x-2}{3x+6}$

5. $\dfrac{6x^2-11x+4}{2x^2+x-1} \div (3x-4)$

6. $\left(\dfrac{3a}{2b^3}\right)^{-2} a^{\frac{2}{3}} b^{-\frac{2}{3}}$

7. $2(3x+2)^{-3} - 3(2-x)(3x+2)^{-4}$

8. $5(x^2-2)^{-\frac{1}{2}} + 4x^2(x^2-2)^{-\frac{3}{2}}$

Exercises 9–10:   Indicate true or false. Justify your answer.

9. If $a > 0$, $b > 0$, and $c > 0$, then $\dfrac{a^{-1}+b^{-2}}{c^{-3}} = \dfrac{c^3}{a+b^2}$

10. $-5^2 = 25$

Exercises 11–18:   Factor the polynomial.

11. $12x^2 - 5x - 2$ 12. $x^2 - 9$ 13. $x^3 - 1$ 14. $x^3 + 125$

15. $2x^3 + x^2 - 3x$ 16. $6x^3 + 7x^2 - 1$ 17. $3x^3 + x^2 - 6x - 2$ 18. $4 - x^4$

Cumulative Review Exercises 71

Exercises 19–21: Given $1 \text{ m} = 3.28 \text{ ft}$, $1 \text{ gm} = 0.0353 \text{ oz}$, $1 \text{ } l = 0.264 \text{ gal.}$, $1 \text{ gal.} = 4 \text{ qt}$, $1 \text{ qt} = 32 \text{ oz.}$

19. Convert 10 gal./min to l/sec.   20. Convert 100 lb/in.^{3} to kg/m^3.

21. How many 2-liter bottles of soda would be needed at a birthday party attended by 30 people, if each person drinks two 8-ounce glasses of soda?

Exercises 22–23: Determine the equation of the line

22. through the points $(1, -2)$ and $(3, -21)$.

23. parallel to $2y - 3x = 4$ and through the point $(1,4)$.

Exercises 24–28: Solve each of the following.

24. $x - \dfrac{1}{3} = \dfrac{5x + 2}{6} - 3$ 25. $3 - 4x > 2x - 8$ 26. $(x - 2)(2x + 3) = x$

27. $(3x - 5)^2 = 2$ 28. $\left. \begin{array}{r} 2x + 3y = -5 \\ 4z - 4x = 12 \\ \dfrac{-y - z}{2} = \dfrac{1}{2} \end{array} \right\}$

Exercise 29: Complete the square to solve the equation: $-2x^2 + 8x = 5$.

Exercise 30: (**Fencing**) A rectangular field that is twice as long as it is wide is to be divided into two areas by a fence running parallel to the width. It costs $2 per foot for the outside fencing and $5 per foot for the inner fencing. Find the dimensions of the field, if the total cost of fencing is $1,275.

Exercise 31: (**Mixtures**) A gallon container is filled to capacity with a solution that is 10% alcohol and 90% water. How many ounces of the solution should be removed and replaced with a 70% alcohol solution for the final mixture to be 40% alcohol? (1 gal. = 4 qt, 1 qt = 32 oz.)

Exercise 32: (**Free Fall**) Near the surface of the earth, the height $s(t)$ above the ground (in feet) of an object subject only to the force of gravity is given by

$$s(t) = -16t^2 + v_0 t + s_0$$

where t is time in seconds, v_0 is initial velocity and s_0 is the initial height of the object. Velocity is positive when the object is going up, and negative when it is coming down.

A ball is thrown upward from the ground at a speed of 50 ft/sec. How long will it take the ball to first reach a height of twenty-eight and one-half feet?

72

CHAPTER 2
AN INTRODUCTION TO FUNCTIONS AND GRAPHS

Roughly speaking, a **function** is a rule that assigns to each element of one set (collection of objects) exactly one element of another set. There is the age function, for example, which assigns to each individual his or her age; the enrollment function, which associates to a course the number of students enrolled in that course; the grade function, which associates a grade to each student in the course; the profit versus production function; the temperature function; and the list goes on. In short, they are so important that we simply could not function without them. And the same can be said for their pictorial representations, or graphs. Just open any newspaper or textbook and you will find graphs which compactly represent relationships, or functions: graphs that depict the rise and fall of new housing starts in recent years; graphs representing stock values over time; graphs representing the unemployment rate, population growth, and so on.

The concept of a function is introduced in Section 1, and graphs are discussed in Section 2. Section 3 discusses a particularly important class of functions, the so-called one-to-one function, and their inverses.

§1. GENERAL CONCEPTS

We will be primarily concerned with functions, f, which assign a real number $f(x)$ to a given real number x. Such functions can often be described by mathematical expressions; as with:

$$f(x) = \frac{x^2}{x + 1}$$

To **evaluate** a function at $x = c$ is to find the value of $f(c)$. For example:

$$f(5) = \frac{5^2}{5 + 1} = \frac{25}{6} \quad \text{and} \quad f(-2) = \frac{(-2)^2}{-2 + 1} = \frac{4}{-1} = -4$$

We want to emphasize the fact that the variable x is a placeholder; a "box" that can hold any meaningful expression. For example:

$$f(x) = 2x + 5$$
$$f\boxed{} = 2\boxed{} + 5$$
$$f(3) = 2 \cdot 3 + 5 = 11$$
$$f(c) = 2 \cdot c + 5 = 2c + 5$$
$$f(3t) = 2 \cdot 3t + 5 = 6t + 5$$
$$f(x^2 + 3) = 2(x^2 + 3) + 5 = 2x^2 + 11$$

Answers:

(a) -11 (b) $3t - 2$

(c) $-6x - 2$ (d) $-\frac{6}{x} - 5$

CHECK YOUR UNDERSTANDING 2.1

For $f(x) = 3x - 5$, determine:

(a) $f(-2)$ (b) $f(t + 1)$ (c) $f(-2x + 1)$ (d) $f\left(\frac{-2}{x}\right)$

74   Chapter 2 An Introduction to Functions and Graphs

We interrupt our discussion of functions momentarily in order to introduce some useful set notation:

INTERVAL NOTATION

A set may be defined by listing its elements inside braces, as in:

$$\{3, \sqrt{7}, -14\}$$

or by specifying the elements by means of some property or condition, as in:

$$\{x \,|\, 1 < x < 5\}$$

read: *such that*

which represents the set of real numbers, x, which are greater than 1 and less than 5.

EMPTY SET

The set which contains no elements is called the **empty set** and is denoted by the symbol $\varnothing$.

Figure 2.1 illustrates interval notation and its corresponding representation on the number line:

	Interval Notation	**Geometrical Representation**
All real numbers strictly between 1 and 5 (not including 1 or 5)	$(1, 5) = \{x \,\vert\, 1 < x < 5\}$ excluding 1 and 5	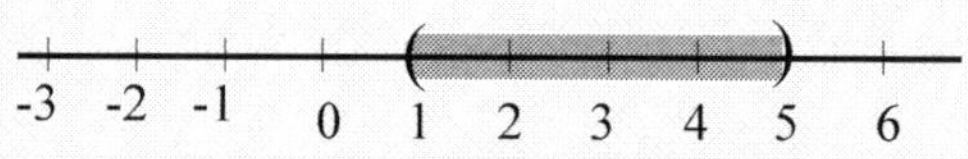
All real numbers between 1 and 5, including both 1 and 5.	$[1, 5] = \{x \,\vert\, 1 \le x \le 5\}$ including 1 and 5	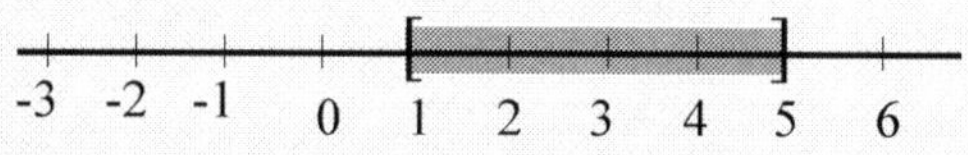
All real numbers between 1 and 5, including 1 but not 5.	$[1, 5) = \{x \,\vert\, 1 \le x < 5\}$ including 1 and excluding 5	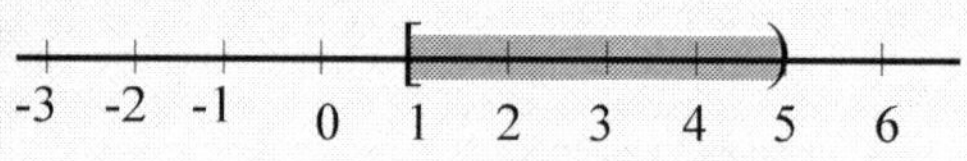
All real numbers between 1 and 5, including 5 but not 1.	$(1, 5] = \{x \,\vert\, 1 < x \le 5\}$ excluding 1 and including 5	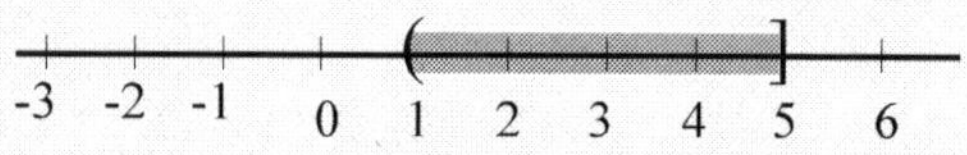
All real numbers greater than 1.	$(1, \infty) = \{x \,\vert\, x > 1\}$ the infinity symbol	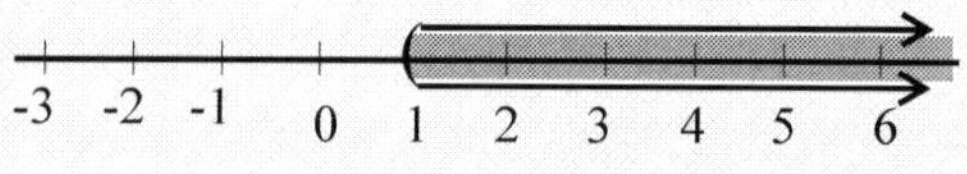
All real numbers greater than or equal to 1.	$[1, \infty) = \{x \,\vert\, x \ge 1\}$	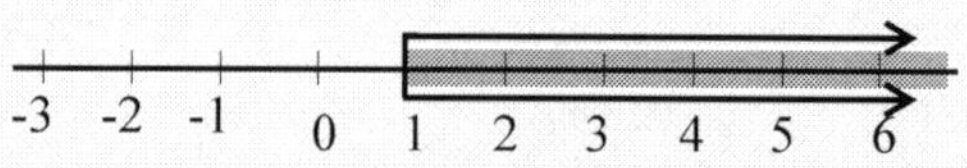
All real numbers strictly less than 5.	$(-\infty, 5) = \{x \,\vert\, x < 5\}$	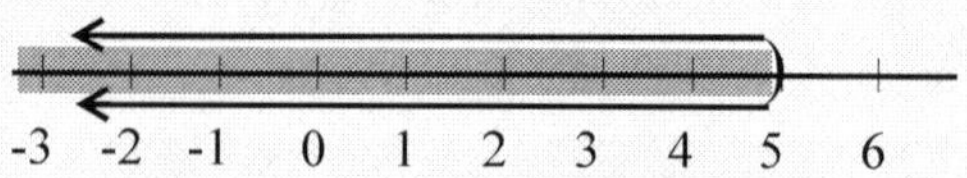
All real numbers less than or equal to 5.	$(-\infty, 5] = \{x \,\vert\, x \le 5\}$	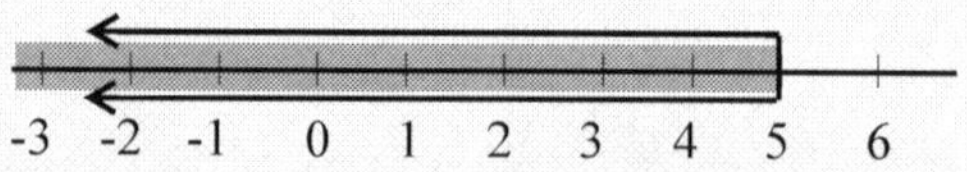
The set of all real numbers.	$(-\infty, \infty) = \{x \,\vert\, -\infty < x < \infty\}$	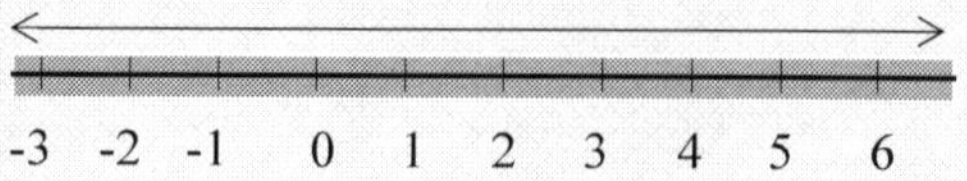

Figure 2.1

The interval (−4, 1) which does not contain its endpoints −4 and 1 is said to be an **open interval**, while [2, 6] which contains its endpoints is a **closed interval**. The intervals [−5, 0) and (1, 5] are **half-open** (or **half-closed**) intervals.

For any two sets A and B, the **union** of A and B is defined to be that set, denoted by $A \cup B$, which consists of all elements that are either in A or in B, including the elements in both A and B. That is:

$$A \cup B = \{x | x \text{ in } A \text{ \textbf{OR} } x \text{ in } B\}$$

The **intersection** of A and B, written $A \cap B$, is the set consisting of the elements common to both A and B. That is:

$$A \cap B = \{x | x \text{ in } A \text{ \textbf{AND} } x \text{ in } B\}$$

For example:

$(-2, 0) \cup [-1, 2] \cup [3, 5]$ $= (-2, 2] \cup [3, 5]$	-3 -2 -1 0 1 2 3 4 5 6
$(-2, 2) \cap [0, 5] = [0, 2)$	-3 -2 -1 0 1 2 3 4 5 6

THE DOMAIN AND RANGE OF A FUNCTION

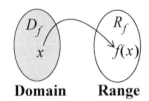

Domain   **Range**

In general, the **domain** of a function f is the set, D_f, on which f "acts," and its **range** is the set R_f of the function values (see margin).

When not specified, the domain of a function defined by an expression is **understood** to be the set of all numbers for which the given expression is defined. For example:

Since x^2 is defined for all numbers, the domain of $f(x) = x^2$ is the set of all numbers: $D_f = (-\infty, \infty)$. Since the function can not assume negative values, and assumes every nonnegative value: $R_f = [0, \infty)$.

Since you can not take the square root of a negative number (in the real number system), the domain of the function $g(x) = \sqrt{x}$ is the set of all numbers greater than or equal to 0: $D_g = [0, \infty)$, with $R_f = [0, \infty)$ (the function can only assume nonnegative values).

Since division by zero is not defined, the domain of the function $h(x) = \dfrac{1}{x-1}$ is the set of all numbers except 1: $D_h = (-\infty, 1) \cup (1, \infty)$.

The range of h is not so easy to determine at this point. You will be able to do so once you know how to graph such a function.

76   Chapter 2 An Introduction to Functions and Graphs

Answers:

(a)

$D_f = [-3, \infty); R_f = [0, \infty)$

(b) $D_g = (-\infty, -1) \cup (-1, 2)$

$\cup (2, \infty)$

CHECK YOUR UNDERSTANDING 2.2

(a) Determine the domain and range of $f(x) = \sqrt{x + 3}$.

(b) Determine the domain of $g(x) = \dfrac{1}{(x + 1)(x - 2)}$.

PIECEWISE-DEFINED FUNCTIONS

Functions such as $f(x) = x^2$ and $g(x) = x + 1$ are described by a single algebraic expression. But there are many situations where different rules apply at different times. A company's profit versus production may take one form up to a certain level of production, and then another form to reflect, say, overtime-pay, or hiring of additional personnel, or the necessity of expanding facilities.

For whatever reasons, one may wish to consider a function such as the function h which acts like $f(x) = x^2$ for $-2 \le x < 0$ and like $g(x) = x + 1$ for $x \ge 0$. Such a function is said to be a **piecewise-defined** function and is represented in the following manner:

$$h(x) = \begin{cases} x^2 & \text{if } -2 \le x < 0 \\ x + 1 & \text{if } x \ge 0 \end{cases}$$

Combining the "if" parts of the definition of h, we see that the domain is $D_h = [-2, \infty)$. To evaluate h at a particular x you must first determine which of the two rules applies. For example:

$$h(-1) = (-1)^2 = 1, \quad \text{and} \quad h(9) = 9 + 1 = 10$$

top rule since $-2 \le -1 < 0$ bottom rule since $9 \ge 0$

EXAMPLE 2.1 Evaluate the function:

$$f(x) = \begin{cases} 3x - 5 & \text{if } x < 0 \\ \dfrac{1}{x + 1} & \text{if } x \ge 0 \end{cases}$$

at $x = -1$, at $x = 0$, and at $x = 3$.

SOLUTION: Since -1 is less than 0, it falls under the jurisdiction of the formula on the top line, and we have:

$$f(-1) = 3(-1) - 5 = -3 - 5 = -8$$

Both 0 and 3 are greater than or equal to 0, thus:

$$f(0) = \frac{1}{0 + 1} = \frac{1}{1} = 1 \quad \text{and} \quad f(3) = \frac{1}{3 + 1} = \frac{1}{4}$$

2.1 General Concepts **77**

CHECK YOUR UNDERSTANDING 2.3

Evaluate the function:

$$f(x) = \begin{cases} -4x + 1 & \text{if } x < 0 \\ x^2 & \text{if } 0 \leq x < 5 \\ -2x & \text{if } 5 \leq x < 10 \end{cases}$$

at: $x = -1$, $x = 1$, $x = 5$, $x = 7$. Is $f(10)$ defined? If not, why not?

Answer:
$f(-1) = 5, f(1) = 1$
$f(5) = -10, f(7) = -14$;
No, 10 is not in the domain of the function.

A particularly important piecewise-defined function is the absolute value function, $f(x) = |x|$, where:

DEFINITION 2.1

ABSOLUTE VALUE

The **absolute value** of a number a is that number $|a|$ given by:

$$|a| = \begin{cases} a & \text{if } a \geq 0 \\ -a & \text{if } a < 0 \end{cases}$$

IN WORDS: The absolute value of a **nonnegative** number is itself. The absolute value of a **negative** number is the negative of that number (a positive number).

You can, and should, interpret $|a|$ as representing the distance (number of units) between the numbers a and 0 on the number line. For example, both 5 and -5 are 5 units from the origin, and we have:

$$|5| = 5 \text{ and } |{-5}| = 5$$

When you subtract one number from another the result is either **plus or minus** the distance (number of units) between those numbers on the number line. For example, $7 - 2 = 5$ while $2 - 7 = -5$. In either case, the absolute value of the difference is 5, the distance between the two numbers:

$$|7 - 2| = |5| = 5 \text{ and } |2 - 7| = |{-5}| = 5$$

This observation leads us to the following definition:

DEFINITION 2.2

DISTANCE

The **distance** between a and b on the number line is given by

$$|a - b|$$

note the negative sign ⬑

In particular, $|(-8) - (-4)| = |{-8} + 4| = 4$ is the distance between -8 and -4, $|(-3) - (4)| = |{-3} - 4| = 7$ is the distance between -3 and 4, and $|2 - 7| = 5$ is the distance between 2 and 7:

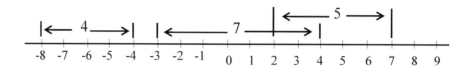

Answers:
(a)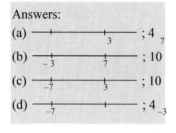

CHECK YOUR UNDERSTANDING 2.4

Plot the two numbers on the number line and determine the distance between them; both visually and by using Definition 2.2.

(a) 3 and 7   (b) –3 and 7   (c) 3 and –7   (d) –3 and –7

EXAMPLE 2.2

WORKER'S PAY

A worker gets paid $18 per hour for the first 8 hours worked, and $22 per hour for overtime, for a maximum of 4 hours of overtime.

(a) Express the worker's pay (in dollars), $P(h)$, as a function of h, the number of hours worked.

(b) What will the worker be paid for 4 hours?

(c) What will the worker be paid for 10 hours?

SOLUTION:

(a) For the first 8 hours, we have to multiply the hours worked, h, by the $18 per hour pay rate (top line):

$$P(h) = \begin{cases} 18h & \text{if } 0 \leq h \leq 8 \\ 144 + 22(h-8) & \text{if } 8 < h \leq 12 \end{cases} \quad (*)$$

If the worker puts in more than 8 hours, then he will be paid a total of $18 \cdot 8 = \$144$ for the first 8 hours worked, plus $22 for every hour **over 8** worked [the $22(h-8)$ in (*)].

(b) Since 4 is less than 8, we have:
$$P(4) = 18 \cdot 4 = 72$$
The worker will be paid $72 for 4 hours.

(c) Since 10 is greater than 8, we have:
$$P(\mathbf{10}) = 144 + 22(\mathbf{10}-8) = 144 + 22 \cdot 2 = 188$$
The worker will be paid $188 for 10 hours.

2.1 General Concepts   79

CHECK YOUR UNDERSTANDING 2.5

The following piecewise-defined function is taken from the IRS Tax Form 1040:

Schedule X—Use if your filing status is **Single**

If the amount on Schedule J, line 15, is: Over—	But not over—	Enter on Schedule J, line 16	*of the amount over—*
$0	$25,350	..........15%	$0
25,350	61,400	$3,802.50+ 28%	25,350
61,400	128,100	13,896.50+ 31%	61,400
128,100	278,450	34,573.50+ 36%	128,100
278,450	....	83,699.50+ 39.6%	278,450

What should be entered on Schedule J, line 16, if the amount on Schedule J, line 15 is:

(a) $20,000   (b) $40,500   (c) $138,100   (d) $205,100   (e) $351,500

Answers::
(a) $3,000
(b) $8,044.50
(c) $38,173.50
(d) $62,293.50
(e) $112,627.30

THE ARITHMETIC OF FUNCTIONS

The following definition is the natural extension of addition, subtraction, multiplication, and division of numbers to functions:

DEFINITION 2.3
COMBINING FUNCTIONS

The sum, difference, product, and quotient of two functions f and g are defined as follows:

$$(f+g)(x) = f(x) + g(x)$$
$$(f-g)(x) = f(x) - g(x)$$
$$(fg)(x) = f(x)g(x)$$
$$\left(\frac{f}{g}\right)(x) = \frac{f(x)}{g(x)}$$

And if c is any number: $(cf)(x) = cf(x)$

Noting that the functions $f+g$, $f-g$, and fg can be evaluated at x if and only if **both** f and g can be evaluated at x, we see that the domain of those three functions coincide with the intersection of the domain of f with that of g:

$$D_{f+g} = D_{f-g} = D_{fg} = D_f \cap D_g$$

In determining the domain of $\dfrac{f}{g}$ one must also exclude those x's where $g(x) = 0$:

$$D_{\frac{f}{g}} = \{x \mid x \text{ is in } (D_f \cap D_g) \text{ and } g(x) \neq 0\}$$

Finally, the domain of the function cf is the same as that of f:

$$D_{cf} = D_f$$

EXAMPLE 2.3 For $f(x) = \sqrt{x}$ and $g(x) = x - 1$, determine the functions $f+g$, $f-g$, fg, $5g$, and $\frac{f}{g}$, and their domains.

SOLUTION: Appealing to Definition 2.3, we have:

$$(f+g)(x) = f(x) + g(x) = \sqrt{x} + x - 1$$

$$(f-g)(x) = f(x) - g(x) = \sqrt{x} - (x-1) = \sqrt{x} - x + 1$$

$$(fg)(x) = f(x)g(x) = \sqrt{x}(x-1) = x^{\frac{3}{2}} - \sqrt{x}$$

$$\left(\frac{f}{g}\right)(x) = \frac{f(x)}{g(x)} = \frac{\sqrt{x}}{x-1}$$

$$5g(x) = 5(x-1) = 5x - 5$$

Noting that $D_f = [0, \infty)$ and $D_g = (-\infty, \infty)$ we conclude that:

$$D_{f+g} = D_{f-g} = D_{fg} = D_f \cap D_g = [0, \infty) \cap (-\infty, \infty) = [0, \infty)$$

$$D_{\frac{f}{g}} = \{x \mid x \text{ is in } [0, \infty) \text{ and } x - 1 \neq 0\} = [0, 1) \cup (1, \infty)$$

$$D_{5g} = D_g = (-\infty, \infty)$$

Answer:

$(f+g)(x) = \frac{x^2 - 6x + 10}{x - 3}$

$(f-g)(x) = \frac{x^2 - 6x + 8}{x - 3}$

$(fg)(x) = \frac{x-3}{x-3} = 1$ if $(x \neq 3)$

$(5g)(x) = \frac{5}{x-3}$

$\left(\frac{f}{g}\right)(x) = x^2 - 6x + 9$;

all domains are $(-\infty, 3) \cup (3, \infty)$

CHECK YOUR UNDERSTANDING 2.6

For $f(x) = x - 3$ and $g(x) = \frac{1}{x-3}$, determine the functions $f+g$, $f-g$, fg, $5g$, and $\frac{f}{g}$, and their domains.

COMPOSITION OF FUNCTIONS

If $h(x) = (x-3)^2$ then $h(8) = 25$. You get that answer by first subtracting 3 from 8: $(x-3) = (8-3) = \mathbf{5}$, and then squaring the result: $(\mathbf{5})^2 = 25$. In other words, you **first** apply the function $f(x) = x - 3$, **and then** apply the function $g(x) = x^2$ to that result. This operation of first performing one function, and then another on that result, is called composition, and is denoted by $(g \circ f)(x)$:

COMPOSITION

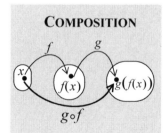

DEFINITION 2.4 The **composition** $(g \circ f)(x)$ is given by:

$$(g \circ f)(x) = g(f(x))$$

↑↑ **first** apply f
  and **then** apply g

[Assuming $f(x)$ is in the domain of g]

Note that composition is **not** a commutative operation:

$$(g \circ f)(2) \neq (f \circ g)(2)$$

EXAMPLE 2.4 Determine $(g \circ f)(2)$ and $(f \circ g)(2)$ for:

$$f(x) = x^2 + 1 \text{ and } g(x) = 2x - 3$$

SOLUTION:

$$(g \circ f)(2) = g(f(2)) = g(2^2 + 1) = g(5) = 2 \cdot 5 - 3 = 7$$

$$(f \circ g)(2) = f(g(2)) = f(2 \cdot 2 - 3) = f(1) = 1^2 + 1 = 2$$

EXAMPLE 2.5 Determine $(f \circ g)(x)$ and $(g \circ f)(x)$ for:

$$f(x) = 3x - 1 \text{ and } g(x) = -2x^2 - 3x + 1$$

SOLUTION:

Definition 2.4 $f(x) = 3x - 1$

$$(f \circ g)(x) = f(g(x)) = f(-2x^2 - 3x + 1) = 3(-2x^2 - 3x + 1) - 1$$
$$= -6x^2 - 9x + 3 - 1$$
$$= -6x^2 - 9x + 2$$

$$(g \circ f)(x) = g(f(x)) = g(3x - 1) = -2(3x - 1)^2 - 3(3x - 1) + 1$$
$$= -2(9x^2 - 6x + 1) - 9x + 3 + 1$$
$$= -18x^2 + 3x + 2$$

EXAMPLE 2.6 Express the function $h(x) = \sqrt{x^2 - 1} + 5$ as a composition $h = g \circ f$.

An alternate solution:
$h(x) = (g \circ f)(x)$ where:
$f(x) = \sqrt{x^2 - 1}$
and $g(x) = x + 5$

SOLUTION: There are many choices for f and g. Perhaps the most natural one is to first do: $f(x) = x^2 - 1$, then take the square root of that result and add 5: $g(x) = \sqrt{x} + 5$. In other words:

$$h(x) = (g \circ f)(x) \text{ where } f(x) = x^2 - 1 \text{ and } g(x) = \sqrt{x} + 5$$

CHECK YOUR UNDERSTANDING 2.7

Answers:
(a-i) 13   (a-ii) $16x^2 + 32x + 13$
(b) One possible answer:

$f(x) = x^2$ and $g(x) = \dfrac{x}{x+3}$

(a) For $f(x) = x^2 + 2x - 2$ and $g(x) = 4x + 3$:

   (i) Evaluate $(f \circ g)(-2)$ (ii) Determine $(f \circ g)(x)$

(b) Express $h(x) = \dfrac{x^2}{x^2 + 3}$ as a composition $h = g \circ f$.

Let's analyze the stated problem a bit. *The heart of the problem is: As time changes, so does the temperature of the food, and as the temperature of the food changes, so does the number of bacteria.* What does (*) tell us? For one thing, that at a temperature of 0 degrees Fahrenheit the food has no bacteria present. Why? Because:

$N(0) = 32 \cdot 0^2 - 2 \cdot 0 = 0$

What does the 50 in (**) represent? The temperature of the food when it is removed from refrigeration. Why? Because $T(t)$ represents the temperature t minutes after being removed from refrigeration, and:

$T(0) = 4 \cdot 0 + 50 = 50$

What is the purpose of the condition $0 \leq t \leq 10$ in (**)? It is an attempt to add credence to the problem. Without it, the temperature of the food would increase to infinity (not to mention that for some time prior to being removed from the refrigerator the food would have a negative number of bacteria).

EXAMPLE 2.7
BACTERIA

Assume that the number N of bacteria per cubic millimeter of a certain food as a function of the Fahrenheit temperature T of the food is given by:

$$N(T) = 32T^2 - 2T \qquad (*)$$

Assume, also, that the temperature T of the food as a function of the time that the food has been removed from refrigeration is given by:

$$T(t) = 4t + 50, \text{ for } 0 \leq t \leq 10 \qquad (**)$$

where t denotes time, in minutes.

(a) Determine the composite function $N \circ T$ and indicate what it represents.

(b) How many bacteria are present per cubic millimeter of food 2 minutes after the food has been removed from refrigeration?

SOLUTION:

(a) $(N \circ T)(t) = N(T(t)) = N(4t + 50) = 32(4t+50)^2 - 2(4t+50)$
$= 512t^2 + 12{,}792t + 79{,}900$

represents the number of bacteria per cubic milimeter of food t minutes after the food was removed from refrigeration.

(b) Using (a):

$(N \circ T)(2) = 512 \cdot 2^2 + 12{,}792 \cdot 2 + 79{,}900 = 107{,}532$ bacteria

Another approach: Use (**) to find the temperature T of the food at $t = 2$: $T(2) = 4 \cdot 2 + 50 = 58$. Then substitute this temperature into (*) and calculate the corresponding bacteria count per cubic millimeter: $N(58) = 32(58)^2 - 2 \cdot 58 = 107{,}532$. Both approaches are depicted below:

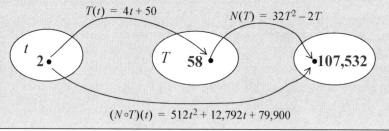

CHECK YOUR UNDERSTANDING 2.8

The number of monthly housing starts, H, in a certain county, as a function of the 30-year mortgage rate, r, are given by:

$$H(r) = 350 + 1000(0.1 - r), \text{ for } 0.04 \leq r \leq 0.1$$

Assume that the mortgage rate is given by $r = i + 0.02$, where i denotes the prime interest rate.

(a) Express the number of monthly housing starts as a function of the prime interest rate.

(b) Determine the number of housing starts in April, if the prime rate in April is $i = 0.043$.

Answers:
(a) $430 - 1000i$; $0.02 \leq i \leq 0.08$
(b) 387

Composition is not shackled to functions that assume numerical values. Consider, for example, the function M which assigns to an individual i his or her (biological) mother, and the function F which assigns to an individual i his or her (biological) father. Then, the function $F \circ M$ is that function which assigns to an individual his or her (biological) maternal grandfather:

$$(F \circ M)(i) = F(M(i))$$

i's mother

the father of i's mother

CHECK YOUR UNDERSTANDING 2.9

Referring to the above, what do the following compositions represent?

(a) $(M \circ F)(i)$ (b) $(M \circ M)(i)$ (c) $(F \circ M \circ M)(i)$

Answers:
(a) paternal grandmother
(b) maternal grandmother
(c) great-grandfather (twice maternal)

84   Chapter 2 An Introduction to Functions and Graphs

EXERCISES

Exercises 1–6: Determine $f(0)$, $f(4)$, and $f(-2)$ for the given function, if the value exists.

1. $f(x) = 2x^2 - 3x - 9$

2. $f(x) = \sqrt{x+5}$

3. $f(s) = \dfrac{4}{s} + 5$

4. $f(t) = \dfrac{t^2 - t}{t^2 + 1}$

5. $f(x) = \dfrac{(x-1)(x-4)}{\sqrt{x}(x+2)}$

6. $f(x) = \dfrac{x-2}{x^{\frac{3}{2}}}$

Exercises 7–9: For the given function

7. $f(x) = 3x - 5$

8. $f(x) = -x^2 + x + 2$

9. $f(x) = \dfrac{1 - 2x}{x}$

determine:

(a) $f(6x)$ (b) $f(7x-4)$ (c) $f(x^2)$ (d) $f(x+h)$ (e) $\dfrac{f(x+h) - f(x)}{h}$, $h \neq 0$

Exercises 10–15: Represent the given interval on the number line.

10. $(-1, 5)$ 11. $[0, 3]$ 12. $[1, 2)$ 13. $(-4, -2]$ 14. $(-\infty, -1)$ 15. $(-1, \infty)$

Exercises 16–19: Express the given set of numbers in interval notation, and represent the interval(s) on the number line. All numbers

16. greater than 2 and less than or equal to 8

17. at least 5 but not more than 9

18. outside the interval $[-2, 4]$

19. strictly between -3 and -5

Exercises 20–25: Simplify, by expressing as one interval. (Suggestion: First represent each of the component intervals on the number line.)

20. $(1, 5) \cup [3, 9)$

21. $(1, 5] \cup [3, 9]$

22. $(-\infty, 5) \cap (3, 9)$

23. $(-\infty, 5) \cup (3, 9)$

24. $(2, 5) \cup [-1, 4) \cup [0, \infty)$

25. $\left(-\infty, \frac{9}{2}\right) \cap (2, 6) \cap [-2, 4]$

Exercises 26–37: Determine the domain of the function.

26. $f(x) = 4x^3 - 5$

27. $f(x) = \dfrac{1}{5x - 3}$

28. $h(x) = \dfrac{3x^2 - 5x - 2}{4 - 3x}$

29. $k(x) = \dfrac{x-1}{2x-2}$

30. $g(t) = \dfrac{\sqrt{1-5t}}{4}$

31. $g(s) = \dfrac{\sqrt{s}}{s^2 + 2s - 3}$

32. $h(r) = 2 - \sqrt[3]{r+1}$

33. $k(x) = \dfrac{1}{\sqrt{3x-5}}$, for $x \leq 3$

34. $g(x) = \begin{cases} -1 & \text{if } x < 4 \\ 2 - x^2 & \text{if } x \geq 4 \end{cases}$

35. $f(x) = \begin{cases} 3x & \text{if } -1 < x \leq 2 \\ x - 1 & \text{if } 2 < x \leq 3 \end{cases}$

$$36.\ h(x) = \begin{cases} x & \text{if } x \le 0 \\ 2x-1 & \text{if } 0 < x < 2 \\ -x^3 & \text{if } x \ge 3 \end{cases} \qquad 37.\ k(x) = \begin{cases} 5x+4 & \text{if } x < -1 \\ -x & \text{if } 0 \le x \le 4 \\ x^3-1 & \text{if } x > 6 \end{cases}$$

Exercises 38–43: Determine the domain and range of the function.

38. $f(x) = 4$

39. $g(x) = 2x$

40. $k(x) = -x^2$

41. $g(t) = \sqrt{t-1}$

42. $h(t) = \sqrt{5t}$

43. $f(x) = \sqrt{x}-1$

Exercises 44–47: Determine $f(-1)$, $f(0)$ and $f(5)$, if possible.

$$44.\ f(x) = \begin{cases} 3x+1 & \text{if } x \le 5 \\ 9-x & \text{if } x > 5 \end{cases} \qquad 45.\ f(x) = \begin{cases} 2 & \text{if } x < 0 \\ -x^2+2x+1 & \text{if } x \ge 0 \end{cases}$$

$$46.\ f(x) = \begin{cases} x-1 & \text{if } 0 < x \le 1 \\ 7x+3 & \text{if } 1 < x < 9 \\ x^2-12x-4 & \text{if } x \ge 9 \end{cases} \qquad 47.\ f(x) = \begin{cases} 0 & \text{if } -3 < x < 1 \\ 5-4x^2 & \text{if } 1 \le x \le 3 \\ x^3-5x^2 & \text{if } 3 < x < 5 \end{cases}$$

Exercises 48–53: Plot the numbers on the number line, and determine the distance between them; both visually and by using Definition 2.2.

48. -2 and -5

49. -2 and 5

50. 2 and -5

51. 2 and 5

52. $-\frac{1}{4}$ and $-\frac{1}{2}$

53. -6 and 0

Exercises 54–58: For $f(x) = 3x+5$ and $g(x) = x-1$, determine the rule and the domain of the function.

54. $(g-f)(x)$

55. $(f+g)(x)$

56. $\left(\dfrac{f}{g}\right)(x)$

57. $(fg)(x)$

58. $\left(\dfrac{f+g}{f-g}\right)(x)$

Exercises 59–66: For $f(x) = 2x^2-1$, $g(x) = \dfrac{1}{x}$, and $h(x) = \sqrt{x+1}$ determine the rule of the function.

59. $(f+g+h)(x)$

60. $(f-g)(x)$

61. $(fgh)(x)$

62. $\left(\dfrac{f}{g}\right)(x)$

63. $\left(\dfrac{g}{h}\right)(x)$

64. $\left(\dfrac{h}{g}\right)(x)$

65. $\left(\dfrac{g}{f}-h\right)(x)$

66. $\left(\dfrac{fh}{g}\right)(x)$

Exercises 67–72: For $f(x) = -3x+2$ and $g(x) = \dfrac{1}{x-2}$, determine:

67. $(g \circ f)(3)$

68. $(f \circ g)(3)$

69. $(f \circ g)(x)$

70. $(g \circ g)(x)$

71. $(g \circ f)(x)$

72. $(f \circ f)(x)$

Exercises 73–77: Given $f(x) = x-1$, $g(x) - x^2$, and $h(x) = \dfrac{1}{x}$, determine:

73. $(f \circ g \circ h)(2)$

74. $(g \circ f \circ h)(2)$

75. $(g \circ h \circ f)(2)$

76. $(f \circ g \circ h)(x)$

77. $(h \circ g \circ f)(x)$

86   Chapter 2 An Introduction to Functions and Graphs

Exercises 78–81: Suppose f and g have the following values:
$$f(1) = 4, \quad f(2) = -3, \quad f(5) = 2, \quad g(1) = 5, \quad g(4) = 0, \quad g(5) = -1$$
Determine

78. $(f \circ g)(1)$ 79. $(g \circ f)(1)$ 80. $(f \circ f)(5)$ 81. $(g \circ g)(1)$

Exercises 82–85: Find functions f and g such that h can be expressed as their composition, $h(x) = (g \circ f)(x)$.

82. $h(x) = \sqrt{x - 5}$ 83. $h(x) = (x^2 + 1)^{-3}$ 84. $h(x) = (x^3 + x - 2)^{\frac{5}{3}}$ 85. $h(x) = \sqrt{\dfrac{2x}{3x - 1}}$

Exercise 86: **(Viral Infection)** A viral infection is spreading through a herd of sheep. The approximate number of diseased sheep is predicted by the function
$$f(t) = 0.06t^3 + 4, \qquad 0 \le t \le 12$$
where t is the number of days since the virus was first detected. Approximately how many sheep were infected when the virus was first detected? After a week? After 10 days?

Exercise 87: **(Population)** Suppose the population of a certain city, in thousands of people, is given approximately by
$$P(t) = 2t + 15, \qquad -6 \le t \le 11$$
where t is time measured in years, with $t = 0$ representing the year 2004.

  (a) What was the population in 1998?

  (b) What will the population be in the year 2013?

  (c) In the year 2020? (Rethink your last answer.)

Exercise 88: **(Average Grade)** Two exams and a final exam are administered in a course. Each of the two exams constitute 30% of the final grade, and the final exam is 40% of the final grade.

  (a) Express the final grade as a function of the three exam grades. (Define all variables.)

  (b) Use your function in (a) to determine the final grade, if the student's grades are 75 and 83 on the two exams and 91 on the final.

Exercise 89: **(Profit)** The monthly cost and revenue functions for producing and selling x Zip-Drives at Zip-Di-Do are given, resp., by
$$C(x) = 45x + 12,500 \quad \text{and} \quad R(x) = 65x$$
The company manufactured and sold 4000 drives in December, and 3000 drives in January.

  (a) Find the December profit for the company. (Profit $=$ Revenue $-$ Cost)

  (b) Find the total profit for the two months.

  (c) What was the combined per unit profit for the two months?

2.1 General Concepts   87

Exercise 90:   (Price and Tax) Let $P(x)$ denote the price of an item x. Let $T(A) = 0.08A$ denote the state tax levied on a purchase price of A.

(a) What does the function $T \circ P$ represent?

(b) What is the domain of the function $T \circ P$?

(c) Evaluate $(T \circ P)$(TI-83), given that the TI-83 calculator is on sale for $72.50.

(d) What does P(TI-83)$+(T \circ P)$(TI-83) represent?

(e) Evaluate P(TI-83)$+(T \circ P)$(TI-83), given that the TI-83 calculator is on sale for $72.50.

Exercise 91:   (Profit) On sunny summer Saturdays, both the attendance and soda sales at a little league game depend on the Fahrenheit temperature, T. Specifically, the attendance, $A(T)$, and the number of sodas sold, $S(T)$, are approximated by the functions

$$A(T) = \begin{cases} 250 & \text{if } T \leq 86° \\ 250 - 2(T - 87) & \text{if } T > 86° \end{cases}$$

$$S(T) = A(T) + 5(T - 70), \text{ for } T \geq 70°$$

Assuming that a profit of 25 cents is made from the sale of each soda, determine the day's profit at the given temperature.

(a) $T = 75°$ (b) $T = 95°$ (c) $T = 95.3°$

Exercise 92:   (Worker's Pay) A worker gets paid $17/hr if he works at most 8 hours, $30/hr for up to 5 hours of overtime, and $40/hr for any additional hours of overtime. Express the worker's pay (in dollars), $P(h)$, as a function of h, the number of hours worked.

Exercise 93:   (Special Delivery) Special Delivery postal rates (in dollars) for first class mail, are $7.65 for up to two pounds, $7.95 for up to ten pounds, and $8.55 for a package weighing between ten and fifteen pounds, inclusive.

(a) Express the Special Delivery cost $C(x)$ as a (piecewise-defined) function of the weight x of the package.

(b) Special Delivery rates for other classes of mail are given as $8.05 for at most 2 pounds, $8.65 for at most 10 pounds, and $9.30 for packages over 10 pounds. Write the Special Delivery cost function $C(x)$ for this class of mail.

Exercise 94:   Given

$$f(x) = \begin{cases} 2x + 1 & \text{if } x \leq -1 \\ x^2 & \text{if } x > -1 \end{cases} \quad \text{and} \quad g(x) = \begin{cases} -x + 1 & \text{if } x < 2 \\ 3x & \text{if } x \geq 2 \end{cases}$$

determine

(a) $(g \circ f)(3)$ (b) $(g \circ f)(-2)$ (c) $(f \circ g)(x)$

(d) $(f + g)(x)$ (e) $(f - g)(x)$ (f) $(fg)(x)$

§2. THE GRAPH OF A FUNCTION

The **Cartesian Plane** is formed by two number lines intersecting at right angles at their origins. It is customary to call the horizontal axis the *x*-axis, and the vertical axis the *y*-axis. As is illustrated in Figure 2.2, each point in the plane can be described by an (ordered) pair of numbers.

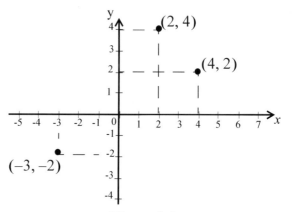

Figure 2.2

The **graph** of a function f is the set of points (x, y) in the plane, with x in the domain of f and $y = f(x)$. The relationship between the graph of a function f, its domain D_f, and its range R_f, is illustrated in Figure 2.3. The domain, being the set of *x*-coordinates of points on the graph of f, is the projection of that graph onto the *x*-axis (the "shadow" the graph casts on the *x*-axis). The range, being the set of *y*-coordinates, is the projection onto the *y*-axis (the "shadow" the graph casts onto the *y*-axis).

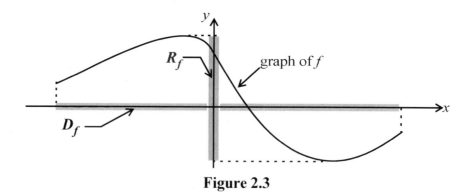

Figure 2.3

EXAMPLE 2.8 Find the domain of the function: $f(x) = \sqrt{x}$ and sketch its graph. Determine the range of f.

SOLUTION: Since no domain was specified, it is understood to be the set of numbers for which $\sqrt{x}$ is defined, namely where x is nonnegative:

$$D_f = [0, \infty)$$

After calculating and plotting a few of the points on the graph of f, we were able to anticipate and sketch the graph of the function (see Figure 2.4). The graph reveals the range of f: $R_f = [0, \infty)$ (the shadow the graph casts onto the y-axis)

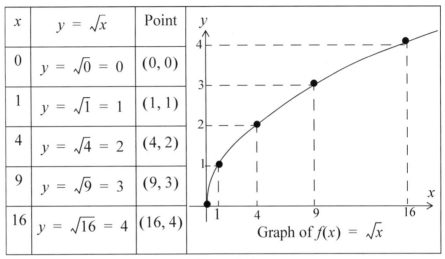

Figure 2.4

EXAMPLE 2.9 Sketch the graph of the absolute value function

$$f(x) = |x|.$$

SOLUTION: From its very definition, the graph of the absolute value function

$$f(x) = |x| = \begin{cases} x & \text{if } x \geq 0 \\ -x & \text{if } x < 0 \end{cases}$$

must coincide with the line $y = x$ for $x \geq 0$, and with the line $y = -x$ for $x < 0$:

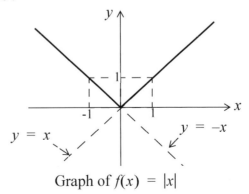

Graph of $f(x) = |x|$

Figure 2.5

INTERCEPTS

X-INTERCEPT

Y-INTERCEPT

The x-coordinate of a point where the graph of a function f intersects the x-axis is said to be an **x-intercept** of that graph [Figure 2.6(a)]. Moreover, if 0 is in the domain of f, then the function value $f(0)$ is said to be the **y-intercept** of the graph of f (where the graph intersects the y-axis) [Figure 2.6(b)]. Although the graph of a function can have more than one x-intercept [as is the case with the function in Figure 2.6(a)], it can not have more than one y-intercept (why not?).

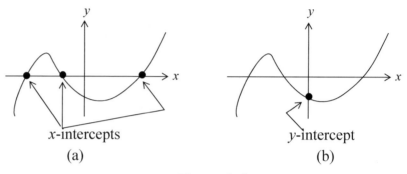

Figure 2.6

CHECK YOUR UNDERSTANDING 2.10

Use the adjacent graph of the function f to approximate the values of:

(a) $f(-1)$

(b) the x-intercepts

(c) the y-intercept

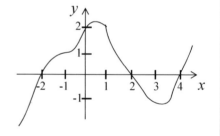

Answers: (a) ≈ 1
(b) $-2, 2, 4$ (c) 2

In the next example, we illustrate how you can generate the graph of the function $h(x) = -2|x-1| + 3$ from that of the "basic" function $f(x) = |x|$. [The same procedure can be used to generate graphs of $h(x) = af(x - x_0) + y_0$ from that of $f(x)$.]

EXAMPLE 2.10 Sketch the graph of:
$$h(x) = -2|x-1| + 3$$
Determine its range, and its x- and y-intercepts.

SOLUTION: We use a a four-step process: two distortion steps and two shifting steps. The first step is easy to take: Reflect the graph of $y = |x|$ about the x-axis (to go from $y = |x|$ to $y = -|x|$). Then stretch the resulting graph by a factor of **2** (to get to the graph of $y = -2|x|$):

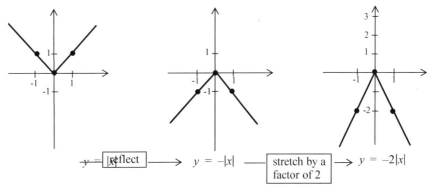

In the third step, "pick up" the graph of $y = -2|x|$ and move it 1 unit to the right: the -1 $-2|x-1|$ (see margin). Finally, move the graph 3 units up to go from $y = -2|x-1|$ to $-2|x-1|+3$:

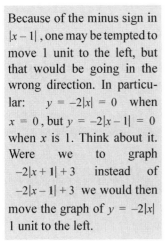

Because of the minus sign in $|x-1|$, one may be tempted to move 1 unit to the left, but that would be going in the wrong direction. In particular: $y = -2|x| = 0$ when $x = 0$, but $y = -2|x-1| = 0$ when x is 1. Think about it. Were we to graph $-2|x+1|+3$ instead of $-2|x-1|+3$ we would then move the graph of $y = -2|x|$ 1 unit to the left.

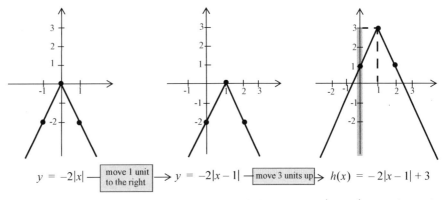

We can easily read off the range of $h(x) = -2|x-1|+3$ from its graph (the shadow the graph casts onto the y-axis): $R_h = (-\infty, 3]$.

The y-intercept of $h(x) = -2|x-1|+3$ is $h(0)$:

$$h(0) = -2|0-1|+3 = -2|-1|+3 = -2+3 = 1$$

To determine the x-intercepts, we solve the equation $h(x) = 0$:

$$-2|x-1|+3 = 0$$
$$|x-1| = \frac{3}{2} \quad (*)$$

Here are two methods that can be used to solve (*):

Geometric Method	Analytical Method
To say that $\lvert x-1 \rvert = \frac{3}{2}$ is to say that the distance between x and 1 is $\frac{3}{2}$. It follows that $x = -\frac{1}{2}$ or $x = \frac{5}{2}$: 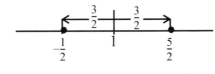	To say that $\lvert x-1 \rvert = \frac{3}{2}$ is to say that: $x-1 = \frac{3}{2}$ OR $x-1 = -\frac{3}{2}$ $x = \frac{5}{2}$ OR $x = -\frac{1}{2}$

Answers:
(a) See page C-14.
(b) $D_h = [-2, \infty)$
$R_h = (-\infty, 4]$
(c) x-int: -1, y-int: $-4\sqrt{2} + 4$

CHECK YOUR UNDERSTANDING 2.11

(a) Follow the above four-step procedure to generate the graph of the function $h(x) = -4\sqrt{x+2} + 4$ from that of the function $f(x) = \sqrt{x}$ appearing in Figure 2.4.
(b) Determine the domain and range of h.
(c) What are the x- and y intercepts of h?

EVEN AND ODD FUNCTIONS

Evaluating the function $f(x) = x^2$ at -2 and at 2 yields the same result. Indeed, for any number x:

$$f(-x) = (-x)^2 = x^2 = f(x) \qquad (*)$$

On the other hand, the value of the function $g(x) = x^3$ at -2 is the negative of the value of the function at 2. Indeed, for any number x:

$$g(-x) = (-x)^3 = -x^3 = -g(x) \qquad (**)$$

As you will see, there are other functions satisfying (*) or (**). Appropriately enough, they are called **even** and **odd** functions, respectively:

DEFINITION 2.5
EVEN AND ODD FUNCTIONS

Suppose that for all x in the domain D_f of a function f, $-x$ is also in D_f. Then f is said to be:

EVEN if: $f(-x) = f(x)$

ODD if: $f(-x) = -f(x)$

for all x in D_f.

Even functions have very characteristic graphs. If f is even, then $f(-x) = f(x)$, and this tells us that both (x, y) and $(-x, y)$ are on the graph of f. To put it geometrically:

The graph of an even function is symmetric with respect to the y-axis [see Figure 2.7(a)].

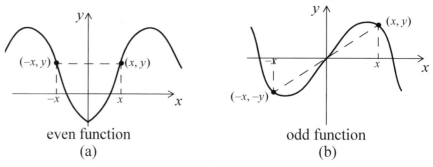

even function
(a)

odd function
(b)

Figure 2.7

Graphs of odd functions also have distinctive features. If f is odd, then $f(-x) = -f(x)$, and this tells us that both (x, y) and $(-x, -y)$ are on the graph of f. To put it geometrically:

> The graph of an odd function is symmetric with respect to the origin [see Figure 2.7(b)].

CHECK YOUR UNDERSTANDING 2.12

Complete the graph in the adjacent figure so that it is the graph of:

(a) an even function

(b) an odd function.

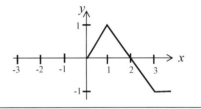

Answers: See page C-15.

NOT ALL CURVES ARE FUNCTIONS

The curve in Figure 2.8(a) is the graph of a function, since no more than one y is assigned to any x. On the other hand, the curve in Figure 2.8(b) does not represent a function, since to some x there is assigned more than one y. Geometrically speaking:

VERTICAL LINE TEST No vertical line can intersect the graph of a function at more than one point.

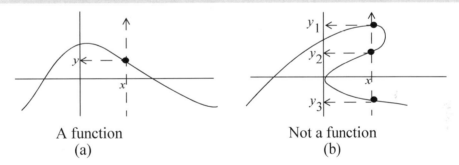

A function
(a)

Not a function
(b)

Figure 2.8

CHECK YOUR UNDERSTANDING 2.13

Indicate whether or not the given curve is the graph of a function.

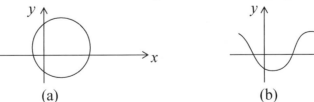

(a)   (b)

Answers: (a) no   (b) yes

EQUATIONS, INEQUALITIES, AND GRAPHS OF FUNCTIONS

Graphs of functions can be used to find approximate solutions of equations and inequalities. Since the graph of the function f is the set of points (x, y) where $y = f(x)$, the solutions of the equation $f(x) = 0$ (called the **zeros** of the function f) are the x-coordinates of the points $(x, 0)$ on the graph of f; that is:

$$f(x) = 0$$
at the x-intercepts of the graph of f [see Figure 2.9]

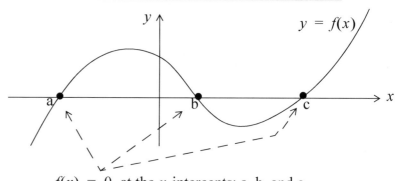

$f(x) = 0$ at the x-intercepts: a, b, and c.

Figure 2.9

The solution set of the inequality $f(x) < 0$ consists of the x-coordinates of the points (x, y) on the graph of f where $y = f(x) < 0$, that is:

$$f(x) < 0$$
Where the graph of f lies **below** the x-axis [see Figure 2.10(a)]

The solution set of the inequality $f(x) > 0$ consists of the x-coordinates of the points (x, y) on the graph of f where $y = f(x) > 0$, that is:

$$f(x) > 0$$
Where the graph of f lies **above** the x-axis [see Figure 2.10(b)]

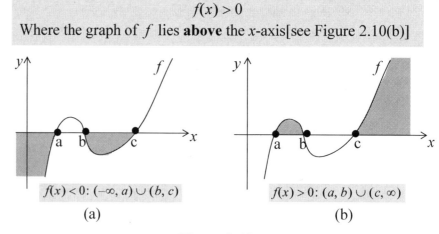

$f(x) < 0: (-\infty, a) \cup (b, c)$ $f(x) > 0: (a, b) \cup (c, \infty)$

(a)   (b)

Figure 2.10

The solution set of $f(x) \leq 0$ is the union of the solution sets of the equation $f(x) = 0$ and the inequality $f(x) < 0$. Likewise, the solution set of $f(x) \geq 0$ is the union of the solution sets of $f(x) = 0$ and $f(x) > 0$. In particular, referring to the function of Figure 2.10, we have:

$$f(x) \leq 0: (-\infty, a] \cup [b, c] \quad \text{and} \quad f(x) \geq 0: [a, b] \cup [c, \infty)$$

Turning our attention to two functions, f and g, we have:

The solutions of $f(x) = g(x)$ are the x-coordinates of the points of intersection of the graphs of f and g; or, equivalently, the x-intercepts of the graph of the function $h(x) = f(x) - g(x)$.

Please make sure you see the connection between these three statements and the situation depicted in Figure 2.11

The solution set of the inequality $f(x) < g(x)$ consists of those x's at which the graph of f lies below that of g; or, equivalently, where the graph of the function $h(x) = f(x) - g(x)$ lies below the x-axis.

The solution set of the inequality $f(x) > g(x)$ consists of those x's at which the graph of f lies above that of g; or, equivalently, where the graph of the function $h(x) = f(x) - g(x)$ lies above the x-axis.

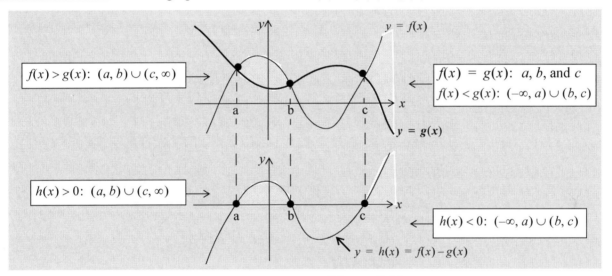

Figure 2.11

CHECK YOUR UNDERSTANDING 2.14

(a) Use the adjacent graph of the function f to solve:
 (i) $f(x) = 0$
 (ii) $f(x) < 0$
 (iii) $f(x) \geq 0$

(b) Use the adjacent graph of the functions f and g to solve:
 (i) $f(x) = g(x)$
 (ii) $f(x) < g(x)$
 (iii) $f(x) \geq g(x)$

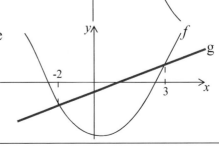

Answers: (a-i) a, b, c
(a-ii) $(a, b) \cup (c, \infty)$
(a-iii) $(-\infty, a] \cup [b, c]$
(b-i) -2, 3
(b-ii) $(-2, 3)$
(b-iii) $(-\infty, -2] \cup [3, \infty)$

APPROXIMATE SOLUTIONS OF EQUATIONS AND INEQUALITIES

In later chapters, we will carefully select equations and inequalities which can be solved analytically: "by hand." Here, that won't be necessary, as we will be using a graphing calculator to find **approximate** solutions.

EXAMPLE 2.11 Use a graphing calclulator to approximate the solutions of the equation:

$$x^4 + x^3 - 2x^2 - 1 = 0$$

SOLUTION: The first step is to enter the function:

$$f(x) = x^4 + x^3 - 2x^2 - 1$$

into the graphing calculator (see [A] in Figure 2.12). We then selected an "appropriate" window size [B] (see margin). Next, we instructed the unit to position the cursor at the left x-intercept (one solution): $x \approx -2.075$ [C], and then at the right x-intercept (another solution): $x \approx 1.212$ [D].

With the Confinement Theorem of chapter 6 (Theorem 6.4 page 266) you will be able to determine that all of the solutions of the given equation are necessarily contained in the interval $[-3, 3]$, and we defined our x-range accordingly. Since our concern is with the x-intercepts of the graph, any "small" y-range which displays the x-axis will do.

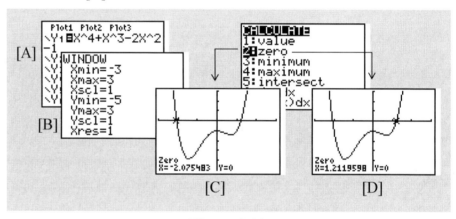

Figure 2.12

CHECK YOUR UNDERSTANDING 2.15

Use a graphing calculator to find an approximate solution set of:

$$x^3 - 3x^2 - 10x + 3 = 0$$

Use a $[-6, 6]$ by $[-30, 30]$ window.

Answer:
$\approx -2.19,\ 0.28,\ 4.91$

EXAMPLE 2.12 Use a graphing calculator to find an approximate solution set for the inequality:.

$$x^3 - 9x^2 - 38x + 33 < 0$$

(Use a $[-10, 20] \times [-400, 150]$ window)

SOLUTION: We begin by entering the function:

$$f(x) = x^3 - 9x^2 - 38x + 33 \qquad \text{(see [A] of Figure 2.13)}$$

and the given window (see [B]). The rest of the figure speaks for itself, and tells us that the solution set of: $x^3 - 9x^2 - 38x + 33 < 0$ is approximately: $(-\infty, -3.7) \cup (0.75, 11.95)$.

The importance of having an adequate window size:

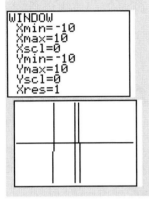

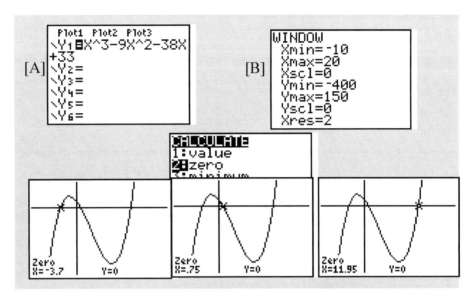

Figure 2.13

CHECK YOUR UNDERSTANDING 2.16

Use a graphing calculator to find an approximate solution set (to two decimal places) of the inequality:

$$x^3 + 5x^2 - 7x - 3 > 0$$

Use a $[-10, 6]$ by $[-20, 50]$ window.

Answer:
$(-6.07, -0.35) \cup (1.42, \infty)$

EXERCISES

Exercises 1–3: Find the domain of the function. Sketch the graph by plotting points. Determine the range of the function from the graph.

1. $f(x) = x^2$
2. $k(x) = x^3$
3. $g(x) = \dfrac{1}{x}$

Exercise 4: Use the graph of the function f
   (a) to determine
   (b) to find x for which

   (i) $f(-8)$
   (ii) $f(1)$
   (iii) $f(5)$

   (i) $f(x) = 0$
   (ii) $f(x) = 1$
   (iii) $f(x) = 6$

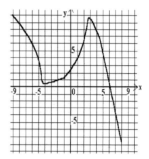

Exercise 5: Use the graph of the function f
   (a) to determine
   (b) to find x for which

   (i) $f(-5)$
   (ii) $f(3)$
   (iii) $f(7)$

   (i) $f(x) = -7$
   (ii) $f(x) = -5$
   (iii) $f(x) = 0$

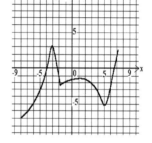

Exercises 6–7: Determine, approximately, the domain, range, and x-and y-intercepts from the graph of f.

6.

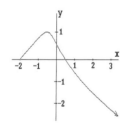

7.

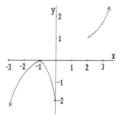

Exercises 8–11: Sketch the graph of the function by reflecting, stretching, and shifting the basic graph of $y = |x|$, indicating the intercepts. Show each step on a separate set of axes.

8. $F(x) = |x + 4|$
9. $k(x) = 2|x - 4| - 1$
10. $f(x) = \dfrac{2}{3}|3 - x| + 2$
11. $K(x) = -\dfrac{1}{2}\left|x + \dfrac{1}{4}\right| + \dfrac{1}{2}$

Exercises 12–21: Using as the basic graph one of the functions $y = \sqrt{x}$ or a function in Exercises 1–3 above, sketch the graph of the function by reflecting, stretching, and shifting the basic graph, and indicate only the y-intercept. Show each step on a separate set of axes.

12. $f(x) = 3\sqrt{x+4} - 1$

13. $k(x) = 5\sqrt{x-3} - 8$

14. $h(x) = -\frac{1}{2}\sqrt{x+8} + 2$

15. $g(x) = -2\sqrt{x-1} + 3$

16. $s(x) = (x-1)^2 + 1$

17. $t(x) = \dfrac{1}{x-3}$

18. $G(x) = -(x-2)^3$

19. $K(x) = -x^2 + 9$

20. $g(x) = -\dfrac{3}{x+4}$

21. $f(x) = (x+4)^3 - 2$

Exercises 22–27: Indicate if the graph appears to represent the graph of an even function, an odd function, or neither.

22.

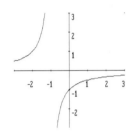

23.

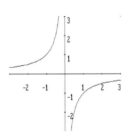

24.

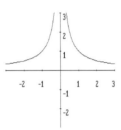

25.

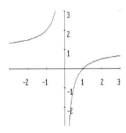

26.

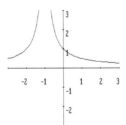

27.

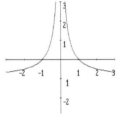

Exercises 28–29: Complete the graph of $f(x)$ below, so that the function f is (a) an even function, (b) an odd function, and (c) neither an even function nor an odd function.

28.

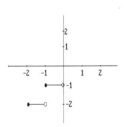

29.

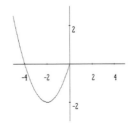

100   Chapter 2 An Introduction to Functions and Graphs

Exercises 30–32: Show that the function is even, by determining $f(-x)$ and verifying that $f(-x) = f(x)$.

30. $f(x) = -3x^2$ 31. $f(x) = -\dfrac{1}{x^2}$ 32. $f(x) = 2x^2 + x^4 + 1$

Exercises 33–35: Show that the function is odd, by determining $f(-x)$ and verifying that $f(-x) = -f(x)$.

33. $f(x) = -x$ 34. $f(x) = 5x$ 35. $f(x) = -x^5$

Exercises 36–41: Which of the following graphs represents a function?

36.
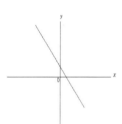

37.
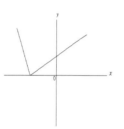

38.
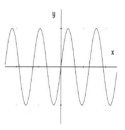

39.
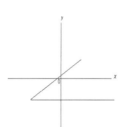

40.
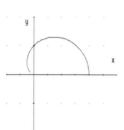

41.
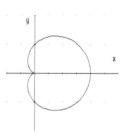

Exercises 42–45: From the graph of the function f, determine, (a) the x-intercepts, (b) the y-intercept, (c) the solutions of $f(x) = 0$, (d) the solution set of $f(x) < 0$, and (e) the solution set of $f(x) > 0$.

42.
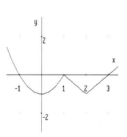

43.
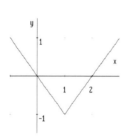

44.
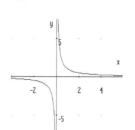

45.
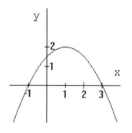

Exercises 46–48: From the given graphs of f and g, determine the solution of
(a) $f(x) = g(x)$, (b) $f(x) < g(x)$, and (c) $f(x) > g(x)$.

46.

47.

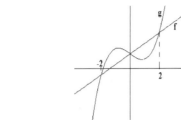

48.

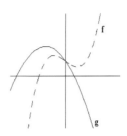

Exercises GC49–GC52: Use a graphing calculator to approximate the solutions of the equation (to two decimal places).

49. $-6x^3 + 8x + 1 = 0$, (window: $[-5, 5]$ by $[-5, 5]$)

50. $2x^3 - 4x + 1 = 0$, (window: $[-5, 5]$ by $[-5, 5]$)

51. $-4x^4 + 3x^2 + 10 = 0$, (window: $[-5, 5]$ by $[-10, 15]$)

52. $5x^4 + 2x^3 - 3x - 10 = 0$, (window: $[-5, 5]$ by $[-15, 10]$)

Exercises GC53-GC56: Use a graphing calculator to approximate the solution set (to two decimal places) of the inequality.

53. $2x^3 - 8x + 3 > 0$, (window: $[-5, 5]$ by $[-10, 10]$)

54. $-2x^2 + 5x + 4 > 0$, (window: $[-5, 5]$ by $[-3, 10]$)

55. $\dfrac{1 - x^2}{2 + x^2} < 0$, (window: $[-5, 5]$ by $[-1, 1]$)

56. $(x + 4)^{\frac{1}{3}} < 0$, (window: $[-10, 5]$ by $[-3, 3]$)

Exercises GC57–GC60: Use a graphing calculator to determine approximately (to two decimal places) the solution set of: (a) $f(x) < g(x)$, and (b) $f(x) > g(x)$, by graphing the function $h(x) = f(x) - g(x)$.

57. $f(x) = -4x + 2$ and $g(x) = 5x - 1$, (window: $[-3, 3]$ by $[-5, 5]$)

58. $f(x) = (x - 2)^2$ and $g(x) = 2x - 1$, (window: $[-5, 10]$ by $[-5, 15]$)

59. $f(x) = x^3 - 2$ and $g(x) = -5x^2 + 3x + 1$, (window: $[-8, 8]$ by $[-15, 30]$)

60. $f(x) = x^3 + 2x + 10$ and $g(x) = -8x + 20$, (window: $[-3, 3]$ by $[-20, 30]$)

102   Chapter 2 An Introduction to Functions and Graphs

Exercise GC61: (**Cost**) The monthly cost function (in hundreds of dollars) for a company is given by

$$C(x) = 9x - \frac{5x^3}{x^2 + 100} + 3000, \quad \text{for} \quad 0 \le x \le 200$$

where x denotes the number of units produced. Use a graphing calculator to approximate the production levels, x, at which the monthly cost exceeds \$350,000.

Exercise GC62: (**Break-Even Point**) The monthly revenue function (in hundreds of dollars) for the company of Exercise GC61 is given by

$$R(x) = 35x - \frac{9x^2}{x + 100}$$

Use a graphing calculator to approximate the production level, x, at which monthly revenue equals monthly cost (called the break-even point).

Exercise GC63: (**Profit**) Referring to the company of Exercises GC61 and GC62, use a graphing calculator to approximate the production levels for which the monthly profit exceeds \$40,000. (Profit = Revenue − Cost)

Exercises 64–65: Sketch the graph of the piece-wise defined function by plotting points, and determine the range from the graph.

64. $f(x) = \begin{cases} 2x + 1, & \text{if } x < 0 \\ 3, & \text{if } x \ge 0 \end{cases}$

65. $f(x) = \begin{cases} -3x, & \text{if } x \le 1 \\ x^2, & \text{if } x > 1 \end{cases}$

Exercises 66–71: Sketch the graph of the piecewise-defined function.

66. $f(x) = \begin{cases} x - 3, & \text{if } x < 2 \\ 4, & \text{if } x \ge 2 \end{cases}$

67. $f(x) = \begin{cases} -2, & \text{if } x \le 1 \\ x + 2, & \text{if } x > 1 \end{cases}$

68. $f(x) = \begin{cases} 2x + 4, & \text{if } x \le 0 \\ -x, & \text{if } x > 0 \end{cases}$

69. $f(x) = \begin{cases} -3x + 1, & \text{if } x < 4 \\ 4x, & \text{if } x \ge 4 \end{cases}$

70. $f(x) = \begin{cases} 1 - x, & \text{if } -1 < x < 0 \\ 2x, & \text{if } 0 \le x \le 2 \\ 4x - 1, & \text{if } x > 2 \end{cases}$

71. $f(x) = \begin{cases} 2x - 3, & \text{if } 0 < x < 2 \\ -2x + 3, & \text{if } 2 \le x \le 4 \\ 3, & \text{if } x > 4 \end{cases}$

Exercises 72–74: Sketch the graph of a function with domain $[-5, 5]$ satisfying the given conditions:

72. with range $[0, 12]$.

73. as in Exercise 72 and with x-intercepts at -1 and at 1.

74. as in Exercise 73 and with y-intercept at 2.

Exercise 75: Explain why there is no function as in Exercise 72 with y-intercepts at 2 and at 4.

Exercises 76–79: Define a function f having the given graph.

76.   77.   78.   79.

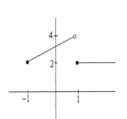 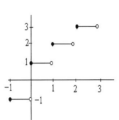 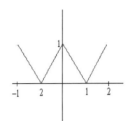 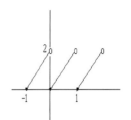

§3. One-To-One Functions and their Inverses

Though a function $y = f(x)$ can not assign more than one value of y to each x in its domain, it can assign the same y-value to different x's. The function $f(x) = x^2$, for example, assigns the number 4 to both 2 and -2, and we say that f **maps** 2 and -2 onto 4.

Of particular interest are those functions that map different values of x onto different values of y:

DEFINITION 2.6 A function f is **one-to-one** if for all a and b in D_f:
ONE-TO-ONE

If $f(a) = f(b)$ then $a = b$

Equivalently: If $a \neq b$ then $f(a) \neq f(b)$

The function f, represented in Figure 2.14(a), **is** one-to-one since no two elements in its domain, $\{1, 2, 3, 4\}$, are "mapped" onto the same element in its range $\{0, 2, 5, 6\}$. The function g, of Figure 2.14(b), is **not** one-to-one since 2 and 3 are both mapped onto 5.

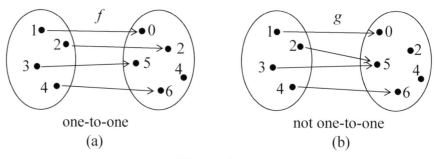

one-to-one   not one-to-one
    (a)   (b)

Figure 2.14

Looking at more traditional graphs of functions, we see that the function f of Figure 2.15(a) is one-to-one (no two xs map onto the same y), while the function g of Figure 2.15(b) is not (different xs map onto the same y).

Horizontal Line Test: No horizontal line can intersect the graph of a one-to-one function at more than one point.

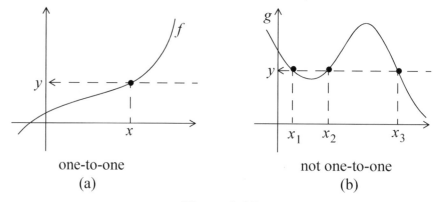

one-to-one   not one-to-one
    (a)   (b)

Figure 2.15

CHECK YOUR UNDERSTANDING 2.17

Which of the following curves: Is not the graph of a function? Is the graph of a one-to-one function? Is the graph of a function that is not one-to-one?

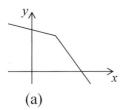
(a)

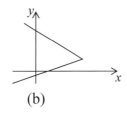
(b)

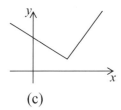
(c)

Answer: (b) is not the graph of a function.
(a) is the graph of a one-to-one function.
(c) is the graph of a function that is not one-to-one.

EXAMPLE 2.13
Show that the function $f(x) = \dfrac{x}{5x+2}$ is one-to-one.

SOLUTION: Appealing to Definition 2.6, we begin with $f(a) = f(b)$, and show that this can only hold if $a = b$:

$$f(a) = f(b)$$
$$\frac{a}{5a+2} = \frac{b}{5b+2}$$
$$a(5b+2) = b(5a+2)$$
$$5ab + 2a = 5ab + 2b$$
$$2a = 2b$$
$$a = b$$

$f(a) = f(b)$
$\Downarrow$
$a = b$

CHECK YOUR UNDERSTANDING 2.18

Show that the function $f(x) = \dfrac{x}{x+1}$ is one-to-one.

Answer: See page C-17.

EXAMPLE 2.14
Show that the given function is not one-to-one:
(a) $f(x) = x^2 - x - 6$
(b) $f(x) = x^3 + 3x^2 - x + 75$

SOLUTION: (a) Noting that:
$$x^2 - x - 6 = 0$$
$$(x-3)(x+2) = 0$$
$$x = 3 \text{ or } x = -2$$

We conclude that
$$f(3) = f(-2) = 0$$

Since two different x's (3 and -2) are mapped to the same y (0), the function is not one-to-one.

(b) We first consider the eqaution $x^3 + 3x^2 - x = 0$:

$$x^3 + 3x^2 - x = 0$$
$$x(x^2 + 3x - 1) = 0$$
$$x = 0, x = \frac{-3 \pm \sqrt{13}}{2}$$

↑ using the quadratic formula

and conclude that the function $g(x) = x^3 + 3x^2 - x$ is not one-to-one: it maps three different x's to zero: 0, $\frac{-3+\sqrt{13}}{2}$, and $\frac{-3-\sqrt{13}}{2}$. It follows that function $f(x) = x^3 + 3x^2 - x + 75$ will map those same three x's to 75, and is therefore not one-to-one.

CHECK YOUR UNDERSTANDING 2.19

Show that the function $f(x) = \dfrac{x^2 - 9}{-x + 2}$ is not one-to-one.

One possible answer:
$f(-3) = f(3) = 0$

Graphing utilitites can often suggest whether or not a function is one-to-one. Sometimes, however, a graph can be misleading. The graph of $f(x) = \dfrac{x^2 + 11x + 31}{x^2 - 4x + 54}$ in [A] of Figure 2.16 suggests that f is one-to-one. Changing the viewing window from $[-10, 10]$ by $[-10, 10]$ to $[-25, 25]$ by $[0, 3]$ shows that we can, in fact, find two different x's which map onto 2 (see [B]):

$$\frac{x^2 + 11x + 31}{x^2 - 4x + 54} = 2$$
$$x^2 + 11x + 31 = 2x^2 - 8x + 108$$
$$x^2 - 19x + 77 = 0$$

Using the quadratic formula: $x = \dfrac{19 \pm \sqrt{19^2 - 4(1)(77)}}{2(1)} = \dfrac{19 \pm \sqrt{53}}{2}$

Since $f\left(\dfrac{19 + \sqrt{53}}{2}\right) = f\left(\dfrac{19 - \sqrt{53}}{2}\right)$ (both numbers map onto 2), f is not one-to-one.

We could have chosen any y-value at which a horizontal line intersects the graph at more than one point.

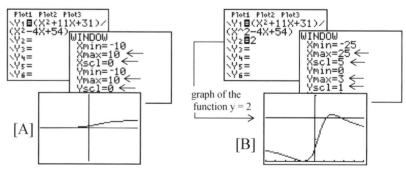

Figure 2.16

INVERSE FUNCTIONS

An attempt to reverse the direction of the arrows in Figure 2.17(a), representing the action of the non-one-to-one function g, would not yield a function. As is shown in Figure 2.17(b), the number 5 would be mapped onto two numbers, 2 and 3, and a function can not assign more than one value to each number in its domain.

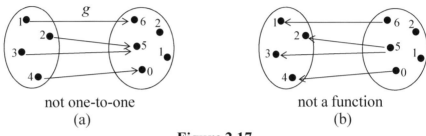

not one-to-one   not a function
(a)   (b)

Figure 2.17

Reversing the arrows of the one-to-one function f of Figure 2.18(a) **does** lead to a function [see Figure 2.18(b)]. That function is called the **inverse** of f and is denoted by the symbol f^{-1}. As you can see, the domain of f^{-1} is the range of f: $\{0, 1, 5, 6\}$, and the range of f^{-1} is the domain of f: $\{1, 2, 3, 4\}$.

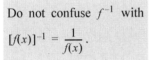

Do not confuse f^{-1} with $[f(x)]^{-1} = \dfrac{1}{f(x)}$.

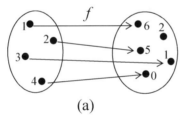 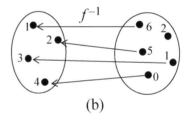

(a)   (b)

Figure 2.18

The relationship between the functions f and f^{-1} depicted in the margin reveals the fact that each function "undoes" the work of the other. For example:

$$(f^{-1} \circ f)(2) = f^{-1}(f(2)) = f^{-1}(5) = 2$$

and

$$(f \circ f^{-1})(5) = f(f^{-1}(5)) = f(2) = 5$$

In general:

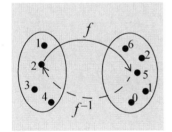

Only one-to-one functions have inverses (see Figure 2.17).

DEFINITION 2.7

INVERSE FUNCTIONS

The inverse of a one-to-one function f with domain D_f and range R_f is that function f^{-1} with domain R_f and range D_f such that:

$$(f^{-1} \circ f)(x) = x \text{ for every } x \text{ in } D_f$$

and

$$(f \circ f^{-1})(x) = x \text{ for every } x \text{ in } R_f$$

108   Chapter 2 An Introduction to Functions and Graphs

EXAMPLE 2.15 Find the inverse of the one-to-one function:

$$f(x) = \frac{x}{5x + 2}$$

SOLUTION: We offer two methods for your consideration.

Start with:

$$f(f^{-1}(x)) = x$$

For notational convenience, substitute t for $f^{-1}(x)$:

$$f(t) = x$$

Since $f(x) = \frac{x}{5x + 2}$:

$$\frac{t}{5t + 2} = x$$

Solve for t:

$$t = (5t + 2)x$$
$$t = 5tx + 2x$$
$$t - 5tx = 2x$$
$$t(1 - 5x) = 2x$$
$$t = \frac{2x}{1 - 5x}$$

Substituting $f^{-1}(x)$ back for t:

$$f^{-1}(x) = \frac{2x}{1 - 5x}$$

To say that $y = f(x)$ is to say that $f^{-1}(y) = x$.

So, start with:

$$y = \frac{x}{5x + 2}$$

And solve for x in terms of y:

$$(5x + 2)y = x$$
$$5xy + 2y = x$$
$$5xy - x = -2y$$
$$x(5y - 1) = -2y$$
$$x = \frac{-2y}{5y - 1} = \frac{2y}{1 - 5y} = f^{-1}(y)$$

To obtain the inverse function expressed in terms of the variable x (instead of y), interchange x and y:

$$y = \frac{2x}{1 - 5x} = f^{-1}(x)$$

As a check, we will verify that $(f^{-1} \circ f)(x) = x$:

Since $f(x) = \frac{x}{5x + 2}$

Since $f^{-1}(x) = \frac{2x}{1 - 5x}$

$$(f^{-1} \circ f)(x) = f^{-1}(f(x)) \stackrel{?}{=} f^{-1}\left(\frac{x}{5x + 2}\right) \stackrel{?}{=} \frac{2\left(\dfrac{x}{5x + 2}\right)}{1 - 5\left(\dfrac{x}{5x + 2}\right)}$$

Multiply numerator and denominator by $5x + 2$:

$$= \frac{2x}{5x + 2 - 5x} = \frac{2x}{2} = x$$

CHECK YOUR UNDERSTANDING 2.20

Determine the inverse of the one-to-one function:

$$f(x) = \frac{x}{x + 1}$$

Answer: $f^{-1}(x) = \frac{x}{1 - x}$

Verify, directly, that $(f \circ f^{-1})(x) = x$ and that $(f^{-1} \circ f)(x) = x$.

GRAPH OF AN INVERSE FUNCTION

We end this section with a result which relates the graph of a one-to-one function with that of its inverse. In that endeavor, we will use the following facts:

THEOREM 2.1 For any number a:
$$|a|^2 = a^2$$

PROOF: Even if $|a|$ and a differ by a negative sign, their squares will certainly be the same.

THEOREM 2.2
DISTANCE BETWEEN TWO POINTS

The distance D between the points (x_1, y_1) and (x_2, y_2) in the plane is given by:
$$D = \sqrt{(x_1 - x_2)^2 + (y_1 - y_2)^2}$$

PROOF:

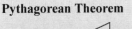
Pythagorean Theorem

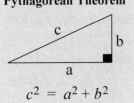
$c^2 = a^2 + b^2$

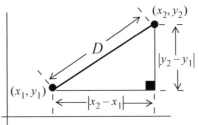

Pythagorean Theorem:
$$D^2 = |x_2 - x_1|^2 + |y_2 - y_1|^2$$
Theorem 2.1: $= (x_2 - x_1)^2 + (y_2 - y_1)^2$
or: $D = \sqrt{(x_2 - x_1)^2 + (y_2 - y_1)^2}$

THEOREM 2.3 The graph of f^{-1} is the reflection of the graph of f about the line $y = x$.

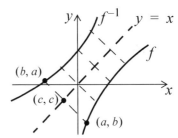

PROOF: Since $f(a) = b$ if and only if $f^{-1}(b) = a$, to say that (a, b) is on the graph of f, is to say that (b, a) is on the graph of f^{-1}. Since the slope of the line joining (b, a) and (a, b) is $\dfrac{a-b}{b-a} = -1$, that line segment is perpendicular to the line $y = x$ (which has slope 1). Moreover, using Theorem 2.2, we see that the point (c, c) on that line segment is equidistant from (b, a) and (a, b):

distance between (b, a) and (c, c) distance between (a, b) and (c, c)

$$\sqrt{(b-c)^2 + (a-c)^2} = \sqrt{(a-c)^2 + (b-c)^2}$$

EXAMPLE 2.16 (a) Sketch the graph of the function $f(x) = \sqrt{x-3} + 2$. Specify its domain and range.
(b) Use Theorem 2.3 to obtain the graph of its inverse $f^{-1}(x)$. Specify its domain and range.
(c) Find $f^{-1}(x)$.

SOLUTION: (a) The graph of the function $y = \sqrt{x}$ of Figure 2.4, page 89, appears in Figure 2.19(a). Moving that graph 3 units to the right and 2 units up brings us to the graph of $f(x) = \sqrt{x-3} + 2$ in Figure 2.19(b). From that graph, we see that the domain of f is $[3, \infty)$, and that its range is $[2, \infty)$.

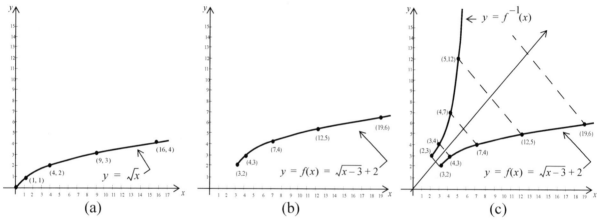

Figure 2.19

(b) Reflecting the graph of f about the line $y = x$ we arrive at the graph of f^{-1} [Figure 2.19(c)]. Note that the domain of f^{-1} is the range of f, namely: $[2, \infty)$; and that the range of f^{-1} is the domain of f, namely: $[3, \infty)$.

(c) Preceding as in Example 2.15 (right-hand side), we find $f^{-1}(x)$:

Start with: $\qquad y = \sqrt{x-3} + 2$

Solve for x in terms of y: $\sqrt{x-3} = y - 2$

$$x - 3 = (y-2)^2$$
$$x - 3 = y^2 - 4y + 4$$
$$x = y^2 - 4y + 7$$

To obtain the inverse function expressed in terms of the variable x (instead of y), interchange x and y: $y = x^2 - 4x + 7 = f^{-1}(x)$

with domain: $[2, \infty)$

2.3 One-To-One Functions and their Inverses 111

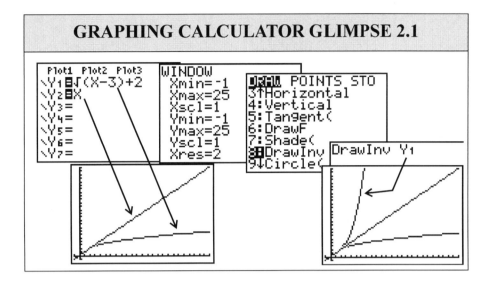

GRAPHING CALCULATOR GLIMPSE 2.1

Answer: For one-to-one, see page C-18.

$D_f = [0, \infty), R_f = [-2, \infty)$

$f^{-1}(x) = (x+2)^2, x \geq -2$

$D_{f^{-1}} = R_f = [-2, \infty)$

$R_{f^{-1}} = D_f = [0, \infty)$

For graph, see page C-18.

CHECK YOUR UNDERSTANDING 2.21

Show that the function $f(x) = \sqrt{x} - 2$ is one-to-one. Indicate its domain and range. Find its inverse and indicate its domain and range. Sketch the graph of both functions on the same set of axes.

EXERCISES

Exercises 1–4: Decide whether or not the function is one-to-one.

1.   2.   3.   4.

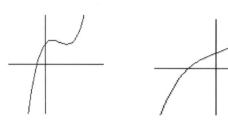
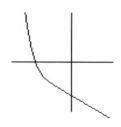

Exercises 5–15: Show that the function is one-to-one.

5. $f(x) = -5x - 1$ 6. $g(x) = 6x + 5$ 7. $h(x) = x^3 - 1$ 8. $p(x) = (x+1)^3$

9. $g(x) = \sqrt{x-2} + 1$ 10. $f(x) = \dfrac{2}{\sqrt{x+1}}$ 11. $g(x) = -\dfrac{1}{x}$ 12. $h(x) = \dfrac{4}{2x-3}$

13. $r(x) = \dfrac{3x}{2x+1}$ 14. $k(x) = \dfrac{x-2}{4-3x}$ 15. $f(x) = \dfrac{2x+3}{3x+1}$

Exercises 16–24: Show that the function is not one-to-one: Find two different values of x, a and b, such that $f(a) = f(b)$.

16. $f(x) = 2$ 17. $f(x) = -\dfrac{1}{x^2}$ 18. $f(x) = \dfrac{3}{x^2+1}$ 19. $f(x) = \dfrac{x^2-16}{3x+4}$

20. $f(x) = \dfrac{x^3 - 3x^2}{x+1}$ 21. $f(x) = 5x^2 + 12x + 4$ 22. $f(x) = (3-2x)(4x-7)(x+1)$

23. $f(x) = 3x^3 - 7x^2 + 2x - 4$ 24. $f(x) = 2x^3 + 9x^2 + 4x + 17$

Exercises GC25–GC28: Show that the function is not one-to-one: Use a graphing calculator to suggest a value that is assumed more than once. Then use that value to show that the function is not one-to-one.

25. $g(x) = \dfrac{x^2 + 3x + 3}{x^2 - x + 2}$,   (window: $[-5, 5]$ by $[-1, 5]$)

26. $f(x) = \dfrac{x^2 + x + 10}{x^2 + 3x + 6}$,   (window: $[-10, 10]$ by $[-2, 10]$)

27. $h(x) = \dfrac{x^2 - 4x + 5}{x^2 + 2x + 3}$,   (window: $[-5, 5]$ by $[-1, 6]$)

28. $k(x) = \dfrac{-x^2 + 5x - 8}{x^2 + x + 2}$,   (window: $[-5, 5]$ by $[-8, 1]$)

2.3 One-To-One Functions and their Inverses   113

Exercises 29–32: Find the inverse of the given one-to-one function, and the domain and range of the inverse function. Verify that $(f \circ f^{-1})(x) = x$ and $(f^{-1} \circ f)(x) = x$.

29. $f(x) = 2x - 3$ 30. $f(x) = -x + 1$ 31. $f(x) = 2\sqrt{x + 3}$ 32. $f(x) = \sqrt{1 - x}$

Exercises 33–42: Find the inverse of the given one-to-one function, and verify that $(f \circ f^{-1})(x) = x$ and $(f^{-1} \circ f)(x) = x$.

33. $f(x) = 8 - x^3$ 34. $f(x) = x^3 + 1$ 35. $f(x) = \dfrac{1}{x}$ 36. $f(x) = -\dfrac{2}{3x}$

37. $f(x) = \dfrac{1}{x - 1}$ 38. $f(x) = \dfrac{3}{2x - 5}$ 39. $f(x) = -\dfrac{4x}{3x + 4}$ 40. $f(x) = \dfrac{3x}{2 + 7x}$

41. $f(x) = \dfrac{3x + 2}{1 - x}$ 42. $f(x) = \dfrac{1 - 2x}{6 - 3x}$

Exercise 43: (**Temperature Conversion**) The conversion equations between degrees Centigrade (Celsius) (C) and degrees Fahrenheit (F) are

$$(1) \quad F = \tfrac{9}{5} C + 32 \quad \text{and} \quad (2) \quad C = \tfrac{5}{9}(F - 32)$$

(a) Use (2) to convert 50°F to Centigrade.

(b) Use the value of C you determined in (a) to find F using equation (1).

(c) What does (b) suggest about the relationship between the functions

$$F(x) = \tfrac{9}{5} x + 32 \quad \text{and} \quad C(x) = \tfrac{5}{9}(x - 32)$$

(d) Verify your hypothesis in (c) by showing that

$$(F \circ C)(x) = x \quad \text{and} \quad (C \circ F)(x) = x$$

Exercises 44–52: Sketch the graph of the one-to-one function f, and then use Theorem 2.3 to sketch the graph of f^{-1}, on the same set of axes. [Do <u>not</u> find the rule for $f^{-1}(x)$.]

44. $f(x) = 3x + 5$ 45. $f(x) = -4x - 2$ 46. $f(x) = 2x - 1$

47. $f(x) = -3x + 2$ 48. $f(x) = \tfrac{1}{2} x - \tfrac{1}{3}$ 49. $f(x) = \tfrac{2}{3} x + \tfrac{1}{4}$

50. $f(x) = \sqrt{1 - x} + 3$ 51. $f(x) = \sqrt{x - 2} - 1$ 52. $f(x) = \sqrt{x + 4} + 2$

114   Chapter 2 An Introduction to Functions and Graphs

Exercises GC53–GC56: Use a graphing calculator to graph $f(x)$ in the given window. Then graph $f^{-1}(x)$ in the same window, as in Calculator Glimpse 2.1.

53. $f(x) = (x+1)^{\frac{5}{2}} - 2,$ (window: $[-5, 5]$ by $[-5, 5]$)

54. $f(x) = (1-x)^{\frac{3}{2}} + 3,$ (window: $[-5, 10]$ by $[-5, 10]$)

55. $f(x) = (x-1)^{\frac{1}{3}} + 1,$ (window: $[-2, 3]$ by $[-1, 3]$)

56. $f(x) = 5(x+3)^{\frac{5}{3}} - 8,$ (window: $[-10, 10]$ by $[-10, 10]$)

Exercises GC57–GC59: Find a viewing window in which you can see that the given function is not one-to-one.

57. $f(x) = x^3 - 3x^2 + 2x + 100$ 58. $g(x) = \dfrac{1}{x^2 + 40x + 401}$ 59. $h(x) = x^2 + 40x + 401$

CHAPTER SUMMARY

INTERVAL NOTATION	For any two numbers a and b with $a < b$: $$(a, b) = \{x \mid a < x < b\} \qquad (a, \infty) = \{x \mid x > a\}$$ $$[a, b] = \{x \mid a \leq x \leq b\} \qquad [a, \infty) = \{x \mid x \geq a\}$$ $$[a, b) = \{x \mid a \leq x < b\} \qquad (-\infty, b) = \{x \mid x < b\}$$ $$(a, b] = \{x \mid a < x \leq b\} \qquad (-\infty, b] = \{x \mid x \leq b\}$$ $(-\infty, \infty)$ denotes the set of all real numbers
FUNCTIONS	A function, f, assigns to each number x in its domain, one and only one number $f(x)$ in its range.
Vertical Line Test	No vertical line can intersect the graph of a function at more than one point: 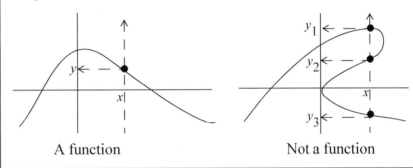 A function Not a function
DOMAIN	Unless otherwise specified, the domain of a function given in terms of an expression is understood to be the set of all numbers for which the expression is defined. For example, the (understood) domain of the function $f(x) = \sqrt{x-5}$ is the set $D_f = [5, \infty)$.
Relation between the graph of a function f, its domain D_f and its range R_f	The domain of f is the projection of the graph of f onto the x-axis. 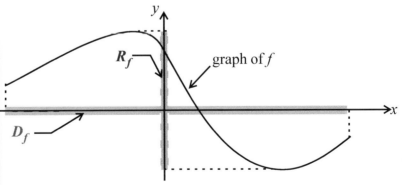 The range of f is the projection of the graph of f onto the y-axis.

116   **Chapter 2 An Introduction to Functions and Graphs**

ABSOLUTE VALUE FUNCTION	The **absolute value** of a number a is that number $\|a\|$ given by: $$\|x\| = \begin{cases} x & \text{if} \quad x \geq 0 \\ -x & \text{if} \quad x < 0 \end{cases}$$ $y = \|x\|$
DISTANCE	The **distance** between a and b on the number line is given by $$\|a - b\|$$ note the negative sign
COMPOSITE FUNCTIONS $g \circ f$	The **composition** $(g \circ f)(x)$ is given by: $$(g \circ f)(x) = g(f(x))$$ **first** apply f and **then** apply g [Assuming $f(x)$ is in the domain of g] In general: $(g \circ f)(x) \neq (f \circ g)(x)$
Graphing the function $f(x) = a(x - x_0)^n + y_0$	**Step 1:** Stretch/Shrink the graph by the magnitude of a, and also reflect the graph about the x-axis if $a < 0$. **Step 2:** Shift the resulting graph $\|y_0\|$ units upward if $y_0 > 0$, or downward if $y_0 < 0$. In addition, shift the graph $\|x_0\|$ units to the right if $x_0 > 0$, or to the left if $x_0 < 0$.
EVEN AND ODD FUNCTIONS	EVEN if: $f(-x) = f(x)$ ODD if: $f(-x) = -f(x)$ (for all x in the domain of f)
DISTANCE BETWEEN POINTS IN THE PLANE	$$D = \sqrt{(x_1 - x_2)^2 + (y_1 - y_2)^2}$$

GRAPHS, EQUATIONS, AND INEQUALITIES	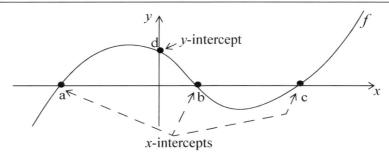
$f(x) = 0$ At the x-intercepts of the graph of f: a, b, and c
$f(x) < 0$ Where the graph of f lies below the x-axis: $(-\infty, a) \cup (b, c)$
$f(x) > 0$ Where the graph of f lies above the x-axis: $(a, b) \cup (c, \infty)$
WHEN TWO FUNCTIONS ARE INVOLVED:
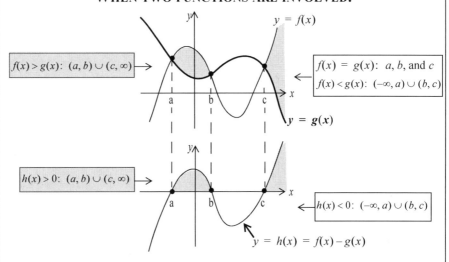
ONE-TO-ONE FUNCTIONS **HORIZONTAL LINE TEST**	A function f is **one-to-one** if for all a and b in D_f:
$$\text{If } f(a) = f(b) \text{ then } a = b$$
Equivalently: If $a \neq b$ then $f(a) \neq f(b)$
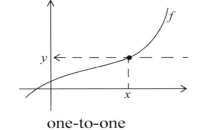   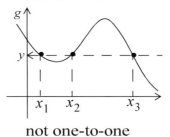
one-to-one   not one-to-one
INVERSE FUNCTIONS (only one-to-one functions have inverses)	The inverse of a one-to-one function f with domain D_f and range R_f is that function f^{-1} with domain R_f and range D_f such that:
$$(f^{-1} \circ f)(x) = x \text{ for every } x \text{ in } D_f$$
and
$$(f \circ f^{-1})(x) = x \text{ for every } x \text{ in } R_f$$

118   **Chapter 2 An Introduction to Functions and Graphs**

Finding f^{-1}	Substitute t for $f^{-1}(x)$ in the equation $f(f^{-1}(x)) = x$, and solve for t. **OR** Solve $y = f(x)$ for x in terms of y, and then interchange x and y.
GRAPH OF INVERSE FUNCTION	The graph of f^{-1} is the reflection of the graph of f about the line $y = x$.

		SOME COMMON PITFALLS	
WRONG:	$f^{-1}(x) = \dfrac{1}{f(x)}$ (The above does hold for some functions, but not for all functions)	**Because:**	For $f(x) = 2x$: $f^{-1}(x) = \dfrac{1}{2}x$ while $\dfrac{1}{f(x)} = \dfrac{1}{2x}$
WRONG:	$f(a+b) = f(a) + f(b)$ (The above does hold for some functions, but not for all functions)	**Because:**	For $f(x) = x^2$: $f(2+3) = f(5) = 5^2 = 25$ while: $f(2) + f(3) = 2^2 + 3^3 = 13$
WRONG:	$f(ax) = af(x)$ (The above does hold for some functions, but not for all functions)	**Because:**	For $f(x) = x^2$: $f(2x) = (2x)^2 = 4x^2$ while: $2f(x) = 2x^2$
WRONG:	Every function is either even or odd	**Because:**	The function $f(x) = x + 1$ is neither even nor odd. (Indeed, most functions are neither even nor odd.)

Chapter Review Exercises   119

CHAPTER REVIEW EXERCISES

Exercises 1–4: Simplify, by expressing as one interval. (Suggestion: First represent each of the component intervals on the number line.)

1. $[3,6) \cap (4,7]$

2. $[3,6) \cup (4,7]$

3. $(-\infty, 4) \cup [1,5] \cup (4,\infty)$

4. $(-\infty, 5) \cap [3,5] \cap (4,\infty)$

Exercises 5–12: Find the domain of the function.

5. $f(x) = 2x - \frac{1}{3}$

6. $g(x) = \dfrac{2x}{x^2 + x}$

7. $h(x) = \sqrt{4 + 3x} - \dfrac{1}{x}$

8. $k(x) = \dfrac{3 - x}{5x\sqrt{2x - 1}}$

9. $F(x) = \sqrt[3]{x + 1}$

10. $G(x) = \dfrac{4x^2 - 9}{2x^2 + x - 3}$

11. $K(x) = 1 - x^{-\frac{3}{2}}$

12. $H(x) = \dfrac{x^{\frac{2}{3}}}{x^2 + x + 1}$

Exercise 13: Determine the domain of the function, and the values $f(1)$, $f(3)$, and f(4).

$$f(x) = \begin{cases} -2x - 3 & \text{if } -\infty < x \le 3 \\ 2x^2 - 1 & \text{if } 3 < x < 5 \end{cases}$$

Exercises 14–20: For $f(x) = 2x^2 - 3x - 1$, determine:

14. $f(3)$

15. $f(-2)$

16. $f\left(\dfrac{1}{x}\right)$

17. $f(-3x)$

18. $f(x^2)$

19. $f(x + h)$

20. $\dfrac{f(x + h) - f(x)}{h}$, for $h \ne 0$

Exercises 21–27: Repeat Exercises 14–20 for $f(x) = \dfrac{1}{x + 1}$

Exercises 28–37: For $f(x) = 3x + 5$, $g(x) = \dfrac{x}{x + 4}$, and $h(x) = 4 - 7x$, determine :

28. $(f - h)(x)$

29. $\left(\dfrac{f}{g}\right)(x)$

30. $\left(\dfrac{h}{f}\right)(x)$

31. $(2h)(x)$

32. $(fg)(x)$

33. $((f + h)g)(x)$

34. $(f \circ g)(x)$

35. $(g \circ h)(x)$

36. $(h \circ h)(x)$

37. $(g \circ g)(x)$

Exercises 38–41: Find functions f and g such that h can be expressed as their composition, $h(x) = (g \circ f)(x)$.

38. $h(x) = \left(\dfrac{2x}{3x - 1}\right)^{\frac{2}{3}}$

39. $h(x) = (3x^3 + 2x - 1)^{10}$

40. $h(x) = \sqrt{x - 5} + 2$

41. $h(x) = \sqrt[4]{2x + 7}$

Exercise 42: (**Commission**) At the end of each month, a salesperson receives a 4% commission on total sales of up to $3,000. An additional 1% is awarded on the portion of total sales that is over $3,000, and another 1% on the portion over $5,000. Determine

(a) the piece-wise commission function, C(x), as a function of monthly sales, x.

(b) the commission on $2,500. (c) the commission on $4,000

(d) the commission on $7,500

Exercises 43–48: Find the domain of the function. Sketch the graph by reflecting, stretching, and shifting the basic graph. Determine the y-intercept and the range of the function.

43. $g(x) = |x+5| - 2$ 44. $f(x) = -|x-8| + 4$ 45. $K(x) = 1 - x^3$

46. $h(x) = x^2 - 3$ 47. $g(x) = -\sqrt{x-5} + 3$ 48. $G(x) = \frac{1}{4}\sqrt{x-4} - 1$

Exercises 49–51: Indicate if the graph appears to represent the graph of an even function, an odd function, or neither.

49.

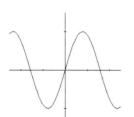

50.

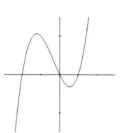

51.

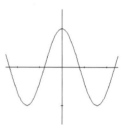

Exercises 52–57: Sketch the graph by reflecting, stretching, and shifting the basic graph, and from the graph estimate the solution set of (a) $f(x) = 0$, (b) $f(x) > 0$ and (c) $f(x) < 0$.

52. $f(x) = 2|x-1|$ 53. $f(x) = |x| - 1$ 54. $f(x) = -|x| + 1$

55. $f(x) = -2|x+3| + 8$ 56. $f(x) = 4|x+3| - 6$ 57. $f(x) = 9\sqrt{x-5}$

Exercises 58–60: From the graph of the function f, determine, (a) the x-intercepts, (b) the y-intercept, (c) the solution set of $f(x) = 0$, (d) the solution set of $f(x) < 0$, and (e) the solution set of $f(x) > 0$.

58.

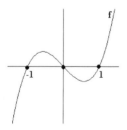

59.

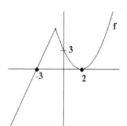

60.

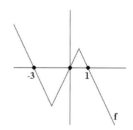

Exercises 61–63: From the given graphs of f and g, determine the solution set of (a) $f(x) = g(x)$, (b) $f(x) < g(x)$, and (c) $f(x) > g(x)$.

61.

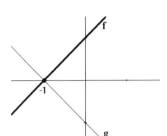

62.

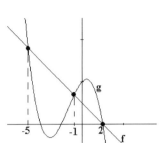

63.

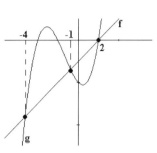

Exercises GC64–GC67: Use a graphing calculator to find the approximate solution set (to two decimal places).

64. $2x^3 + 3x^2 - 5x - 1 = 0$, (window: $[-5, 5]$ by $[-5, 10]$)

65. $-x^4 + 2x^3 - 3x + 5 = 0$, (window: $[-5, 5]$ by $[-5, 10]$)

66. $(x^2 + 3x)^3 - 5 < 0$, (window: $[-5, 5]$ by $[-50, 50]$)

67. $\dfrac{x^2 - x - 1}{x^2 + x + 2} > 0$, (window: $[-5, 5]$ by $[-1, 1]$)

Exercise GC68: (**Salary**) An individual's yearly salary (in thousands of dollars) as a function of years worked is given by $S(y) = 37 + \dfrac{2y^2}{y + 10}$.

(a) What is the starting salary?

(b) Use a graphing calculator to determine how many years the individual has to work before earning twice his starting salary.

Exercises 69–71: Decide whether or not the function is one-to-one.

69.

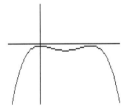

70.

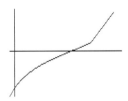

71.

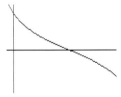

122   Chapter 2 An Introduction to Functions and Graphs

Exercises GC72–GC73: Show that the function is not one-to-one: (a) Use a graphing calculator to suggest a value of the function that is assumed more than once. (b) Use the value determined in (a) to show that the function is not one-to-one.

72. $f(x) = \dfrac{2x^2 - x + 5}{x^2 + x + 5}$,   (window: $[-10, 10]$ by $[-2, 3]$)

73. $g(x) = \dfrac{-2x^2 + x - 4}{x^2 + x + 4}$,   (window: $[-6, 6]$ by $[-4, 1]$)

Exercises 74-79: Determine whether or not the given function is one-to-one. If it is, find the inverse function.

74. $f(x) = -5x + 3$ 75. $f(x) = \frac{2}{3}x + 4$ 76. $f(x) = -x^2 - 4$

77. $f(x) = \dfrac{4}{x}$ 78. $f(x) = \dfrac{x}{4}$ 79. $f(x) = \dfrac{2}{1 + x^2}$

Exercises 80–83: Find the inverse of the given one-to-one function, and verify that $(f \circ f^{-1})(x) = x$ and $(f^{-1} \circ f)(x) = x$.

80. $f(x) = \dfrac{1}{x - 2}$ 81. $f(x) = \dfrac{3}{x + 4}$ 82. $f(x) = \dfrac{x - 4}{1 - 5x}$ 83. $f(x) = -\dfrac{5x + 3}{3x + 1}$

Exercises 84–86: Find the inverse of the given one-to-one function f, and sketch the graphs of f and f^{-1} on the same set of axes.

84. $f(x) = -3x + 2$ 85. $f(x) = \sqrt{2x + 1}$ 86. $f(x) = \sqrt{x - 3} - 3$

Exercises GC87–GC88: Use a graphing calculator to graph $f(x)$ and $f^{-1}(x)$ in the given window.

87. $f(x) = (x - 2)^{\frac{3}{2}} - 1$,   (window: $[-1, 15]$ by $[-2, 20]$)

88. $f(x) = (x + 1)^{\frac{2}{3}} + 2$,   (window: $[-15, 15]$ by $[-20, 20]$)

C U M U L A T I V E   R E V I E W   E X E R C I S E S

Exercises 1–3: Simplify each of the following.

1. $\dfrac{2x}{x - 3} - \dfrac{3x - 5}{2x + 1} + 2$ 2. $\dfrac{(-3x)^2 \sqrt{x}}{3(x^2)^3}$ 3. $\dfrac{(3x + 4)^2 \cdot 5 - (5x - 2) \cdot 2(3x + 4) \cdot 3}{(3x + 4)^4}$

Exercises 4–5: Solve.

4. $(3x + 1)(x - 2) = 4$ 5. $(5x - 1)^2 = 9$

Exercise 6: Convert 3 ft^3/min to m^3/hr, given 3.28 ft = 1 m.

Exercises 7–8: Determine $f(3)$, $f(t)$, $f(2t)$, $f(x-3)$, and $f\left(\dfrac{1}{x}\right)$, for the given function.

7. $f(x) = \dfrac{x+3}{x}$

8. $f(x) = \sqrt{x-2} + 2x$

Exercises 9–14: If $f(x) = 2x + 5$ and $g(x) = \dfrac{1}{x-3}$ determine:

9. $\left(\dfrac{f}{g}\right)(x)$ 10. $(g-f)(x)$ 11. $(f \circ f)(2)$ 12. $(f \circ g)(x)$ 13. $(g \circ g)(x)$ 14. $(g \circ f)(x)$

Exercises 15–16: Determine the domain of g. Sketch the graph of g by reflecting, stretching, and shifting the basic graph. Determine the intercepts and the range.

15. $g(x) = \sqrt{x-6}$

16. $g(x) = -10|x+1| + 5$

Exercise 17: (**Discount Pricing**) The price of a 4 gigabyte flash drive at Computer Supplies is $8.50. If the school buys at least 10 flash drives, the price per flash drive is reduced to $8.00, and if the school buys 20 or more flash drives, the price is further reduced to $7.50 per flash drive.

(a) Determine the piece-wise cost function, C(x), for the purchase of x flash drives.

(b) Compare the cost of buying 9 flash drives with that of buying 10 flash drives.

(c) Compare the cost of buying 19 flash drives with that of buying 20 flash drives.

Exercise 18: Complete the graph in the adjacent figure so that it is the graph of

(a) an even function   (b) an odd function

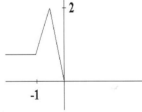

Exercise 19: Establish that $f(x) = \dfrac{1}{3x-2}$ is a one-to-one function, and find its inverse. Show that
$$(f \circ f^{-1})(x) = x \quad \text{and} \quad (f^{-1} \circ f)(x) = x$$

Exercise 20: Show that the function $f(x) = x^2 - x$ is **not** one-to-one.

Exercise 21: Find the inverse of the one-to-one function $f(x) = \sqrt{x-4}$, and sketch the graphs of f and f^{-1} on the same set of axes.

Exercise 22: From the adjacent graph of the function f, determine,

(a) the x-intercepts

(b) the y-intercept

(c) the solution set of $f(x) = 0$

(d) the solution set of $f(x) < 0$

(e) the solution set of $f(x) > 0$

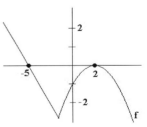

Exercise 23: From the given graphs of f and g, determine the approximate solution set of

(a) $f(x) = g(x)$ (b) $f(x) < g(x)$ (c) $f(x) > g(x)$

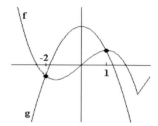

Exercise GC24: Use a graphing calculator to view the graph of $h(x) = f(x) - g(x)$ for $f(x) = 4x^4 - 1$ and $g(x) = 3 - x^2$ in the window: $[-6, 6]$ by $[-6, 6]$. Determine where, approximately (to two decimal places),

(a) $f = g$ (b) $f < g$ (b) $f > g$

Exercise GC25: Use a graphing calculator to find the approximate solution set (to two decimal places) of

$$-x^3 + 4x^2 + 8x - 4 > 0 \quad (\text{window}: [-5, 10] \text{ by } [-20, 40])$$

Exercise GC26: (**Profit**) The weekly profit function (in dollars) for a company is given by

$$P(x) = 6x - \frac{7x^2}{x + 2500} - 1200$$

where x denotes the number of units produced. Use a graphing calculator to find the production levels, x:

(a) at which weekly profit is positive.

(b) at which weekly profit exceeds $3,000.

Exercise 27: (**Mixture**) Beaker A contains a 35% acid solution and beaker B contains 10 ounces of a 50% acid solution. The contents of both beakers are poured into an empty beaker C, resulting in a 42% acid solution. How many ounces of solution did beaker A contain?

CHAPTER 3

TRIGONOMETRIC FUNCTIONS OF ANGLES

The Greek astronomer Hipparchus (180-125 BC) compiled the first trigonometric tables, and is often called the *father of trigonometry*. While the word trigonometry is itself a derivation of the Greek words *trigonon* (triangle) and *metron* (measure), the origin of the discipline can be traced to earlier times and cultures. As early as 1500 BC, in the Ahmes Papyrus, one already finds problems involving similar triangles, and there is ample evidence that trigonometry was used by the Chinese in 1100 BC to calculate distances and heights.

Throughout the centuries trigonometry continued to be a valuable tool for measurement, especially in astronomy, but it did not begin to develop into a formal mathematical discipline until the fifteenth century, with the work of the German mathematician Johan Muller (1436-1476).

In 1748, the great Swiss mathematician, Leonard Euler (1707-1783) published his *Introduction to Infinite Analysis*, in which the trigonometric functions were defined in terms of ratios of sides of right triangles. Because of its practical and geometrical nature, the importance of that approach has not diminished through the years.

The historical and geometrical right triangle approach is used in Section 1 in defining the trigonometric functions of acute angles. Applications of these functions are featured in Section 2. Oriented angles are introduced in Section 3, and the trigonometric functions are extended to accommodate them in Section 4.

§1. TRIGONOMETRIC FUNCTIONS OF ACUTE ANGLES

An **angle** is formed by two line segments having a common endpoint. The line segments are the **sides** of the angle and the common endpoint is the **vertex** [Figure 3.1(a)]. Lower case Greek letters will be used to denote angles; particularly the letters α (alpha), β (beta), γ (gamma), and θ (theta).

A **central angle** of a circle is an angle whose vertex is at the center of a circle. Referring to the central angle θ of Figure 3.1(b), we say that the arc AB **subtends the angle** θ, or that θ is **subtended by that arc**.

> A degree is further subdivided into 60 equal units called minutes, and each minute into 60 seconds. For example, one writes $\theta = 73°15'30''$ to mean that the angle θ has a measure of 73 degrees, 15 minutes, 30 seconds. Decimal degrees, such as 73.4° are also used.

The most commonly used angle measurement comes from the Greeks. Influenced by the base-60 Babylonian number system, they divided the circumference of a circle into 360 equal parts. Each part subtends a central angle of the circle whose measure is defined to be one **degree** (denoted 1°) [Figure 3.1(c)]. A 90° angle is said to be a **right angle**, an angle strictly between 0° and 90° is an **acute angle**, and an angle strictly between 90° and 180° is an **obtuse angle**.

126   Chapter 3 Trigonometric Functions of Angles

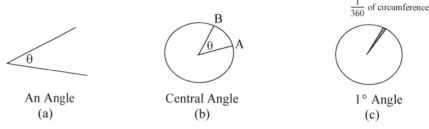

An Angle
(a)

Central Angle
(b)

1° Angle
(c)

Figure 3.1

A triangle containing a right angle is a **right triangle**, and any other triangle is an **oblique triangle**. In a right triangle, the side opposite the right angle is the **hypotenuse**, and the other two sides are the **legs** of the triangle.

The following important result is attributed to the Pythagoreans, circa 500 B.C.

THEOREM 3.1   The sum of the angles in any triangle is $180°$.

Two triangles are said to be **similar** if the angles of one of them are the same as those of the other:

> In general, we label the angles in a triangle with the Greek letters α, β, and γ (alpha, beta, and gamma, respectively). The lengths of the sides opposite α, β, and γ are labeled a, b, and c, respectively.

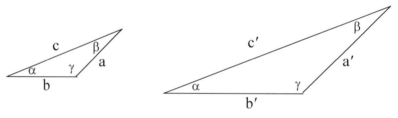

Figure 3.2

While similar triangles have the same shape, they need not be of the same size; however:

THEOREM 3.2   The ratio of corresponding sides of similar triangles are equal.

In particular, referring to the similar triangles of Figure 3.2, we have:

$$\frac{a}{a'} = \frac{b}{b'} = \frac{c}{c'}$$

Since the sum of the angles in any triangle equals $180°$, if two angles in one triangle equal two angles in another, then the two triangles are similar. In particular, any right triangle with acute angle θ has to be similar to any other right triangle with the same acute angle θ. This enables us to define the trigonometric functions of an acute angle θ in terms of ratios of lengths of sides of **any** right triangle containing θ:

3.1 Trigonometric Functions of Acute Angles   127

DEFINITION 3.1
TRIGONOMETRIC FUNCTIONS

Let θ be an acute angle. The functions sine, cosine, tangent, cosecant, secant, and cotangent of θ (abbreviated sin, cos, tan, csc, sec, and cot, respectively), are defined as follows:

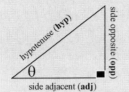

$$\sin\theta = \frac{\text{opp}}{\text{hyp}}, \quad \cos\theta = \frac{\text{adj}}{\text{hyp}}, \quad \tan\theta = \frac{\text{opp}}{\text{adj}}$$

$$\csc\theta = \frac{\text{hyp}}{\text{opp}}, \quad \sec\theta = \frac{\text{hyp}}{\text{adj}}, \quad \cot\theta = \frac{\text{adj}}{\text{opp}}$$

where opp, adj, and hyp are the lengths of the opposite side, adjacent side and hypotenuse, respectively.

> Note that by virtue of Theorem 3.2, any right triangle containing θ can be selected in the definition process

EXAMPLE 3.1 Determine the values of the remaining trigonometric functions of the acute angle θ, if $\sin\theta = \frac{2}{7}$

SOLUTION: The given information:

$$\sin\theta = \frac{\text{opp}}{\text{hyp}} = \frac{2}{7}$$

leads us to form the adjacent right triangle in which the ratio of the side opposite the angle θ to the hypotenuse is $\frac{2}{7}$.

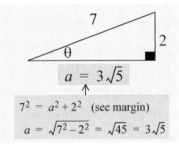

$7^2 = a^2 + 2^2$ (see margin)
$a = \sqrt{7^2 - 2^2} = \sqrt{45} = 3\sqrt{5}$

Referring to that triangle, we see that:

$$\sin\theta = \frac{\text{opp}}{\text{hyp}} = \frac{2}{7}, \quad \cos\theta = \frac{\text{adj}}{\text{hyp}} = \frac{3\sqrt{5}}{7}, \quad \tan\theta = \frac{\text{opp}}{\text{adj}} = \frac{2}{3\sqrt{5}}$$

$$\csc\theta = \frac{\text{hyp}}{\text{opp}} = \frac{7}{2}, \quad \sec\theta = \frac{\text{hyp}}{\text{adj}} = \frac{7}{3\sqrt{5}}, \quad \cot\theta = \frac{\text{adj}}{\text{opp}} = \frac{3\sqrt{5}}{2}$$

> **Pythagorean Theorem**
> In any right triangle:
>
> $c^2 = a^2 + b^2$

> Answer: $\sin\theta = \frac{1}{5\sqrt{2}}$,
> $\cos\theta = \frac{7}{5\sqrt{2}}$, $\tan\theta = \frac{1}{7}$,
> $\sec\theta = \frac{5\sqrt{2}}{7}$, $\csc\theta = 5\sqrt{2}$

CHECK YOUR UNDERSTANDING 3.1

Determine the values of the remaining trigonometric functions of the acute angle θ, if $\cot\theta = 7$.

128   Chapter 3 Trigonometric Functions of Angles

> **FUNDAMENTAL IDENTITIES**
>
> An **identity** is an equation that holds for every value of the variable(s) for which both sides of the equation are defined.
>
> The following basic identities express the tangent, cosecant, secant, and cotangent functions in terms of the sine and cosine functions:

Once the sine and the cosine are known, so are the rest.

THEOREM 3.3
BASIC IDENTITIES

$$\tan\theta = \frac{\sin\theta}{\cos\theta} \qquad \csc\theta = \frac{1}{\sin\theta}$$

$$\sec\theta = \frac{1}{\cos\theta} \qquad \cot\theta = \frac{\cos\theta}{\sin\theta} = \frac{1}{\tan\theta}$$

Each of these identities follows immediately from the definition of the trigonometric functions. For example:

$$\tan\theta = \frac{\text{opp}}{\text{adj}} = \frac{\frac{\text{opp}}{\text{hyp}}}{\frac{\text{adj}}{\text{hyp}}} = \frac{\sin\theta}{\cos\theta}$$

The generic term **trig θ** will be used to represent any of the six trigonometric functions of θ. When n is a positive integer, the nth power of trig θ is typically written as $\text{trig}^n\theta$ instead of $(\text{trig}\,\theta)^n$. In particular, $\sin^2\theta$ is simply another way of writing $(\sin\theta)^2$.

This result is appropriately named, as it follows directly from the Pythagorean Theorem.

THEOREM 3.4
PYTHAGOREAN IDENTITY

$$\sin^2\theta + \cos^2\theta = 1$$

PROOF:

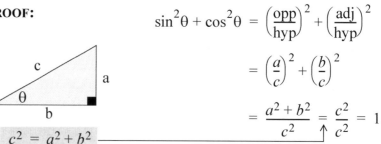

$$\sin^2\theta + \cos^2\theta = \left(\frac{\text{opp}}{\text{hyp}}\right)^2 + \left(\frac{\text{adj}}{\text{hyp}}\right)^2$$

$$= \left(\frac{a}{c}\right)^2 + \left(\frac{b}{c}\right)^2$$

$$= \frac{a^2 + b^2}{c^2} = \frac{c^2}{c^2} = 1$$

If θ is an acute angle in a right triangle, then the other acute angle is $90° - \theta$, and the two angles are said to be **complementary**.

THEOREM 3.5

COMPLEMENTARY IDENTITIES $\sin(90° - \theta) = \cos\theta$, and $\cos(90° - \theta) = \sin\theta$

PROOF:

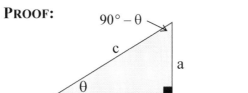

$$\sin(90° - \theta) = \frac{b}{c} = \cos\theta$$

$$\cos(90° - \theta) = \frac{a}{c} = \sin\theta$$

b (side opposite $90° - \theta$, and side adjacent to θ)

CHECK YOUR UNDERSTANDING 3.2

(a) Establish the basic identity:
$$\csc\theta = \frac{1}{\sin\theta}$$

(b) Determine the complementary identity (similar to those of Theorem 3.5) that involves the tangent function.

Answers: (a) See page C-19.
(b) $\tan(90° - \theta) = \cot\theta$
$\cot(90° - \theta) = \tan\theta$

TWO IMPORTANT RIGHT TRIANGLES

If one acute angle of a right triangle measures 45°, then so does the other: $90° - 45° = 45°$. This means that the legs of the triangle are equal in length. Such a triangle is said to be **isosceles**. Since all isosceles right triangles are similar, any one of them can be used to compute the values of the trigonometric functions of a 45° angle. The one in Figure 3.3(a), with legs of length 1 unit and, consequently, with hypotenuse of length $\sqrt{1^2 + 1^2} = \sqrt{2}$ units, is perhaps the easiest to remember, and we will refer to it as the **45° reference triangle**.

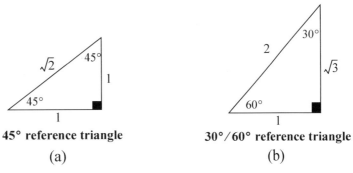

45° reference triangle   30°/60° reference triangle
         (a)   (b)

Figure 3.3

The 30°/60° reference triangle of Figure 3.3(b) was obtained by folding the adjacent **equilateral triangle** (all sides of equal length) in half along the dashed line. We chose the sides to be 2 units long so that, after folding, the halved side would be 1 unit. Using the Pythagorean Theorem, we then found the length of the leg opposite the 60° angle:

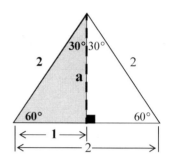

$$2^2 = 1 + a^2, \text{ or } a = \sqrt{3}$$

From the reference triangles of Figure 3.3 you can easily determine the values of the trigonometric functions of a 45°, 30°, or 60° angle. For example:

$$\cos 45° = \frac{\text{adj}}{\text{hyp}} = \frac{1}{\sqrt{2}}$$

$$\csc 30° = \frac{1}{\sin 30°} = \frac{1}{\frac{\text{opp}}{\text{hyp}}} = \frac{1}{\frac{1}{2}} = 2$$

$$\tan 60° = \frac{\text{opp}}{\text{adj}} = \frac{\sqrt{3}}{1} = \sqrt{3}$$

CHECK YOUR UNDERSTANDING 3.3

Complete the table of values:

θ	sin θ	cos θ	tan θ	csc θ	sec θ	cot θ
30°				2		
45°		$\frac{1}{\sqrt{2}}$				
60°			$\sqrt{3}$			

Answer: See page C-19.

SOLVING RIGHT TRIANGLES

To solve a triangle is to find the lengths of its sides and the measures of its angles. The following examples show that any right triangle can be solved once you know two sides, or one side and an acute angle.

As is the case in this example, the letter c is used to label the side opposite the right angle in any right triangle:

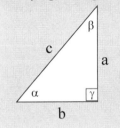

EXAMPLE 3.2 Solve the right triangle with $\alpha = 51°$ and $a = 12$.

SOLUTION: The first step is to sketch the triangle and label its known and unknown parts (see adjacent figure).

Since $\beta = 90° - \alpha$, $\beta = 90° - 51° = 39°$

To determine b and c, we look to the equations:

$$\sin 51° = \frac{\text{opp}}{\text{hyp}} = \frac{12}{c} \quad \text{or} \quad c = \frac{12}{\sin 51°} \approx 15.4$$

$$\tan 51° = \frac{\text{opp}}{\text{adj}} = \frac{12}{b} \quad \text{or} \quad b = \frac{12}{\tan 51°} \approx 9.7$$

We have found: $\beta = 39°$, $b \approx 9.7$, and $c \approx 15.4$.

EXAMPLE 3.3 Solve the right triangle with hypotenuse 6.2 units long and a leg 2.3 units long.

SOLUTION: Using the Pythagorean Theorem, we can readily find the value of a in the adjacent triangle:

$$a^2 + (2.3)^2 = (6.2)^2$$

$$a = \sqrt{(6.2)^2 - (2.3)^2} \approx 5.8$$

We also see that $\cos\alpha = \frac{\text{adj}}{\text{hyp}} = \frac{2.3}{6.2}$

With a calculator (see margin), we obtain the approximation $\alpha \approx 68°$.

Knowing α, the value of β is easily determined:

$$\beta = 90° - \alpha \approx 90° - 68° = 22°$$

We have found: $a \approx 5.8$, $\alpha \approx 68°$, and $\beta \approx 22°$

As you can see, we made sure that our calculator was in its DEGree mode. We then employed the inverse cosine function, $\cos^{-1}$, discussed in detail in the next chapter, to arrive at an approximation for α:

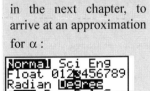

Answer: (a) $a = 7\sqrt{3}$, $b = 7$, $\beta = 30°$
(b) For $a = \sqrt{3}$, $b = 3.25$, $\alpha \approx 28°$, $\beta \approx 62°$, $c \approx 3.68$

CHECK YOUR UNDERSTANDING 3.4

(a) Without a calculator, solve the right triangle with $\alpha = 60°$ and $c = 14$.

(b) With a calculator, solve the right triangle with legs of length 3.25 and $\sqrt{3}$ units.

EXERCISES

Exercises 1–2: Determine the values of x and y in each pair of similar triangles.

1.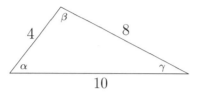

2.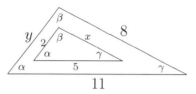

Exercise 3: Find the values of α, β, and x in the similar triangles in the figure.

Exercise 4: Find the values of w, z, x, y, and s in the similar triangles in the figure.

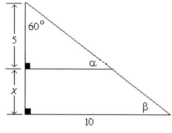
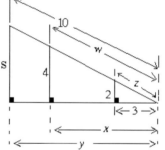

Exercises 5–16: Determine the values of the remaining trigonometric functions of the acute angle θ, given

5. $\sin\theta = \frac{2}{3}$ 6. $\sin\theta = \frac{12}{13}$ 7. $\cos\theta = \frac{1}{4}$ 8. $\cos\theta = \frac{3}{5}$ 9. $\tan\theta = 2$ 10. $\tan\theta = \frac{1}{3}$

11. $\cot\theta = \frac{3}{4}$ 12. $\cot\theta = 5$ 13. $\sec\theta = 3$ 14. $\sec\theta = \frac{5}{4}$ 15. $\csc\theta = \frac{13}{5}$ 16. $\csc\theta = \frac{4}{3}$

Exercise 17: Find the exact value of $\cos\theta$, if $\tan\theta = \frac{2}{x}$ and $\sin\theta = \frac{4}{3x}$.

Exercise 18: Explain why, for any acute angle θ,
$$0 < \sin\theta < 1, \quad 0 < \cos\theta < 1, \quad \csc\theta > 1, \quad \text{and} \quad \sec\theta > 1$$

Exercise 19: Explain why $\tan\theta$ and $\cot\theta$ assume all values from zero to infinity as θ ranges over all possible acute angles.

3.1 Trigonometric Functions of Acute Angles   **133**

Exercises 20–24: By direct computation of the value of each trigonometric function, verify each of the following.

20. $\cos(90° - 30°) = \sin 30°$ 21. $\sin(90° - 30°) = \cos 30°$ 22. $\cos^2 30° + \sin^2 30° = 1$

23. $\cos^2 45° + \sin^2 45° = 1$ 24. $\cos^2 60° + \sin^2 60° = 1$

Exercises 25–36: By direct computation of the value of each trigonometric function, show that

25. $\cos(60° - 30°) \neq \cos 60° - \cos 30°$ 26. $\sin(60° - 30°) \neq \sin 60° - \sin 30°$

27. $\tan(60° - 30°) \neq \tan 60° - \tan 30°$ 28. $\csc(60° - 30°) \neq \csc 60° - \csc 30°$

29. $\sec(60° - 30°) \neq \sec 60° - \sec 30°$ 30. $\cot(60° - 30°) \neq \cot 60° - \cot 30°$

31. $\sin 2(30°) \neq 2\sin 30°$ 32. $\cos 2(30°) \neq 2\cos 30°$ 33. $\tan 2(30°) \neq 2\tan 30°$

34. $\csc 2(30°) \neq 2\csc 30°$ 35. $\sec 2(30°) \neq 2\sec 30°$ 36. $\cot 2(30°) \neq 2\cot 30°$

Exercises 37–42: Solve the right triangle, without a calculator.

37. $\beta = 45°$ and $a = 3$ 38. $\alpha = 30°$ and $a = \sqrt{2}$ 39. $\alpha = 60°$ and $c = 12$

40. $\alpha = 45°$ and $c = 6$ 41. $a = 5\sqrt{3}$ and $c = 10$ 42. $b = \frac{1}{2}$ and $c = \frac{\sqrt{2}}{2}$

Exercises 43–48: Solve the right triangle, with a calculator.

43. $a = 7$ and $b = 5$ 44. $b = 5$ and $c = 13$ 45. $\alpha = 53°$ and $b = 2$

46. $\beta = 12°$ and $c = 4.7$ 47. $\beta = 21°$ and $b = 3.8$ 48. $\beta = 37°$ and $c = 9$

Exercise 49:   **(Clock Hands)** Approximate the degree measure of the angle between the hour and minute hands of a clock at 3:35 (to the nearest degree). (Don't forget to take into account the fact that the hour hand points to 3 only at precisely 3 o'clock.)

Exercises 50–51: Use the Pythagorean Identity, $\sin^2 \theta + \cos^2 \theta = 1$, to establish the given identity.

50. $1 + \tan^2 \theta = \sec^2 \theta$ 51. $1 + \cot^2 \theta = \csc^2 \theta$

Exercises 52–60:   Indicate True or False. Justify your answer. (Assume all angles are acute.)

52. $\sin \alpha < \sin \beta$ if and only if $\alpha < \beta$.   53. $\cos \alpha < \cos \beta$ if and only if $\alpha < \beta$.

54. $\tan \alpha < \tan \beta$ if and only if $\alpha < \beta$.   55. If $\sin \theta = \frac{5}{6}$, then $\tan \theta < \frac{5}{6}$

56. If $\sin \theta = \frac{5}{6}$, then $\cos \theta < \frac{5}{6}$ 57. If $\sin \theta = \frac{5}{6}$, then $\cos \theta \geq \frac{5}{6}$

58. If $\theta < 45°$, then $\sin \theta < \cos \theta$ 59. If $\theta < 45°$, then $\sin \theta < \tan \theta$

60. If $\theta < 45°$, then $\tan \theta < \sec \theta$

§2. RIGHT TRIANGLE APPLICATIONS

Each problem in this section reduces to finding an unknown angle or side of a right triangle. In Chapter 5 we will consider applications involving more general triangles.

We begin by recalling two useful results from plane geometry:

Of course, if two alternate angles are equal, then so are several other pairs of angles equal; for example:

THEOREM 3.6
ALTERNATE INTERIOR ANGLES

When parallel lines are cut by a transversal, alternate interior angles are equal.

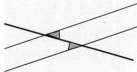

THEOREM 3.7
VERTICAL ANGLES

Vertical angles of intersecting lines are equal.

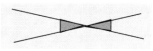

ANGLE OF ELEVATION/DEPRESSION

The **angle of elevation** of an object is the angle formed by a horizontal ray and the line of sight to that object.

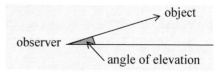

The **angle of depression** of an object is the an angle formed by a horizontal ray and the line of sight to that object.

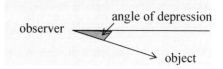

EXAMPLE 3.4
HEIGHT OF A BUILDING

The angle of elevation of the top of a building, from a point on the ground that is 170 feet from the base of the building, is 30°. How high is the building?

SOLUTION:

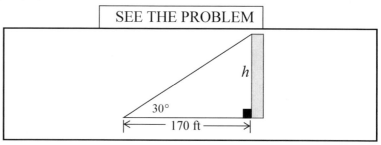

SEE THE PROBLEM

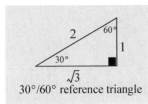

30°/60° reference triangle

To find the height of the building, h, we turn to the tangent function, because it relates h with a known length and a known angle:

$$\tan 30° = \frac{h}{170 \text{ ft}}$$

$$h = (170 \text{ ft})(\tan 30°) = (170 \text{ ft})\left(\frac{\text{opp}}{\text{adj}}\right) = (170 \text{ ft})\left(\frac{1}{\sqrt{3}}\right) \approx 98.1 \text{ ft}$$

EXAMPLE 3.5
DISTANCE OF BUOY

From the top of a vertical cliff, 75 meters above the ocean, the angle of depression to a buoy is 71°. How far is the buoy from the top of the cliff?

SOLUTION:

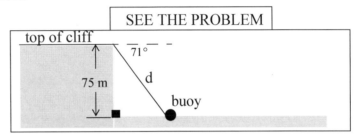

Extracting the right triangle from the above figure, and using the fact that alternate interior angles are equal, we arrive at the adjacent triangle. From that triangle, we see that:

$$\sin 71° = \frac{\text{opp}}{\text{hyp}} = \frac{75 \text{ m}}{d}$$

$$d = \frac{75 \text{ m}}{\sin 71°} \approx 79 \text{ m}$$

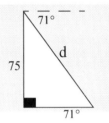

EXAMPLE 3.6
HEIGHT OF PYRAMID

From a certain point near the base of a pyramid, the angle of elevation to the top of the pyramid is 35°. When 97 feet closer, the angle of elevation is 47°. What is the height of the pyramid?

SOLUTION:

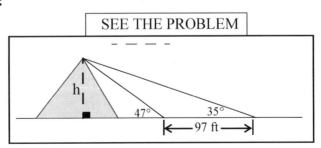

We see two right triangles in the above figure, but neither of them carries enough information to enable us to solve for h. Adding a new variable, x, (adjacent figure), enables us to derive two equations in two unknowns; namely:

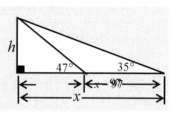

From the larger right triangle: (1) $\tan 35° = \dfrac{h}{x}$

From the smaller right triangle: (2) $\tan 47° = \dfrac{h}{x - 97}$

From (1) we see that $x = \dfrac{h}{\tan 35°}$, and substituting this expression for x in (2) we obtain an equation involving only the variable h:

Keep in mind the fact that $\tan 47°$ and $\tan 35°$ are simply numbers.

$$\tan 47° = \dfrac{h}{\dfrac{h}{\tan 35°} - 97}$$

Multiply numerator and denominator on the right by $\tan 35°$:

$$\tan 47° = \dfrac{h \tan 35°}{h - 97 \tan 35°}$$

Cross multiply: $(\tan 47°)(h - 97 \tan 35°) = h \tan 35°$

$$h \tan 47° - 97 \tan 47° \tan 35° = h \tan 35°$$

$$h \tan 47° - h \tan 35° = 97 \tan 47° \tan 35°$$

$$h(\tan 47° - \tan 35°) = 97 \tan 47° \tan 35°$$

$$h = \dfrac{97 \tan 47° \tan 35°}{\tan 47° - \tan 35°} \approx 196$$

We find that the pyramid is approximately 196 feet high.

CHECK YOUR UNDERSTANDING 3.5

From the top of a 200 foot cliff, the angle of depression to a point A on the far side of a river is 62°, and to a point B on the near side it is 71°. How wide is the river?

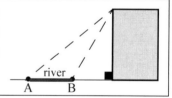

Answer: ≈ 37.5 ft.

BEARING

Another bearing is discussed in Chapter 5.

A bearing is used to describe the direction of an object from a given point P. One method of establishing a **bearing** of an object Q from a point P is to specify the acute angle between the north-south line through P and the line joining Q to P. For example, the bearing N40°E indicates the direction 40° east of due north [see Figure 3.4(a)], while S57°W denotes the direction that is 57° west of due south [see Figure 3.4(b)].

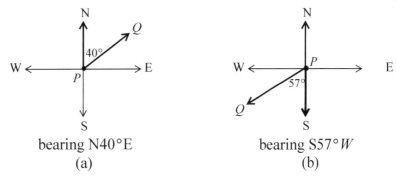

bearing N40°E
(a)

bearing S57°W
(b)

Figure 3.4

EXAMPLE 3.7

DISTANCE OF SHIP FROM COASTLINE

At 2:00 P.M., a ship leaves the port of New York at a bearing of S30°E. Assuming that the coastline is a straight north-south line, how far is the ship from the coastline at 4:00 P.M., if the ship averaged 35 nautical miles per hour during that time?

SOLUTION:

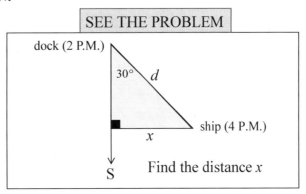

At 4:00 P.M., which is two hours after the time the ship leaves New York, the distance d between the ship and the dock is:

$$d = 35 \frac{\text{nautical miles}}{\text{hr}} \cdot 2\text{hr} = 70 \text{ nautical miles}$$

Then:

$$\sin 30° = \frac{\text{opp}}{\text{hyp}} = \frac{x}{70 \text{ nautical miles}}$$

$$\frac{1}{2} = \frac{x}{70 \text{ nautical miles}}$$

$$x = 35 \text{ nautical miles}$$

(see margin):
$$= (35 \text{ nautical miles})\left(\frac{6076 \text{ ft}}{1 \text{ nautical mile}}\right)\left(\frac{1 \text{ mi}}{5280 \text{ ft}}\right)$$

$$= \frac{35 \cdot 6076}{5280} \text{ mi} \approx 40 \text{ mi}$$

> In the U.S., the nautical mile used to be 6,080.27 feet, which is the length of the subtending arc of one minute of a great circle on the earth. It has been replaced by the international nautical mile of 6,076.11549 feet. We will use the approximation of 6,076 feet.

EXAMPLE 3.8

FOREST FIRE

Ranger A spots a forest fire at a bearing of S15°E. Ranger B, who is 37 miles away from Ranger A, at a bearing of N75°E from A, spots the same fire at a bearing of S19°W. How far is the fire from Ranger A?

SOLUTION:

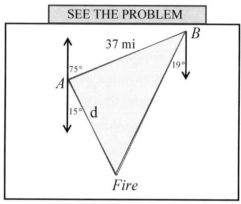

SEE THE PROBLEM

We are to determine the distance d. As luck would have it, the triangle turns out to be a right triangle:

$\angle BAF = 180° - 75° - 15° = 90°$

↑ denotes the angle with vertex at A

Since the line segment AB is a transversal of the two parallel dotted lines, we have:

$\alpha + 19° = 75°$
$\alpha = 75° - 19° = 56°$

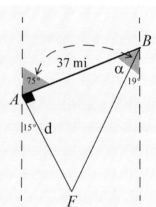

Then:

$$\tan 56° = \frac{\text{opp}}{\text{adj}} = \frac{d}{37 \text{ mi}}$$

$$d = (37 \text{ mi})(\tan 56°) \approx 55 \text{ mi}$$

Similar problems that do not lead to right triangles are considered in Chapter 5.

CHECK YOUR UNDERSTANDING 3.6

Billow is due south of Bingly, and Brier is 34 miles east of Bingly. What is the distance between Billow and Brier, if Billow is at a bearing of S23°W from Brier?

Answer: ≈ 87.0 mi.

3.2 Right Triangle Applications

INDEX OF REFRACTION

When a pencil is partly inserted in water, it appears to bend downward at the surface of the water. This phenomenon is due to **refraction**: the physical bending of light waves as light moves from one medium to another.

The following result, attributed to the French mathematician, physicist, and philosopher, Rene Descartes (1596-1650), describes refraction mathematically. Actually, the result was discovered twenty years earlier by the Dutch mathematician and physicist, Willebrord Snell (1591-1626). Descartes, living in Holland at the time, is certain to have heard of it.

THEOREM 3.8
REFRACTION

Let a denote the speed of light in medium A (say air) and b denote the speed of light in medium B (say water). If α is the acute angle between a light ray and the vertical in medium A (called the angle of incidence), and β is the acute angle between the ray and the vertical in medium B (called the angle of refraction), then:

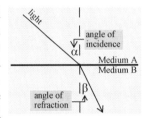

$$\frac{\sin\alpha}{\sin\beta} = \frac{a}{b}$$

At least Snell is credited with introducing the concept of the **index of refraction**, I, which is the ratio of the speed of light in a vacuum ($c \approx 2.998 \times 10^8 \frac{m}{sec}$) to the speed of light in another medium (such as air or water). In particular, if a represents the speed of light in medium A, and c the speed of light in a vacuum, then the index of refraction of medium A is given by:

$$I_a = \frac{c}{a}$$

Returning to Theorem 3.8, we then have:

THEOREM 3.9
SNELL'S LAW

If I_a and I_b denote the indices of refraction of mediums A and B, respectively, then:

$$\frac{I_b}{I_a} = \frac{\sin\alpha}{\sin\beta}$$

PROOF:

$$\frac{\sin\alpha}{\sin\beta} = \frac{a}{b} = \frac{\frac{a}{c}}{\frac{b}{c}} = \frac{\frac{1}{I_a}}{\frac{1}{I_b}} = \frac{I_b}{I_a}$$

EXAMPLE 3.9

ANGLE OF REFRACTION

The index of refraction of air is 1.0003, and that of water is 1.3330. Assume that a light shines on a lake with an angle of incidence of 22.312°. Determine the angle of refraction.

SOLUTION:

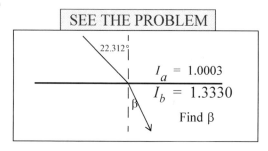

Applying Snell's Law, we obtain an equation from which the value of $\sin\beta$, and therefore of β, can be determined:

$$\frac{I_b}{I_a} = \frac{\sin\alpha}{\sin\beta}$$

$$\frac{1.3330}{1.0003} = \frac{\sin 22.312°}{\sin\beta}$$

$$\sin\beta = \frac{1.0003}{1.3330}\sin 22.312°$$

$$\beta \approx 16.553°$$

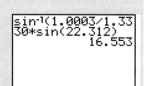

CHECK YOUR UNDERSTANDING 3.7

What is the index of refraction of liquid X, if a light shining in air on the liquid at an angle of incidence of 22.135° is refracted to an angle of 18.712°?

Answer: ≈ 1.1748

EXERCISES

Exercise 1: (**Ladder**) A ladder leans against the side of a building with its foot 12 feet from the building. How far from the ground is the top of the ladder and how long is the ladder, if it makes an angle of 60° with the ground?

Exercise 2: (**Angle of Depression**) From the top of a vertical cliff, the angle of depression to a boat 100 meters from the base of the cliff is 30°. What is the height of the cliff?

Exercise 3: (**Angle of Elevation**) The angle of elevation to the top of a flagpole from a point 75 feet from the base of the pole is 40°. What is the height of the pole?

Exercise 4: (**Angle of Elevation**) An airplane flies directly over point A at an altitude of 25,000 feet, and a minute later the angle of elevation from A to the plane is 45°. Find the speed of the plane.

Exercise 5: (**Kite**) A kite is 87 meters above the ground, and the kite's string forms an angle of 50° with the horizontal. What is the length of the extended string if it is held at a point 3 feet above the ground? (1m = 3.28 ft)

Exercise 6: (**Kite**) A kite string makes an angle of 39° with the horizontal. How high is the kite above the point at which the string is held, if 103 feet of string is paid out?

Exercise 7: (**River Width**) To find the width of a river, a surveyor set up her transit at C on one bank of the river and sighted directly across to a point B on the opposite bank; then turning through an angle of 90°, she laid off a distance CA equal to 225 m. Finally setting the transit at A, she measured angle CAB as 48°. Find the width (BC) of the river.

Exercise 8: (**Tree Height**) Find the height of a tree if the angle of elevation of its top changes from 20° to 40° as the observer advances 75 ft toward its base?

Exercise 9: (**Broken Tree**) A tree broken over by the wind forms a right triangle with the ground. If the top of the tree now 20 ft from the bottom makes an angle of 50° with the ground, how tall was the tree?

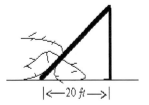

Exercise 10: (**Angle of Depression**) In the adjacent figure, the observer on top of the 95 ft cliff notes that the angle of depression of the boat is 35°, while the angle of depression of the buoy is 50°. How far from the buoy is the boat?

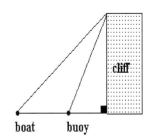

142   Chapter 3 Trigonometric Functions of Angles

Exercise 11: (**Angle of Elevation**) Two buildings with flat roofs are 60 meters apart. From the roof of the shorter building, 40 meters high, the angle of elevation to the edge of the roof of the taller building is 40°. How high is the taller building?

Exercise 12: (**Angle of Elevation**) Find the height of two buildings that are 190 feet apart, if the angle of depression from the top of the taller building to the top of the shorter building is 25° and the angle of elevation from the bottom of the shorter building to the top of the taller building is 51°.

Exercise 13: (**Angle of Elevation**) The angle of elevation to the top of a pyramid is 37°. When 40 feet closer to the pyramid, the angle of elevation is 43°. What is the height of the pyramid?

Exercise 14: (**Bearing**) A boat travels 25 km at a bearing of N23°W. How far north and how far west has it traveled?

Exercise 15: (**Bearing**) A car travels at 50 mph for 15 minutes at a bearing of S45°E. It then travels at 30 mph at a bearing of S45°W. How long does the car travel at S45°W, if it ends up at a point whose bearing from the starting point is S10°W?

Exercise 16: (**Bearing**) A ship is sailing due east when a light is observed at a bearing of N62°E. After the ship has traveled 2250 m further, the light's bearing is N48°E. If the course is continued, what is the closest the ship will ever be to the light?

Exercises 17–21: (**Refraction**) In each of the following, use the fact that the speed of light in a vacuum is 2.998×10^8 m/sec.

17. Compute the speed of light in air, given its index of refraction is 1.0003.

18. A ray of light is incident on a plane surface separating two transparent substances of refractive indices 1.60 and 1.40. The angle of incidence is 30° and the ray originates in the medium of higher index. Compute the angle of refraction.

19. The speed of light of wavelength 656 nanometers in heavy flint glass is 1.60×10^8 m/sec. What is the index of refraction of this glass?

20. What is the speed of light of wavelength 500 nm (in a vacuum) in glass whose index of refraction at this wavelength is 1.50?

21. A ray of light traveling in air strikes a glass plate 3.00 cm thick at a 50° angle of incidence. The angle of refraction is 29.6°. Calculate (a) the index of refraction of the glass, and (b) the speed of light in the glass.

Exercise 22: (**Polygon**) A polygon is a plane figure bounded by three or more line segments, meeting in an equal number of points. When each of the interior angles of a polygon is less than 180°, the polygon is said to be convex. It can be shown that the sum of the interior angles of a convex polygon of n sides is $(n-2) \times 180°$. A regular polygon is a convex polygon with equal sides and equal angles.

Steve, Basil, and Shaun want to extend the rectangular porch in the adjacent figure to include the shaded region, which represents $\frac{3}{8}$ of an octagon (8 sided regular polygon). Unlike the Greeks, who studied ruler and compass constructions, carpenters today depend on T-squares (to mark off right angles and for straight edges) and tape measures (to measure lengths) in their construction process. Indicate approximate measurements of x and y in the adjacent figure which will enable those three master carpenters to build the addition.

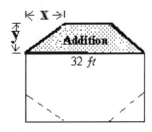

§3. RADIAN MEASURE AND ORIENTED ANGLES

Though degree measure is used in such fields as surveying and navigation, another measure is typically used in mathematics and the sciences. We are referring to **radian** measure, which we now introduce.

RADIAN MEASURE

Consider the two concentric circles of Figure 3.5(a), where arcs of lengths s and s' are depicted. Since each arc subtends an angle of $90° = \frac{1}{4}(360°)$, the length of each arc is one-fourth the circumference of its corresponding circle:

$$s = \tfrac{1}{4}(2\pi r) = \tfrac{\pi}{2}r \quad \text{and} \quad s' = \tfrac{1}{4}(2\pi r') = \tfrac{\pi}{2}r'$$

Although $s \neq s'$, the ratios of s and s' to their corresponding radii **are** the same:

$$\frac{s}{r} = \frac{\tfrac{\pi}{2}r}{r} = \frac{\pi}{2} \quad \text{and} \quad \frac{s'}{r'} = \frac{\tfrac{\pi}{2}r'}{r'} = \frac{\pi}{2}$$

This observation holds for any central angle [see Figure 3.5(b)].

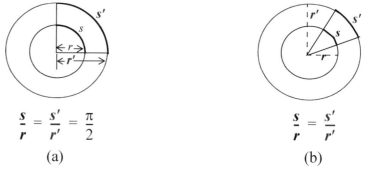

Figure 3.5

Since the ratio of the length of an arc subtending a central angle to the radius of the circle is independent of the radius of the circle, that ratio can be used as a measure of the angle:

DEFINITION 3.2

RADIAN MEASURE

The **radian measure** of a central angle of a circle is the ratio of the length of the arc subtending that angle to the radius of the circle.

Radian measure, being the ratio of two lengths, is a real number and is **not** associated with a unit (like degrees). Nonetheless, when referring to radian measure the word "radian" is often used to refer to that measure; and just as we understand the expression $\theta = 90°$ to mean that the measure of the angle θ is $90°$, the expression $\theta = \frac{\pi}{2}$ (or $\theta = \frac{\pi}{2}$ radians) means that the angle θ has a radian measure of $\frac{\pi}{2}$.

EXAMPLE 3.10 (a) Convert $\theta = 45°$ to radian measure.

(b) Convert $\theta = \dfrac{3\pi}{2}$ to degree measure.

SOLUTION: From Figure 3.5(a), we know that:

$$90° = \frac{\pi}{2} \text{ radians}$$

$$\text{or:} \quad 180° = \pi \text{ radians} \quad (*)$$

In the conversion process, it is useful to treat "radian" as if it were a unit of measure, and use (*) as if it were a conversion bridge:

(a) $\theta = 45° = 45° \cdot \dfrac{\pi \text{ radians}}{180°} = \dfrac{\pi}{4} \text{ radians}$

(b) $\theta = \dfrac{3\pi}{2} = \left(\dfrac{3\pi}{2} \text{ radians}\right)\left(\dfrac{180°}{\pi \text{ radians}}\right) = 270°$

CHECK YOUR UNDERSTANDING 3.8

(a) Convert $\theta = 120°$ to radians.

(b) Convert $\theta = \dfrac{\pi}{6}$ radians to degrees.

Answers: (a) $\dfrac{2\pi}{3}$ (b) $30°$

ORIENTED ANGLES

It is often useful to think of an angle θ as evolving from the following dynamic process:

> A fixed ray or half-line, called the **initial side** of the angle, is rotated about an endpoint O, called the **vertex** of the angle, to a final destination, called the **terminal side** of the angle. If the rotation is counterclockwise, then the angle is said to be **positive**, and if clockwise then it is **negative**. Because of the sign associated with it, such angles are said to be **oriented angles**.

Typically, one positions an angle in the Cartesian plane, with its vertex at the origin and its initial side along the positive x-axis. In such a setting, the angle is said to be in **standard position.** Two angles in standard position are depicted in Figure 3.6. The one in Figure 3.6(a) is positive (counterclockwise rotation) and that in (b) is negative (clockwise rotation).

146   Chapter 3 Trigonometric Functions of Angles

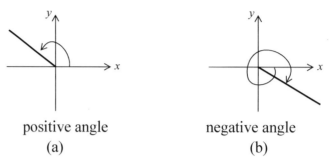

positive angle   negative angle
      (a)   (b)

Figure 3.6

When the terminal sides of two angles in standard position coincide, then the angles are said to be **coterminal**. The two angles depicted in the adjacent figure are coterminal (one positive and the other negative).

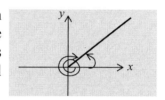

The Cartesian plane is divided into four quadrants QI through QIV (see adjacent figure). When the terminal side of an angle in standard position lies in one of the four quadrants, the angle is said to **lie in that quadrant**. In particular, the angle of Figure 3.6(a) lies in the second quadrant (QII), the angle in Figure 3.6(b) lies in the fourth quadrant (QIV), and the angle in 3.6(c) lies in QI.

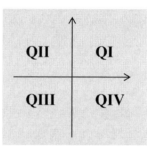

No quadrant is associated with an angle whose terminal side lies on a coordinate axis, such as $90°$ or $\frac{3\pi}{2}$ radian. Such angles are called **quadrantal** angles.

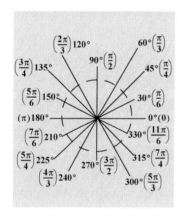

Henceforth, unless otherwise specified, "angle" will be understood to refer to an oriented angle in standard position.

As you will see in the next section, it is often necessary to locate the quadrant in which an angle θ lies. To do so, simply add to or subtract from θ an integer multiple of $360°$ (or 2π), in order to arrive at a coterminal angle between $-360°$ and $360°$ (or, -2π and 2π). Consider the following examples.

EXAMPLE 3.11   Determine in which quadrant $\theta = -475°$ lies.

SOLUTION: Since:

$$-475° + 360° = -115°$$

θ is coterminal with $-115°$, and therefore lies in the third quadrant.

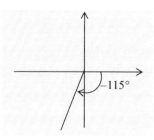

EXAMPLE 3.12
In which quadrant does $\theta = \dfrac{15\pi}{4}$ lie?

SOLUTION: Since:

$$\frac{15\pi}{4} - 2\pi = \frac{7\pi}{4}$$

θ is coterminal with $\dfrac{7\pi}{4}$, and therefore lies in the fourth quadrant.

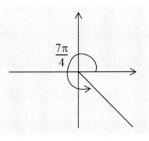

CHECK YOUR UNDERSTANDING 3.9

Determine in which quadrant θ lies.

(a) $\theta = 999°$ (b) $\theta = -\dfrac{23\pi}{7}$

Answers: (a) QIV (b) QII

ARC LENGTH

From the definition of radian measure $\left(\theta = \dfrac{s}{r}\right)$, we have:

Let s denote the length of the arc subtending a central angle of θ radians in a circle of radius r. Then:

$$s = r\theta$$

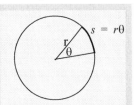

EXAMPLE 3.13
PULLEY

Referring to the adjacent figure, find the angle through which the 5 inch pulley turns, if the 12 inch wheel is rotated through an angle of 3 radians.

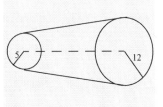

SOLUTION: Letting θ_5 and θ_{12} denote subtending angles in the 5 inch and 12 inch wheels respectively, we have:

148   Chapter 3 Trigonometric Functions of Angles

A length *l* of the belt remains constant through the rotation process on either wheel. In other words the arclength *l* on the larger wheel must correspond to an equal arclength on the smaller wheel.

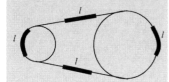

Applying $s = r\theta$ to the larger wheel: $(12 \text{ in.})\theta_{12} = (12 \text{ in.}) \cdot 3 = 36 \text{ in.}$

Applying $s = r\theta$ to the smaller wheel: $(5 \text{ in.})\theta_5 = 36 \text{ in.}$ (see margin)

$$\theta_5 = \frac{36 \text{ in.}}{5 \text{ in.}} = \frac{36}{5} \text{ (radians)}$$

EXAMPLE 3.14

WINCH

Referring to the winch in the adjacent figure, determine the number of revolutions required to lift the object 3 feet, if the wheel has a diameter of 1 ft.

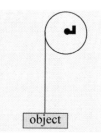

SOLUTION: The 3 feet rise of the object calls for the wheel turning through an arclength of 3 feet. Since $r\theta = s$ we have:

diameter = 1 ft $(\frac{1}{2} \text{ ft})\theta = 3 \text{ ft}$

$$\theta = 6 \text{ (radians)}$$

$$= (6 \text{ radians})\left(\frac{1 \text{ revolution}}{2\pi \text{ radians}}\right) = \frac{3}{\pi} \text{ revolutions}$$

(almost one complete revolution)

CHECK YOUR UNDERSTANDING 3.10

Referring to Example 3.14, determine the distance the object will be lifted if the winch is rotated through an angle of 405°.

Answer: $\frac{9\pi}{8}$ ft.

3.3 Radian Measure and Oriented Angles   149

EXERCISES

Exercises 1–15: Convert to radian measure.

1. $\theta = 270°$ 2. $\theta = 60°$ 3. $\theta = -30°$ 4. $\theta = 135°$ 5. $\theta = -120°$

6. $\theta = 210°$ 7. $\theta = 180°$ 8. $\theta = 720°$ 9. $\theta = -75°$ 10. $\theta = -480°$

11. $\theta = 495°$ 12. $\theta = 1000°$ 13. $\theta = 225°$ 14. $\theta = -780°$ 15. $\theta = -315°$

Exercises 16–29: Convert to degree measure.

16. $\theta = \frac{\pi}{2}$ 17. $\theta = \frac{5\pi}{3}$ 18. $\theta = 3\pi$ 19. $\theta = \frac{11\pi}{6}$ 20. $\theta = -\frac{3\pi}{2}$

21. $\theta = -\frac{3\pi}{4}$ 22. $\theta = -4\pi$ 23. $\theta = \frac{5\pi}{6}$ 24. $\theta = -\frac{8\pi}{3}$ 25. $\theta = \frac{5\pi}{4}$

26. $\theta = 4$ 27. $\theta = 7.3$ 28. $\theta = 90$ 29. $\theta = -22.4$

Exercises 30–45: Represent the angle in standard position in the xy-plane.

30. $\theta = 540°$ 31. $\theta = -270°$ 32. $\theta = -225°$ 33. $\theta = 225°$ 34. $\theta = 315°$ 35. $\theta = 630°$

36. $\theta = 990°$ 37. $\theta = -675°$ 38. $\theta = \frac{5\pi}{4}$ 39. $\theta = \frac{17\pi}{6}$ 40. $\theta = \frac{13\pi}{2}$ 41. $\theta = -5\pi$

42. $\theta = \frac{5\pi}{3}$ 43. $\theta = \frac{11\pi}{3}$ 44. $\theta = \frac{5\pi}{6}$ 45. $\theta = -\frac{13\pi}{3}$

Exercises 46–60: Determine the quadrant in which the angle lies.

46. $\theta = 462°$ 47. $\theta = 110°$ 48. $\theta = -1488°$ 49. $\theta = -500°$ 50. $\theta = 623°$

51. $\theta = -75°$ 52. $\theta = 430°$ 53. $\theta = -327°$ 54. $\theta = \frac{11\pi}{3}$ 55. $\theta = \frac{5\pi}{6}$

56. $\theta = -\frac{17\pi}{6}$ 57. $\theta = -\frac{3\pi}{4}$ 58. $\theta = \frac{23\pi}{6}$ 59. $\theta = \frac{8\pi}{3}$ 60. $\theta = -\frac{23\pi}{6}$

Exercises 61–65: Find a coterminal angle between $-360°$ and $360°$.

61. $\theta = -448°$ 62. $\theta = -610°$ 63. $\theta = 785°$ 64. $\theta = 690°$ 65. $\theta = -4000°$

Exercises 66–69: Find a coterminal angle between -2π and 2π.

66. $\theta = -\frac{31\pi}{4}$ 67. $\theta = \frac{15\pi}{4}$ 68. $\theta = \frac{25\pi}{6}$ 69. $\theta = -\frac{29\pi}{3}$

150 Chapter 3 Trigonometric Functions of Angles

Exercises 70–74: Indicate the degree measure and the radian measure of the given quadrantal angle.

70.
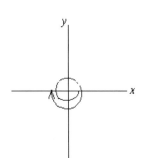

71.
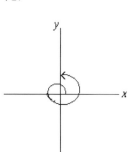

72.
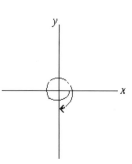

73.
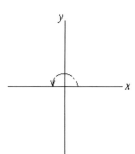

74.
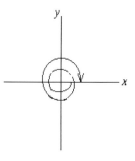

Exercises 75–82: (Arc Length) In each of the following, use the arc length formula $s = r\theta$.

75. **(Clock Hand)** How far does the tip of an hour hand measuring 2 centimeters move in three hours?

76. **(Latitude/Longitude)** The adjacent figure defines Longitude and Latitude. The Prime Meridian is distinguished from all other meridians, or great circles (circles passing through the North and South poles), as that meridian passing through Greenwich, England (G). The longitude of a point P can be any angle from 0° to 180°, designated as east or west longitude, depending on whether P is east or west of the prime meridian. Similarly the latitude of a point can be any angle from 0° to 90°, designated as north or south of the equator.

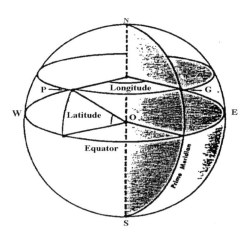

Using essentially the same principles as have been used more recently to approximate the circumference of the earth, but without benefit of sophisticated instrumentation, the Greek mathematician Eratosthenes, in about 235 B.C., was able to arrive at a very close approximation of the circumference of the earth. (The actual circumference is approximately 24,900 miles.)

He assumed that the two cities of Alexandria and Syene (now Aswan), which are 500 miles apart, lie along the same longitude line (they do approximately). At the time when the sun was directly overhead in Syene, Erastosthenes measured the sun's rays to be 7.2° from the vertical in Alexandria (see figure below). Assuming that the sun's rays are parallel when they reach the earth, he concluded that the rays from the center of the earth to Syene and Alexandria subtended a central angle also of 7.2°.

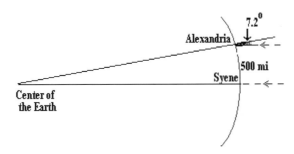

Using this information, he was able to approximate the circumference of the earth. Do the calculation.

77. (**Latitude/Longitude**) The cities of Bukittinggi, Sumatra and Pontianak, Borneo are located approximately on the equator. Bukittinggi lies at 100°E longitude, and Pontianak is at 108°E longitude. Approximate the distance between the two cities, given the radius of the earth is approximately 3963 miles. (See Exercise #76.)

78. (**Latitude/Longitude**) Find the distance between two points north of the equator at the same longitude, if one is at latitude 35° and the other is at latitude 12°. Assume the diameter of the earth to be 7,926 miles. (See Exercise #76.)

79. (**Nautical Miles**) A nautical mile was defined in Section 3.2 as the distance on the surface of the earth subtended by a central angle of measure 1 minute = $\frac{1}{60}$th of a degree (see the adjacent figure).

    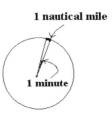

    Given that the diameter of the earth is approximately 7,926 miles, calculate the number of feet in a nautical mile, directly from the formula $s = r\theta$.

80. (**Message in a Bottle**) A message in a bottle traveled 1000 nautical miles. What percentage of a complete revolution about the globe did the bottle travel? (See Exercise #79.)

81. Daffy Duck swam 50 nautical miles and Pluto ran 60 miles. Who traveled further, and by how many feet? (See Exercise #79.)

82. **(Bicycle)** Consider the bicycle pedal and wheel configuration in the figure.

    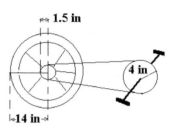

    (a) How many times will the back wheel revolve for every complete pedal revolution?

    (b) How far will the bicycle move, if the pedals move through 6 complete revolutions?

Exercises 83–86: **(Centesimal System)** In the centesimal system of angle measurement, a right angle is divided into 100 equal parts called grads. Develop formulas for converting R radians to G grads, and G grads to R radians.

83. Convert 2π radians to grads.

84. Convert $\dfrac{5\pi}{6}$ radians to grads.

85. Convert 180 grads to radians.

86. Convert 500 grads to radians.

Exercise 87: **(Area)** A sector of a circle is the region bounded by a central angle and its subtending arc (see the adjacent figure). Determine a formula for the area of a sector in terms of r and θ.

Exercise 88: **(Area)** Using the formula from the previous exercise, and the formula for the area of a triangle ($A = \frac{1}{2}bh$), find a formula for the area of the shaded region in the adjacent figure, in terms only of r and θ.

Exercises 89–94: **(Linear/Angular Speed)** For an object traveling at a constant speed along a circle of radius r, the ratio $\dfrac{s}{t}$ is called the linear speed of the object, where s is the distance traveled and t is time. The angular speed of the object, ω, is the ratio $\dfrac{\theta}{t}$. Since $s = \theta r$, the linear speed

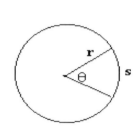

$$v = \frac{s}{t} = \frac{\theta r}{t} = r\omega$$

that is, the linear speed, v, is r times the angular speed, ω.

89. **(Bicycle)** The wheels of a bike are 14 inches in diameter, and they are revolving at 600 radians/min. What is the linear speed of the bike?

90. **(Bicycle)** The wheels of a bike are 14 inches in diameter, and are rotating at a speed of 300 rpm (revolutions per minute).

    (a) What is the linear speed of the bike?

    (b) Will doubling the rpm result in doubling the linear speed? Justify your answer.

3.3 Radian Measure and Oriented Angles   **153**

91. (**Vinyl Records**) As the terms indicate, 45 rpm and $33\frac{1}{3}$ rpm records revolve at the rate of 45 and $33\frac{1}{3}$ revolutions **per minute**, respectively.

    (a) Compare the linear speed of points on the rims of the two types of records, if 45 rpm records have a diameter of 7 inches and the $33\frac{1}{3}$ rpm records have a diameter of 12 inches.

    (b) Compare the linear speed of a point that is one inch from the rim of a 45 rpm record with the linear speed of a point that is two inches from the rim of the record.

    (c) Determine d_1 and d_2 such that a point d_1 inches from the center of a revolving 45 rpm record and a point d_2 inches from the center of a revolving $33\frac{1}{3}$ rpm record have the same linear speed.

92. (**Earth**) The planet Earth moves in an approximately circular orbit about the sun at an average distance from the sun of about 93,000,000 miles. Approximate the linear speed of this point, Earth.

93. (**Earth**) The diameter of the earth is estimated to be 7,926 miles. Approximate the linear speed of kings and fools on the Earth's equator that results from the Earth's rotation about its own axis.

94. (**Second Hand of a Watch**) Determine the angular speed of the second hand of a watch. Do the same for the hour hand.

§4. TRIGONOMETRIC FUNCTIONS OF ORIENTED ANGLES

The sine and cosine functions of acute angles were defined in terms of ratios of lengths of sides of a right triangle. We now extend those definitions to include all oriented angles, and do so in terms of the coordinates of points on **the unit circle**, the circle of radius 1 centered at the origin:

DEFINITION 3.3
SINE AND COSINE FUNCTIONS

For any angle θ, $(\cos\theta, \sin\theta)$ is the point of intersection of the terminal side of θ with the **unit** circle.

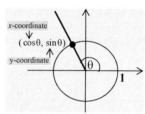

In other words:

For any angle θ:

$\cos\theta$ is the *x*-coordinate of the point of intersection of the terminal side of θ with the unit circle.

$\sin\theta$ is the *y*-coordinate of the point of intersection of the terminal side of θ with the unit circle.

To see that this definition coincides with the previous one when θ is an acute angle, consider the right triangle in the adjacent figure:

> The hypotenuse has length 1 as it coincides with the radius of the circle.

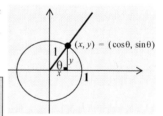

From right-triangle definition:		From Definition 3.3:
$\sin\theta = \dfrac{\text{opp}}{\text{hyp}} = \dfrac{y}{1}$	same ↔	$y = \sin\theta$
$\cos\theta = \dfrac{\text{adj}}{\text{hyp}} = \dfrac{x}{1}$	same ↔	$x = \cos\theta$

The remaining four trigonometric functions, tangent, cotangent, secant, and cosecant, are defined in accordance with the basic identities of Theorem 3.3, page 128:

DEFINITION 3.4
TANGENT, COSECANT, SECANT, COTANGENT

For any θ:

$$\tan\theta = \frac{\sin\theta}{\cos\theta} \qquad \csc\theta = \frac{1}{\sin\theta}$$

$$\sec\theta = \frac{1}{\cos\theta} \qquad \cot\theta = \frac{\cos\theta}{\sin\theta}$$

> Note that:
> $$\cot\theta = \frac{1}{\tan\theta}$$
> **except** when $\cos\theta = 0$.

Since the trigonometric functions of an angle depend solely on the terminal side of that angle, we have:

THEOREM 3.10 If α and β are coterminal, then:
$$\text{trig } \alpha = \text{trig } \beta$$

For example, since $-\frac{7\pi}{4} = \frac{\pi}{4} - 2\pi$, $-\frac{7\pi}{4}$ and $\frac{\pi}{4}$ are coterminal, and so:

$$\sin\left(-\frac{7\pi}{4}\right) = \sin\frac{\pi}{4} = \frac{1}{\sqrt{2}} \quad \text{and} \quad \tan\left(-\frac{7\pi}{4}\right) = \tan\frac{\pi}{4} = 1$$

> We remind you that the expression trig θ is generically used to represent any one of the six trigonometric functions.
>
>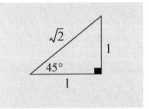
>
> Answers: (a) $\frac{1}{2}$ (b) 2

CHECK YOUR UNDERSTANDING 3.11

Evaluate:

(a) $\cos 420°$

(b) $\csc\left(-\frac{11\pi}{6}\right)$

TRIGONOMETRIC VALUES OF QUADRANTAL ANGLES

Figure 3.7 shows the four points of intersection of the unit circle and the *x*- and *y*-axes: $(1, 0)$, $(0, 1)$, $(-1, 0)$, and $(0, -1)$. These points lie on the terminal side of quadrantal angles (angles whose terminal side lies on an axis). The sine and cosine of such angles are easily determined, as is illustrated in the following examples.

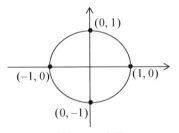

Figure 3.7

EXAMPLE 3.15 Determine the value of $\sin 270°$.

SOLUTION: Placing the angle in standard position and reading off the *y*-coordinate of the intersection of its terminal side with the unit circle, we find that:

$$\sin 270° = -1$$

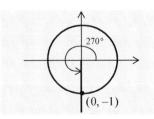

156 Chapter 3 Trigonometric Functions of Angles

EXAMPLE 3.16 Evaluate $\tan(-3\pi)$.

SOLUTION: Placing the angle in standard position we see that $\sin(-3\pi) = 0$ (y-coordinate), and that $\cos(-3\pi) = -1$ (x-coordinate). Consequently:

$$\tan(-3\pi) = \frac{\sin(-3\pi)}{\cos(-3\pi)} = \frac{0}{-1} = 0$$

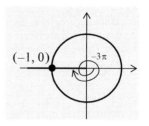

CHECK YOUR UNDERSTANDING 3.12

Complete the following table:

θ degrees	θ radians	sin θ	cos θ	tan θ	csc θ	sec θ	cot θ
0°	0	0	1	0			undef
90°							
	π						
270°							

Answer: See page C-22.

TRIGONOMETRIC VALUES OF NON-QUADRANTAL ANGLES

We now turn our attention to that of determining trig θ for certain non-quadrantal angles θ. This will involve two steps: finding (1) the sign of trig θ and (2) the magnitude of trig θ.

THE SIGN OF Trig θ

To determine the signs of trig θ you need only remember that (cos θ, sin θ) is the point of intersection of the terminal side of θ with the unit circle:

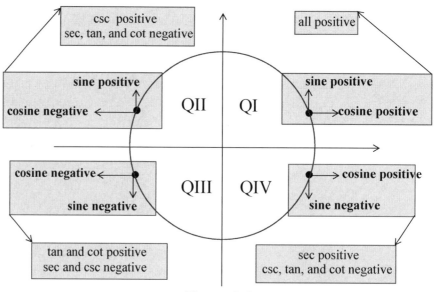

Figure 3.8

3.4 Trigonometric Functions of Oriented Angles   157

EXAMPLE 3.17 Determine the sign of the six trigonometric functions of $\theta = 825°$

SOLUTION: Since: $825° = 2(360°) + 105°$, $825°$ is coterminal with $105°$, and therefore lies in the second quadrant, where the sine is positive and the cosine is negative. Consequently:

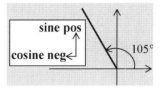

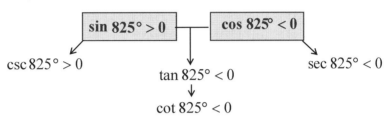

Answer: cosine and secant are positive, the others are negative.

CHECK YOUR UNDERSTANDING 3.13

Determine the sign of the six trigonometric functions of $\theta = -\dfrac{29\pi}{7}$.

THE MAGNITUDE OF Trig θ

As you have seen, determining the sign of trig θ is an easy matter, once the quadrant of θ has been determined. Our next concern is with the magnitude of trig θ, and we begin by defining the **reference angle**, θ_r, of any non-quadrantal angle θ, to be that **acute angle** formed by the terminal side of θ and the *x*-axis. The angles $\theta = \dfrac{5\pi}{6}$, $\theta = -\dfrac{7\pi}{4}$, and $\theta = 330°$, and their corresponding reference angles appear in Figure 3.9.

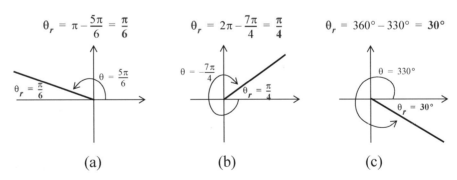

Figure 3.9

Let θ be any angle with reference angle θ_r. The terminal side of θ is one of the four depicted in the adjacent figure. By symmetry, the coordinates of the points of intersection of the terminal sides with the unit circle only differ in sign:

$(x, y), (-x, y), (-x, -y), (x, -y)$

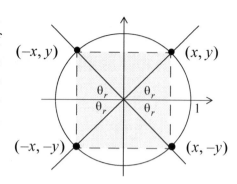

Thus $\cos\theta = x$ or $-x$, and $\sin\theta = y$ or $-y$, depending on the quadrant in which θ lies. Noting that $(x, y) = (\cos\theta_r, \sin\theta_r)$, we have:

$\left.\begin{array}{l} |\cos\theta| = \cos\theta_r \\ |\sin\theta| = \sin\theta_r \end{array}\right\}$ consequently: $\boxed{|\text{trig}\,\theta| = \text{trig}\,\theta_r}$

We can now find the exact value of $\text{trig}\,\theta$ for any θ whose reference angle is 30°, 45°, or 60°. It is a two-step process:

Step 1. Determine the sign of trig θ. Locate the quadrant in which θ lies, and then refer to Figure 3.8.

Step 2. Determine the magnitude of trig θ. Find the reference angle θ_r, and then use the fact that $|\text{trig}\,\theta| = \text{trig}\,\theta_r$.

EXAMPLE 3.18   Evaluate $\cos 570°$.

SOLUTION:

STEP 1: Noting that $570° - 360° = 210°$, we conclude that $\theta = 570°$ lies in QIII, where the cosine is **negative**.

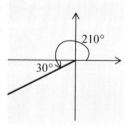

Step 2: Noting that the reference angle of θ is $\theta_r = 210° - 180° = 30°$, we conclude that $|\cos 570°| = \cos 30°$. Thus:

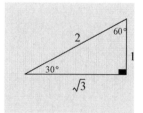

$$\cos 570° = \overset{\text{Step 1}}{\underset{}{-}} \overset{\text{Step 2}}{\cos 30°} = -\frac{\sqrt{3}}{2}$$

see margin

EXAMPLE 3.19

Evaluate $\cot\left(-\dfrac{17\pi}{4}\right)$.

SOLUTION: STEP 1: Noting that $-\dfrac{17\pi}{4} = -4\pi - \dfrac{\pi}{4}$, we conclude that $\theta = -\dfrac{17\pi}{4}$ lies in QIV, where the cotangent is **negative** ($\cot\theta = \dfrac{\cos\theta}{\sin\theta}$).

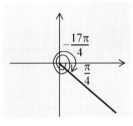

Step 2: Noting that the reference angle of θ is $\theta_r = \dfrac{\pi}{4}$, we conclude that $\left|\cot\left(-\dfrac{17\pi}{4}\right)\right| = \cot\dfrac{\pi}{4}$. Thus:

$$\cot\left(-\dfrac{17\pi}{4}\right) \overset{\text{Step 1}}{=} -\underset{\text{see margin}}{\cot\dfrac{\pi}{4}} \overset{\text{Step 2}}{=} -1$$

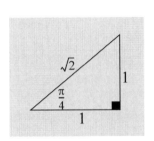

Answers: (a) $-\dfrac{\sqrt{3}}{2}$ (b) -1 (c) $\dfrac{2}{\sqrt{3}}$

CHECK YOUR UNDERSTANDING 3.14

Determine the exact value of:

(a) $\sin(-840°)$ (b) $\cot\dfrac{11\pi}{4}$ (c) $\sec\left(-\dfrac{25\pi}{6}\right)$

EXAMPLE 3.20

Given that $\cos\theta = -\dfrac{3}{5}$ and that $\sin\theta > 0$, determine $\sin\theta$ and $\tan\theta$.

SOLUTION: We know that

$$|\cos\theta| = \cos\theta_r = \dfrac{3}{5} = \dfrac{\text{adj}}{\text{hyp}}$$

leading us to the adjacent reference triangle for θ_r, where the Pythagorean Theorem was used to determine the length of the side opposite θ_r. From that triangle, we can easily read off the **magnitude** of $\sin\theta$ and $\tan\theta$:

$$|\sin\theta| = \sin\theta_r = \dfrac{4}{5} \quad \text{and} \quad |\tan\theta| = \tan\theta_r = \dfrac{4}{3}$$

160   Chapter 3 Trigonometric Functions of Angles

Since we were given that $\sin\theta$ is positive, $\sin\theta = \frac{4}{5}$.

Since the cosine is negative ($\cos\theta = -\frac{3}{5}$) and the sine is positive, the tangent is negative, $\tan\theta = -\frac{4}{3}$.

Answer: $\sin\theta = -\frac{\sqrt{33}}{7}$, $\cos\theta = -\frac{4}{7}$, $\tan\theta = \frac{\sqrt{33}}{4}$, $\csc\theta = -\frac{7}{\sqrt{33}}$, $\cot\theta = \frac{4}{\sqrt{33}}$

CHECK YOUR UNDERSTANDING 3.15

Determine the values of the remaining five trigonometric functions of θ, if:

$$\sec\theta = -\frac{7}{4} \quad \text{and} \quad \sin\theta < 0$$

For any angle θ, it is also possible to determine trigθ from the coordinates of **any** point (x, y) (other than the origin) on the terminal side of θ (in standard position). The point need not lie on the unit circle. To see how this is done, we call your attention to Figure 3.10 (where $r = \sqrt{x^2 + y^2}$):

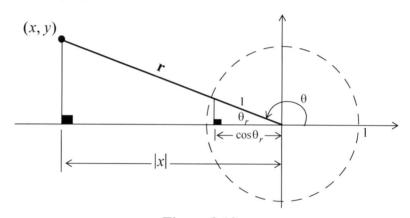

Figure 3.10

Since the two right triangles are similar: $\quad \dfrac{\cos\theta_r}{1} = \dfrac{|x|}{r}$

Since $\cos\theta_r = |\cos\theta|$: $\quad |\cos\theta| = \dfrac{|x|}{r}$

Since $\cos\theta$ and x are always of the same sign: $\quad \cos\theta = \dfrac{x}{r}$

In a similar manner, one can show that $\sin\theta = \dfrac{y}{r}$ (Exercise 100).

From the definition of the remaining four trigonometric functions in terms of sine and cosine, we can also express those functions in terms of x, y, and r, and we have:

3.4 Trigonometric Functions of Oriented Angles   161

THEOREM 3.11   Let $(x, y) \neq (0, 0)$ be any point on the terminal side of θ, and let $r = \sqrt{x^2 + y^2}$. Then, when defined:

$$\sin\theta = \frac{y}{r} \qquad \cos\theta = \frac{x}{r} \qquad \tan\theta = \frac{y}{x}$$

$$\csc\theta = \frac{r}{y} \qquad \sec\theta = \frac{r}{x} \qquad \cot\theta = \frac{x}{y}$$

This theorem holds for all angles, including quadrantal angles.

EXAMPLE 3.21   Evaluate the six trigonometric functions of θ, if the point $(4, -3)$ lies on the terminal side of θ.

SOLUTION: Appealing to Theorem 3.11, with $r = \sqrt{4^2 + (-3)^2} = 5$, we have:

$$\sin\theta = \frac{y}{r} = \frac{-3}{5} = -\frac{3}{5} \qquad \cos\theta = \frac{x}{r} = \frac{4}{5} \qquad \tan\theta = \frac{y}{x} = \frac{-3}{4} = -\frac{3}{4}$$

$$\csc\theta = \frac{r}{y} = -\frac{5}{3} \qquad \sec\theta = \frac{r}{x} = \frac{5}{4} \qquad \cot\theta = \frac{x}{y} = -\frac{4}{3}$$

Answer: $\sin\theta = \dfrac{1}{\sqrt{5}}$,

$\cos\theta = -\dfrac{2}{\sqrt{5}}$, $\tan\theta = -\dfrac{1}{2}$,

$\sec\theta = -\dfrac{\sqrt{5}}{2}$, $\csc\theta = \sqrt{5}$

$\cot\theta = -2$

CHECK YOUR UNDERSTANDING 3.16

Find the exact values of the trigonometric functions of θ, if the point $(-2, 1)$ lies on the terminal side of θ.

162   Chapter 3 Trigonometric Functions of Angles

EXERCISES

Exercises 1–4: (a) Express each of the following in terms of a coterminal angle of $30°\left(\frac{\pi}{6}\right)$, $60°\left(\frac{\pi}{3}\right)$, or $45°\left(\frac{\pi}{4}\right)$, and (b) use part (a) to evaluate the given expression.

1. $\sin 390°$
2. $\tan \frac{11\pi}{3}$
3. $\sec(-300°)$
4. $\cot\left(-\frac{9\pi}{4}\right)$

Exercises 5–10: Find the point on the unit circle that lies on the terminal side of the quadrantal angle in standard position.

5. $\theta = 540°$
6. $\theta = -\frac{13\pi}{2}$
7. $\theta = 630°$
8. $\theta = 28\pi$
9. $\theta = -\frac{3\pi}{2}$
10. $\theta = -180°$

Exercises 11–16: Evaluate the six trigonometric functions of the quadrantal angle.

11. $\theta = 5\pi$
12. $\theta = 810°$
13. $\theta = -\frac{7\pi}{2}$
14. $\theta = 28\pi$
15. $\theta = -\frac{\pi}{2}$
16. $\theta = -180°$

Exercises 17–28: Determine the sign (only) of the given expression.

17. $\tan(-99°)$
18. $\cos 236°$
19. $\sec \frac{8\pi}{3}$
20. $\sin \frac{11\pi}{4}$
21. $\cos(-11\pi)$
22. $\sin(-320°)$

23. $\tan \frac{9\pi}{4}$
24. $\csc \frac{2\pi}{5}$
25. $\cot 905°$
26. $\sec \pi$
27. $\csc\left(-\frac{23\pi}{2}\right)$
28. $\cot 655°$

Exercises 29–34: In what quadrant(s) is

29. $\cos \theta < 0$ and $\sin \theta > 0$
30. $\tan \theta > 0$ and $\sec \theta < 0$
31. $\csc \theta > 0$ and $\cot \theta > 0$

32. $\sin \theta < 0$ and $\csc \theta < 0$
33. $\cot \theta < 0$ and $\cos \theta < 0$
34. $\sec \theta > 0$ and $\csc \theta < 0$

Exercises 35–42: Find the reference angle, θ_r, of θ.

35. $\theta = 120°$
36. $\theta = 255°$
37. $\theta = \frac{7\pi}{8}$
38. $\theta = \frac{10\pi}{7}$

39. $\theta = -210°$
40. $\theta = 310°$
41. $\theta = -\frac{2\pi}{3}$
42. $\theta = -\frac{\pi}{6}$

Exercises 43–51: Express each of the following in terms of the reference angle, but do not evaluate. (For example: $\sin 300° = -\sin 60°$).

43. $\sin 145°$
44. $\sin(-200°)$
45. $\sec 195°$
46. $\cot 610°$
47. $\sec 325°$

48. $\cos\left(-\frac{5\pi}{3}\right)$
49. $\cos \frac{5\pi}{4}$
50. $\tan \frac{9\pi}{7}$
51. $\csc \frac{4\pi}{5}$

Exercises 52–76: Determine the exact value (without a calculator).

52. $\tan \frac{\pi}{4}$
53. $\cos 0$
54. $\sin 60°$
55. $\csc\left(-\frac{\pi}{6}\right)$
56. $\sec 0°$

57. $\tan 0$
58. $\sec\left(-\frac{7\pi}{6}\right)$
59. $\cos\left(-\frac{\pi}{4}\right)$
60. $\cot 45°$
61. $\sin 0°$

62. $\sec \frac{9\pi}{2}$
63. $\tan \frac{\pi}{6}$
64. $\sin \pi$
65. $\tan 270°$
66. $\cos(-150°)$

67. $\cot 210°$
68. $\csc 390°$
69. $\sin \frac{15\pi}{4}$
70. $\sin \frac{7\pi}{3}$
71. $\tan 150°$

72. $\sin(-225°)$
73. $\tan(-225°)$
74. $\cos 225°$
75. $\sin 300°$
76. $\cot\left(-\frac{5\pi}{4}\right)$

3.4 Trigonometric Functions of Oriented Angles   163

Exercises 77–82: Determine the exact values of the remaining five trigonometric functions of θ or explain why that is not possible, given

77. $\csc \theta = \frac{1}{4}$ and θ lies in QII

78. $\cos \theta = -\frac{2}{3}$ and $\tan \theta > 0$

79. $\sin \theta = -\frac{24}{25}$ and $\cos \theta < 0$

80. $\tan \theta = -4$ and $\csc \theta < 0$

81. $\sec \theta = \frac{13}{5}$ and θ lies in QIV

82. $\sin \theta = \frac{1}{4}$ and $\cos \theta > 0$

Exercises 83–87: Evaluate the remaining five trigonometric functions of θ, for each possible quadrant in which θ could lie. (For example, if $\sin \theta = \frac{1}{2}$, then θ lies in the first or second quadrants, and there are two sets of values to be determined.)

83. $\sec \theta = -2$ 84. $\cot \theta = \frac{12}{5}$ 85. $\tan \theta = -\frac{1}{2}$ 86. $\csc \theta = -\frac{25}{7}$ 87. $\cos \theta = \frac{3}{5}$

Exercises 88–99: Evaluate the six trigonometric functions of θ, if the given point lies on the terminal side of θ.

88. $(3, 4)$ 89. $(-3, 4)$ 90. $(2, 0)$ 91. $(2, 4)$ 92. $(-5, 12)$ 93. $(0, -6)$

94. $(1, -3)$ 95. $(-12, -5)$ 96. $(-4, -3)$ 97. $(12, -5)$ 98. $\left(0, \frac{3}{4}\right)$ 99. $(-3, -2)$

Exercise 100:   Refer to Figure 3.10, and show that $\sin \theta = \frac{y}{r}$.

Exercises 101–103: Find the degree measure of a pair of angles α and β if

101. $\sin(\alpha + \beta) = -\frac{1}{2}$ and $\cos(\alpha - \beta) = \frac{1}{\sqrt{2}}$

102. $\sin(\alpha + \beta) = -\cos(\alpha + \beta)$ and $\cos(\alpha - \beta) = \frac{1}{\sqrt{2}}$

103. $\sin(2\alpha) = \frac{\sqrt{3}}{2}$ and $\tan(\alpha + 2\beta) = 1$

Chapter Summary

Right triangle definition of trigonometric functions of an acute angle θ	$\sin\theta = \dfrac{\text{opp}}{\text{hyp}}$, $\cos\theta = \dfrac{\text{adj}}{\text{hyp}}$, $\tan\theta = \dfrac{\text{opp}}{\text{adj}}$ $\csc\theta = \dfrac{\text{hyp}}{\text{opp}}$, $\sec\theta = \dfrac{\text{hyp}}{\text{adj}}$, $\cot\theta = \dfrac{\text{adj}}{\text{opp}}$ 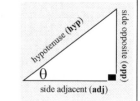
Some Fundamental Identities for acute angles θ	**Pythagorean:** $\sin^2\theta + \cos^2\theta = 1$ **Basic:** $\tan\theta = \dfrac{\sin\theta}{\cos\theta}$ $\csc\theta = \dfrac{1}{\sin\theta}$ $\sec\theta = \dfrac{1}{\cos\theta}$ $\cot\theta = \dfrac{\cos\theta}{\sin\theta} = \dfrac{1}{\tan\theta}$ **Complementary identities:** $\sin(90° - \theta) = \cos\theta$, and $\cos(90° - \theta) = \sin\theta$
Two important reference triangles	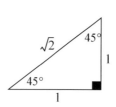 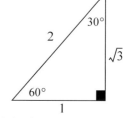 **45° reference triangle** **30°/60° reference triangle**
Alternate Interior Angles	When parallel lines are cut by a transversal, alternate interior angles are equal. 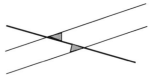
Vertical Angles	Vertical angles of intersecting lines are equal. 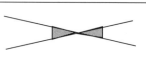
Angles of elevation and depression	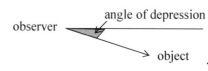

BEARING	The **bearing** of an object Q from a point P can be given in terms of the acute angle between the north-south line through P and the line segment joining P to Q. For example: bearing N40°E bearing S57°W
REFRACTION	Let a denote the speed of light in medium A (say, air) and b denote the speed of light in medium B (say, water). If α is the acute angle between a light ray and the vertical medium A (called the angle of incidence), and β is the acute angle between the ray and the vertical in medium B (called the angle of refraction), then: $$\frac{\sin\alpha}{\sin\beta} = \frac{a}{b}$$
INDEX OF REFRACTION	The **index of refraction** of a medium A (such as air or water), is the ratio of the speed of light in a vacuum to the speed of light in A.
Snell's Law	If I_a and I_b denote the indices of refraction of mediums A and B, respectively, then: $$\frac{I_b}{I_a} = \frac{\sin\alpha}{\sin\beta}$$
RADIAN MEASURE	The **radian measure** of a central angle of a circle is the ratio of the length of the arc subtending that angle to the radius of the circle. (Radian measure, being the ratio of two lengths, is a real number and is **not** associated with a unit (like degrees). Nonetheless, one can use an expression of the form $\theta = \pi$ radians to indicate that the angle θ as radian measure π. Alternatively, one may simply write $\theta = \pi$.)
Conversions	Since an angle of 180° has radian measure π, one can use the conversion bridge: $$180° = \pi \text{ radians}$$ to convert from degree to radian measure, and vice versa.

ORIENTED ANGLE	It is often useful to think of an angle θ as evolving from the following dynamic process: A fixed ray or half-line, called the **initial side** of the angle, is rotated about an endpoint O, called the **vertex** of the angle, to a final destination, called the **terminal side** of the angle. If the rotation is counterclockwise, then the angle is said to be **positive**, and if clockwise then it is **negative**. Because of the sign associated with it, such angles are said to be **oriented angles**.
STANDARD POSITION	An angle in the Cartesian plane is said to be in **standard position when** its vertex is at the origin and its initial side is along the positive x-axis.
Arc length	Let s denote the length of the arc subtending a central angle of θ radians in a circle of radius r. Then: $$s = r\theta$$
UNIT CIRCLE DEFINITION OF TRIGONOMETRIC FUNCTIONS	For any angle θ, $(\cos\theta, \sin\theta)$ is the point of intersection of the terminal side of θ with the **unit** circle. The remaining four trigonometric functions are defined in terms of the sine and cosine functions: $$\tan\theta = \frac{\sin\theta}{\cos\theta} \qquad \csc\theta = \frac{1}{\sin\theta}$$ $$\sec\theta = \frac{1}{\cos\theta} \qquad \cot\theta = \frac{\cos\theta}{\sin\theta}$$
Quadrantal Angles	Angles whose terminal side lies on an axis. The sine and cosine of such angles are easily determined from the adjacent figure.
Coterminal Angles	If α and β are coterminal, then trig α = trig β.
REFERENCE ANGLE	The **reference angle**, θ_r, of any non-quadrantal angle θ, is that **acute angle** formed by the terminal side of θ and the x-axis.

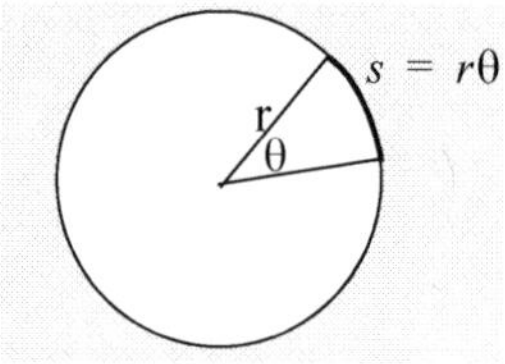

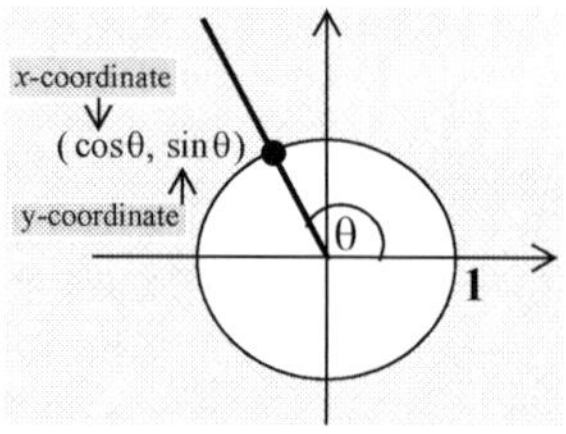

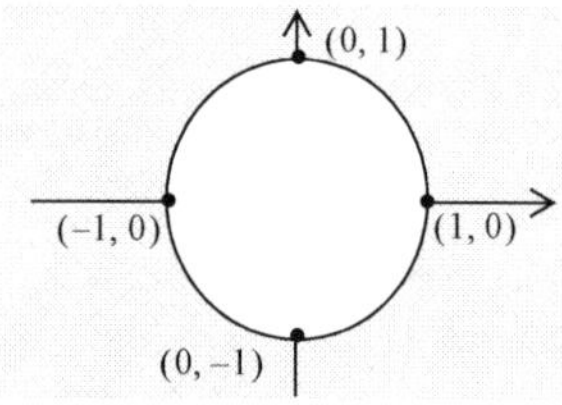

Evaluating trigonometric functions of oriented angles	Step 1. Locate the angle θ. If it is a quadrantal angle, then refer to the *Quadrantal Angles*-Figure above. If not: Step 2. Determine the **sign** of trig θ from the unit circle and: Step 3. Determine the **magnitude** of trig θ: Find the reference angle θ_r, and use the fact that $\lvert \text{trig } \theta \rvert = \text{trig } \theta_r$.
Using a point on the terminal side of θ to determine trigθ	Let $(x, y) \neq (0, 0)$ be any point on the terminal side of θ, and let $r = \sqrt{x^2 + y^2}$. Then, when defined: $$\sin\theta = \frac{y}{r} \qquad \cos\theta = \frac{x}{r} \qquad \tan\theta = \frac{y}{x}$$ $$\csc\theta = \frac{r}{y} \qquad \sec\theta = \frac{r}{x} \qquad \cot\theta = \frac{x}{y}$$

168   **Chapter 3 Trigonometric Functions of Angles**

CHAPTER REVIEW EXERCISES

Exercises 1–16:   Represent the angle in standard position in the xy-plane, indicate the quadrant in which it lies, and find a coterminal angle between 0° and 360°, or between 0 and 2π, inclusive.

1. $-240°$ 2. $-888°$ 3. $-742°$ 4. $1100°$ 5. $921°$ 6. $-\pi°$

7. $\dfrac{2500°}{3}$ 8. $-587.8°$ 9. $-\dfrac{31\pi}{4}$ 10. 9π 11. $\dfrac{11\pi}{3}$ 12. $\dfrac{20\pi}{7}$

13. $-\dfrac{5\pi}{12}$ 14. -2 15. $\dfrac{13\pi}{5}$ 16. $-\dfrac{53\pi}{11}$

Exercises 17–20:   Convert each of the given angles to radian measure.

17. $-225°$ 18. $315°$ 19. $210°$ 20. $-150°$

Exercises 21–24:   Convert each of the given angles to degree measure.

21. $-\dfrac{\pi}{6}$ 22. $-\dfrac{2\pi}{3}$ 23. $\dfrac{7\pi}{3}$ 24. $\dfrac{3\pi}{4}$

Exercises 25–34:   Find the exact value of each of the following.

25. $\sin\dfrac{11\pi}{2}$ 26. $\sec 0$ 27. $\cos(-\pi)$ 28. $\cot\dfrac{3\pi}{2}$ 29. $\sin 135°$

30. $\sin 150°$ 31. $\tan(-135°)$ 32. $\tan\left(-\dfrac{7\pi}{3}\right)$ 33. $\sec\left(-\dfrac{\pi}{3}\right)$ 34. $\tan 300°$

Exercises 35–38:   Solve the right triangle (without a calculator).

35. $\alpha = 45°$ and $a = \sqrt{2}$ 36. $\beta = 60°$ and $b = 2$

37. $\beta = 30°$ and $c = 3$ 38. $a = 5$ and $b = \dfrac{5\sqrt{3}}{3}$

Exercises 39–40:   Solve the right triangle, approximately, with a calculator.

39. $\alpha = 32°$ and $a = \sqrt{2}$ 40. $a = 3$ and $b = \dfrac{5\sqrt{5}}{2}$

Exercises 41–42:   Determine the exact values of the trigonometric functions of θ.

41. $\sin\theta = -\dfrac{1}{4}$ and $\tan\theta > 0$ 42. $\cos\theta = \dfrac{9}{10}$ and θ lies in the fourth quadrant

Exercises 43–44:   Determine the exact values of the trigonometric functions of θ for each possible quadrant in which θ could lie.

43. $\cot\theta = -\dfrac{4}{3}$ 44. $\csc\theta = 3$

Exercise 45:   (Angle of Depression) From the top of a lighthouse, 120 meters above the sea, the angle of depression of a boat is 15°. How far is the boat from the lighthouse?

Exercise 46:   (Angle of Depression) The angle of depression from the top of a building to the far side of a 25 meter street is 58°. The angle of depression to the near side is 70°.

(a) How tall is the building?   (b) How far is the street from the building?

Exercise 47: (**Bearing**) Two forest rangers, one directly north of the other, are thirty miles apart. A fire is observed to be S39°W of one ranger, and N51°W of the other. How far is each ranger from the fire?

Exercise 48: (**Bearing**) A motor boat moves in the direction N40°E for 3 hours at 20 mph. How far north and how far east does the boat travel?

Exercise 49: (**River Width**) Standing on the bank of a straight river, Mary measures an angle of 27° from the bank of the river to a tree downstream on the opposite bank. From a point 50 feet downstream, closer to the tree which is still downstream from her, the angle from the bank of the river to the tree is 35°. How wide is the river?

Exercise 50: (**Refraction**) A beam of light traveling in air is incident at an angle of 41° with the normal (perpendicular) to a surface of transparent plastic. It is refracted at the interface and makes an angle of 26.5° with the normal in the plastic. Calculate

(a) the index of refraction of the plastic.   (b) the speed of light in the plastic.

Exercise 51: (**Pulley**) A belt moving at a speed of 45 feet per second drives a pulley at a speed of 600 revolutions per minute. What is the radius of the pulley?

C U M U L A T I V E   R E V I E W   E X E R C I S E S

Exercises 1–2: Simplify.

1. $\dfrac{(-2b)^2(ab)^3}{\sqrt{ab}}$

2. $\dfrac{x^{-\frac{1}{2}}(x-3) + \sqrt{x}(x-3)^{-1}}{x^2 - 5x + 9}$

Exercises 3–4: Solve the equation.

3. $-2x^2 + 6x + 1 = 0$

4. $x(3x - 5) = 4 + 6x$

Exercise 5: (**Unit Conversion**) The speed of light is approximately 3×10^8 meters per second. Convert this to miles per hour. (1 m = 3.28 ft, 1mi = 5280 ft)

Exercise 6:   Suppose $f(x) = -\dfrac{x}{3x + 4}$ and $g(x) = x^2 - 1$.

(a) Determine the domain of f.

(b) Determine $(f \circ g)(x)$.

(c) Evaluate $(g \circ f)(-1)$.

(d) Show that f is one-to-one.

(e) Find $f^{-1}(x)$, the inverse of the function f.

(f) Verify that $(f \circ f^{-1})(x) = x$

170   Chapter 3 Trigonometric Functions of Angles

Exercise 7:   For the function $g(x) = -\sqrt{x+4}$,

(a) determine the domain of g

(b) determine the intercepts of the graph of g

(c) sketch the graph of g by reflecting, stretching, and shifting the basic graph

(d) determine the range of g

Exercises 8–11:   Find the exact value of each of the following.

8. $\sin 30° + \tan 45°$

9. $\sin 30° \cos 60° + \cos 30° \sin 60°$

10. $\sin 0° + 3 \cot 90° + 5 \sec 180° - 4 \cos 270°$

11. $3 \sin \pi + 4 \cos 0 - 3 \cos \pi + \csc \frac{\pi}{2}$

Exercises 12–17:   Determine the exact values of the six trigonometric functions of the given angle.

12. $\theta = 225°$ 13. $\theta = -240°$ 14. $\theta = 690°$ 15. $\theta = \frac{3\pi}{2}$ 16. $\theta = -\frac{7\pi}{4}$ 17. $\theta = \frac{5\pi}{6}$

Exercises 18–19:   Find the exact values of the six trigonometric functions of θ.

18. $\sin \theta = \frac{7}{25}$, $\tan \theta < 0$

19. $\cos \theta = -\frac{4}{5}$, $\csc \theta < 0$

Exercise 20: (**Bearing**) Ship A is 350 miles east of ship B, and ship C is due north of B at a bearing of N32°W from A.

(a) How far is ship A from ship C?   (b) What is the bearing of ship A from ship C?

CHAPTER 4
TRIGONOMETRIC FUNCTIONS OF A REAL VARIABLE

Though it may be comfortable to think of the trigonometric functions as acting on angles, a different interpretation is called for in the calculus where one is concerned with functions defined on real numbers. The trigonometric functions in this chapter are real-valued functions of a real variable.

In Section 1, we define the trigonometric functions of a real variable, construct their graphs, and discuss some of their basic properties. Trigonometric identities are established in Sections 2 and 3. The inverse trigonometric functions are defined and investigated in Section 4, and are then used to solve trigonometric equations in Section 5.

---

§1. THE TRIGONOMETRIC FUNCTIONS

As you can see from the following definition, the transition from trigonometric functions of angles to trigonometric functions of numbers hinges on the fact that the radian measure of an angle, being a ratio of two lengths (Definition 3.2, page 144), is actually a real number.

DEFINITION 4.1
TRIGONOMETRIC FUNCTIONS OF A REAL VARIABLE

Let x be a real number and let θ be the angle with **radian measure** x. Then:

$$\underset{\underset{\text{a number}}{\uparrow}}{\text{trig } x} = \underset{\underset{\text{an angle}}{\uparrow}}{\text{trig } \theta}$$

While one generally finds that the number π makes an appearance in the "x" of trig x, this need not be so. There is nothing wrong with an expression of the form $\sin(17.3)$, where 17.3 can represent the radian measure of an angle, or a real number.

For example, since the sine of the **angle** $\theta = \dfrac{\pi}{4}$ (radians) equals $\dfrac{1}{\sqrt{2}}$, the sine of the **real number** $\dfrac{\pi}{4}$ is also $\dfrac{1}{\sqrt{2}}$:

$$\sin\frac{\pi}{4} = \frac{1}{\sqrt{2}}$$

Evaluating our "new" trigonometric functions offers no new challenge. Indeed, the following Check Your Understanding box could just as well have been placed in the previous chapter, except that now we are applying the trigonometric functions to numbers rather than geometric objects (angles).

Answers:

(a) $\sin\dfrac{5\pi}{6} = \dfrac{1}{2}, \quad \cos\dfrac{5\pi}{6} = -\dfrac{\sqrt{3}}{2}$

(b) $\sin\left(-\dfrac{11\pi}{3}\right) = \dfrac{\sqrt{3}}{2}, \quad \cos\left(-\dfrac{11\pi}{3}\right) = \dfrac{1}{2}$

(c)
$\sin\left(\dfrac{\pi}{4} + 361\pi\right) = \cos\left(\dfrac{\pi}{4} + 361\pi\right) = -\dfrac{1}{\sqrt{2}}$

CHECK YOUR UNDERSTANDING 4.1

Evaluate $\sin x$ and $\cos x$ for the real number x:

(a) $x = \dfrac{5\pi}{6}$ (b) $x = -\dfrac{11\pi}{3}$ (c) $x = \dfrac{\pi}{4} + 361\pi$

THE SINE AND COSINE FUNCTIONS AND THEIR GRAPHS

Since the sine and cosine functions are defined on all angles, and every real number is the radian measure of some angle, the domain of both functions is the set of all numbers: $(-\infty, \infty)$.

Since the sine and cosine functions are the *x*- and *y*-coordinates, respectively, of points on the **unit** circle, the range of both functions is $[-1, 1]$.

We have already noted that the sine and cosine of coterminal angles are equal. In particular:

$$\sin(x + 2\pi) = \sin x$$
$$\cos(x + 2\pi) = \cos x$$

> A function *f* is **periodic** if there exists a positive number *p* such that:
> $$f(x + p) = f(x)$$
> for every *x* in the domain of *f*. The smallest such *p* is said to be the **period** of the function.

for all *x*, and no smaller positive value that 2π added to *x* satisfies either of the above two equations. For this reason we say that the sine and cosine functions are **periodic function**, of period 2π.

The sine of an angle is the **y-coordinate** of the point on the unit circle that also lies on the terminal side of that angle. Consequently, as the radian measure of an angle increases from 0 to $\frac{\pi}{2}$, the sine of the angle increases from 0 to 1. It follows that as the number *x* increases from 0 to $\frac{\pi}{2}$, the function $f(x) = \sin x$ increases from 0 to 1. Continuing in this fashion, we arrive at the graph of $f(x) = \sin x$ depicted in Figure 4.1(a). From the periodicity of the sine function, the graph of $f(x) = \sin x$ over any interval $[2n\pi, 2n\pi + 2\pi]$ must coincide with that of Figure 4.1(a), as is depicted in Figure 4.1(b).

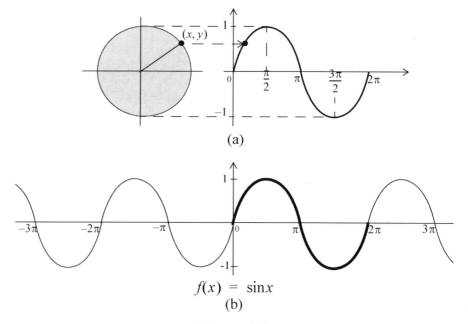

Figure 4.1

The fact that the sine function has domain $(-\infty, \infty)$ and range $[-1, 1]$, with period 2π, is revealed in Figure 4.1. In addition, the graph suggests that the sine function is an odd function $[\sin(-x) = -\sin x$ for all $x]$, a fact that you are asked to establish in the exercises.

To obtain the graph of the cosine function, we proceed as we did with the sine function, except that we now focus our attention of the **x-coordinate** of the point where the terminal side of the angle of radian measure x intersects the unit circle (see Figure 4.2).

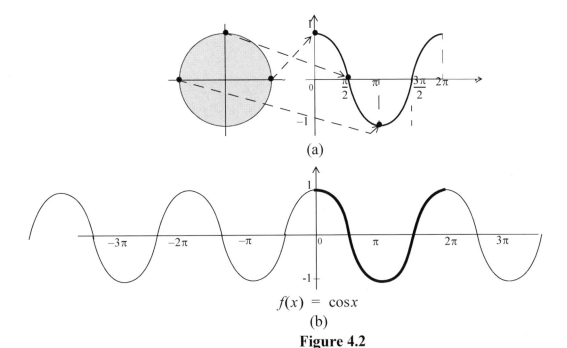

(a)

$f(x) = \cos x$

(b)

Figure 4.2

The fact that the cosine function has domain $(-\infty, \infty)$ and range $[-1, 1]$, with period 2π is revealed in Figure 4.2. In addition, the graph suggests that the cosine function is an even function $[\cos(-x) = \cos x$ for all $x]$, a fact that you are asked to establish in the exercises.

GRAPHS OF $f(x) = a \sin b(x-c) + d$ **AND**
$f(x) = a \cos b(x-c) + d$, $b > 0$.

As is illustrated in our first example, the method of pages 90-91 of Section 2.2 can be used to graph such functions when $b = 1$.

EXAMPLE 4.1 Sketch the graph of the function:
$$f(x) = -2\sin(x - \pi) + 3$$

174 Chapter 4 Trigonometric Functions of a Real Variable

Compare this solution with that of Example 2.10, page 90.

SOLUTION: First, reflect the graph of $y = \sin x$ about the x-axis (to go from $y = \sin x$ to $y = -\sin x$), and then stretch the resulting graph by a factor of **2** (to get to the graph of $y = -2\sin x$):

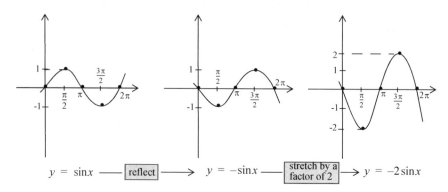

In the third step, "pick up" the graph of $y = -2\sin x$ and move it π units to the right: the $-\pi$ in $-2\sin(x - \pi)$. Finally, move the graph **3** units up to go from $y = -2\sin(x - \pi)$ to $-2\sin(x - \pi) + 3$:

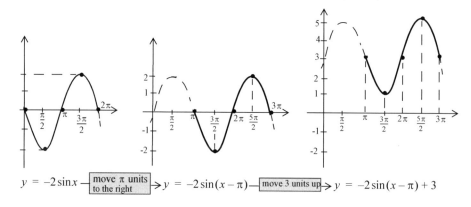

CHECK YOUR UNDERSTANDING 4.2

Sketch the graph of:
$$f(x) = -4\cos\left(x - \frac{\pi}{2}\right) + 3$$

Answer: See page C-25.

The graph of a periodic function restricted to any one period is called a **cycle**. The next example is a bit more ambitious than the previous one, in that it will involve a "cycle-compression."

EXAMPLE 4.2 Sketch the graph of the function:
$$f(x) = 3\cos(4x - \pi) - 2$$

> You can also use the process of this example to graph functions like the one in Example 4.1.

SOLUTION: The graph of the function $y = 3\cos x$ appears in Figure 4.3(a). To arrive at the graph of $f(x) = 3\cos(4x - \pi)$, we reason as follows:

Since one cycle of $y = 3\cos\square$ starts when $\square$ is 0 and ends when $\square$ is 2π, one cycle of $y = 3\cos(4x - \pi)$ begins when $4x - \pi = 0$ and ends when $4x - \pi = 2\pi$:

$$0 \le 4x - \pi \le 2\pi$$

$$\pi \le 4x \le 3\pi$$

start of cycle → $\dfrac{\pi}{4} \le x \le \dfrac{3\pi}{4}$ ← end of cycle

We find that the period of the $f(x)$ is $\dfrac{3\pi}{4} - \dfrac{\pi}{4} = \dfrac{\pi}{2}$. Dividing the period into 4 equal parts of length $\dfrac{1}{4} \cdot \dfrac{\pi}{2} = \dfrac{\pi}{8}$ enables us to easily sketch one cycle by shifting the 5 points in Figure 4.3(a) to the right to:

$x = \dfrac{\pi}{4} = \dfrac{2\pi}{8}, \dfrac{2\pi}{8} + \dfrac{\pi}{8} = \dfrac{3\pi}{8}, \dfrac{3\pi}{8} + \dfrac{\pi}{8} = \dfrac{4\pi}{8}$, etc. [see Figure 4.3(b)].

Finally, lower the graph in Figure 4.3(b) to arrive at the graph of $f(x) = 3\cos(4x - \pi) - 2$ [Figure 4.3(c)]

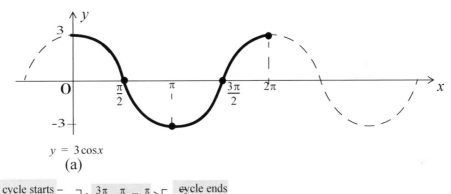

$y = 3\cos x$
(a)

> Since the start of the cycle is shifted to the right by $\dfrac{\pi}{4}$ units one says that a **phase shift** of $\dfrac{\pi}{4}$ has occurred

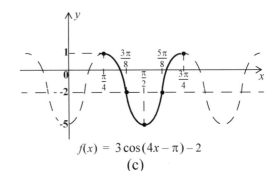

$y = 3\cos(4x - \pi)$
(b)

$f(x) = 3\cos(4x - \pi) - 2$
(c)

Figure 4.3

The graphs of the sine and cosine functions have the same shape, and both are called **sine waves**. The **amplitude** of a sine wave is defined to be one-half of the difference between the largest and smallest values of the function (see margin). The function $f(x) = 3\cos(4x - \pi) - 2$ of the previous example has amplitude 3, and the function $f(x) = -2\sin(x - \pi) + 3$ of Example 4.1 has amplitude 2.

176　Chapter 4　Trigonometric Functions of a Real Variable

> **CHECK YOUR UNDERSTANDING 4.3**
>
> Sketch the graph of:
> $$f(x) = -\sin\left(\frac{1}{2}x + \frac{\pi}{2}\right) + 1$$
> Specify its period, amplitude, and phase shift.

Answer: See page C-26.

THE TANGENT FUNCTION AND ITS GRAPH

By definition, the tangent function is the quotient function:

$$\tan x = \frac{\sin x}{\cos x}$$

Since the numerator of $\tan x = \dfrac{\sin x}{\cos x}$, is 0 at multiples of π [Figure 4.4(a)], its graph has x-intercepts at those points [Figure 4.4(b)].

Since the denominator, $\cos x$, is zero at odd multiples of $\dfrac{\pi}{2}$ [Figure 4.4(a)], the tangent function is **not defined** at those points. As you **approach** those points, the magnitude of the sine function tends to 1 while the cosine function is shrinking to zero. It follows that the graph of $\tan x = \dfrac{\sin x}{\cos x}$ must tend to plus or minus infinity as x approaches an odd multiple of $\dfrac{\pi}{2}$ — it will tend to $+\infty$ if both the sine and cosine are of the same sign), and it will tend to $-\infty$ if they are of opposite signs (shaded regions of Figure 4.4).

The domain of the tangent function is all real numbers except odd multiples of $\dfrac{\pi}{2}$.

A **vertical asymptote** for the graph of a function is represented by a dashed vertical line about which the graph tends to either plus or minus infinity. In particular the graph of the tangent function has vertical asymptotes at odd multiples of $\dfrac{\pi}{2}$.

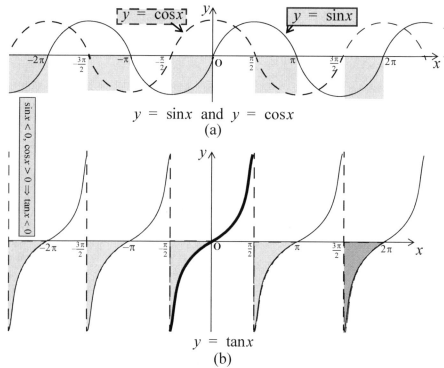

$y = \sin x$ and $y = \cos x$
(a)

$y = \tan x$
(b)

Figure 4.4

We have already noted that the domain of the tangent function consists of all numbers excluding odd multiples of $\frac{\pi}{2}$. The above graph tells us that its range is $(-\infty, \infty)$ and its period is π. The graph also suggests that the tangent function is an odd function, a fact which you are invited to establish in the exercises.

Answers: (a) All real numbers except for those of the form $n\pi$ for any integer n.
Graph: See page C-27.
(b) Range: $(-\infty, \infty)$
   Period: π
(c) Odd.

CHECK YOUR UNDERSTANDING 4.4

(a) Determine the domain of $f(x) = \cot x$ and sketch its graph.
(b) What is the range and period of the cotangent function.
(c) Is the cotangent function an even or an odd function?

THE COSECANT FUNCTION AND ITS GRAPH

The cosecant function is the reciprocal of the sine function:

$$\csc x = \frac{1}{\sin x}$$

As such, the cosecant function is undefined at the zeros of the sine: multiples of π. It follows that the domain of the cosecant function is all real numbers except multiples of π.

Since the cosecant function $\csc x = \frac{1}{\sin x}$ in never zero, its graph has no x-intercepts. It does have vertical asymptotes where the denominator, $\sin x$, is zero: at multiples of π (Figure 4.5). As is depicted in Figure 4.5, the cosecant is positive where the sine function is positive, and negative where the sine function is negative, and assumes a value of ± 1 wherever the sine function assumes those values (at multiples of $\frac{\pi}{2}$).

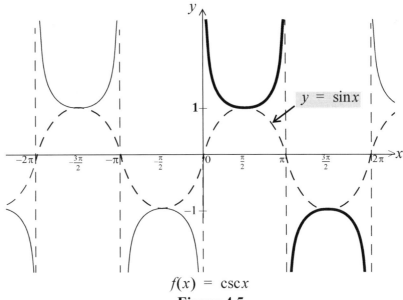

$f(x) = \csc x$
Figure 4.5

178   Chapter 4 Trigonometric Functions of a Real Variable

We have already noted that the domain of the cosecant function consists of all numbers excluding odd multiples of π. The above graph tells us that its range is $(-\infty, -1] \cup [1, \infty)$, and its period is 2π. In addition, the graph suggests that the cosecant function is an odd function [$\cos(-x) = -\cos x$ for all x], a fact which you are invited to establish in the exercises.

Answers: (a) All real numbers except for those of the form $(2n+1)\frac{\pi}{2}$ for any integer n.
Graph: See page C-28.
(b) Range:
$(-\infty, -1] \cup [1, \infty)$
Period: 2π.
(c) Even.

CHECK YOUR UNDERSTANDING 4.5

(a) Determine the domain of $f(x) = \sec x$ and sketch its graph.

(b) What is the range and period of the secant function.

(c) Is the secant function an even or an odd function?

4.1 The Trigonometric Functions   179

EXERCISES

Exercises 1–4: Evaluate the six trigonometric functions for the real number x.

1. $x = \frac{7\pi}{6}$
2. $x = \frac{2\pi}{3}$
3. $x = -\frac{5\pi}{4}$
4. $x = \frac{5\pi}{6} - 423\pi$

Exercises 5–12: Determine the amplitude and sketch the graph of the function.

5. $f(x) = 4\cos x - 2$
6. $f(x) = -4\sin x + 4$
7. $f(x) = 2\cos(x - \pi)$

8. $f(x) = \sin\left(x + \frac{\pi}{2}\right)$
9. $f(x) = 3\cos(x + 3\pi) - 4$
10. $f(x) = \frac{1}{2}\cos\left(x + \frac{3\pi}{2}\right) - 2$

11. $f(x) = -2\sin\left(x - \frac{\pi}{2}\right) + 1$
12. $f(x) = \frac{4}{3}\sin\left(x - \frac{3\pi}{2}\right) + 1$

Exercises 13–20: Determine the amplitude and period, and sketch the graph of the function.

13. $f(x) = 2\cos 4x$
14. $k(x) = -5\sin 2x$
15. $g(x) = \frac{5}{2}\cos 8x$

16. $h(x) = -10\cos 10x$
17. $G(x) = 3\sin 3x$
18. $F(x) = 3\cos\frac{1}{2}x$

19. $H(x) = -\sin\frac{2}{3}x$
20. $K(x) = \frac{1}{2}\sin 5x$

Exercises 21–28: Determine the amplitude, period, and phase shift, and sketch the graph of the function.

21. $f(x) = \frac{1}{3}\sin\left(x - \frac{3\pi}{4}\right) - 1$
22. $g(x) = -5\sin 2x + 1$

23. $h(x) = \sin\left(\frac{1}{2}x - \pi\right) + 2$
24. $f(x) = \frac{5}{2}\cos(8x - \pi)$

25. $k(x) = -3\cos(4x - 2\pi) - 3$
26. $g(x) = 3\sin\left(3x + \frac{\pi}{2}\right) + 4$

27. $k(x) = -10\cos(10x + \pi) - 10$
28. $h(x) = -2\sin\left(\frac{3}{2}x - 2\pi\right) + 3$

Exercises 29–40: Determine the vertical asymptotes and intercepts of the graph of the function for $-2\pi \le x \le 2\pi$.

29. $f(x) = \tan 2x$
30. $f(x) = \tan(x - \pi)$
31. $f(x) = \tan x - 1$

32. $f(x) = \cot 4x$
33. $f(x) = \cot\left(x + \frac{\pi}{2}\right)$
34. $f(x) = \cot\left(\frac{1}{2}x - \frac{\pi}{2}\right)$

35. $f(x) = \sec 2x$
36. $f(x) = \sec x + 1$
37. $f(x) = -3\sec x + 3$

38. $f(x) = \csc x - 1$
39. $f(x) = 2\csc x + 2$
40. $f(x) = 4\csc 4(x + \pi) + 3$

Exercises 41–52: Find the period and sketch the graph of the function.

41. $f(x) = \tan 2x$
42. $f(x) = \tan(x - \pi)$
43. $f(x) = \tan x - 1$

44. $f(x) = \cot 4x$
45. $f(x) = \cot\left(x + \frac{\pi}{2}\right)$
46. $f(x) = \cot\left(\frac{1}{2}x - \frac{\pi}{2}\right)$

47. $f(x) = \sec x + 1$
48. $f(x) = \sec 2x$
49. $f(x) = -3\sec x + 3$

50. $f(x) = 2\csc x + 2$
51. $f(x) = \csc x - 1$
52. $f(x) = 4\csc 4(x + \pi) + 3$

Exercises 53–57: Show that the two functions are not equal, by determining a particular value of x, a, for which $f(a) \neq g(a)$.

53. $f(x) = 2\sin x$, $g(x) = \sin 2x$

54. $f(x) = \sec 3x$, $g(x) = 3\sec x$

55. $f(x) = \cos(-x)$, $g(x) = -\cos x$

56. $f(x) = \cos(x + \pi)$, $g(x) = \cos x + \pi$

57. $f(x) = \tan\left(x - \frac{\pi}{4}\right)$, $g(x) = \tan x - \frac{\pi}{4}$

Exercises 58–61: Prove that

58. the sine function is an odd function.

59. the cosine function is an even function.

60. the tangent function is an odd function.

61. the cosecant function is an odd function.

Exercise 62: Prove that the graph of $f(x) = a\sin b(x - c) + d$ or $f(x) = a\cos b(x - c) + d$ with $b > 0$ has amplitude $|a|$, period $\frac{2\pi}{b}$, and phase shift c.

Exercises 63–65: Find a function $f(x) = a\cos b(x - c)$ with the given graph.

63.
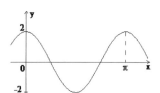

64.
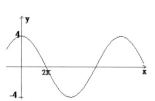

65.
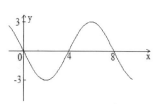

Exercises 66–68: Find a function $f(x) = a\sin b(x - c)$ with the given graph.

66.
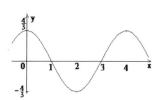

67.
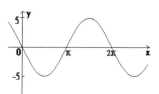

68.
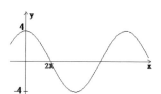

Exercise GC69: Use a graphing calculator to view the graph of $h(x) = \tan^2 x - \sec^2 x$, and, from that graph, formulate an identity involving the tangent and secant functions.

Exercise GC70: Repeat the previous exercise for the function $h(x) = \cot^2 x - \csc^2 x$.

Exercises 71–74: Construct a tangent or secant function satisfying the given conditions.

71. It has a period of 3π.

72. It has a period of 3π and a vertical asymptote at π.

73. It has a period of 4π, a vertical asymptote at π, and a local minimum at 2π.

74. It has a period of 4π, a vertical asymptote at π, and a local minimum value of 9 at 2π.

Exercise GC75: (**Factorials**) The symbol $n!$, read n-factorial, represents the product of the first n positive integers. For example,

$$4! = 1 \cdot 2 \cdot 3 \cdot 4 = 24$$

For each positive integer n, let

$$f_n(x) = x - \frac{x^3}{3!} + \frac{x^5}{5!} - \frac{x^7}{7!} + \cdots + \frac{(-1)^{n-1} x^{2n-1}}{(2n-1)!}$$

For instance, $f_1(x) = x$, $f_2(x) = x - \frac{x^3}{3!}$, $f_3(x) = x - \frac{x^3}{3!} + \frac{x^5}{5!}$, etc. Use a graphing calculator to graph the functions $g(x) = \sin x$ and $f_n(x)$ for $n = 3$, first on the interval $[-2\pi, 2\pi]$ and then on $[-\pi, \pi]$. Repeat this with $n = 4$, then with $n = 5$. Discuss your observation, and formulate a hypothesis related to that observation.

Exercises 76–79: Functions other than trigonometric functions can be periodic. For example,

$$f(x) = \begin{cases} 1, & \text{if } 2n < x \leq 2n+1 \quad \text{for all integers } n \\ -1, & \text{if } 2n+1 < x \leq 2n+2 \quad \text{for all integers } n \end{cases}$$

has period 2 (See Figure below.)

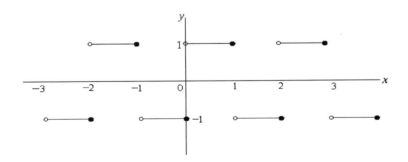

Define a nonconstant function f, that is not a trigonometric function, which is

76. periodic of period 9.

77. periodic of period 9 such that $f(x) \geq 0$ where defined.

78. periodic of period 9 such that $f(x) \geq 0$ where defined, and with y-intercept 3.

79. periodic of period 9 such that $f(x) \geq 0$ where defined, with y-intercept 3, and an x-intercept at 3.

Exercises 80–82: (Simple Harmonic Motion) Motion described by an equation of the form
$$y = a\sin(bt + c) \quad \text{or} \quad y = a\cos(bt + c)$$
is said to be simple harmonic motion. (This concept is illustrated in these exercises.)

80. Ignoring friction, when a mass suspended on a spring is pulled or compressed a units from its equilibrium position and then released, its resulting motion is simple harmonic
$$d = a\cos\omega t$$
where the displacement d is relative to the equilibrium position.

    A mass on a spring is displaced 5 inches from the equilibrium point and then released. Find the equation of motion of the mass, if it first passes through the equilibrium point $\frac{1}{15}$th of a second later.

81. The simple harmonic motion equation of the previous exercise has to be modified to take into account the damping effect of friction on the system. The modified equation
$$d = ae^{kt}\cos\omega t$$
where k is a negative constant that depends on the particular spring, is an example of damped harmonic motion.

    (a) Referring to the spring of the previous exercise, find the value of k if the maximum displacement of the spring is reduced to 3 inches after 4 seconds.

    (b) Suppose the spring in (a) is compressed a distance a from its equilibrium point and then released. How long will it take until the maximum displacement of the resulting motion is reduced by 90%?

82. The electrical circuit in the adjacent figure consists of a coil and a capacitor. The charge Q on the capacitor t seconds after the circuit is closed is given by

    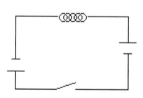

    $$Q = Q_0 \sin\left[\pi\left(\frac{1}{\sqrt{LC}}t + \frac{1}{2}\right)\right]$$

    where Q_0 is the initial charge (in coulombs), L is the inductance of the coil (in henrys), and C is the capacitance of the capacitor (in farads).

4.1 The Trigonometric Functions   183

(a) What is the maximum charge on the condenser?

(b) Can negative charge occur?

(c) What is the frequency (cycles per unit time) of oscillation of the current?

(d) Find the frequency, if $L = 0.30$ henrys and $C = 9.25 \times 10^{-4}$ farads.

(e) Find the initial charge if $L = 0.30$ henrys, $C = 9.25 \times 10^{-4}$ farads, and if the charge on the capacitor is 7.34×10^{-3} coulombs when one fifth of the first cycle has just occurred.

Exercises 83–93:   Indicate True or False. Where appropriate, use a graphing calculator to check your answer. Justify your answer.

83. The graph of the product function $f(x) = \sin x \cos x$ never assumes values greater than 1.

84. The graph of the quotient function $g(x) = \dfrac{\sin x}{\cos x}$ never assumes values greater than 1.

85. The function $k(x) = -2 + \sin x + \cos x$ does not take on any positive values.

86. The function $h(x) = 1 + \sin x + \cos x$ does not take on any negative values.

87. The functions $f(x) = \dfrac{\tan x}{\cot x}$ and $g(x) = \tan^2 x$ are equal.

88. The function $f(x) = 2 + \tan x$ is never negative.

89. Where defined, $\tan x + 1 > \cot x - 1$.

90. Where defined, $\tan 2x = 2 \tan x$.

91. For any nonzero polynomial $p(x)$, the graph of the function $f(x) = p(x) \tan x$ has infinitely many x-intercepts.

92. For any nonzero polynomial $p(x)$, the graph of the function $f(x) = p(x) \tan x$ has infinitely many vertical asymptotes.

93. The graph of the function $f(x) = \tan x + \cot x$ has no vertical asymptotes.

§2. Trigonometric Identities

Trigonometric Identities enable us to simplify trigonometric expressions or write them in alternate forms. As you will see later on in this chapter, identities also play an important role in solving trigonometric equations.

The following identities are particularly important in that they are the building blocks for many other trigonometric identities. You would do well to commit them to memory:

THEOREM 4.1 For all numbers x and y:

Pythagorean Identity (i) $\sin^2 x + \cos^2 x = 1$

Addition Identities

(ii) $\cos(x+y) = \cos x \cos y - \sin x \sin y$
$\cos(x-y) = \cos x \cos y + \sin x \sin y$

(iii) $\sin(x+y) = \sin x \cos y + \cos x \sin y$
$\sin(x-y) = \sin x \cos y - \cos x \sin y$

PROOF: (i). Let x be any number, and let θ be the angle of radian measure x. Applying the Pythagorean Theorem to the point $(\cos\theta, \sin\theta)$ on the **unit circle**, we have:

$$\sin^2\theta + \cos^2\theta = 1$$

Definition 4.1: $\sin^2 x + \cos^2 x = 1$

(ii) We begin by establishing the identity:

$$\cos(x-y) = \cos x \cos y + \sin x \sin y$$

Let α and β be angles of radian measure x and y, respectively. Consider the chord of length L in Figure 4.6(a) subtending the angle $\alpha - \beta$, and the same situation in Figure 4.6(b) but with $\alpha - \beta$ in standard position.

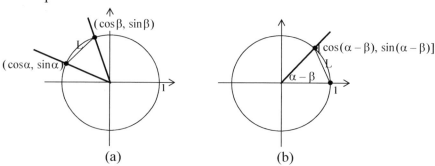

Figure 4.6

Applying the formula for the distance between two points (Theorem 2.2, page 109), we obtain two expressions for L^2.

From Figure 4.6(a):

$$L^2 = (\cos\alpha - \cos\beta)^2 + (\sin\alpha - \sin\beta)^2$$

$$= (\cos^2\alpha - 2\cos\alpha\cos\beta + \cos^2\beta) + (\sin^2\alpha - 2\sin\alpha\sin\beta + \sin^2\beta)$$

$$= (\cos^2\alpha + \sin^2\alpha) + (\cos^2\beta + \sin^2\beta) - 2\cos\alpha\cos\beta - 2\sin\alpha\sin\beta$$

$$= 1 + 1 - 2(\cos\alpha\cos\beta + \sin\alpha\sin\beta) = \mathbf{2 - 2(\cos\alpha\cos\beta + \sin\alpha\sin\beta)}$$

Pythagorean Identity

From Figure 4.6(b)

$$L^2 = [\cos(\alpha - \beta) - 1]^2 + \sin[(\alpha - \beta) - 0]^2$$

$$= [\cos^2(\alpha - \beta) - 2\cos(\alpha - \beta) + 1] + \sin^2(\alpha - \beta)$$

$$= \cos^2(\alpha - \beta) + \sin^2(\alpha - \beta) - 2\cos(\alpha - \beta) + 1$$

$$= 1 - 2\cos(\alpha - \beta) + 1 = \mathbf{2 - 2\cos(\alpha - \beta)}$$

Pythagorean Identity

Equating the two expressions for L^2, we find that:

$$\mathbf{2 - 2\cos(\alpha - \beta) = 2 - 2(\cos\alpha\cos\beta + \sin\alpha\sin\beta)}$$

$$\cos(\alpha - \beta) = \cos\alpha\cos\beta + \sin\alpha\sin\beta$$

Definition 4.1: $\quad \mathbf{\cos(x - y) = \cos x \cos y + \sin x \sin y} \quad (*)$

We now use (*) to establish the identity for the cosine of the sum of two numbers (or angles):

> Any trigonometric identity can be formulated in terms of real numbers or in terms of angles (see Definition 4.1, page 171).

from (*)

$$\mathbf{\cos(x + y)} = \cos[x - (-y)] = \cos x \cos(-y) + \sin x \sin(-y)$$

since the cosine function is an even function and the sine function is an odd function:

$$= \cos x \cos y + \sin x(-\sin y)$$

$$= \mathbf{\cos x \cos y - \sin x \sin y}$$

The proof of Theorem 4.1(iii) is relegated to the exercises.

CHECK YOUR UNDERSTANDING 4.6

(a) Establish the following "variations" of the Pythagorean Identity:

$\quad$ (i) $1 + \tan^2 x = \sec^2 x$ $\qquad\qquad$ (ii) $1 + \cot^2 x = \csc^2 x$

[Hint: Divide both sides of the Pythagorean identity by $\sin^2 x$ or $\cos^2 x$.]

(b) Use Theorem 4.1(ii) and (iii) and the definition of the tangent function to establish the following addition identities for the tangent:

$\quad$ (i) $\tan(x + y) = \dfrac{\tan x + \tan y}{1 - \tan x \tan y}$ $\qquad$ (ii) $\tan(x - y) = \dfrac{\tan x - \tan y}{1 + \tan x \tan y}$

Answers: See page C-29.

186 Chapter 4 Trigonometric Functions of a Real Variable

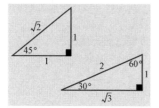

EXAMPLE 4.3 Determine the exact value of $\sin 15°$

SOLUTION:

$$\sin 15° = \sin(45° - 30°)$$

Theorem 4.1(iii): $\qquad = \sin 45° \cos 30° - \cos 45° \sin 30°$

See margin: $\qquad = \dfrac{1}{\sqrt{2}} \cdot \dfrac{\sqrt{3}}{2} - \dfrac{1}{\sqrt{2}} \cdot \dfrac{1}{2} = \dfrac{\sqrt{3}-1}{2\sqrt{2}} = \dfrac{\sqrt{6}-\sqrt{2}}{4}$

CHECK YOUR UNDERSTANDING 4.7

Use the fact that $\dfrac{\pi}{12} = \dfrac{\pi}{3} - \dfrac{\pi}{4}$ to find the exact value of $\cos\dfrac{\pi}{12}$.

Answer: $\dfrac{\sqrt{6}+\sqrt{2}}{4}$

EXAMPLE 4.4 Determine the value of $\cos(\alpha + \beta)$, if $\tan\alpha = -\dfrac{2}{3}$ and $\csc\beta = \dfrac{5}{3}$, and both α and β lie in the second quadrant.

SOLUTION: From Theorem 4.1(ii), we have:

$$\cos(\alpha + \beta) = \cos\alpha\cos\beta - \sin\alpha\sin\beta$$

The given information that $\tan\alpha = -\dfrac{2}{3}$ and $\csc\beta = \dfrac{5}{3}$ for α and β in the second quadrant, leads us to the following reference triangles:

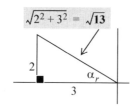   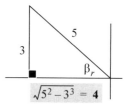

$\tan\alpha = -\dfrac{2}{3} \Rightarrow \tan\alpha_r = \dfrac{2}{3}\begin{smallmatrix}\leftarrow \text{opp}\\ \leftarrow \text{adj}\end{smallmatrix}$ $\qquad \sin\beta = \dfrac{3}{5}\begin{smallmatrix}\leftarrow \text{opp}\\ \leftarrow \text{hyp}\end{smallmatrix}$

Since the sine is positive in the second quadrant and the cosine is negative:

$\sin\alpha = \sin\alpha_r = \dfrac{2}{\sqrt{13}}\begin{smallmatrix}\leftarrow \text{opp}\\ \leftarrow \text{hyp}\end{smallmatrix}$ $\qquad \sin\beta = \sin\beta_r = \dfrac{3}{5}$

$\cos\alpha = -\cos\alpha_r = -\dfrac{3}{\sqrt{13}}\begin{smallmatrix}\leftarrow \text{adj}\\ \leftarrow \text{hyp}\end{smallmatrix}$ $\qquad \cos\beta = -\cos\beta_r = -\dfrac{4}{5}$

Bringing us to:

$$\cos(\alpha + \beta) = \cos\alpha\cos\beta - \sin\alpha\sin\beta$$
$$= \left(-\dfrac{3}{\sqrt{13}}\right)\left(-\dfrac{4}{5}\right) - \left(\dfrac{2}{\sqrt{13}}\right)\left(\dfrac{3}{5}\right) = \dfrac{6}{5\sqrt{13}}$$

CHECK YOUR UNDERSTANDING 4.8

Answer: $\dfrac{4-6\sqrt{2}}{15}$

Determine the value of $\sin(\alpha - \beta)$, if $\cos\alpha = -\dfrac{1}{3}$ and $\tan\beta = \dfrac{4}{3}$, where α is in the third quadrant and β is in the first quadrant.

DOUBLE ANGLE IDENTITIES

The following two identities follow readily from those of Theorem 4.1:

THEOREM 4.2

DOUBLE ANGLE IDENTITIES

For all numbers x:

(i) $\sin 2x = 2\sin x \cos x$

(ii) $\cos 2x = \cos^2 x - \sin^2 x$

$\qquad\qquad = 1 - 2\sin^2 x$

$\qquad\qquad = 2\cos^2 x - 1$

PROOF:

(i) $\sin 2x = \sin(x + x) \overset{\text{Theorem 4.1(iii), with } x = y}{=} \sin x \cos x + \cos x \sin x = 2\sin x \cos x$

(ii) $\cos 2x = \cos(x + x) \overset{\text{Theorem 4.1(ii), with } x = y}{=} \cos x \cos x - \sin x \sin x = \cos^2 x - \sin^2 x$

Then:

$$\cos 2x = \underset{\uparrow}{\cos^2 x} - \sin^2 x = \underset{\big|}{(1 - \sin^2 x)} - \sin^2 x = 1 - 2(\sin^2 x)$$

$$\sin^2 x + \cos^2 x = 1$$

$$\cos 2x = \cos^2 x - \underset{\downarrow}{\sin^2 x} = \cos^2 x - (1 - \cos^2 x) = 2\cos^2 x - 1$$

CHECK YOUR UNDERSTANDING 4.9

Establish the following double angle identity for the tangent function:

$$\tan 2x = \frac{2\tan x}{1 - \tan^2 x}$$

Answer: See page C-30.

EXAMPLE 4.5 Determine $\sin 2x$, given that $\tan x = \dfrac{2}{5}$ and $\sin x < 0$.

SOLUTION: From Theorem 4.1(i), we have:
$$\sin 2x = 2\sin x \cos x$$

To determine $\sin x$ and $\cos x$, we think in terms of angles, and let θ denote the angle of radian measure x.

Since $\tan \theta = \frac{2}{5}$ is positive and $\sin \theta$ is negative, θ must lie in the third quadrant. Looking at the adjacent reference triangle, and noting that the cosine is also negative in QIII, we conclude that:

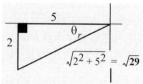

$$\sin 2x = 2\sin x \cos x = 2\left(-\frac{2}{\sqrt{29}}\right)\left(-\frac{5}{\sqrt{29}}\right) = \frac{20}{29}$$

CHECK YOUR UNDERSTANDING 4.10

Determine $\cos 2x$, given that $\tan x = -\frac{1}{2}$ and $\cos x > 0$.

Answer: $\frac{3}{5}$

EXAMPLE 4.6 Simplify:
$$\frac{(\sin x + \cos x)^2 - 1}{\sin 2x}$$

SOLUTION:

$$\frac{(\sin x + \cos x)^2 - 1}{\sin 2x} = \frac{(\sin^2 x + 2\sin x \cos x + \cos^2 x) - 1}{2\sin x \cos x}$$

Theorem 4.2(i):
$$= \frac{(\sin^2 x + \cos^2 x) + 2\sin x \cos x - 1}{2\sin x \cos x}$$

Theorem 4.1(i):
$$= \frac{1 + 2\sin x \cos x - 1}{2\sin x \cos x} = 1$$

CHECK YOUR UNDERSTANDING 4.11

Simplify:
$$\frac{\sin^2 x}{\cos x} - \sec x$$

Answer: $-\cos x$

4.2 Trigonometric Identities   189

The following identities play an important role in the calculus:

THEOREM 4.3

POWER REDUCING IDENTITIES

For all numbers x:

(i) $\sin^2 x = \dfrac{1 - \cos 2x}{2}$

(ii) $\cos^2 x = \dfrac{1 + \cos 2x}{2}$

PROOF: Applying Theorem 4.2(ii):

$$\cos 2x = \cos^2 x - \sin^2 x = \begin{cases} 1 - 2\sin^2 x \\ 2\cos^2 x - 1 \end{cases}$$

we have:

(i): $\cos 2x = 1 - 2\sin^2 x$

$2\sin^2 x = 1 - \cos 2x$

$\sin^2 x = \dfrac{1 - \cos 2x}{2}$

(ii): $\cos 2x = 2\cos^2 x - 1$

$2\cos^2 x = 1 + \cos 2x$

$\cos^2 x = \dfrac{1 + \cos 2x}{2}$

HALF-ANGLE IDENTITIES

The next two identities are a consequence of Theorem 4.3:

THEOREM 4.4

HALF-ANGLE IDENTITIES

For all numbers x:

(i) $\sin \dfrac{x}{2} = \pm\sqrt{\dfrac{1 - \cos x}{2}}$

(ii) $\cos \dfrac{x}{2} = \pm\sqrt{\dfrac{1 + \cos x}{2}}$

The "$\pm$" means that **either** the plus **or** the minus is to be chosen, **depending** on the quadrant in which the angle of radian measure $\frac{x}{2}$ lies.

(Here, $\pm$ cannot mean both plus and minus, since functions can only assign one value.)

PROOF:

(i):

From Theorem 4.3(i): $\sin^2 x = \dfrac{1 - \cos 2x}{2}$

Replace x with $\frac{x}{2}$: $\sin^2\left(\dfrac{x}{2}\right) = \dfrac{1 - \cos x}{2}$

Take the square root: (there are two of them) $\sin\dfrac{x}{2} = \pm\sqrt{\dfrac{1 - \cos x}{2}}$

(ii):

From Theorem 4.3(ii): $\cos^2 x = \dfrac{1 + \cos 2x}{2}$

Replace x with $\frac{x}{2}$: $\cos^2\left(\dfrac{x}{2}\right) = \dfrac{1 + \cos x}{2}$

Take the square root: (there are two of them) $\cos\dfrac{x}{2} = \pm\sqrt{\dfrac{1 + \cos x}{2}}$

190   Chapter 4   Trigonometric Functions of a Real Variable

CHECK YOUR UNDERSTANDING 4.12

Use the results of Theorem 4.4 to establish the following half-angle formula for the tangent function:

$$\tan\frac{x}{2} = \pm\sqrt{\frac{1-\cos x}{1+\cos x}}$$

Answer: See page C-31.

EXAMPLE 4.7 Determine the value of $\sin\frac{x}{2}$, given that $\tan x = -\frac{3}{4}$ and $\frac{3\pi}{2} < x < 2\pi$.

SOLUTION: Dividing $\frac{3\pi}{2} < x < 2\pi$ by 2 gives:

$$\frac{3\pi}{4} < \frac{x}{2} < \pi$$

which shows that the angle with radian measure $\frac{x}{2}$ lies in the second quadrant. This being the case, we choose the positive sign in the identity for $\sin\frac{x}{2}$:

$$\sin\frac{x}{2} = +\sqrt{\frac{1-\cos x}{2}}$$

All that remains is to determine the value of $\cos x$.

The given information, leads us to the adjacent reference triangle in QIV, where θ is the angle of radian measure x. Since the cosine is positive in the fourth quadrant, $\cos x = \cos\theta = \cos\theta_r$, and so:

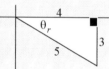

$$\sin\frac{x}{2} = \sqrt{\frac{1-\cos x}{2}} = \sqrt{\frac{1-\cos\theta_r}{2}} = \sqrt{\frac{1-\frac{4}{5}}{2}} = \sqrt{\frac{5-4}{10}} = \frac{1}{\sqrt{10}}$$

CHECK YOUR UNDERSTANDING 4.13

Given that $\cot x = -\frac{24}{7}$ and $\frac{3\pi}{2} \le x \le 2\pi$, determine the value of $\sec\frac{x}{2}$.

Answer: $-\frac{5\sqrt{2}}{7}$

4.2 Trigonometric Identities

VERIFYING IDENTITIES

One way to establish that an equation is an identity (called verifying the identity), is to start with one side of the equation and, using previously established identities, mold it into the other side (see Example 4.8 below). A less formal way is to rewrite each side of the equation separately, until a common expression is reached (see Example 4.9 below).

EXAMPLE 4.8 Verify the identity:
$$1 + \cos^2 x = \frac{1 - \cos^4 x}{\sin^2 x}$$

SOLUTION: Since the right side of the equation is more complex, we will work with it and transform it into the left side:

$$\frac{1 - \cos^4 x}{\sin^2 x} \stackrel{?}{=} \frac{(1 + \cos^2 x)(1 - \cos^2 x)}{\sin^2 x} \quad [a^2 - b^2 = (a+b)(a-b)]$$

$$= \frac{(1 + \cos^2 x)\sin^2 x}{\sin^2 x} = 1 + \cos^2 x \quad [\sin^2 x + \cos^2 x = 1]$$

EXAMPLE 4.9 Verify the identity:
$$\cot x \sin 2x = 1 + \cos 2x$$

SOLUTION: Since both sides of the equation involve a double angle, we worked with both sides until a common expression was reached. We began working on the left side, and stopped when we got to the form $2\cos^2 x$. We then worked on the right side and were able to arrive at the same expression:

$$\cot x \sin 2x \qquad\qquad 1 + \cos 2x$$
$$\left(\frac{\cos x}{\sin x}\right)(2\sin x \cos x) \quad\text{[Theorem 4.2(i)]} \qquad 1 + (2\cos^2 x - 1) \quad\text{[Theorem 4.2(ii)]}$$
$$2\cos^2 x \longleftarrow \qquad\qquad 2\cos^2 x$$

A more formal solution can be obtained by: starting at the top of one of the above columns work down to the end of that columnn cross over to the bottom of the other column and then work up to its top (see margin).

$\cot x \sin 2x$
$= \left(\frac{\cos x}{\sin x}\right)(2\sin x \cos x)$
$= 2\cos^2 x$
$= 1 + (2\cos^2 x - 1)$
$= 1 + \cos 2x$

EXAMPLE 4.10 Verify the identity:
$$\cos 2x = \frac{1 - \tan^2 x}{1 + \tan^2 x}$$

192 Chapter 4 Trigonometric Functions of a Real Variable

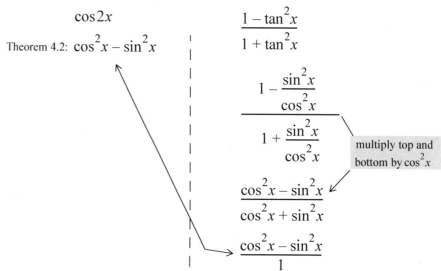

SOLUTION:

$$\cos 2x$$

Theorem 4.2: $\cos^2 x - \sin^2 x$

$$\dfrac{1 - \tan^2 x}{1 + \tan^2 x}$$

$$\dfrac{1 - \dfrac{\sin^2 x}{\cos^2 x}}{1 + \dfrac{\sin^2 x}{\cos^2 x}}$$ (multiply top and bottom by $\cos^2 x$)

$$\dfrac{\cos^2 x - \sin^2 x}{\cos^2 x + \sin^2 x}$$

$$\dfrac{\cos^2 x - \sin^2 x}{1}$$

If you prefer:

right side ↓

$$\dfrac{1 - \tan^2 x}{1 + \tan^2 x} = \dfrac{1 - \dfrac{\sin^2 x}{\cos^2 x}}{1 + \dfrac{\sin^2 x}{\cos^2 x}}$$

$$= \dfrac{\cos^2 x - \sin^2 x}{\cos^2 x + \sin^2 x}$$

$$= \cos^2 x - \sin^2 x$$

$$= \cos 2x$$

↑ left side

EXAMPLE 4.11 Verify the identity:

$$\dfrac{1 - \cos^4 x}{\sin x \cos x + \sin x \cos^3 x} = \dfrac{\sec x(1 + \cot x)}{\csc x} - 1$$

SOLUTION:

$$\dfrac{1 - \cos^4 x}{\sin x \cos x + \sin x \cos^3 x}$$

$$\dfrac{(1 + \cos^2 x)(1 - \cos^2 x)}{\sin x \cos x (1 + \cos^2 x)}$$

$$\dfrac{1 - \cos^2 x}{\sin x \cos x}$$

$$\dfrac{\sin^2 x}{\sin x \cos x}$$

$$\mathbf{\dfrac{\sin x}{\cos x}}$$

$$\dfrac{\sec x(1 + \cot x)}{\csc x} - 1$$

$$\dfrac{\dfrac{1}{\cos x}\left(1 + \dfrac{\cos x}{\sin x}\right)}{\dfrac{1}{\sin x}} - 1$$

$$\dfrac{\dfrac{1}{\cos x} + \dfrac{1}{\sin x}}{\dfrac{1}{\sin x}} - 1$$

$$\sin x\left(\dfrac{1}{\cos x} + \dfrac{1}{\sin x}\right) - 1$$

$$\dfrac{\sin x}{\cos x} + 1 - 1$$

$$\mathbf{\dfrac{\sin x}{\cos x}}$$

CHECK YOUR UNDERSTANDING 4.14

Verify the identity.

(a) $\dfrac{\csc x + \sec x}{1 + \tan x} = \csc x$

(b) $\dfrac{1 + \cos 2x}{\sin 2x} = \cot x$

Answers: See page C-32.

4.2 Trigonometric Identities   193

E X E R C I S E S

Exercises 1–3: Show, by means of a specific example that, in general,

1. $\sin(x+y) \neq \sin x + \sin y$ 2. $\cos(x+y) \neq \cos x + \cos y$ 3. $\tan(x+y) \neq \tan x + \tan y$

Exercises 4–6: Express in a form involving only $\sin x$, $\sin y$, $\cos x$, $\cos y$, and, where appropriate, $\sin z$ and $\cos z$.

4. $\sin(x+y+z)$ 5. $\sin(x-y)\sin(x+y)$ 6. $\cos(x+y+z)$

Exercises 7–8: (**Complementary Identities**) Show that the Complementary Identities hold for all x:

7. $\cos\left(\frac{\pi}{2} - x\right) = \sin x$

   [Suggestion:Apply Theorem 4.1(ii)]

8. $\sin\left(\frac{\pi}{2} - x\right) = \cos x$

   [Suggestion:Substitute $\frac{\pi}{2} - x$ for x
    in Exercise 7]

Exercises 9–10: Establish the Addition Identities, Theorem 4.1(iii).
[Suggestion: Apply the Complementary Identities (Exercises 7 and 8), and then apply
   Theorem 4.1(ii)].

9. $\sin(x+y) = \sin x \cos y + \cos x \sin y$ 10. $\sin(x-y) = \sin x \cos y - \cos x \sin y$

Exercise 11–14: Write the angle as a sum or difference of angles whose trigonometric functions can be evaluated exactly, and then find the exact value of the given expression. (Note that $\frac{\pi}{12} = 15°$.)

11. $\sin 75°$ 12. $\tan \frac{5\pi}{12}$ 13. $\tan \frac{7\pi}{12}$ 14. $\cos 255°$

Exercises 15–18: Evaluate $\sin(\alpha - \beta)$, $\cos(\alpha - \beta)$, $\tan(\alpha - \beta)$, given

15. $\cos \alpha = \frac{12}{13}$, $\cos \beta = -\frac{3}{5}$, $\frac{3\pi}{2} < \alpha < 2\pi$, $\pi < \beta < \frac{3\pi}{2}$

16. $\sin \alpha = \frac{8}{17}$, $\sin \beta = \frac{5}{13}$, $0 < \alpha < \frac{\pi}{2}$, $0 < \beta < \frac{\pi}{2}$

17. $\tan \alpha = \frac{3}{4}$, $\cot \beta = -\frac{2}{3}$, $\pi < \alpha < \frac{3\pi}{2}$, $\frac{\pi}{2} < \beta < \pi$

18. $\sin \alpha = -\frac{7}{25}$, $\csc \beta = -\frac{2}{\sqrt{3}}$, $\pi < \alpha < \frac{3\pi}{2}$, $\frac{3\pi}{2} < \beta < 2\pi$

Exercises 19–22: Evaluate $\sin 2x$, $\cos 2x$, $\tan 2x$, given

19. $\sin x = \frac{3}{5}$,   and $\frac{\pi}{2} < x < \pi$ 20. $\cos x = -\frac{12}{13}$,   and $\pi < x < \frac{3\pi}{2}$

21. $\cot x = \frac{3}{5}$,   and $\sec x > 0$ 22. $\csc x = -\frac{5}{4}$,   and $\tan x > 0$

194   Chapter 4 Trigonometric Functions of a Real Variable

Exercises 23–28: Rewrite each expression as a single trigonometric function of an angle.

23. $\cos 31° \cos 48° - \sin 31° \sin 48°$

24. $\sin 75° \cos 28° - \cos 75° \sin 28°$

25. $2 \sin 75° \cos 75°$

26. $1 - 2 \sin^2 37°$

27. $\sin 87° \cos 87°$

28. $2 \cos^2 151° - 1$

Exercises 29–33: Use trigonometric identities to simplify the expression.

29. $\dfrac{(\sin x - \cos x)^2 - 1}{\sin 2x}$

30. $\dfrac{\sin^2 x - \cos^2 x}{2 \cos 2x}$

31. $\dfrac{\tan x + \cot x}{\csc 2x}$

32. $\cos^3 x \sin x + \sin^3 x \cos x$

33. $(\sec x + \tan x)^2 (\sec x - \tan x)$

Exercises 34–36: Use trigonometric identities to simplify the expression, ultimately writing it in terms of $\tan x$ only.

34. $\dfrac{\sin^2 x}{\sin 2x}$

35. $\dfrac{\sin^2 \frac{x}{2} \cos^2 \frac{x}{2}}{1 + \cos 2x}$

36. $\dfrac{\tan 2x \, \tan^3 x}{\sec^2 x}$

Exercises 37–39: Express each of the following in terms only of first powers of cosine functions.

37. $\sin^4 x$

38. $\sin^2 x \cos^2 y - \cos^2 x \sin^2 y$

39. $1 + \tan^2 2x$

Exercise 40: Express $\sin 3x$ in terms only of $\sin x$. [Suggestion: $3x = x + 2x$.]

Exercises 41–42: Express each of the following in terms only of $\cos x$. [Suggestion: $3x = x + 2x$, $4x = 2(2x)$.]

41. $\cos 4x$

42. $\cos 3x$

Exercise 43: Use trigonometric identities to simplify the expression.

$$\cos^4 x - \frac{1}{8} \cos 4x - \frac{1}{2} \cos 2x$$

Exercise 44: Determine the exact value of $\sin 15°$ by a half-angle identity. Compare your answer with that obtained in Example 4.3.

Exercise 45: Find the exact value of $\sin \frac{\pi}{8}$, by a half-angle identity.

Exercises 46–48: Evaluate the six trigonometric functions of x using the half-angle or double-angle identities, given

46. $\sin 2x = \frac{4}{5}$, and $0 < x < \frac{\pi}{4}$

47. $\cos 2x = -\frac{3}{4}$, and $\pi < x < \frac{3\pi}{2}$

48. $\sin \frac{x}{2} = \frac{4}{5}$, and $\frac{\pi}{2} < x < \pi$

Exercises 49–52: Evaluate the six trigonometric functions of $\frac{x}{2}$, given

49. $\sin x = \frac{4}{5}$, and $\frac{\pi}{2} < x < \pi$

50. $\cot x = \frac{2}{\sqrt{5}}$, and $-\pi < x < -\frac{\pi}{2}$

51. $\csc x = -\frac{5}{3}$, and $-\frac{\pi}{2} < x < 0$

52. $\sec x = 4$, and $0 < x < \frac{\pi}{2}$

4.2 Trigonometric Identities   **195**

Exercises 53–56:   (**Product-to-Sum Identities**) These identities express products of sine and cosine functions in terms of sums or differences of sine and cosine functions:

(i) $\sin x \cos y = \frac{1}{2}[\sin(x+y) + \sin(x-y)]$,   (ii) $\cos x \sin y = \frac{1}{2}[\sin(x+y) - \sin(x-y)]$,

(iii) $\cos x \cos y = \frac{1}{2}[\cos(x+y) + \cos(x-y)]$,   (iv) $\sin x \sin y = \frac{1}{2}[\cos(x-y) - \cos(x+y)]$.

   Express as a sum of sine and/or cosine functions.

53. $\cos 5x \cos 4x$ 54. $\sin 2x \sin 10x$ 55. $\sin(4x-1)\cos(3x+2)$ 56. $\cos \frac{2x}{3} \sin \frac{x}{3}$

Exercise 57:   Establish the Product-to-Sum Identity of Exercises 53–56, by applying Theorem 4.1 (ii) or (iii) to the right side of the equation.

(a) (i)   (b) (ii)   (c) (iii)   (d) (iv)

Exercise 58–61:   (**Sum-to-Product Identities**) These identities express sums and differences of sine and cosine functions in terms of products of sine and cosine functions:

(i) $\sin x + \sin y = 2\sin\left(\frac{x+y}{2}\right)\cos\left(\frac{x-y}{2}\right)$,   (ii) $\sin x - \sin y = 2\cos\left(\frac{x+y}{2}\right)\sin\left(\frac{x-y}{2}\right)$,

(iii) $\cos x + \cos y = 2\cos\left(\frac{x+y}{2}\right)\cos\left(\frac{x-y}{2}\right)$,   (iv) $\cos x - \cos y = -2\sin\left(\frac{x+y}{2}\right)\sin\left(\frac{x-y}{2}\right)$.

   Express as a product of sine and/or cosine functions.

58. $\sin 5x + \sin 8x$ 59. $\cos 6x - \cos 2x$ 60. $\sin 3x^2 - \sin 7x^2$ 61. $\cos 9x + \cos 6x$

Exercise 62:   Establish the Sum-to-Product Identity of Exercises 58–61, by reversing the appropriate equation of Exercises 53–56 and substituting $\frac{x+y}{2}$ for x and $\frac{x-y}{2}$ for y.

(a) (i)   (b) (ii)   (c) (iii)   (d) (iv)

Exercises 63–66:   Make the substitution, and use an identity to simplify the resulting expression. Assume $0 \le \theta < \frac{\pi}{2}$ and $a > 0$.

63. $\sqrt{x^2 - a^2}$,   let $x = a \sec \theta$ 64. $\sqrt{a^2 + x^2}$,   let $x = a \tan \theta$

65. $\sqrt{a^2 - x^2}$,   let $x = a \sin \theta$ 66. $\sqrt{a^2 + 4x^2}$,   let $x = \frac{1}{2} a \tan \theta$

Exercises GC67–GC69:   Express each function as either $f(x) = a\sin(x+c)$ or $f(x) = a\cos(x+c)$. Then use a graphing calculator to graph both versions of the function and check your answer.

67. $f(x) = \cos x - \sin x$ 68. $f(x) = -\sin x + \sqrt{3}\cos x$ 69. $f(x) = \cos x + \frac{\sqrt{3}}{3}\sin x$

Exercise 70:   Show that every expression of the form $a\sin x \pm b\cos x$ can be written in the form $c\sin(x \pm d)$, for some constants c and d depending only on a and b.

Exercises 71–109:   Verify the identity.

71. $(1 - \cos x)(1 + \sec x)\cot x = \sin x$ 72. $1 - \dfrac{\cos^2 x}{1 + \sin x} = \sin x$

73. $\dfrac{\sin x + \tan x}{\cot x + \csc x} = \sin x \tan x$ 74. $\dfrac{\sin x \cos x}{\cos^2 x - \sin^2 x} = \dfrac{\tan x}{1 - \tan^2 x}$

196   Chapter 4 Trigonometric Functions of a Real Variable

75. $\sin x = 2\sin\frac{x}{2}\cos\frac{x}{2}$

76. $\tan^2 x \csc x = \dfrac{\sin x}{1-\sin^2 x}$

77. $\sin x \tan x = \sec x - \cos x$

78. $\sin\left(\frac{\pi}{6}+x\right)+\cos\left(\frac{\pi}{3}+x\right)=\cos x$

79. $\cos\left(x+\frac{\pi}{4}\right)=-\sin\left(x-\frac{\pi}{4}\right)$

80. $\sin\left(\frac{\pi}{4}+x\right)-\sin\left(\frac{\pi}{4}-x\right)=\sqrt{2}\sin x$

81. $\dfrac{\sec x - \csc x}{\tan x + \cot x}=\dfrac{\tan x - \cot x}{\sec x + \csc x}$

82. $\cos 6x = 1 - 2\sin^2 3x$

83. $\sin^2\frac{1}{2}x = \frac{1}{2}(1-\cos x)$

84. $\cos^2\frac{1}{2}x = \frac{1}{2}(1+\cos x)$

85. $\sin x = \pm\sqrt{\dfrac{1-\cos 2x}{2}}$

86. $\sin(x+y)\sin(x-y)=\sin^2 x - \sin^2 y$

87. $\cos 2x = \cos^4 x - \sin^4 x$

88. $\dfrac{\sin x}{1+\cos x}+\dfrac{1+\cos x}{\sin x}=2\csc x$

89. $\dfrac{\sin x}{1+\cos x}=\dfrac{1-\cos x}{\sin x}$

90. $1+\cos 2x = (\cot^2 x)(1-\cos 2x)$

91. $2\csc 2x = \tan x + \cot x$

92. $\cot^2 x - \cos^2 x = \cot^2 x \cos^2 x$

93. $\dfrac{1+\cot x}{\csc x}=\dfrac{1+\tan x}{\sec x}$

94. $\dfrac{1-\cot^2 x}{1+\cot^2 x}=\sin^2 x - \cos^2 x$

95. $\dfrac{\tan x}{\tan^2 x - 1}=\dfrac{1}{\tan x - \cot x}$

96. $\sin 3x = 3\sin x - 4\sin^3 x$

97. $\dfrac{\csc x - \sec x}{\csc x + \sec x}=\dfrac{\cos 2x}{1+\sin 2x}$

98. $\tan x \sin 2x = 2\sin^2 x$

99. $\dfrac{\sin 4x}{1+\cos 2x}\cdot\dfrac{\cos 2x}{1+\cos 4x}=\tan x$

100. $\dfrac{\cos x + \cos y}{\sin x - \sin y}+\dfrac{\sin x + \sin y}{\cos x - \cos y}=0$

101. $\dfrac{\tan x \tan y + 1}{1-\tan x \tan y}=\dfrac{\cos(x-y)}{\cos(x+y)}$

102. $\dfrac{\cos^3 x + \sin^3 x}{\cos x + \sin x}=1-\frac{1}{2}\sin 2x$

103. $\dfrac{\sin(x-y)}{\sin x \sin y}=\cot y - \cot x$

104. $\dfrac{2\tan x}{1+\tan^2 x}=\sin 2x$

105. $\dfrac{1+\sin 2x + \cos 2x}{1+\sin 2x - \cos 2x}=\cot x$

106. $\cos x \sin(x+y)=\sin x \cos(x+y)+\sin y$

107. $\cos^2 x - \cos x \cos\left(\frac{\pi}{3}+x\right)+\sin^2\left(\frac{\pi}{6}-x\right)=\frac{3}{4}$

108. $\dfrac{\sin(x+h)-\sin x}{h}=(\sin x)\left(\dfrac{\cos h - 1}{h}\right)+(\cos x)\left(\dfrac{\sin h}{h}\right)$

109. $\dfrac{\cos(x+h)-\cos x}{h}=(\cos x)\left(\dfrac{\cos h - 1}{h}\right)-(\sin x)\left(\dfrac{\sin h}{h}\right)$

4.2 Trigonometric Identities   **197**

Exercise 110:   Show that the equation $\dfrac{1}{(1 - \tan x)^2} = \dfrac{1}{\sec^2 x - 2\tan x}$ is an identity, and specify the values of x for which equality holds.

Exercises 111–112:   Show that the equation is **not** an identity by exhibiting a specific value of x for which the two sides are unequal.

111. $\dfrac{\cos x + \sin x}{\cos x \sin x} = \dfrac{1}{\sin x + \cos x}$

112. $\sec x + \sec 2x = \sec 3x$

Exercises 113–114:   Prove that if α, β, and γ are angles in a triangle, then

113. $\sin \alpha \cos \beta + \cos \alpha \sin \beta = \sin \gamma$

114. $\cos \alpha \cos \beta - \sin \alpha \sin \beta = -\cos \gamma$

Exercises GC115–GC118:   Use a graphing calculator to help you decide whether or not the equation is an identity. Justify your claim, by verifying the identity, or by exhibiting a value of x for which equality does not hold.

115. $\sqrt{1 - \sin^2 \frac{x}{10}} = \cos \frac{x}{10}$

116. $\tan x - \cot x = \dfrac{1}{\sin x \cos x}$

117. $\dfrac{1 - \cos x}{1 + \cos x} = \left(\dfrac{\sin x}{1 + \cos x} \right)^2$

118. $\dfrac{\sin x}{\cos x - \sin x} = \dfrac{11 \tan x}{10 - 11 \tan x}$

Exercises 119–121:   Establish the identity, and interpret it geometrically. For example, the geometric interpretation of the identity in Exercise 119 is that shifting the graph of the sine function $\frac{\pi}{2}$ units to the left yields the graph of the cosine function.

119. $\sin \left(x + \frac{\pi}{2} \right) = \cos x$

120. $2 \cos \left(x - \frac{\pi}{4} \right) = \sqrt{2}(\cos x + \sin x)$

121. $\cos \left(x + \frac{\pi}{2} \right) = -\sin x$

§3. INVERSE TRIGONOMETRIC FUNCTIONS

Since trigonometric functions are not one-to-one, they do not have inverses (see page 107). We can, however, restrict the domain of each trigonometric function to an interval on which it is one-to-one, and then consider the inverse of the resulting **restricted** function. Convention dictates that the interval chosen is the largest interval containing 0 and some positive values, on which the function is one-to-one.

Restricting the domain of a function to produce a one-to-one function is not new to us. Indeed, by restricting the non-one-to-one function $f(x) = x^2$ to $x \geq 0$ we obtained a one-to-one function $g(x)$ (see figure), with inverse $g^{-1}(x) = \sqrt{x}$.

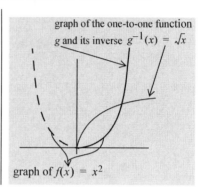

graph of the one-to-one function g and its inverse $g^{-1}(x) = \sqrt{x}$

graph of $f(x) = x^2$

THE INVERSE SINE FUNCTION

Restricting the sine function to the interval $\left[-\frac{\pi}{2}, \frac{\pi}{2}\right]$ produces a one-to-one function [see Figure 4.7(a)]. The inverse of this restricted function is called the **inverse sine** function, and is denoted by $\sin^{-1}x$. Reflecting the graph of the restricted sine function about the line $y = x$ yields the graph of the inverse sine function in Figure 4.7(b) (see Theorem 2.3, page 109).

> In spite of its notation and name, $\sin^{-1}x$ is **not** the inverse of the sine function. It can't be, since the sine function, not being one-to-one, has no inverse. $\sin^{-1}x$ is the inverse of the sine function **restricted** to the interval $\left[-\frac{\pi}{2}, \frac{\pi}{2}\right]$. We note that the inverse sine function is also called the **arcsine** function, and can be denoted by Arcsin x. These alternative choices are now essentially extinct, thanks to graphing calculators which have a $\sin^{-1}$ button.

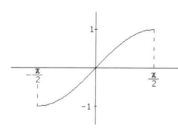

$y = \sin x$
Restricted to $\left[-\frac{\pi}{2}, \frac{\pi}{2}\right]$
(a)

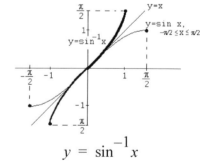

$y = \sin^{-1}x$
Domain: $[-1, 1]$, Range: $\left[-\frac{\pi}{2}, \frac{\pi}{2}\right]$
(b)

Figure 4.7

In summary:

DEFINITION 4.2
INVERSE SINE FUNCTION

The inverse sine function, denoted by $\sin^{-1}x$, has domain $[-1, 1]$ and is given by:

$$y = \sin^{-1}x \text{ if } \sin y = x \text{ with } -\frac{\pi}{2} \leq y \leq \frac{\pi}{2}.$$

Condition (2):

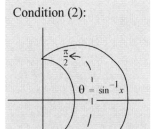

Thinking in terms of angles (which we highly recommend that you do) we have:

For $-1 \leq x \leq 1$, $\sin^{-1} x$ is that angle θ such that: (*)
(1) $\sin\theta = x$ and (2) $-\dfrac{\pi}{2} \leq \theta \leq \dfrac{\pi}{2}$

From condition (1), it follows that for any $-1 \leq x \leq 1$ whatsoever,

$$\sin(\sin^{-1} x) = x$$

If you take the sine of the angle whose sine is x, you will certainly get x.

For example:

$$\sin\left(\sin^{-1}\frac{2}{7}\right) = \frac{2}{7} \quad \text{and} \quad \sin\left[\sin^{-1}\left(-\frac{3}{5}\right)\right] = -\frac{3}{5}$$

You **CANNOT** for example have: $\sin^{-1}\left[\sin\dfrac{7\pi}{5}\right] = \dfrac{7\pi}{5}$ since $\dfrac{7\pi}{5}$ is **NOT** in $\left[-\dfrac{\pi}{2}, \dfrac{\pi}{2}\right]$.

An analogous situation: You **CANNOT** have:.

$$\sqrt{(-4)^2} = -4$$

since -4 is **NOT** in $[0, \infty)$

On the other hand, $\sin^{-1}(\sin\alpha)$ **need not** equal α. The reason is that $\sin^{-1} x$ is not just any angle whose sine is x, it is that angle **contained in the interval** $\left[-\dfrac{\pi}{2}, \dfrac{\pi}{2}\right]$ whose sine is x.

To determine $\sin^{-1}(\sin\alpha)$ for any angle α, we first note that $\sin\alpha$ is the "x" in (*) above. It follows that we are to find that angle $-\dfrac{\pi}{2} \leq \theta \leq \dfrac{\pi}{2}$ for which $\sin\theta = \sin\alpha$; for then:

$$\sin^{-1}[\sin\alpha] = \sin^{-1}[\sin\theta] = \theta.$$

For example, to find $\sin^{-1}\left[\sin\left(\dfrac{7\pi}{5}\right)\right]$ we seek $-\dfrac{\pi}{2} \leq \theta \leq \dfrac{\pi}{2}$ such that $\sin\theta = \sin\left(\dfrac{7\pi}{5}\right)$. As $\sin\left(\dfrac{7\pi}{5}\right) = -\sin\left(\dfrac{2\pi}{5}\right)$ is negative, θ lies in the fourth quadrant with a reference angle of $\dfrac{2\pi}{5}$. Consequently $\theta = -\dfrac{2\pi}{5}$ (see margin). Hence:

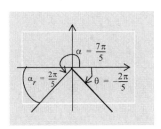

$$\sin^{-1}\left[\sin\left(\frac{7\pi}{5}\right)\right] = \sin^{-1}\left[\sin\left(-\frac{2\pi}{5}\right)\right] = -\frac{2\pi}{5}$$

CHECK YOUR UNDERSTANDING 4.15

Evaluate:

(a) $\sin^{-1}\left(\sin\dfrac{5\pi}{8}\right)$ (b) $\sin\left(\sin^{-1}\dfrac{5}{9}\right)$

Answers: (a) $\dfrac{3\pi}{8}$ (b) $\dfrac{5}{9}$

EXAMPLE 4.12 Evaluate:

$$\sin^{-1}\frac{\sqrt{3}}{2} \quad \text{and} \quad \sin^{-1}\left(-\frac{\sqrt{3}}{2}\right)$$

SOLUTION: In the first instance, we want the angle θ **in the interval** $\left[-\frac{\pi}{2}, \frac{\pi}{2}\right]$ whose sine is $\frac{\sqrt{3}}{2}$, and in the second instance we want the angle θ in the interval $\left[-\frac{\pi}{2}, \frac{\pi}{2}\right]$ whose sine is $-\frac{\sqrt{3}}{2}$.

In either case, $\sin\theta_r = \frac{\sqrt{3}}{2}$, and we are dealing with a reference angle of radian measure $\frac{\pi}{3}$. Since the sine function is positive in the first quadrant and negative in the fourth, we conclude that:

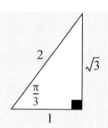

Note it would be **WRONG** to say that:

$$\sin^{-1}\left(-\frac{\sqrt{3}}{2}\right) = \frac{5\pi}{3}$$

$$\sin^{-1}\frac{\sqrt{3}}{2} = \frac{\pi}{3} \quad \text{and}$$

$$\sin^{-1}\left(-\frac{\sqrt{3}}{2}\right) = -\frac{\pi}{3}$$

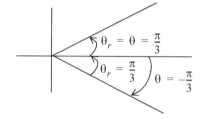

EXAMPLE 4.13 Evaluate:

$$\cos\left[\sin^{-1}\left(-\frac{3}{5}\right)\right] \quad \text{and} \quad \tan\left[\sin^{-1}\left(-\frac{3}{5}\right)\right]$$

SOLUTION: Let $\theta = \sin^{-1}\left(-\frac{3}{5}\right)$. Then $\sin\theta = -\frac{3}{5}$ and $-\frac{\pi}{2} \leq \theta \leq \frac{\pi}{2}$, so that θ lies in the fourth quadrant. From the adjacent reference triangle for θ_r, we see that $|\cos\theta| = \cos\theta_r = \frac{4}{5}$ and $|\tan\theta| = \tan\theta_r = \frac{3}{4}$. Since in the fourth quadrant the cosine function is positive and the tangent function is negative, we have:

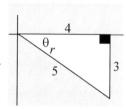

$$\cos\left[\sin^{-1}\left(-\frac{3}{5}\right)\right] = \cos\theta = \frac{4}{5} \quad \text{and} \quad \tan\left[\sin^{-1}\left(-\frac{3}{5}\right)\right] = \tan\theta = -\frac{3}{4}$$

4.3 Inverse Trigonometric Functions   201

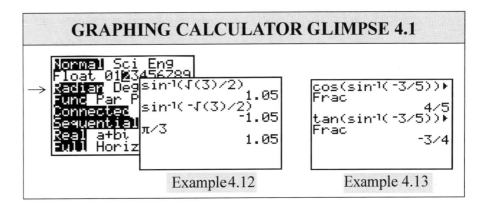

Example 4.12   Example 4.13

CHECK YOUR UNDERSTANDING 4.16

Evaluate:

Answers: (a) $-\frac{\pi}{6}$ (b) $-\frac{\sqrt{5}}{2}$

(a) $\sin^{-1}\left(-\frac{1}{2}\right)$ (b) $\cot\left[\sin^{-1}\left(-\frac{2}{3}\right)\right]$

THE INVERSE COSINE FUNCTION

Restricting the cosine function to the interval $[0, \pi]$ produces a one-to-one function [see Figure 4.8(a)]. The inverse of this restricted function is called the **inverse cosine** function, and is denoted by $\cos^{-1} x$. Reflecting the graph of the restricted cosine function about the line $y = x$ yields the graph of the inverse cosine function in Figure 4.8(b).

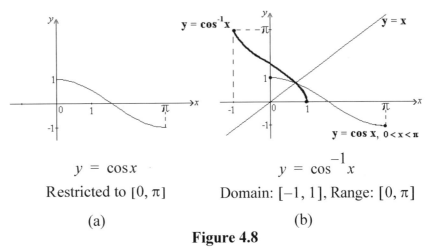

$y = \cos x$ $y = \cos^{-1} x$
Restricted to $[0, \pi]$ Domain: $[-1, 1]$, Range: $[0, \pi]$

(a)   (b)

Figure 4.8

In summary:

DEFINITION 4.3

INVERSE COSINE FUNCTION

The inverse cosine function, denoted by $\cos^{-1} x$, has domain $[-1, 1]$ and is given by:

$y = \cos^{-1} x$ if $\cos y = x$ with $0 \le y \le \pi$.

202 Chapter 4 Trigonometric Functions of a Real Variable

Condition (2):

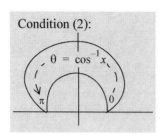

Thinking in terms of angles we have:

For $-1 \leq x \leq 1$, $\cos^{-1}x$ is that angle θ such that: (*)

(1) $\cos\theta = x$ and (2) $0 \leq \theta \leq \pi$

As was the case with the inverse sine function, while $\cos(\cos^{-1}x) = x$ holds for any $-1 \leq x \leq 1$, $\cos^{-1}(\cos\alpha)$ **need not equal** α. Why not? Because $\cos\alpha$ is playing the role of "x" in (*), and $\cos^{-1}(\cos\alpha)$ must be that angle $0 \leq \theta \leq \pi$ for which $\cos\theta = \cos\alpha$.

For example, to find $\cos^{-1}\left[\cos\left(\frac{7\pi}{5}\right)\right]$ we seek $0 \leq \theta \leq \pi$ such that

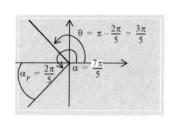

$\cos\theta = \cos\left(\frac{7\pi}{5}\right)$. That angle is $\theta = \frac{3\pi}{5}$ (see margin). Hence:

$$\cos^{-1}\left[\cos\left(\frac{7\pi}{5}\right)\right] = \cos^{-1}\left[\cos\left(\frac{3\pi}{5}\right)\right] = \frac{3\pi}{5}$$

CHECK YOUR UNDERSTANDING 4.17

Evaluate:

(a) $\cos^{-1}\left[\cos\left(\frac{12\pi}{7}\right)\right]$ (b) $\cos\left[\cos^{-1}\left(-\frac{1}{\pi}\right)\right]$

Answers: (a) $\frac{2\pi}{7}$ (b) $-\frac{1}{\pi}$

EXAMPLE 4.14 Evaluate:

$$\cos^{-1}\frac{1}{\sqrt{2}} \quad \text{and} \quad \cos^{-1}\left(-\frac{1}{\sqrt{2}}\right)$$

SOLUTION: In the first instance, we want the angle θ **in the interval** $[0, \pi]$ whose cosine is $\frac{1}{\sqrt{2}}$, and in the second instance we want the angle θ in the interval $[0, \pi]$ whose cosine is $-\frac{1}{\sqrt{2}}$.

In either case, $\cos\theta_r = \frac{1}{\sqrt{2}}$, and we are dealing with a reference angle of radian measure $\frac{\pi}{4}$. Since the cosine function is positive in the first quadrant and negative in the second, we conclude that:

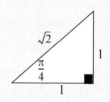

$$\cos^{-1}\left(\frac{1}{\sqrt{2}}\right) = \frac{\pi}{4} \quad \text{and} \quad \cos^{-1}\left(-\frac{1}{\sqrt{2}}\right) = \pi - \frac{\pi}{4} = \frac{3\pi}{4}$$

4.3 Inverse Trigonometric Functions

EXAMPLE 4.15 Evaluate:
$$\sin\left[\cos^{-1}\left(-\frac{2}{3}\right) + \sin^{-1}\frac{1}{5}\right]$$

SOLUTION: Let:
$$\alpha = \cos^{-1}\left(-\frac{2}{3}\right) \quad \text{and} \quad \beta = \sin^{-1}\frac{1}{5} \qquad (*)$$

Then $\cos\alpha = -\frac{2}{3}$, and $\cos\alpha_r = \frac{2}{3}$ with α_r in the second quadrant ($\cos\alpha < 0$). Similarly, $\sin\beta_r = \frac{1}{5}$ with β_r in the first quadrant ($\sin\beta > 0$):

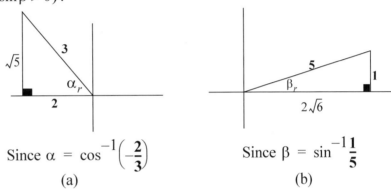

Since $\alpha = \cos^{-1}\left(-\frac{2}{3}\right)$ Since $\beta = \sin^{-1}\frac{1}{5}$
(a)   (b)

Figure 4.9

We then have:

$$\sin\left[\cos^{-1}\left(-\frac{2}{3}\right) + \sin^{-1}\frac{1}{5}\right] = \sin(\alpha + \beta) = \sin\alpha\cos\beta + \cos\alpha\sin\beta$$

(Theorem 4.1(iii), page 184; Figure 4.9(a): $\sin\alpha = \frac{\sqrt{5}}{3}$, $\cos\alpha = -\frac{2}{3}$; Figure 4.9(b): $\cos\beta = \frac{2\sqrt{6}}{5}$, $\sin\beta = \frac{1}{5}$)

Which is to say:

$$\sin\left[\cos^{-1}\left(-\frac{2}{3}\right) + \sin^{-1}\frac{1}{5}\right] = \frac{\sqrt{5}}{3} \cdot \frac{2\sqrt{6}}{5} - \frac{2}{3} \cdot \frac{1}{5} = \frac{2\sqrt{30} - 2}{15}$$

CHECK YOUR UNDERSTANDING 4.18

Evaluate:
(a) $\cos^{-1}\left(-\frac{1}{2}\right)$ (b) $\cos\left[2\cos^{-1}\left(-\frac{4}{5}\right)\right]$

[Suggestion: Use a double angle formula (Theorem 4.2, page 187)]

Answers: (a) $\frac{2\pi}{3}$ (b) $\frac{7}{25}$

THE INVERSE TANGENT FUNCTION

Restricting the tangent function to the open interval $\left(-\frac{\pi}{2}, \frac{\pi}{2}\right)$ produces a one-to-one function [see Figure 4.10(a)]. The inverse of this restricted function is called the **inverse tangent** function, and is denoted by $\tan^{-1} x$. Reflecting the graph of the restricted tangent function about the line $y = x$ yields the graph of the inverse tangent function in Figure 4.10(b).

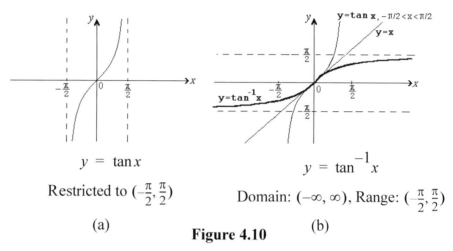

$y = \tan x$
Restricted to $(-\frac{\pi}{2}, \frac{\pi}{2})$
(a)

$y = \tan^{-1} x$
Domain: $(-\infty, \infty)$, Range: $(-\frac{\pi}{2}, \frac{\pi}{2})$
(b)

Figure 4.10

In summary:

DEFINITION 4.4
INVERSE TANGENT FUNCTION

The inverse tangent function, denoted by $\tan^{-1} x$, has domain $(-\infty, \infty)$ and is given by:

$$y = \tan^{-1} x \text{ if } \tan y = x \text{ with } -\frac{\pi}{2} \leq y \leq \frac{\pi}{2}.$$

Thinking in terms of angles we have:

For any x, $\tan^{-1} x$ is that angle θ such that:
(1) $\tan\theta = x$ and (2) $-\frac{\pi}{2} < \theta < \frac{\pi}{2}$

Condition (2):

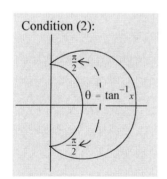

Repeating a common theme, we note that do to the restriction on the range of the inverse tangent function, we find that $\tan^{-1}(\tan x)$ will only equal x if x is in the interval $\left(-\frac{\pi}{2}, \frac{\pi}{2}\right)$, while $\tan(\tan^{-1} x) = x$ for all x.

CHECK YOUR UNDERSTANDING 4.19

Evaluate:

(a) $\tan^{-1}\left(\tan\frac{5\pi}{3}\right)$ (b) $\tan(\tan^{-1} 500)$

Answers: (a) $-\frac{\pi}{3}$ (b) 500

EXAMPLE 4.16
Evaluate:
$$\sin\left[2\tan^{-1}\left(-\frac{3}{4}\right)\right]$$

SOLUTION:

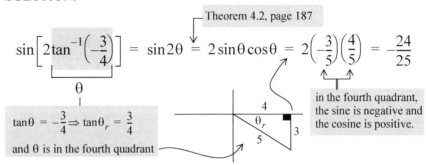

EXAMPLE 4.17
Express:
$$\sin(\tan^{-1}x)$$
as an algebraic function of x.

SOLUTION: Let $\theta = \tan^{-1}x$. Then:
$$\tan\theta = x, \text{ and } -\frac{\pi}{2} < \theta < \frac{\pi}{2}.$$

If $x > 0$, then $x = \tan\theta > 0$, and θ lies in the first quadrant [Figure 4.11(a)]. We then have:
$$\sin\theta = \frac{x}{\sqrt{1+x^2}}$$

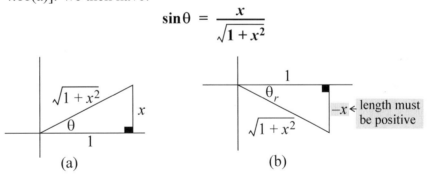

Figure 4.11

If $x < 0$, then $x = \tan\theta < 0$, and θ lies in the fourth quadrant [Figure 4.11(b)]. We have:
$$\sin\theta = -\sin\theta_r = -\frac{-x}{\sqrt{1+x^2}} = \frac{x}{\sqrt{1+x^2}}$$

Finally, if $x = 0$, then $x = \tan\theta = 0$, and $\theta = 0$ (remember that $-\frac{\pi}{2} < \theta < \frac{\pi}{2}$). Once more, we have:
$$\sin\theta = \frac{x}{\sqrt{1+x^2}} \quad \text{(both sides equal 0)}$$

Having exhausted all possible cases, we conclude that for any x:

$$\sin(\tan^{-1}x) = \frac{x}{\sqrt{1+x^2}}$$

CHECK YOUR UNDERSTANDING 4.20

Answer: $\dfrac{2}{\sqrt{x^2+4}}$

Express $\cos\left(\tan^{-1}\dfrac{x}{2}\right)$ as an algebraic function of x.

The inverse cosecant, inverse secant, and inverse cotangent functions are addressed in the exercises.

We began this section by developing the inverse sine function, and then mimicked the process for the inverse cosine and inverse tangent functions. The adjacent figure recalls the range of those inverse trigonometric functions.

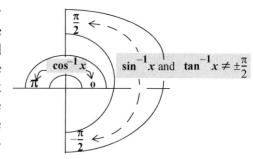

4.3 Inverse Trigonometric Functions   **207**

EXERCISES

Exercises 1–27: Find the exact value of each of the following.

1. $\sin^{-1}(-1)$

2. $\sin^{-1}\frac{\sqrt{2}}{2} - \sin^{-1}\frac{1}{2}$

3. $\sin\left(\sin^{-1}\frac{1}{2}\right)$

4. $\sin^{-1}\left(\sin\frac{3\pi}{5}\right)$

5. $\sin^{-1}\left(\sin\frac{3\pi}{10}\right)$

6. $\cot\left[\sin^{-1}\left(-\frac{\sqrt{2}}{2}\right)\right]$

7. $\cos\left[\sin^{-1}\left(-\frac{12}{13}\right)\right]$

8. $\tan\left(\sin^{-1}\frac{7}{12}\right)$

9. $\cos\left(2\sin^{-1}\frac{3}{5}\right)$

10. $\sin^{-1}\left[\cos\left(\frac{7\pi}{12}\right)\right]$

11. $\sin(\sin^{-1}2)$

12. $\cos^{-1}\frac{1}{2}$

13. $\cos\left(\cos^{-1}\frac{5}{9}\right)$

14. $\cos[\cos^{-1}\sqrt{3})]$

15. $\cos^{-1}\left[\cos\left(-\frac{21\pi}{8}\right)\right]$

16. $\cos^{-1}\left(\cos\frac{3\pi}{5}\right)$

17. $\sin\left[\cos^{-1}\left(-\frac{2}{3}\right)\right]$

18. $\csc\left(\cos^{-1}\frac{3}{5}\right)$

19. $\cos\left(2\cos^{-1}\frac{1}{4}\right)$

20. $\tan^{-1}\sqrt{3}$

21. $\tan^{-1}\left(-\frac{\sqrt{3}}{3}\right)$

22. $\tan^{-1}\left(\tan\frac{5\pi}{6}\right)$

23. $\tan^{-1}\left[\tan\left(-\frac{\pi}{9}\right)\right]$

24. $\tan(\tan^{-1}74)$

25. $\sec\left[\tan^{-1}\left(-\frac{3}{4}\right)\right]$

26. $\sin(\tan^{-1}2)$

27. $\sin(2\tan^{-1}3)$

Exercises GC1–GC27: Use a graphing calculator to evaluate the expressions in Exercises 1–27 as in Graphing Calculator Glimpse 4.1.

Exercises 28–31: Find the exact value of the given expression.

28. $\cos\left[\sin^{-1}\left(-\frac{2}{3}\right) + \cos^{-1}\frac{1}{3}\right]$

29. $\sin\left[\cos^{-1}\left(-\frac{5}{13}\right) + \tan^{-1}\frac{2}{3}\right]$

30. $\tan\left(\frac{1}{2}\sin^{-1}\frac{4}{5}\right)$

31. $\tan\left(\sin^{-1}\frac{4}{5} - \tan^{-1}2\right)$

Exercises 32–33: Show that

32. $\sin\left(\sin^{-1}\frac{5}{13} + \sin^{-1}\frac{4}{5}\right) = \frac{63}{65}$

33. $\cos\left[\tan^{-1}\frac{12}{5} - \sin^{-1}\left(-\frac{1}{4}\right)\right] = \frac{5\sqrt{15} - 12}{52}$

Exercise 34: Show that $\tan^{-1}\frac{1}{2} + \tan^{-1}\frac{1}{3} = \frac{\pi}{4}$.

[Suggestion: Apply the tangent function, and also consider the quadrant in which the angle represented by the left side of the equation lies.]

Exercises 35–41: Write the trigonometric expression as an equivalent algebraic expression in x, for suitable $x > 0$.

35. $\cot(\sin^{-1}x)$

36. $\csc\left(\cos^{-1}\frac{x}{2}\right)$

37. $\tan(\sin^{-1}2x)$

38. $\sec[\tan^{-1}(x+1)]$

39. $\cot(2\cos^{-1}x)$

40. $\cos(2\sin^{-1}x)$

41. $\sin\left(\frac{\pi}{2} + \tan^{-1}x\right)$

Exercise 42: Verify the identity, $\sin^{-1}x + \cos^{-1}x = \frac{\pi}{2}$, for all x in the interval $[-1, 1]$.

208   Chapter 4   Trigonometric Functions of a Real Variable

Exercises 43–44:   Solve for y in terms of x. Indicate suitable values of x.

43. $x + \frac{\pi}{4} = \cos \frac{y}{4}$ 44. $\sin^{-1} 2y = 4x$

Exercises 45–50:   Solve for x.

45. $\sin(\tan^{-1} x) = \frac{\sqrt{3}}{2}$ 46. $\cos^{-1}(x - 2) = \tan^{-1}\left(\tan \frac{5\pi}{4}\right)$

47. $2\cos^{-1}(x + 2) = \sin^{-1} \frac{3}{5}$ 48. $4\tan^{-1} 3x = \frac{3\pi}{4} + \tan^{-1} 1$

49. $\sin(x + 1) = -\frac{1}{2}$ 50. $\sqrt{2} \tan x = 1$

Exercises GC51–GC57:   Use a graphing calculator to approximate the solutions, to three decimal places.

51. $\sin^{-1}(x + 200) - 800 = 4x + \frac{1}{2}$ 52. $\cos^{-1}(x - 600) - 1200 = -2x + \frac{6}{5}$

53. $\sin x = \tan^{-1} x$ 54. $\sin^{-1} x = \tan x$

55. $\sin\left(x + \frac{\pi}{2}\right) = \sin^{-1} x$ 56. $\cos x = \cos^{-1} x$

57. $\cos^{-1} x = \sin^{-1} x$

Exercise 58:   If an object is given an initial velocity of v_0 feet per second at an angle θ with the horizontal, its horizontal displacement s (when it returns to the same height as that at which it started) is given by

$$s = \frac{1}{32} v_0^2 \sin 2\theta$$

(a) Determine what the angle of a rifle barrel from the horizontal should be, if the bullet is to hit a target 3000 feet away at a point at the same level as the muzzle of the gun, given that the muzzle velocity is 400 ft/sec.

(b) What angle will maximize the (horizontal) distance traveled by the bullet?

Exercises 59–61: (Inverse cosecant function) The cosecant function is one-to-one on $\left[-\frac{\pi}{2}, 0\right) \cup \left(0, \frac{\pi}{2}\right]$. It's inverse function is called the inverse cosecant and is denoted by $\csc^{-1} x$.

59. Sketch the graph of $\csc^{-1} x$.

60. Complete the definition (as for the inverse sine function).
The inverse cosecant function, denoted by $\csc^{-1} x$, has domain _____ and range _____. Furthermore, for _____, $y = \csc^{-1} x$ if and only if _____.

61. Find the exact value of each of the following.

     (a) $\csc^{-1} \frac{2}{\sqrt{3}}$ (b) $\csc\left(\csc^{-1} \frac{4}{3}\right)$ (c) $\csc^{-1}\left(\csc \frac{3\pi}{5}\right)$

4.3 Inverse Trigonometric Functions

Exercises 62–64: (**Inverse secant function**) The secant function is one-to-one on $\left[0, \frac{\pi}{2}\right) \cup \left(\frac{\pi}{2}, \pi\right]$. It's inverse function is called the inverse secant and is denoted by $\sec^{-1} x$.

62. Sketch the graph of $\sec^{-1} x$.

63. Complete the definition (as for the inverse sine function).
    The inverse secant function, denoted by $\sec^{-1} x$, has domain _____ and range _____. Furthermore, for _____, $y = \sec^{-1} x$ if and only if _____.

64. Find the exact value of each of the following.
    (a) $\sec^{-1}\left(-\frac{2}{\sqrt{3}}\right)$
    (b) $\sec\left(\sec^{-1} \frac{4}{3}\right)$
    (c) $\sec^{-1}\left(\sec \frac{9\pi}{7}\right)$

Exercises 65–67: (**Inverse cotangent function**) The cotangent function is one-to-one on $(0, \pi)$. It's inverse function is called the inverse cotangent and is denoted by $\cot^{-1} x$.

65. Sketch the graph of $\cot^{-1} x$.

66. Complete the definition (as for the inverse sine function).
    The inverse cotangent function, denoted by $\cot^{-1} x$, has domain _____ and range _____. Furthermore, for _____, $y = \cot^{-1} x$ if and only if _____.

67. Find the exact value of each of the following.
    (a) $\cot^{-1} \sqrt{3}$
    (b) $\cot\left(\cot^{-1} \frac{8}{15}\right)$
    (c) $\cot^{-1}\left(\cot \frac{9\pi}{5}\right)$

Exercise 68: At supersonic speeds, aircraft produce a conical effect of sound (shock) waves (see adjacent figure). The mach number m of supersonic speed, named after the Austrian physicist Ernst Mach (1838-1916), is the ratio of the plane's speed to the speed of sound (approximately 742 mph). The angle at which the waves strike the ground depends on the mach number of the plane, and is given by the equation

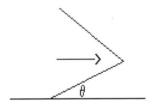

$$m \sin \theta = 1, \quad \text{for } m \geq 1$$

(a) What is the angle θ of the shock wave at mach 1?

(b) One of the following questions is meaningful while the other is not. Answer the question that is meaningful, and explain why the other is not.

　i. What is the speed of the plane if the angle θ of the shock wave is 10% greater than the mach 1 angle?

　ii. What is the speed of the plane if the angle θ of the shock wave is 10% less than the mach 1 angle?

210   Chapter 4 Trigonometric Functions of a Real Variable

Exercises 69–71:   Graph the function. Give the domain and range. Indicate x- and y-intercepts. Then use a graphing utility to check your results.

69. $f(x) = 2\sin^{-1}(x - 1) + 5$

70. $f(x) = -\frac{1}{2}\cos^{-1}(x + 2) - 1$

71. $f(x) = \tan^{-1}(x + 1) - 5$

Exercises 72–74:   Find the domain and range of the function.

72. $f(x) = \sin^{-1}(x^2 - 1)$

73. $f(x) = \dfrac{1}{\cos^{-1} x}$

74. $f(x) = \sin^{-1}(\cos^{-1} x)$

Exercises 75–76:   Solve the system of equations.

75. $\left. \begin{array}{l} 2\sin^{-1} x + \cos^{-1} y = 2\pi \\ \sin^{-1} x - \frac{1}{2}\cos^{-1} y = 0 \end{array} \right\}$

76. $\left. \begin{array}{l} \sin(\sin^{-1} 2x) + y = 5 \\ y^2 + x = 3 \end{array} \right\}$

Exercises 77–83:   Indicate True or False. Justify your answer.

77. The expression $\sin^{-1}(\sin^{-1} \pi)$ represents a unique number.

78. For every x, the expression $\sin^{-1}(\sin^{-1} x)$ represents a number.

79. For every x, the expression $\sin^{-1}(\sin x)$ represents a number.

80. For every x, the expression $\sin(\sin^{-1} x)$ represents a number.

81. For every x, the expression $\sin(\sin x)$ represents a number.

82. If $\sin^{-1} x < 0$, then $x < 0$.

83. $\dfrac{\sin^{-1} x}{\cos^{-1} x} = \tan^{-1} x$ is an identity.

§4. TRIGONOMETRIC EQUATIONS

An equation such as $\cos^2 x - \sin x = 1$ that involves trigonometric functions is said to be a **trigonometric equation**. As is shown in the following examples, both algebraic and trigonometric steps may be called for to solve such an equation.

EXAMPLE 4.18 Find all $0 \leq x < 2\pi$ which satisfy the equation:
$$\csc x = -2$$

SOLUTION: Since $\csc x = \dfrac{1}{\sin x}$, to say that $\csc x = -2$ is to say that $\sin x = -\dfrac{1}{2}$. In terms of angles, $\sin \theta_r = |\sin \theta| = \dfrac{1}{2}$ and $\theta_r = \dfrac{\pi}{6}$.

The sine is negative in the third and fourth quadrants, and from the adjacent figure we see that the equation has two solutions in the interval $[0, 2\pi)$; namely:

$$x = \pi + \frac{\pi}{6} = \frac{7\pi}{6} \quad \text{and} \quad x = 2\pi - \frac{\pi}{6} = \frac{11\pi}{6}$$

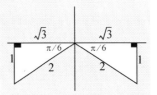

EXAMPLE 4.19 Solve, for $0 \leq x < 2\pi$:
$$\cos^2 x - \sin x = 1$$

SOLUTION:

$$\cos^2 x - \sin x = 1$$

$\sin^2 x + \cos^2 x = 1:$ $\quad (1 - \sin^2 x) - \sin x = 1$

$$-\sin^2 x - \sin x = 0$$

$$-(\sin x)(\sin x + 1) = 0$$

$\sin x = 0$ or $\sin x = -1$

$x = 0, x = \pi \qquad x = \dfrac{3\pi}{2}$

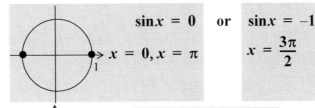
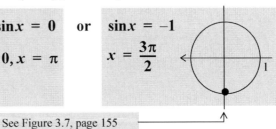

See Figure 3.7, page 155

EXAMPLE 4.20 Find all solutions of:
$$\cos^2 x - \sin x = 1$$

SOLUTION: In the previous example we found that 0, π, and $\frac{3\pi}{2}$ are the solutions of the equation, in the interval $[0, 2\pi)$. Since the sine function is periodic with period 2π, adding multiples of 2π to each of those three numbers will yield all solutions of the equation:

In terms of angles: Adding multiples of 2π to each of the three angles generates all of their coterminal angles.

$$0 + 2n\pi = 2n\pi, \quad \pi + 2n\pi, \quad \frac{3\pi}{2} + 2n\pi, \text{ where } n \text{ is any integer}$$

A nicer representation can be obtained by noting that since $2n\pi$ gives all the even multiples of π, and since $\pi + 2n\pi = (2n+1)\pi$ gives all odd multiples of π, the two of them taken together pick up all multiples of π. Consequently, the solution set of the given equation can be written in the form:

$$\left\{ n\pi, \frac{3\pi}{2} + 2n\pi \,\middle|\, n \text{ is any integer} \right\}$$

EXAMPLE 4.21 Solve for $0 \le x < 2\pi$:
$$\cos 2x - \cos x = 0$$

SOLUTION: We are able to obtain a quadratic equation in $\cos x$ by applying a double angle identity for the cosine:

$$\cos 2x - \cos x = 0$$

Theorem 4.2(ii), page 187: $(2\cos^2 x - 1) - \cos x = 0$

$$2\cos^2 x - \cos x - 1 = 0$$

$$(2\cos x + 1)(\cos x - 1) = 0$$

$$\cos x = -\frac{1}{2} \quad \text{or} \quad \cos x = 1$$

$$x = \frac{2\pi}{3}, x = \frac{4\pi}{3} \qquad x = 0$$

$\cos\theta = -\frac{1}{2} \Rightarrow \cos\theta_r = \frac{1}{2}$ and $\theta_r = \frac{\pi}{3}$. Since $\cos\theta < 0$, θ lies in the second or third quadrant.

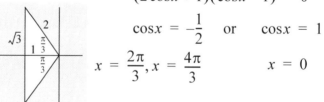

CHECK YOUR UNDERSTANDING 4.21

(a) Solve $2\sin^2 x - 5\sin x - 3 = 0$, for $0 \le x < 2\pi$.
(b) Solve $2\cos x \sin x + \cos x = 2\sin x + 1$.

Answers: (a) $\frac{7\pi}{6}, \frac{11\pi}{6}$
(b) $2n\pi, \frac{7\pi}{6} + 2n\pi, \frac{11\pi}{6} + 2n\pi$

4.4 Trigonometric Equations 213

EXAMPLE 4.22 Solve, for $0 \le x < 2\pi$:
$$\cos^2 2x = \frac{3}{4}$$

SOLUTION:
$$\cos^2 2x = \frac{3}{4}$$

Let $Y = 2x$: $\cos^2 Y = \frac{3}{4}$

$$\cos Y = \pm \frac{\sqrt{3}}{2}$$

The restriction $0 \le x < 2\pi$ imposes the restriction $0 \le 2x < 4\pi$, that is: $0 \le Y < 4\pi$. Consequently:

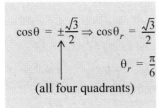
(all four quadrants)

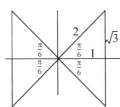

$Y = \frac{\pi}{6}$, $Y = \pi - \frac{\pi}{6} = \frac{5\pi}{6}$, $Y = \pi + \frac{\pi}{6} = \frac{7\pi}{6}$, $Y = 2\pi - \frac{\pi}{6} = \frac{11\pi}{6}$

$Y = \frac{\pi}{6} + 2\pi$ $\quad Y = \frac{5\pi}{6} + 2\pi$ $\quad Y = \frac{7\pi}{6} + 2\pi$ $\quad Y = \frac{11\pi}{6} + 2\pi$

$= \frac{13\pi}{6}$ $\quad = \frac{17\pi}{6}$ $\quad = \frac{19\pi}{6}$ $\quad = \frac{23\pi}{6}$

(Y can go all the way up to 4π) ⟶

Since $Y = 2x$, $x = \frac{Y}{2}$, and the solutions of the equations are seen to be:

$$\frac{\pi}{12}, \frac{5\pi}{12}, \frac{7\pi}{12}, \frac{11\pi}{12}, \frac{13\pi}{12}, \frac{17\pi}{12}, \frac{19\pi}{12}, \text{ and } \frac{23\pi}{12}$$

EXAMPLE 4.23 Solve:
$$2\tan\left(\frac{x}{2} + 1\right)\sec\left(\frac{x}{2} + 1\right) - \tan\left(\frac{x}{2} + 1\right) = 0$$

SOLUTION:
$$2\tan\left(\frac{x}{2} + 1\right)\sec\left(\frac{x}{2} + 1\right) - \tan\left(\frac{x}{2} + 1\right) = 0$$

Let $Y = \frac{x}{2} + 1$: $\quad 2\tan Y \sec Y - \tan Y = 0$

$\tan Y(2\sec Y - 1) = 0$

$\tan Y = 0 \quad$ or $\quad \sec Y = \frac{1}{2}$

$\frac{\sin Y}{\cos Y} = 0 \quad$ or $\quad \cos Y = 2$ **no solution**

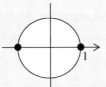

$\left.\begin{array}{l} Y = 0 + 2n\pi \\ Y = \pi + 2n\pi \end{array}\right\} \to Y = n\pi$, for any integer n

Since $Y = \frac{x}{2} + 1$: $\quad \frac{x}{2} + 1 = n\pi$, for any integer n

$x = 2n\pi - 2$, for any integer n

214 Chapter 4 Trigonometric Functions of a Real Variable

Answers: (a) $\frac{\pi}{2}$

(b) $\frac{13\pi}{24} + \frac{n\pi}{2}, \frac{17\pi}{24} + \frac{n\pi}{2}$

CHECK YOUR UNDERSTANDING 4.22

(a) Solve $2\sin\frac{x}{3} - 1 = 0$, for $0 \le x < 2\pi$.

(b) Solve $2\sin^2(4x - \pi) = 5\sin(4x - \pi) + 3$.

We were able to solve the previous trigonometric equations without using any inverse trigonometric function. This is still the case with the first equation in the next example, but not with the second.

EXAMPLE 4.24 Solve, for $0 \le x < 2\pi$:

(a) $2\sin^2 x + \sin x - 1 = 0$

(b) $3\sin^2 x + 2\sin x - 1 = 0$

SOLUTION: We solve both equations, side by side, to highlight their similarities:

$2\sin^2 x + \sin x - 1 = 0$ | $3\sin^2 x + 2\sin x - 1 = 0$
$(2\sin x - 1)(\sin x + 1) = 0$ | $(3\sin x - 1)(\sin x + 1) = 0$
$\sin x = \frac{1}{2}$ or $\sin x = -1$ | $\sin x = \frac{1}{3}$ or $\sin x = -1$

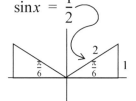 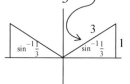

$x = \frac{\pi}{6}, x = \pi - \frac{\pi}{6} = \frac{5\pi}{6}$ $x = \frac{3\pi}{2}$ $x = \sin^{-1}\frac{1}{3}, x = \pi - \sin^{-1}\frac{1}{3}$ $x = \frac{3\pi}{2}$

The main difference between these two solutions stems from the equations:

$\sin x = \frac{1}{2}$, with reference angle: $x = \sin^{-1}\frac{1}{2} = \frac{\pi}{6}$

and: $\sin x = \frac{1}{3}$, with reference angle: $x = \sin^{-1}\frac{1}{3}$ (that's as far as it goes)

EXAMPLE 4.25 Solve, for $0 \le x < 2\pi$:

$2\tan x - \cot x - 1 = 0$

4.4 Trigonometric Equations 215

SOLUTION:

$$2\tan x - \cot x - 1 = 0$$

$$2\tan x - \frac{1}{\tan x} - 1 = 0$$

$$2\tan^2 x - 1 - \tan x = 0$$

Let $y = \tan x$:
$$2y^2 - y - 1 = 0$$

$$(2y + 1)(y - 1) = 0$$

Since $y = \tan x$:
$$y = -\frac{1}{2} \text{ or } y = 1$$

$$\tan x = -\frac{1}{2} \quad \text{or} \quad \tan x = 1$$

Tangent is negative in the second and fourth quadrants.

$$\tan^{-1}\left(\frac{1}{2}\right)$$

$$\tan^{-1}1 = \frac{\pi}{4}$$

Tangent is positive in the first and third quadrant.

$$\tan\theta = -\frac{1}{2}$$

$$\tan\theta_r = \frac{1}{2}$$

$$\theta_r = \tan^{-1}\frac{1}{2}$$

$$x = \pi - \tan^{-1}\left(\frac{1}{2}\right), x = 2\pi - \tan^{-1}\left(\frac{1}{2}\right)$$

$$x = \frac{\pi}{4}, x = \frac{5\pi}{4}$$

CHECK YOUR UNDERSTANDING 4.23

Solve, for $0 \le x < 2\pi$:

$$3\cos^2 x - 2\cos x - 1 = 0$$

Answer: $0, \pi \pm \cos^{-1}\frac{1}{3}$

216   **Chapter 4 Trigonometric Functions of a Real Variable**

E X E R C I S E S

Exercises 1–10: Solve for x in $[0, 2\pi)$.

1. $\sin x = \dfrac{\sqrt{3}}{2}$

2. $\cot x = -1$

3. $\sin x - 2\sin x \cos x = 0$

4. $2\cos^2 x = 1$

5. $2\tan^2 x + \sec^2 x = 2$

6. $\tan^2 x - 3\sec x + 3 = 0$

7. $\sin 2x = 2\cos x$

8. $4\sin^2 x + 2(\sqrt{2} - \sqrt{3})\sin x = \sqrt{6}$

9. $\cos 2x + 2\cos x + 1 = 0$

10. $3\sin x - \sin 3x = \dfrac{1}{2}$

Exercises 11–18: Solve.

11. $\tan x = -1$

12. $2\sin x = 1$

13. $2\sin x - \csc x = 1$

14. $(\sqrt{2}\cos x - 1)(2\cos x + \sqrt{3}) = 0$

15. $\sec^2 x + \tan x = 1 - \sqrt{3}(\tan x + 1)$

16. $\sec^2 x - \tan x = 1$

17. $4\sin^2 x = 3$

18. $2\sin^2 x - \cos x = 1$

Exercises 19–23: Solve for x in $[0, 2\pi)$.

19. $\csc 2x = -\dfrac{2}{\sqrt{3}}$

20. $\sin \dfrac{x}{3} = 0$

21. $2\cos^3 \dfrac{1}{2}x = \cos \dfrac{1}{2}x$

22. $\tan^2 3x - 3 = 0$

23. $2\sin 2x \cos 3x = -\cos 3x$

Exercises 24–28: Solve.

24. $\tan(3x + 5) = -1$

25. $\csc^2 \dfrac{x}{2} = 4$

26. $\sin 2x = \cos 2x$

27. $\sec 3x - \cos 3x = 0$

28. $2\cos^2(4x - 1) + 3\cos(4x - 1) + 1 = 0$

Exercises 29–34: Express all angles satisfying the given conditions in terms of inverse trigonometric functions.

29. $\cos \theta = \dfrac{3}{5}$ and $0 < \theta < \dfrac{\pi}{2}$

30. $\cos \theta = -\dfrac{3}{5}$ and $\dfrac{\pi}{2} < \theta < \pi$

31. $\sin \theta = \dfrac{3}{4}$ and $\dfrac{\pi}{2} < \theta < \pi$

32. $\tan \theta = \dfrac{3}{2}$ and $\pi < \theta < \dfrac{3\pi}{2}$

33. $\tan \theta = -\dfrac{3}{2}$

34. $\sin \theta = -\dfrac{2}{3}$

Exercises 35–45: Solve in $[0, 2\pi)$ expressing some of the solutions in terms of inverse trigonometric functions. Then approximate those values with a calculator.

35. $\tan x = -8$

36. $12\cos^2 x = 2 - 5\cos x$

37. $24\sin^2 x + 2\sin x - 15 = 0$

38. $6\sin 3x = 5$

39. $6\cos^2 x - \sin x - 4 = 0$

40. $3\tan^2 x = 3 + 8\tan x$

41. $4\sin 2x = 3\cos x$

42. $2\cos 2x = -1 - 3\cos x$

43. $8\sin^2 x + 6\sin x - 9 = 0$

44. $9\cos^2 x + 6\cos x = 8$

45. $3\tan x - \cot x = -2$

4.4 Trigonometric Equations **217**

Exercises 46–47: Determine the solution set of the system of equations, for x and y in $[0, 2\pi)$.

46. $\left.\begin{array}{l} \sin x + \cos y = \sin 2x \\[2mm] \sin x - \cos y = 0 \end{array}\right\}$

47. $\left.\begin{array}{l} \sin 2x + y = \cos x \\[2mm] \sin x = \dfrac{y+1}{2} \end{array}\right\}$

Exercises GC48–GC51: Use a graphing calculator to approximate the solutions of the equation in the interval $[0, \pi]$.

48. $3 \tan x - \sec x = 1$

49. $4 \sin x + 3 \cos x = 4$

50. $\cos^2 2x + \cot x = 2 \sin 3x$

51. $5 \cos x = 7 \sin \frac{x}{2} + 1$

Exercise 52: Solve the inequality $\dfrac{\cos x}{\cos^2 x \sin x + \sin^3 x} \geq 1$ in the interval $[0, 2\pi)$.

Exercises GC53–GC55: Use a graphing calculator to approximate the solution set of the inequality in the interval $[0, \pi]$.

53. $1 - \sec x < x$

54. $3 \cos 3x > 2 \sin x$

55. $\cos^2 2x + \cot x > 2 \sin 3x$

Exercise 56: In Exercises 29, 30, 32, and 33, express all angles satisfying the given conditions in terms of the inverse sine function.
(This illustrates the fact that, theoretically, we could have restricted our attention in this section to just one of the inverse trigonometric functions.)

CHAPTER SUMMARY

TRIGONOMETRIC FUNCTIONS OF A REAL VARIABLE	Let x be a real number and let θ be the angle with **radian measure** x. Then: $$\text{trig } x = \text{trig } \theta$$ (a number = an angle)
PERIOD	A function f is **periodic** if there exists a positive number p such that: $$f(x + p) = f(x)$$ The smallest such p is said to be the **period** of the function.
CYCLE	The graph of a periodic function restricted to any one period is called a **cycle**.

GRAPHS OF THE TRIGONOMETRIC FUNCTIONS

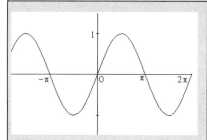

The sine function has period 2π and is an odd function:
$$\sin(-x) = -\sin x$$

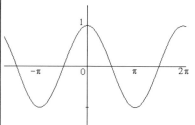

The cosine function has period 2π and is an even function:
$$\cos(-x) = \cos x$$

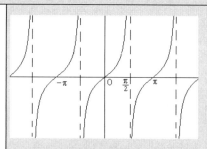

The tangent function has period π and is an odd function:
$$\tan(-x) = -\tan x$$

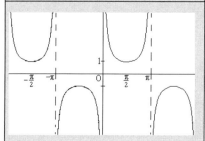

The cosecant function has period 2π and is an odd function:
$$\csc(-x) = -\csc x$$

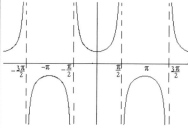

The secant function has period 2π and is an even function:
$$\sec(-x) = \sec x$$

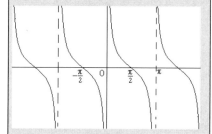

The cotangent function has period π and is an odd function:
$$\cot(-x) = -\cot x$$

SINE WAVE	A sine wave is a curve with shape that of the sine or cosine function. The **amplitude** of the sine wave is defined to be one-half of the difference between the largest and smallest values of the function:

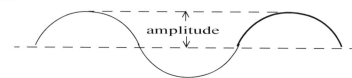

$a \sin b(x - c) + d$ and $a \cos b(x - c) + d$ have amplitude $|a|$.

Chapter 4 Trigonometric Functions of a Real Variable

TRIGONOMETRIC IDENTITIES	
PYTHAGOREAN IDENTITIES	$\sin^2 x + \cos^2 x = 1$ $1 + \tan^2 x = \sec^2 x$ $1 + \cot^2 x = \csc^2 x$
ADDITION IDENTITIES	$\sin^2 x + \cos^2 x = 1$ $\cos(x+y) = \cos x \cos y - \sin x \sin y$ $\cos(x-y) = \cos x \cos y + \sin x \sin y$ $\sin(x+y) = \sin x \cos y + \cos x \sin y$ $\sin(x-y) = \sin x \cos y - \cos x \sin y$ $\tan(x+y) = \dfrac{\tan x + \tan y}{1 - \tan x \tan y}$ $\tan(x-y) = \dfrac{\tan x - \tan y}{1 + \tan x \tan y}$
DOUBLE-ANGLE IDENTITIES	$\sin 2x = 2 \sin x \cos x$ $\cos 2x = \cos^2 x - \sin^2 x$ $\quad\quad = 1 - 2\sin^2 x$ $\quad\quad = 2\cos^2 x - 1$ $\tan 2x = \dfrac{2 \tan x}{1 - \tan^2 x}$
HALF-ANGLE IDENTITIES	$\sin\left(\dfrac{x}{2}\right) = \pm\sqrt{\dfrac{1 - \cos x}{2}}$ $\cos\left(\dfrac{x}{2}\right) = \pm\sqrt{\dfrac{1 + \cos x}{2}}$ $\tan\left(\dfrac{x}{2}\right) = \pm\sqrt{\dfrac{1 - \cos x}{1 + \cos x}}$ The choice of sign (+ or -) is dependent on the quadrant in which the angle of radian measure $\frac{x}{2}$ lies.

	INVERSE TRIGONOMETRIC FUNCTIONS	
INVERSE SINE FUNCTION	For $-1 \leq x \leq 1$: $\sin^{-1} x$ is that angle θ such that $\sin\theta = x$, and $-\frac{\pi}{2} \leq \theta \leq \frac{\pi}{2}$.	
INVERSE COSINE FUNCTION	For $-1 \leq x \leq 1$: $\cos^{-1} x$ is that angle θ such that $\cos\theta = x$, and $0 \leq \theta \leq \pi$.	
INVERSE TANGENT FUNCTION	For any x: $\tan^{-1} x$ is that angle θ such that $\tan\theta = x$, and $-\frac{\pi}{2} < \theta < \frac{\pi}{2}$.	
RANGE OF THE INVERSE TRIGONOMETRIC FUNCTIONS		

GRAPHS OF THE INVERSE TRIGONOMETRIC FUNCTIONS

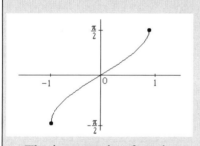

The inverse sine function
Domain: $[-1, 1]$
Range: $\left[-\frac{\pi}{2}, \frac{\pi}{2}\right]$

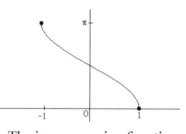

The inverse cosine function
Domain: $[-1, 1]$
Range: $[0, \pi]$

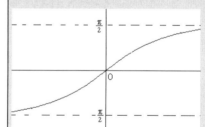

The inverse tangent function
Domain: $(-\infty, \infty)$
Range: $\left(-\frac{\pi}{2}, \frac{\pi}{2}\right)$

Chapter Review Exercises   **221**

C H A P T E R   R E V I E W   E X E R C I S E S

Exercises 1–12: Sketch the graph of the function, and determine the period, y-intercept, and (where appropriate) the amplitude, phase shift, and vertical asymptotes.

1. $f(x) = \sin \pi x + \pi$ 2. $f(x) = -\frac{1}{2}\sin(\pi - x) - \frac{1}{2}$ 3. $f(x) = 4\cos\left(2x - \frac{\pi}{2}\right) + 2$

4. $f(x) = \cos 3x - 1$ 5. $f(x) = \frac{1}{2}\tan\left(x + \frac{\pi}{2}\right)$ 6. $f(x) = -\tan\left(x - \frac{\pi}{4}\right)$

7. $f(x) = -2\csc x$ 8. $f(x) = 3\csc\left(\frac{1}{2}x - \frac{\pi}{4}\right)$ 9. $f(x) = \cot(-2x)$

10. $f(x) = 2\cot(x + 3\pi)$ 11. $f(x) = 1 - \sec 4x$ 12. $f(x) = 2\sec(x - \pi)$

Exercises GC13–GC16: Use a graphing calculator to graph each pair of functions f and g, and verify that they are **not** equal. Illustrate analytically that they are not equal, by determining a particular value of x, $x = a$, and showing that $f(a) \neq g(a)$.

13. $f(x) = 3\sin x, \quad g(x) = \sin 3x$ 14. $f(x) = \cos^3 x, \quad g(x) = \cos x^3$

15. $f(x) = \tan^{-1} x, \quad g(x) = \dfrac{1}{\tan x}$ 16. $f(x) = \sec x - 1, \quad g(x) = \sec(x - 1)$

Exercises 17–18: Use trigonometric identities to simplify the given expression.

17. $\left(\dfrac{\cot x}{\cos x}\right)\left(\dfrac{\sin x}{1 - \cos^2 x} - \cot x\right)$ 18. $\dfrac{\cot x \cos x + \sin x}{2\csc x}$

Exercises 19–30: Verify the identity.

19. $(\csc x - \cot x)^2 = \dfrac{1 - \cos x}{1 + \cos x}$ 20. $\cot x \cos x + \sin x = \csc x$

21. $\dfrac{\tan x}{1 - \tan^2 x} = \dfrac{\cos x}{\csc x - 2\sin x}$ 22. $\dfrac{\sin x - \cos x}{\sec x - \csc x} = \dfrac{\cos x}{\csc x}$

23. $\sin^4 x = \frac{1}{8}\cos 4x - \frac{1}{2}\cos 2x + \frac{3}{8}$ 24. $\sin^4 x = \frac{1}{4} - \frac{1}{2}\cos 2x + \frac{1}{4}\cos^2 2x$

25. $\cos^4 x = \frac{3}{8} + \frac{1}{2}\cos 2x + \frac{1}{8}\cos 4x$ 26. $\sin 3x = 3\sin x \cos^2 x - \sin^3 x$

27. $\cos 3x = 4\cos^3 x - 3\cos x$ 28. $\dfrac{1 - \tan x}{1 + \tan x} = \dfrac{1 - \sin 2x}{\cos 2x}$

29. $\dfrac{1 - \sin x}{1 + \sin x} = (\sec x - \tan x)^2$ 30. $\dfrac{\tan x}{1 + \cos x} = \dfrac{1 - \cos x}{\sin x \cos x}$

Exercises 31–32: Evaluate $\sin(\alpha + \beta)$, $\cos(\alpha - \beta)$, $\sin 2\alpha$, and $\cos 2\beta$, given

31. $\sin \alpha = \frac{4}{5}, \quad \cos \beta = -\frac{12}{13}, \quad \tan \alpha < 0, \quad \csc \beta < 0$

32. $\cos \alpha = -\frac{1}{2}, \quad \cot \beta = -\frac{2}{\sqrt{5}}, \quad \pi < \alpha < \frac{3\pi}{2}, \quad \frac{3\pi}{2} < \beta < 2\pi$

Exercises 33–34: Evaluate the six trigonometric functions of θ, given

33. $\cos 2\theta = -\frac{15}{17}, \quad$ and $\quad 0 < \theta < \frac{\pi}{2}$ 34. $\cos \frac{\theta}{2} = \frac{2}{3}, \quad$ and $\quad \frac{\pi}{2} < \theta < \pi$

Exercises 35–36: Determine the value of

35. $\tan \frac{x}{2}$, if $\csc x = -3$ and $\pi < x < \frac{3\pi}{2}$ 36. $\sec 3x$, if $\cos 6x = -\frac{5}{13}$ and $\frac{\pi}{12} < x < \frac{\pi}{6}$

Exercise 37: Express $\sin 5x$ in terms only of $\sin x$.

Exercises 38–43: Find the exact value.

38. $\sin^{-1}\left(\tan \frac{3\pi}{4}\right)$ 39. $\tan(\sin^{-1} 0)$ 40. $\sin\left[\sin^{-1}\left(-\frac{1}{2}\right)\right]$

41. $\tan^{-1} \frac{\sqrt{3}}{3}$ 42. $\sin\left[\cos^{-1}\left(-\frac{1}{5}\right)\right]$ 43. $\cos\left[\sin^{-1}\frac{1}{2} - \tan^{-1}\left(-\frac{3}{4}\right)\right]$

Exercises 44–47: Write the trigonometric expression as an equivalent algebraic expression in x, for $0 < x < 1$.

44. $\tan(\cos^{-1} x)$ 45. $\csc(\tan^{-1} x)$ 46. $\sec\left(\sin^{-1} \frac{x}{2}\right)$ 47. $\cot(\sin^{-1} 2x)$

Exercises 48–50: Solve the equation in $[0, 2\pi)$.

48. $\sin x = -\frac{\sqrt{2}}{2}$ 49. $\sec x = 2 \sin x$

50. $\sin(x-1)\tan(x-1) + \tan(x-1) = \sqrt{3}\sin(x-1) + \sqrt{3}$

Exercises 51–53: Solve the equation.

51. $\tan^2 x + \cot^2 x = 2$ 52. $\csc^2 x = 4$ 53. $\cos 3x = -1$

Exercises 54–56: Express the solutions in $[0, 2\pi)$ in terms of inverse trigonometric functions. Then approximate those values with a calculator.

54. $\sin x = -\frac{3}{5}$ 55. $12\cos^2 x + \cos x - 6 = 0$ 56. $20\sin^2 x + 3 \sin x = 2$

Exercises GC57–GC60: Use a graphing calculator to approximate the solutions of the equation in the interval $[0, \pi]$.

57. $\tan(3 \sin x) = \cos x$ 58. $2 \sin x - 5 \cos x = 3$

59. $\frac{\sin x}{x} = 1$ (Think about your answer)   60. $3\cos\left(\frac{1}{x+1}\right) = x + 1$

Exercises GC61–GC62: Use a graphing calculator to approximate the solutions of the equation.

61. $\sin^{-1}(x+400) = -\cos(x-3) + 2$ 62. $\cos^{-1}(x+850) - 2500 = 3x + 50$

Exercise 63: Express the angle of elevation, θ, to the top of a structure H units high from an h units high vantage point whose horizontal distance to the taller structure is d units, in terms of H, h and d.

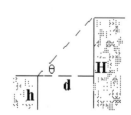

Exercise 64: A mass suspended from a string is a simple pendulum, and its movement approximates simple harmonic motion given by the equation

$$\theta = 0.15 \cos\left(t\sqrt{\frac{g}{L}}\right)$$

where g is the gravitational constant and L is the length of the string.

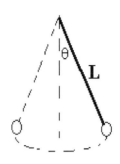

Taking g to be 32 ft/sec^2, determine L, the length of the pendulum of a grandfather clock, if it is to take exactly one second to swing from its maximal displacement on one side of the vertical to its maximal displacement on the other side of the vertical.

Exercises 65–69: Indicate True or False. Justify your answer. Where appropriate, use a graphing calculator to check your answer.

65. The graph of the function $f(x) = \tan x + \cot x$ has no x-intercepts.

66. Where defined, $|(\tan x + \sec x)\cos x| \leq 2$.

67. If $0 \leq x \leq 1$, then $(\sin^{-1} x)^2 + (\cos^{-1} x)^2 = 1$.

68. $[\sin^{-1}(\sin x)]^2 + [\cos^{-1}(\cos x)]^2 = 2x^2$.

69. $\sin^2(\sin^{-1} x) + \cos^2(\cos^{-1} x) = 2x^2$, if $-1 \leq x \leq 1$.

CUMULATIVE REVIEW EXERCISES

Exercises 1–4: Simplify

1. $\dfrac{(2x-4)^4}{(\sqrt{2}\,x)^2} \cdot \dfrac{5x+10}{x^2-4}$

2. $x^{\frac{1}{3}} \left(\dfrac{x+1}{x^{\frac{1}{6}}}\right)^{-1} - x^{-\frac{3}{2}}(x+1)^2$

Exercise 3: For $f(x) = 3x^2 - 4x + 2$, determine $\dfrac{f(x+h) - f(x)}{h}$, for $h \neq 0$.

Exercise 4: Determine the domain and range of the function $g(t) = \sqrt{t+2}$, and sketch the graph by reflecting, stretching, and/or shifting the basic graph.

Exercise 5: Find the exact values of the remaining five trigonometric functions of θ, given $\cot \theta = \frac{24}{7}$, and $\pi < \theta < \frac{3\pi}{2}$.

Exercises 6–7: Solve the equation.

6. $2\sec^2 x - \tan x = 3$

7. $2\cos 2x + 3\sin x - 2 = 0$, in $[0, 2\pi)$

224   Chapter 4 Trigonometric Functions of a Real Variable

Exercise 8:   Verify the identity $\csc 2x + \cot 2x = \cot x$.

Exercise 9:   Express $\sin(2\cos^{-1}3x)$ as an algebraic function of x.

Exercise 10: (**Angle of elevation**) From a point on level ground, the angles of elevation of the top and bottom of a flagpole situated on the top of a hill are measured as 48° and 40°, resp. Find the height of the hill if the height of the flagpole is 115 ft.

Exercise 11:   For $h(x) = 2\sin\left(4x + \frac{\pi}{2}\right) - 3$,

(a) determine the intercepts of the graph of h

(b) determine the amplitude, period, and phase shift

(c) sketch the graph of the function

(d) determine the range of g

Exercises 12–16:   If $\csc\alpha = -\frac{5}{3}$, $\tan\beta = \frac{12}{5}$, $\frac{3\pi}{2} < \alpha < 2\pi$, and $\pi < \beta < \frac{3\pi}{2}$, determine the value of

12. $\sin(\alpha - \beta)$ 13. $\cos(\alpha - \beta)$ 14. $\tan(\alpha - \beta)$ 15. $\sin 2\alpha$ 16. $\cos\frac{\alpha}{2}$

Exercises 17–24:   Find the exact value of each of the following.

17. $2\sin 15°$ 18. $\cos^2\left(\frac{3\pi}{8}\right) - \sin^2\left(\frac{3\pi}{8}\right)$ 19. $\sin^{-1}\left(-\frac{1}{2}\right)$ 20. $\tan^{-1}(-1)$

21. $\sin^{-1}\left(\tan\frac{3\pi}{4}\right)$ 22. $\sin\left[2\sin^{-1}\left(-\frac{12}{13}\right)\right]$ 23. $\cos^{-1}\left(\cos\frac{8\pi}{7}\right)$ 24. $\csc\left[\cos^{-1}\left(-\frac{1}{3}\right)\right]$

Exercise 25:   Evaluate

$$\cos\left[\sin^{-1}\frac{1}{3} + \cos^{-1}\left(-\frac{1}{2}\right)\right]$$

CHAPTER 5

SOLVING OBLIQUE TRIANGLES

> In Chapter 3, we solved right-triangles, and considered some right-triangle applications. We will do this again in this chapter, but for arbitrary triangles.

A triangle consists of six components: three angles and three sides, and to solve a triangle is to determine each of them. In Section 1, you will see that you can solve any triangle, if you are given a side and any two of its other components (three angles won't do, since three angles do not uniquely determine a triangle). In Section 2, the last section of this brief chapter, we turn to some additional applications; applications that involve oblique triangles (as opposed to right triangles).

§1. LAW OF SINES AND LAW OF COSINES

For **any** triangle, we have:

THEOREM 5.1

> Note that the Pythagorean Theorem is a special case of the Law of Cosines, with $\gamma = 90°$, then $\cos\gamma = 0$, and we have:
> $c^2 = a^2 + b^2 - 2ab\cos\gamma$
> $c^2 = a^2 + b^2$

LAW OF SINES: $\dfrac{\sin\alpha}{a} = \dfrac{\sin\beta}{b} = \dfrac{\sin\gamma}{c}$

LAW OF COSINES: $a^2 = b^2 + c^2 - 2bc\cos\alpha$
$b^2 = a^2 + c^2 - 2ac\cos\beta$
$c^2 = a^2 + b^2 - 2ab\cos\gamma$

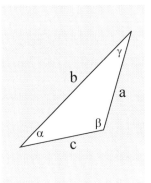

PROOF: Let h denote the altitude of the triangle in Figure 5.1. The coordinates of the indicated point were arrived at by focusing on the shaded right triangle and observing that:

$$\cos\alpha = \frac{x}{b}, \text{ or: } x = b\cos\alpha$$

$$\sin\alpha = \frac{h}{b}, \text{ or: } h = b\sin\alpha$$

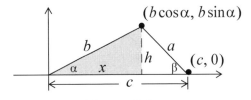

Figure 5.1

FOR THE LAW OF SINES	FOR THE LAW OF COSINES
$\sin\alpha = \dfrac{h}{b}$ and $\sin\beta = \dfrac{h}{a}$	Applying the distance formula to the line segment labeled a:
$h = b\sin\alpha$ and $h = a\sin\beta$	$a = \sqrt{(b\cos\alpha - c)^2 + (b\sin\alpha - 0)^2}$
$b\sin\alpha = a\sin\beta$	Then: $a^2 = b^2\cos^2\alpha - 2bc\cos\alpha + c^2 + b^2\sin^2 a$
$\dfrac{\sin\alpha}{a} = \dfrac{\sin\beta}{b}$	$= b^2(\cos^2\alpha + \sin^2 a) + c^2 - 2bc\cos\alpha$
	$= b^2 + c^2 - 2bc\cos\alpha$

To arrive an any of the other formal expressions of the Law of Sines or Law of Cosines, simply swap letters in the above arguments: b and β with a and α, and so on.

When should you use the Law of Sines, and when the Law of Cosines? First of all, don't bother with either of them if you are dealing with a right triangle, since you have the Pythagorean Theorem and the definition of the trigonometric functions of acute angles (page 127) at your disposal. But if the triangle is not a right triangle, then the decision on which law to use rests on which of them results in an equation involving only one unknown. Here is some advice that you can choose to ignore, since the situation itself will point to the appropriate tool:

Use the Law of Sines when:	
(AAS) You know two (and therefore all three) angles and a side.	**(SSA)** You know two sides and the angle opposite one of them.
Say you have α, β, and a, then:	Say you have a, c, and α, then:
$\dfrac{\sin\alpha}{a} = \dfrac{\sin\beta}{b}$	$\dfrac{\sin\alpha}{a} = \dfrac{\sin\gamma}{c}$
only unknown	only unknown

Use the Law of Cosines when:	
(SAS) You know two sides and their included angle.	**(SSS)** You know all three sides.
Say you have a, b, and γ, then:	Then, for example:
$c^2 = a^2 + b^2 - 2ab\cos\gamma$	$b^2 = a^2 + c^2 - 2ac\cos\beta$
only unknown	only unknown

EXAMPLE 5.1 Solve the triangle with $\alpha = 25°$, $\gamma = 50°$, and $a = 3$.

SOLUTION: First, you should draw and label the triangle with the given information (adjacent figure).

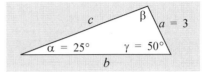

It is an easy matter to determine β:

$$\beta = 180° - 25° - 50° = 105°$$

The Law of Cosines is of no help at this point (try it)

Using the Law of Sines, we are able to determine the length of side c:

$$\frac{\sin 25°}{3} = \frac{\sin 50°}{c}$$

$$c = \frac{3\sin 50°}{\sin 25°} \approx 5.4 \text{ (margin)}$$

At this point, knowing the value of c, we could use the Law of Cosines to find the value of b (margin), but decide to go again with the easier Law of Sines:

$b^2 = a^2 + c^2 - 2ac\cos\beta$
all known

$$\frac{\sin 25°}{3} = \frac{\sin 105°}{b}$$

$$b = \frac{3\sin 105°}{\sin 25°} \approx 6.7$$

EXAMPLE 5.2 Solve the triangle with sides $a = 4$, $b = 3$, and $c = 6$.

SOLUTION: While the Law of Sines is of no initial help (try it), the Law of Cosines can be used to find an angle of the triangle:

$$b^2 = a^2 + c^2 - 2ac\cos\beta$$
$$3^2 = 4^2 + 6^2 - 2(4)(6)\cos\beta$$
$$\cos\beta = \frac{9 - 16 - 36}{-48} = \frac{43}{48}$$
$$\beta = \cos^{-1}\left(\frac{43}{48}\right) \approx 26°$$

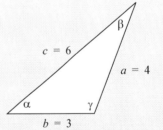

Note that the longer the side of a triangle, the larger is the angle opposite that side. In particular, in the above triangle, the largest angle γ is opposite the longest side, and the smallest angle β is opposite the smallest side.

228 Chater 5 Solving Oblique Triangles

> We know that α is acute since it is not opposite the longest side of the triangle. The advantage of focusing on an acute angle when using the Law of Sines is that $\sin^{-1} x$ cannot be greater that $\frac{\pi}{2}$ (see Definition 4.2, page 198).

Now that we know β, we have the **option** of using the Law of Sines to determine the value of either of the two remaining angles, and we elect to do so, focusing on the acute angle α (see margin):

$$\frac{\sin\alpha}{4} = \frac{\sin\beta}{3}$$

$$\sin\alpha = \frac{4}{3}\sin\beta \approx \frac{4}{3}\sin 26° \approx 0.5845$$

$$\alpha \approx \sin^{-1}(0.5845) \approx 36°$$

Finally: $\gamma = 180° - \beta - \alpha \approx 180° - 26° - 36° = 118°$

Given three sticks, you may not be able to position them so as to form a triangle, for the length of any one of the sticks must be less than the combined lengths of the remaining two (the shortest distance between two points is a straight line). The "three sticks" in the previous example ($a = 4$, $b = 3$, and $c = 6$) did form a triangle, since:

$$4 < 6 + 3, \ 3 < 6 + 4, \text{ and } 6 < 4 + 3$$

However, there can be no triangle with sides $a = 8$, $b = 4$, and $c = 3$, since $8 \not< 4 + 3$. But don't worry if you do not spot this discrepancy, because if you attempt to solve for any angle in the would-be triangle, you will end up with an equation that has no solution. For example:

$$3^2 = 8^2 + 4^2 - 2(8)(4)\cos\gamma$$

$$\cos\gamma = \frac{9 - 64 - 16}{-64} = \frac{71}{64} > 1$$

⌞ Can't be: No solution

Answers:
(a) $\gamma = 130°, b \approx 2.6, a \approx 1.8$
(b) $c \approx 1.9, \alpha \approx 46°, \beta \approx 107°$
(c) $\alpha \approx 21°, \beta \approx 32°, \gamma \approx 127°$

CHECK YOUR UNDERSTANDING 5.1

Solve the triangle (approximating sides to one decimal place, and angles to the nearest degree), given that:

(a) $\alpha = 20°$, $\beta = 30°$, $c = 4$

(b) $a = 3$, $b = 4$, $\gamma = 27°$

(c) $a = 4$, $b = 6$, $c = 9$

THE AMBIGUOUS CASE OF THE LAW OF SINES (SSA)

Given two sides of a triangle, say a and b, and the angle opposite one of those sides, say the angle α, then there are three possibilities:

No triangle [Figure 5.2(a)], one triangle [Figure 5.2(b)], and two triangles [Figure 5.2(c)]

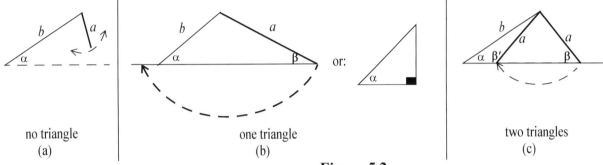

no triangle
(a)

one triangle
(b)

two triangles
(c)

Figure 5.2

It is not difficult to establish rigid rules as to when no triangle, exactly one triangle, or two triangles exist in a SSA situation (Exercise 49). We suggest instead that you simply reason your way through the process in a direct manner, realizing that:

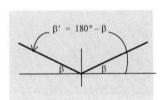

For any given t with $0 < t \leq 1$, there are **two** β's satisfying the equation $\sin\beta = t$ that **can be angles in a triangle**:

$\beta = \sin^{-1} t$ and $\beta' = 180° - \sin^{-1} t$ (see margin)

   ↑   ↑
in first quadrant   in second quadrant

and both must be taken into consideration.

EXAMPLE 5.3 Solve the triangle with:
$$a = 11, \; b = 7, \text{ and } \alpha = 19°$$

SOLUTION: Using the Law of Sines we have:
$$\frac{\sin\alpha}{a} = \frac{\sin\beta}{b}$$
$$\frac{\sin 19°}{11} = \frac{\sin\beta}{7}$$
$$\sin\beta = \frac{7\sin 19°}{11} \quad (*)$$

We apply the inverse sine function to both sides of the above equation. Being a function, it can only provide us with the one value of β.

$$\beta = \sin^{-1}\left(\frac{7\sin 19°}{11}\right) \approx 12°$$

230   Chater 5 Solving Oblique Triangles

As noted earlier, however, we also have to consider the angle

$$\beta' = 180° - \beta \approx 168°$$

for it too satisfies (*), and can also be an angle of a triangle. But, **and here is the deciding factor**:

$$\alpha = 19° \text{ and } \beta' \approx 168° \text{ leaves } \underline{\textbf{no room}} \text{ for a third angle, since}$$

$$\alpha + \beta' \approx 19° + 168° = 187° > 180°$$

We therefore conclude that there is only one triangle. Solving for the remaining angle and side we have:

> Another way to see that $\beta' \approx 168°$ will not do is that since $b < a$ $(7 < 11)$, the angle opposite side b must be smaller than the angle opposite a, and $\beta' \approx 168°$ is not smaller than $\alpha = 19°$

$$\gamma = 180° - \alpha - \beta \approx 180° - 19° - 12° = 149°$$

$$c = \sqrt{a^2 + b^2 - 2ab\cos\gamma} \approx \sqrt{11^2 + 7^2 - 2 \cdot 11 \cdot 7\cos 149°} \approx 17.4$$

EXAMPLE 5.4   Solve the triangle with:

$$a = 4, \ b = 5, \text{ and } \alpha = 30°$$

SOLUTION: Turning to the Law of Sines, we have:

$$\frac{\sin 30°}{4} = \frac{\sin\beta}{5}$$

$$\sin\beta = \frac{5\sin 30°}{4} = \frac{5 \cdot \dfrac{1}{2}}{4} = \frac{5}{8}$$

$$\beta = \sin^{-1}\frac{5}{8} \approx 39°$$

and

$$\beta' \approx 180° - 39° = 141°$$

$$\gamma \approx 180° - 30° - 39° = 111°$$

$$\gamma' \approx 180° - 30° - 141° = 9°$$

enough room for a third angle!

Using the Law of Sines:

$$\frac{\sin 111°}{c} = \frac{\sin 30°}{4}$$

$$c = \frac{4\sin 111°}{\sin 30°} \approx 7.5$$

Using the Law of Sines:

$$\frac{\sin 9°}{c'} = \frac{\sin 30°}{4}$$

$$c' = \frac{4\sin 9°}{\sin 30°} \approx 1.3$$

5.1 Law of Sines and Law of Cosines   **231**

EXAMPLE 5.5   Solve the triangle with:

$$a = 4, \ b = 5, \text{ and } \alpha = 55°$$

SOLUTION:

$$\frac{\sin 55°}{4} = \frac{\sin \beta}{5}$$

$$\sin \beta = \frac{5 \sin 55°}{4} \approx 1.02 > 1$$

Since the sine function cannot assume values greater than one, we conclude that there is no triangle satisfying the given conditions.

CHECK YOUR UNDERSTANDING 5.2

Solve the triangle with:

$$a = 4, \ b = 3, \text{ and } \beta = 15°$$

Answer:

$c \approx 6.6, \alpha \approx 20°, \gamma \approx 145°$

$c' \approx 1.0, \alpha' \approx 160°, \gamma' \approx 5°$

232   Chapter 5 Solving Oblique Triangles

E X E R C I S E S

Exercises 1–6: Use the Law of Sines initially to solve the triangle.

1. $\beta = 33°, \ \gamma = 18°, \ b = 8$

2. $\alpha = 15°, \ \beta = 107°, \ c = 7$

3. $\alpha = 100°, \ \beta = 55°, \ c = 10$

4. $\alpha = 20°, \ \gamma = 95°, \ a = 7$

5. $\beta = 31°, \ \gamma = 24°, \ a = 15$

6. $\alpha = 20°, \ \gamma = 85°, \ a = 7$

Exercises 7–12: Use the Law of Cosines initially to solve the triangle.

7. $a = 2, \ b = 6, \ c = 5$

8. $a = 40, \ c = 131, \ \beta = 9°$

9. $b = 15, \ c = 12, \ \alpha = 50°$

10. $a = 3, \ b = 7, \ c = 4$

11. $a = 3, \ b = 7, \ c = 5$

12. $a = 10, \ b = 8, \ c = 3$

Exercises 13–24: Solve the triangle.

13. $\alpha = 30°, \ b = 7, \ c = 5$

14. $\alpha = 57°, \ \beta = 138°, \ b = 5$

15. $a = 10, \ b = 20, \ c = 35$

16. $a = 9, \ b = 4, \ \gamma = 146°$

17. $\beta = 68°, \ \gamma = 83°, \ b = 5$

18. $\alpha = 10°, \ b = 20, \ \gamma = 55°$

19. $a = 3.5, \ b = 1.2, \ c = 4.0$

20. $a = 49, \ b = 9, \ c = 40$

21. $a = 0.6, \ \alpha = 20°, \ \beta = 123°$

22. $a = 9, \ b = 9, \ c = 5$

23. $a = 14, \ c = 9, \ \beta = 100°$

24. $a = 4, \ b = 5, \ c = 7$

Exercise 25:   Solve the isosceles triangle with $\alpha = 100°,$ and $a = 10$.

Exercises 26–43:   Solve the triangle. There may be no triangle, one triangle or two triangles.

26. $a = 8, \ c = 10, \ \alpha = 32°$

27. $\gamma = 30°, \ a = 4, \ c = 10$

28. $a = 4, \ b = 5, \ \beta = 17°$

29. $a = 7, \ b = 15, \ \alpha = 49°$

30. $b = 10, \ c = 20, \ \beta = 25°$

31. $\alpha = 33°, \ a = 5, \ b = 10$

32. $a = 9, \ c = 40, \ \alpha = 50°$

33. $a = 9, \ c = \dfrac{9}{\sin 50°}, \ \alpha = 50°$

34. $a = 9, \ c = 8, \ \gamma = 42°$

35. $a = 4, \ b = 5, \ \alpha = 27°$

36. $a = 2, \ c = 5, \ \alpha = 80°$

37. $a = 10, \ b = 5, \ \alpha = 31°$

38. $b = 6, \ c = 3, \ \beta = 47°$

39. $b = 6, \ c = 7, \ \beta = 55°$

40. $b = 4, \ c = 6, \ \gamma = 130°$

41. $b = 10, \ c = 9, \ \gamma = 62°$

42. $a = 10, \ b = 11, \ \beta = 53°$

43. $a = 2.3, \ b = 5.6, \ \alpha = 24°$

5.1 Law of Sines and Law of Cosines **233**

Exercise 44: (**Parallelogram**) The length of a diagonal of a parallelogram is 7 inches and it makes an angle of 27° with a side of length 3 inches. What are the dimensions of the parallelogram?

Exercise 45: Show that a triangle is a right triangle if the square of one side equals the sum of the squares of the other two sides (i.e. the Pythagorean Theorem holds only for right triangles).

Exercise 46: The Law of Cosines can be used in the ambiguous case of the Law of Sines. (While this requires more work, it does remove the ambiguity from the situation.) Solve Examples 5.3 and 5.4, using only the Law of Cosines.

Exercise 47: Use the Law of Cosines to prove that the sum of the squares of the lengths of the sides of a parallelogram equals the sum of the squares of the lengths of its two diagonals.

Exercise 48: Use the Law of Cosines to show that there will always exist a unique triangle with given sides a, b, and c, whenever each of the given sides is less than the combined length of the remaining two.

Exercise 49: Verify the following rules determining when there is no triangle, one triangle, or two triangles, given a, b, and angle α, where $h = b\sin\alpha$:

(a) When α is acute,

    i. there is no triangle when $a < h$

    ii. there is one triangle when $a = h$, or $a \geq b$

    iii. there are two triangles when $h < a < b$

(b) When α is obtuse,

    i. there is no triangle when $a \leq b$

    ii. there is one triangle when $a > b$

Exercise 50: Find the smallest a for which there exists a triangle with $b = 6$ and $\alpha = 30°$ [see Figure 5.2(b)].

Exercise 51:

(a) Does there exist a triangle of area 3 with $b = 6$ and $\alpha = 30°$? If so, determine the length of a.

(b) Find the area of another triangle with $b = 6$, $\alpha = 30°$ and the a of part (a).

Exercise 52: If possible, find a such that there exist two triangles with $b = 6$ and $\alpha = 30°$ [see Figure 5.2(c)], and for one of those triangles, $c = 3$.

Exercise 53: Find the range of a for which there exist two triangles with $b = 6$ and $\alpha = 30°$ (see Figure 5.2).

Exercise 54: If possible, find a such that there exist two triangles with $b = 6$ and $\alpha = 30°$ (see Figure 5.2), and for one of those triangles $c = 3$, and for the other $c' = 8$.

Exercise 55:

(a) Find two different numbers c and c', such that there exist two triangles both having $b = 6$, and $\alpha = 30°$, but one triangle has side c while the other has side c'.

(b) Is there only one pair of numbers c and c' satisfying (a)? Justify your answer.

Exercise 56: Find the value of α in the adjacent figure, for which a parallelogram can be constructed.

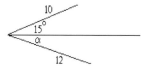

Exercise 57: Can the adjacent figure be constructed? If so? is it a parallelogram? Justify your answer.

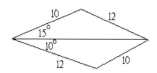

Exercise GC58: Use a graphing calculator to find the approximate dimensions of the triangle of maximum perimeter which is positioned in the first quadrant and has one vertex at the origin, another on the line $y = x$, and another on the line $y = 2x$, and whose side opposite the vertex at the origin has length 2.

§2. OBLIQUE TRIANGLE APPLICATIONS

The applications discussed in Section 3.2 eventually led to finding an unknown angle or side of a triangle. The same is true here, except that the triangles will no longer be right triangles. The Law of Sines or the Law of Cosines will need to be invoked in the solution.

EXAMPLE 5.6 Two guy wires are anchored 83.0 feet apart on opposite sides of a pole. What is the height of the pole, if the angles the guy wires make with the ground are 50° and 62°?

SOLUTION:

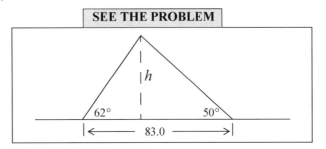

We can easily find the third angle of the triangle: $180° - 62° - 50° = 68°$. Knowing two angles and the side opposite one of them, we now turn to the Law of Sines to find the value of x:

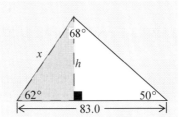

$$\frac{\sin 68°}{83.0} = \frac{\sin 50°}{x}$$

$$x = \frac{83.0 \sin 50°}{\sin 68°} \approx 68.58$$

We can now use the shaded right triangle to determine the height of the pole:

$$\sin 62° = \frac{h}{x}$$

$$h = x \sin 62° \approx 68.58 \sin 62° \approx 60.6$$

We conclude that the height of the pole is approximately 60.6 feet.

EXAMPLE 5.7 A lighthouse is 10.3 km northeast of a dock. A ship leaves the dock at 9 A.M. and steams east 12.2 km/hr. At what time will it be 9.0 km from the lighthouse?

SOLUTION:

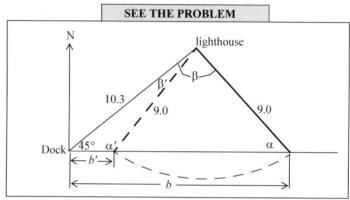

We anticipate that there are two answers to the problem. Applying the Law of Sines (ambiguous case), we have:

$$\frac{\sin \alpha}{10.3} = \frac{\sin 45°}{9.0}$$

$$\sin \alpha = \frac{10.3}{9.0} \cdot \sin 45° = \frac{10.3}{9.0} \cdot \frac{1}{\sqrt{2}}$$

$$\alpha = \sin^{-1}\left(\frac{10.3}{9.0} \cdot \frac{1}{\sqrt{2}}\right) \approx 54° \quad \text{and} \quad \begin{aligned} \alpha' &\approx 180° - 54° = 126° \\ \beta' &\approx 180° - 45° - 126° = 9° \end{aligned}$$

$$\beta = 180° - 45° - 54° = 81°$$

Applying the Law of Cosines to both triangles (we could use the Law of Sines), we obtain the values of b and b':

$$b = \sqrt{10.3^2 + 9.0^2 - 2(10.3)(9.0)\cos 81°} \approx 12.6 \text{ km}$$

and $b' = \sqrt{10.3^2 + 9.0^2 - 2(10.3)(9.0)\cos 9°} \approx 2.0$ km

All that remains is to determine the time it takes the ship to navigate those distances, and to add that time to its time of departure (9 A.M.):

First time ship is 9.0 km from lighthouse:

time = distance / rate

$$9 \text{ A.M.} + \frac{2.0 \text{ km}}{12.2 \frac{\text{km}}{\text{hr}}} = 9 \text{ A.M.} + \frac{2.0}{12.2} \text{hr}$$

$$= 9 \text{ A.M.} + \frac{2.0}{12.2} \text{hr} \cdot \frac{60 \text{ min}}{1 \text{ hr}} \approx 9 \text{ A.M.} + 10 \text{ min} \approx 9:10 \text{ A.M.}$$

Second time ship is 9.0 km from lighthouse:

$$9 \text{ A.M.} + \frac{12.6 \text{ km}}{\frac{12.2 \text{ km}}{\text{hr}}} = 9 \text{ A.M.} + \frac{12.6}{12.2}\text{hr}$$

$$= 9 \text{ A.M.} + \frac{12.6}{12.2}\text{hr} \cdot \frac{60 \text{ min}}{\text{hr}} \approx 9 \text{ A.M.} + 62 \text{ min} \approx 10{:}02 \text{ A.M.}$$

CHECK YOUR UNDERSTANDING 5.3

A dragon, fleeing Iltrode, flew 30 miles due west, and then altered his course by 30° and flew an additional 136 miles to Ghant. How far is Ghant from Iltrode?

Answer: ≈ 162.7 miles

BEARING

As you recall, a bearing is used to describe the direction of an object from a given point. You are already familiar with one type of bearing: the acute angle between the north-south line and the line of sight to the object (see Figure 3.4, page 137). A bearing can also be specified by an angle measured clockwise (not counterclockwise!) from due north. For example, the bearing of 78° in Figure 5.3(a) indicates a direction of 78° clockwise from due North, while the bearing in Figure 5.3(b) indicates a direction of 240° measured clockwise from due North.

The bearing in (a) can also be given as: N 78° E; and that in (b) as: S 60° W.

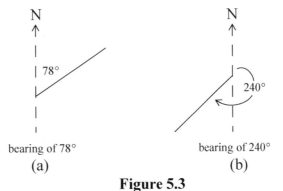

bearing of 78°
(a)

bearing of 240°
(b)

Figure 5.3

EXAMPLE 5.8 Nina and Jared start from the same point at 2 P.M. Nina walks due north at a rate of two miles per hour, while Jared runs at a rate of four miles per hour at a bearing of 330°. How far apart are they at 4 P.M.?

238   Chater 5 Solving Oblique Triangles

SOLUTION:

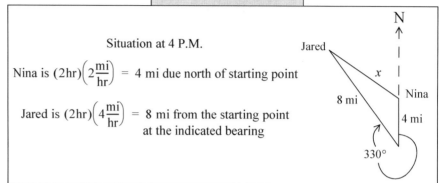

SEE THE PROBLEM

Situation at 4 P.M.

Nina is $(2\text{hr})\left(2\dfrac{\text{mi}}{\text{hr}}\right) = 4$ mi due north of starting point

Jared is $(2\text{hr})\left(4\dfrac{\text{mi}}{\text{hr}}\right) = 8$ mi from the starting point at the indicated bearing

Knowing two sides of the triangle, and the included angle: $360° - 330° = 30°$, we turn to the Law of Cosines to determine the desired distance x:

$$x = \sqrt{8^2 + 4^2 - 2(8)(4)\cos 30°}$$
$$= \sqrt{64 + 16 - 64\left(\dfrac{\sqrt{3}}{2}\right)} \approx 5.0$$

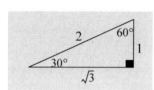

Conclusion: at 4 P.M. Nina and Jared are approximately five miles apart.

EXAMPLE 5.9 Using sonar equipment, a fishing boat detects a school of tuna 3 nautical miles from the boat at a bearing of $37°$. The tuna are swimming at a speed of 5 nautical miles per hour at a bearing of $320°$. In what direction should the boat travel to intercept the tuna, if the boat is to maintain a constant speed of 20 nautical miles per hour?

SOLUTION:

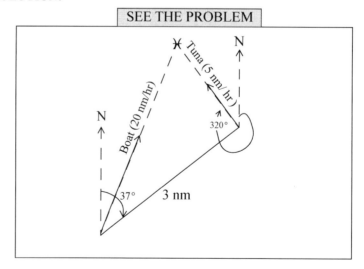

SEE THE PROBLEM

Let x denote the distance the tuna travels before being intercepted by the ship. Since the ship is moving four times faster than the tuna, it will travel $4x$ nautical miles to the point of intersection. This is reflected in the adjacent figure, wherein we are to determine the angle α.

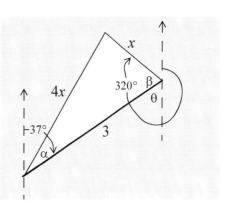

The dark line in the figure is a transversal to the two (parallel) North lines. Consequently: $\theta = 37°$ (Theorem 3.6, page 134). We can now find the angle β:

$$320° = 180° + \theta + \beta$$
$$\beta = 320° - 180° - 37° = 103°$$

Applying the Law of Sines, we have:

$$\frac{\sin 103°}{4x} = \frac{\sin \alpha}{x}$$

$$\sin \alpha = \frac{x \sin 103°}{4x} = \frac{\sin 103°}{4}$$

$$\alpha = \sin^{-1}\left(\frac{\sin 103°}{4}\right) \approx 14°$$

We conclude that in order to intercept the tuna, the boat should navigate at a bearing of approximately $37° - 14° = 23°$.

> We are in the Ambiguous Case of the Law of Sines, but know that α must be an acute angle (how?)
>
> ```
> sin⁻¹(sin(103)/4)
>             14.10
> ```

CHECK YOUR UNDERSTANDING 5.4

A ship sails from Port Pleasant for 112 nautical miles at a bearing of 27°. It then turns and sails at a bearing of 290°. How far is the ship from the port when its bearing from the port is 325°?

Answer: ≈ 193.8 nm.

EXERCISES

Exercise 1: (**A Shadow**) Tipsy Toby, ramrod straight and 6 feet tall, is leaning forward at an angle of 3° from the vertical. What is the length of Toby's shadow if the sun is directly behind him, with an angle of elevation of 39°?

Exercise 2: (**Rolling Marble**) A board is at an incline of 28° with the ground. A marble is released from a point A on the board 25 inches from the base of the board. It comes to rest at a point B on the ground 75 inches from the base of the board. What is the straight line distance between A and B?

Exercise 3: (**Angle of Depression**) An eagle is flying in a straight line at constant altitude toward its nest. At one point, the angle of depression to the nest is 25°. After flying 50 meters farther, the angle of depression is 40°, how many meters is the eagle from its nest at that moment?

Exercise 4: (**Roads**) Two straight roads meet at an angle of 30° at Middletown, one leading to Pleasantville and the other to Daisyville. Pleasantville is 2 miles from Middletown and 3 miles from Daisyville. How far is Daisyville from Middletown?

Exercise 5: (**Angle of Elevation**) A plane traveling in a straight line at constant altitude passes over a stadium and then over a tower that is 5 miles from the stadium. A few minutes later the angle of elevation from the stadium to the plane is 50° and the angle of elevation from the base of the tower to the plane is 71°. How far is the plane from the stadium at that moment?

Exercise 6: (**Bearing-Airplane**) At 1:00 P.M. plane A is directly over plane B. Plane A is flying at 210 mph due north at an altitude of 4200 feet, and plane B is flying 250 mph at an altitude of 3000 feet and at a bearing of 323°. How far apart are the planes at 2:00 P.M.?

Exercise 7: (**Bearing-Ship**) A ship leaves Port Slumber at noon and sails due east for 6 hours at a rate of 15 mph. It then changes its direction to a bearing of 40° and alters its speed to 20 mph for an additional 30 minutes before docking on the Isle of Lethargy. How far is Port Slumber from the Isle of Lethargy?

Exercise 8: (**Baseball**) A baseball diamond forms a square of side 90 feet. The pitching mound is 10 inches high and is located 60.5 feet from home plate, between home plate and second base (see figure). How far is the pitching rubber (situated atop the pitching mound) from first base?

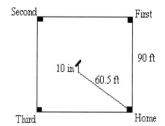

Exercise 9: (**Parallelogram**) One of the vertex angles of a parallelogram is 30°. Find all of its angles and its area, if two of its sides are 3 inches long , and the other two sides are 4 inches long.

Exercise 10: (Post on Hillside) A 25 foot post is to be set vertically on a hillside that is at an incline of 35° with the horizontal (see figure). A cable is to run from the top of the post to an eye-hook positioned on the ground, up the hill from the post. How far from the base of the post is the eye-hook positioned, if the cable is 35 feet long.

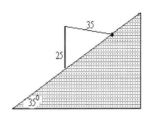

Exercise 11: (Post on Hillside) (Referring to the previous exercise). It turns out that the strongest reinforcement of the post will result when the wire makes an angle of 45° with the hillside. Where should the eye-hook be positioned, and how long a cable will be required?

Exercise 12: (Bearing-Forest Fire) Forest ranger A is positioned 25 miles due south of ranger B. B spots a fire at a bearing of 316°, while A spots the same fire at a bearing of 328°. A chemical drop is to be made over the fire from a helicopter stationed 23 miles due west of A. How long will it take the chopper to reach the fire, if it flies at a speed of 83 mph?

Exercise 13: (Bearing-Fishing) Using sonar equipment, a fishing vessel locates a school of tuna 3 miles away at a bearing of 37°. It finds that the school is traveling at a rate of 6 mph at a bearing of 141°.

(a) Traveling at 25 mph, find the direction the boat should take to intercept the school.

(b) How long will it take the boat to reach the tuna?

Exercise 14: (Bearing-Boating) A boat departs from a dock, A, and travels at a bearing of 23° for 10 minutes at a speed of 27 mph to position B. It then travels at a bearing of 50° for 15 minutes at 30 mph to position C, before changing its course to a bearing of 27°. It travels in that direction at a speed of 25 mph reaching the dock, D, in 12 minutes.

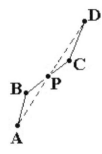

Find the minimum time it would take the boat to go from dock A to dock D at a speed of 35 mph (see adjacent figure).

(Suggestion: Let x be the distance from B to P, and express the straight line distance from A to D as a function of x. Use a graphing utility to find the minimum distance. Then find the minimum time.)

Chapter Summary

LAW OF SINES:	$$\frac{\sin\alpha}{a} = \frac{\sin\beta}{b} = \frac{\sin\gamma}{c}$$ 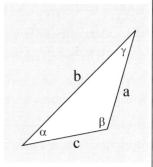
LAW OF COSINES:	$$a^2 = b^2 + c^2 - 2bc\cos\alpha$$ $$b^2 = a^2 + c^2 - 2ac\cos\beta$$ $$c^2 = a^2 + b^2 - 2ab\cos\gamma$$
THE AMBIGUOUS CASE OF THE LAW OF SINES	Given two sides of a triangle, say a and b, and the angle opposite one of those sides, say angle α, then there are three possibilities: 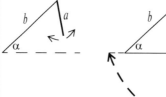 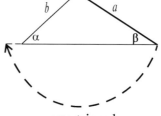 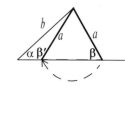 no triangle — one triangle — two triangles
BEARING	A bearing indicates the direction of an object Q from a given point P. One type of bearing uses the acute angle between the north-south line and the line of sight to the object [figure (a) below]. Another bearing uses the angle measured clockwise from due north [figure (b) below]. Bearing of Q from P 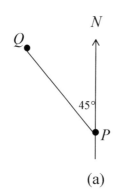 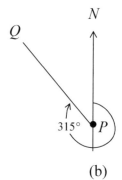 (a) bearing of N 45° W — (b) bearing of 315°

Chapter Review Exercises   243

CHAPTER REVIEW EXERCISES

Exercises 1–15: Solve the triangle.

1. $a = 5$, $b = 10$, $\gamma = 30°$

2. $b = 3$, $c = 7$, $\alpha = 45°$

3. $a = 6$, $c = 4$, $\beta = 115°$

4. $a = 6$, $b = 7$, $c = 4$

5. $a = 6$, $b = 4$, $c = 11$

6. $a = 7$, $b = 25$, $c = 24$

7. $b = 15$, $c = 7$, $\gamma = 20°$

8. $a = 15$, $b = 8$, $\beta = 12°$

9. $a = 14$, $b = 7$, $\alpha = 70°$

10. $a = 10$ $b = 20$, $\alpha = 55°$

11. $b = 37$, $c = 24$, $\gamma = 110°$

12. $a = \frac{1}{2}$, $c = 1$, $\gamma = 80°$

13. $a = 5$ $b = 6$, $c = 3$

14. $a = 4$, $b = 7$, $\gamma = 65°$

15. $a = 3$, $b = 2$, $\beta = 41°$

Exercise 16: (**A Shadow**) A yardstick is set vertically on an inclined plane, and casts a shadow of 4 feet along the plane when the angle of elevation of the sun is 27°. Find the angle of inclination of the plane.

Exercise 17: (**Bearing**) A boat leaves a dock and sails for 3 miles at a bearing of 335°. It then alters its course to a bearing of 205° and sails for 6 miles. How far away is the dock?

Exercise 18: (**Soccer**) Armando, Guiglielmo, and Franco are playing soccer. Armando is 23 meters from Guiglielmo, and 17 meters from Franco. The angle between the line of sight from Armando to Guiglielmo and the line of sight from Armando to Franco is 29°. How far apart are Guiglielmo and Franco?

Exercise 19: (**Angle of Elevation**) Ivan and his son David are 100 feet apart when they observe Adam's parrot at a point between them, at respective angles of elevation of 67° and 39°. How high above them is the parrot?

Exercise 20: (**Longitude/Latitude**) Find the shortest straight-line distance between two points on the surface of the earth that have the same longitude but are separated by 30° of latitude. Use 3950 miles for the radius of the earth.
[See Exercise 76 in Section 3-3 (page 150) for longitude/latitude.]

Exercise 21: (**Leaning Tower of Pisa**) The leaning tower of Pisa is 179 feet tall. How far does the tower lean from its original vertical position, if the angle of elevation to the top of the tower from a point 150 feet from its base (in the direction the tower leans) is 53.3°?

244   Chapter 5 Solving Oblique Triangles

CUMULATIVE REVIEW EXERCISES

Exercises 1–2: Simplify.

1. $\dfrac{[(2a-4)^2 b^3]^5}{(2-a)^3 \sqrt{b}}$

2. $\dfrac{x^{\frac{1}{2}}(3x-4)^2 - x^{-\frac{1}{2}}(3x-4)}{2x(4-3x)}$

Exercise 3:   Determine the equation of the line through the y-intercept of the line $6x = 2y - 3$ and parallel to the line through the points $(6, -8)$ and $(2,8)$.

Exercises 4–6: If $f(x) = \sqrt{x-1}$ and $g(x) = \dfrac{3}{4x}$ determine:

4. $(f \circ g)\left(\dfrac{3}{5}\right)$

5. $(f \circ f \circ f)(26)$

6. $(g \circ g \circ f)(x)$

Exercise 7:   Find the inverse function of $f(x) = \dfrac{x}{x+1}$ and verify that $(f \circ f^{-1})(x) = x$.

Exercise 8:   Sketch the graph of the function $f(x) = |x-2| - 1$ by reflecting, stretching, and shifting the basic graph. Determine and label the intercepts.

Exercise 9:   Verify the identity $\dfrac{\cos x}{1 - \sin x} = \sec x + \tan x$.

Exercise 10:   Solve $3\tan^2 2x - 1 = 0$ for x in $[0, 2\pi)$.

Exercises 11–14: Solve the triangle.

11. $c = 25, \quad \alpha = 35°, \quad \beta = 68°$

12. $a = 132, \quad b = 224, \quad \gamma = 28°$

13. $b = 17, \quad a = 12, \quad \alpha = 24°$

14. $a = 51, \quad c = 41, \quad \gamma = 38°$

Exercise 15: (**Angle of Inclination**) A tower 175 ft high is situated at the top of a hill. At a point 500 ft down the hill the angle between the surface of the hill and the line of sight to the top of the tower is 11°. Find the angle of inclination of the hill.

Exercise 16:   Sketch the graph of $f(x) = -2\cos\left(\dfrac{1}{2}x - \dfrac{\pi}{3}\right)$.

Exercise 17:   Evaluate $\cos 2x$, if $\sin x = \dfrac{1}{4}$.

Exercises 18–21: Evaluate each of the following.

18. $\tan\left(-\dfrac{\pi}{6}\right)$

19. $\sin\left(\dfrac{3\pi}{4}\right)$

20. $\cos \pi$

21. $\sec\left(\dfrac{\pi}{3}\right)$

Exercise 22:   If $\sin\theta = 3x - 1$ and $0 < \theta < \dfrac{\pi}{2}$,

(a) Solve for θ in terms of x

(b) Express $\tan\theta$ as an algebraic function of x

Exercise 23:   Show that

$$\sin^{-1}\frac{5}{13} + \tan^{-1}\frac{7}{24} = \cos^{-1}\frac{253}{325}$$

$\left[\text{Hint: Show that the left side is in } [0, \pi] \text{ and has a cosine equal to } \dfrac{253}{325}.\right]$

CHAPTER 6
Polynomial Functions

A function of the form $p(x) = a_n x^n + a_{n-1} x^{n-1} + \cdots + a_0$, with $a_n \neq 0$ is said to be a **polynomial function** of degree n. Since polynomials are defined everywhere, the domain of any polynomial function is $(-\infty, \infty)$.

Polynomial functions of degree 1, $p(x) = a_1 x + a_0$, or simply $p(x) = ax + b$, are said to be **linear functions**. As was observed in Chapter 1, graphs of linear functions are lines.

Polynomial functions of degree 2 are called **quadratic functions** and, they, along with solutions of quadratic equations, are considered in Section 1. In Section 2, we focus our attention on the solution of higher degree polynomial equations, and then move on to a consideration of polynomial inequalities in Section 3. While there is no disputing the value of graphing utilities, benefits remain to be derived from learning how to graph a function by hand, and graphing techniques are discussed in Section 4.

§1. QUADRATIC FUNCTIONS

The height of a ball (in feet) t seconds after being thrown directly upward is given by the function $s(t) = -16t^2 + v_0 t + s_0$. This is an example of a quadratic function. In general:

DEFINITION 6.1
QUADRATIC FUNCTION

A **quadratic function** is a function which can be expressed in the form:

$$f(x) = ax^2 + bx + c, \text{ with } a \neq 0$$

The graph of a quadratic function can be shown to be a parabola. We will show that the parabola opens upward if $a > 0$ [Figure 6.1(a)], and downward if $a < 0$ [Figure 6.1(b)].

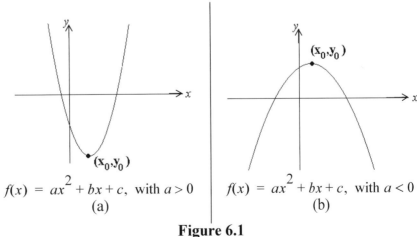

$f(x) = ax^2 + bx + c$, with $a > 0$ $f(x) = ax^2 + bx + c$, with $a < 0$
(a)   (b)

Figure 6.1

246   Chapter 6 Polynomial Functions

The point (x_0, y_0) in Figure 6.1(a) or (b) is called the **vertex** of the parabola. The vertex is the lowest or minimum point if the parabola opens upward [Figure 6.1(a)], and it is the highest or maximum point if the parabola opens downward [Figure 6.1(b)]. In either case, the graph of a parabola is symmetric with respect to the vertical line through its vertex (that line is called the **axis** of the parabola).

The following theorem can be used to determine the vertex of any parabola:

THEOREM 6.1   The vertex of the parabolic graph of the quadratic function $f(x) = ax^2 + bx + c$ occurs at:

$$x = x_0 = -\frac{b}{2a}$$

PROOF: The proof hinges on the completing-the-square method of page 53:

$$f(x) = ax^2 + bx + c$$

Factor out the coefficient of x^2:
$$= a\left(x^2 + \frac{b}{a}x\right) + c$$

Add the square of one-half the coefficient of x inside the brackets, and subtract a times that number so as not to alter the equation:
$$= a\left[x^2 + \frac{b}{a}x + \left(\frac{b}{2a}\right)^2\right] + c - a\left(\frac{b}{2a}\right)^2$$

$$= a\left(x + \frac{b}{2a}\right)^2 + \left(c - \frac{b^2}{4a}\right)$$

For $a > 0$: The term $a\left(x + \frac{b}{2a}\right)^2$ must be greater than or equal to 0, with equality holding at $x = -\frac{b}{2a}$. It follows that $f(x)$ attains its minimum value at $x = -\frac{b}{2a}$ (the x-coordinate of the vertex of a parabola that opens upward).

For $a < 0$: The term $a\left(x + \frac{b}{2a}\right)^2$ must be less than or equal to 0, with equality holding at $x = -\frac{b}{2a}$. It follows that $f(x)$ attains its maximum value at $x = -\frac{b}{2a}$ (the x-coordinate of the vertex of a parabola that opens downward).

We also see from the proof that the vertex of the parabola is at (x_0, y_0), when $f(x)$ is expressed in **standard form**:

$$f(x) = a(x - x_0)^2 + y_0$$

6.1 Quadratic Functions   247

EXAMPLE 6.1   Find the vertex, and sketch the graph of the quadratic function:

$$f(x) = -3x^2 + 6x - 1$$

Indicate the x- and y-intercepts of the graph. Determine the range of the function.

SOLUTION: Since the leading coefficient of $-3x^2 + 6x - 1$ is negative, the parabolic graph of f opens downward, and Theorem 6.1 tells us that the vertex occurs at:

$$x_0 = -\frac{b}{2a} = -\frac{6}{2(-3)} = -\frac{6}{-6} = 1$$

$$\underset{a}{\underset{\uparrow}{-3x^2}} + \underset{b}{\underset{\uparrow}{6x}} - 1$$

Applying the function to the x-coordinate of the vertex, we find its y-coordinate:

$$y_0 = f(x_0) = f(1) = -3(1)^2 + 6(1) - 1 = 2$$

The vertex is at the point $(1, 2)$.

To find the y-intercept, we evaluate the function at $x = 0$: $f(0) = -1$. We need the quadratic formula to find the x-intercepts:

$$f(x) = 0$$

$$-3x^2 + 6x - 1 = 0: \quad x = \frac{-6 \pm \sqrt{6^2 - 4(-3)(-1)}}{2(-3)} = \frac{-6 \pm \sqrt{24}}{-6}$$

$$= \frac{-6 \pm 2\sqrt{6}}{-6} = \frac{-2(3 \pm \sqrt{6})}{-6} = \frac{3 \pm \sqrt{6}}{3}$$

Graph of $f(x) = -3x^2 + 6x - 1$:

From the graph we see that the range of f is $R_f = (-\infty, 2]$

Answer:

Vertex: $(-2, -18)$

y-int: -10; x-ints: -5, 1

range: $[-18, \infty)$

For graph, see page C-39.

CHECK YOUR UNDERSTANDING 6.1

Find the vertex, and sketch the graph of the quadratic function:

$$f(x) = 2x^2 + 8x - 10$$

Indicate the x- and y-intercepts of the graph, and the range of f.

QUADRATIC FUNCTIONS AND EQUATIONS

As noted on page 94, the solutions of an equation $f(x) = 0$ are the x-intercepts of the graph of f. It follows that the graph of $f(x) = ax^2 + bx + c$ has two x-intercepts if the discriminant $b^2 - 4ac$ of the quadratic equation $ax^2 + bx + c = 0$ is positive, one x-intercept if it is zero, and no x-intercept if it is negative (see Figure 6.2).

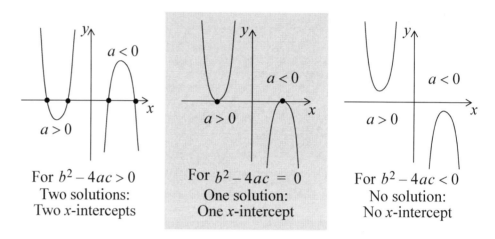

Figure 6.2

CHECK YOUR UNDERSTANDING 6.2

From the discriminant $b^2 - 4ac$ of the quadratic equation $f(x) = 0$ determine the number of x-intercepts of the graph of f.

(a) $f(x) = 3x^2 + 2x + 1$

(b) $f(x) = -x^2 + 8x - 16$

(c) $f(x) = 5x^2 - 4x - 2$

Answers: (a) 0 (b) 1 (c) 2

EXAMPLE 6.2 BREAK-EVEN POINTS

It costs a company $2x^2 + 100x + 3600$ dollars per month to produce x units, for $x \leq 200$. Due to supply and demand, the price per unit drops linearly. If 100 units are manufactured, then each unit will sell for $300, while the price drops to $100 per unit, if 200 units are manufactured. Find the monthly production levels at which break-even points occur.

The **break-even point** of production occurs when total cost of production equals total revenue, or to put it another way: when profit (= revenue - cost) is zero.

SOLUTION: We are given the cost function:

$$C(x) = 2x^2 + 100x + 3600$$

We are also told that there is a linear relationship between the price per unit, u, and the number x of units produced. Knowing that $(100, 300)$ and $(200, 100)$ lie on the graph of that linear function, we have:

$$u = \underbrace{\frac{300-100}{100-200}}_{\text{slope of line}} x + b = -2x + b$$

since $(100, 300)$ is on line: $\quad 300 = -2(100) + b$

$$b = 500$$

Multiplying the unit-price function $u(x) = -2x + 500$ by the number of units produced we arrive at the revenue function:

$$R(x) = (-2x + 500)x = -2x^2 + 500x$$

Setting total cost equal to total revenue, we solve for the break-even point of the company:

$$2x^2 + 100x + 3600 = -2x^2 + 500x$$
$$4x^2 - 400x + 3600 = 0$$
$$x^2 - 100x + 900 = 0$$
$$(x - 90)(x - 10) = 0$$
$$x = 90 \text{ or } x = 10$$

We see that break-even points occur at production levels of 10 units per month and at 90 units per month. Note that these break-even points occur at the x-coordinates of the points of intersection of the graphs of the cost and revenue functions below.

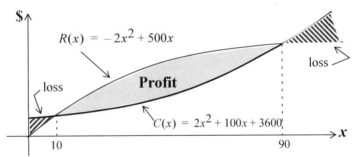

CHECK YOUR UNDERSTANDING 6.3

Referring to the cost function $C(x) = 2x^2 + 100x + 3600$ of Example 6.2, the constant term 3600 represents the **fixed monthly cost** of the company: insurance, taxes, etc. Determine the monthly production levels at which break-even points occur, if the fixed monthly cost is increased to $6,400. (Assume that apart from the fixed-cost, all conditions of Example 6.2 remain in place.)

Answer: 80 units; 20 units

NONLINEAR SYSTEMS OF EQUATIONS

We now turn our attention to systems of equations that do not consist entirely of linear equations.

EXAMPLE 6.3 Solve the system:
$$\left.\begin{array}{r}2x^2 + 3x - y = 1 \\ -x + y = 3\end{array}\right\}$$

SOLUTION: Adding the two equations, we arrive at an equation in x which is easily solved:

$$2x^2 + 3x - y = 1$$
$$\boxed{-x + y = 3}$$
$$2x^2 + 2x = 4 \longrightarrow x^2 + x - 2 = 0$$
$$(x+2)(x-1) = 0$$
$$x = -2 \text{ or } x = 1$$
$$y = x+3: \quad y = 1 \quad\quad y = 4$$

We see that there are two solutions to the given system of equations, $(x = 2, y = 1)$ and $(x = 1, y = 4)$, and they correspond to the points of intersection $(-2, 1)$ and $(1, 4)$, respectively, in the margin figure.

The given equations can easily be written in function form:
$$y = 2x^2 - 3x - 1$$
and: $y = x + 3$

Looking at the graphs of those functions we can see that the system will have two solutions (do you understand why?):

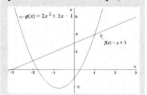

EXAMPLE 6.4 Solve the system:
$$\left.\begin{array}{l}(1): 2x + y = 2 \\ (2): 4x^2 + 2x - 3y = -1\end{array}\right\}$$

SOLUTION:

From (1): $y = 2 - 2x$ \quad (*)

Substitute in (2): $4x^2 + 2x - 3(2 - 2x) = -1$

Solve for x: $\quad 4x^2 + 2x - 6 + 6x = -1$
$$4x^2 + 8x - 5 = 0$$
$$(2x - 1)(2x + 5) = 0$$
$$x = \frac{1}{2} \,\bigg|\, x = -\frac{5}{2}$$

From (*): $y = 2 - 2\left(\frac{1}{2}\right) = 1 \,\bigg|\, y = 2 - 2\left(-\frac{5}{2}\right) = 7$

Solutions: $\left(x = \frac{1}{2}, y = 1\right)$ and $\left(x = -\frac{5}{2}, y = 7\right)$

6.1 Quadratic Functions 251

CHECK YOUR UNDERSTANDING 6.4

(a) Solve:
$$\left.\begin{array}{r} x^2 - 4x - y = 4 \\ x^2 + y = 2 \end{array}\right\}$$

(b) Sketch the parabolas associated with the two given quadratic equations.

Answers:
(a) $(x = 3, y = -7)$
    $(x = -1, y = 1)$
(b) See page C-40.

EXAMPLE 6.5 Find all $0 \leq \alpha < 2\pi$ and $0 \leq \beta < 2\pi$ for which:

$$\left.\begin{array}{l}(1):\ 2\sin\alpha + \cos\beta = 2 \\ (2):\ 4\sin^2\alpha + 2\sin\alpha - 3\cos\beta = -1\end{array}\right\}$$

SOLUTION: Invoking the substitutions:
$$x = \sin\alpha \text{ and } y = \cos\beta \quad (*)$$
the given system takes on a less intimidating form:

$$\left.\begin{array}{l}(1):\ 2x + y = 2 \\ (2):\ 4x^2 + 2x - 3y = -1\end{array}\right\}$$

From Example 6.4:
$$\left(x = \frac{1}{2}, y = 1\right) \text{ and } \left(x = -\frac{5}{2}, y = 7\right)$$

Returning to (*):
$$\left(\sin\alpha = \frac{1}{2}, \cos\beta = 1\right) \text{ and } \left(\sin\alpha = -\frac{5}{2}, \cos\beta = 7\right)$$

Since the magnitude of the sine and cosine functions cannot exceed 1, we discard $\left(\sin\alpha = -\frac{5}{2}, \cos\beta = 7\right)$ and turn our attention to the admissible solution: $\left(\sin\alpha = \frac{1}{2}, \cos\beta = 1\right)$:

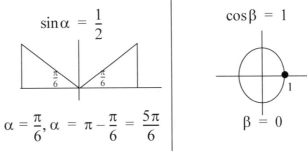

$$\sin\alpha = \frac{1}{2} \qquad\qquad \cos\beta = 1$$

$$\alpha = \frac{\pi}{6},\ \alpha = \pi - \frac{\pi}{6} = \frac{5\pi}{6} \qquad\qquad \beta = 0$$

Conclusion: $\left(\alpha = \frac{\pi}{6}, \beta = 0\right)$ and $\left(\alpha = \frac{5\pi}{6}, \beta = 0\right)$ are the solutions of the given system of equations in the specified region.

Answer:
$\left(\alpha = \frac{3\pi}{2}, \beta = \frac{2\pi}{3}\right), \left(\alpha = \frac{3\pi}{2}, \beta = \frac{4\pi}{3}\right)$

CHECK YOUR UNDERSTANDING 6.5

Find all $0 \leq \alpha < 2\pi$ and $0 \leq \beta < 2\pi$ for which:

$$\left. \begin{array}{r} 3\sin^2\alpha - 2\sin\alpha + \cos\beta = \dfrac{9}{2} \\ 4\sin\alpha + 2\cos\beta = -5 \end{array} \right\}$$

OPTIMIZATION PROBLEMS

Consider the graph of the function f in Figure 6.3. Since $f(a) \geq f(x)$ for every x "near" a, f is said to have a **local maximum** at a, and $f(a)$ is said to be a **local maximum value** of f (another local maximum occurs at c). Since for every x "near" b, $f(b) \leq f(x)$, f is said to have a **local minimum** at b, and $f(b)$ is said to be a **local minimum value** of f (another local minimum occurs at d).

Since the greatest value of f occurs at c, we say that the **absolute maximum** (or simply maximum) occurs at that point, and the maximum (value) of f is $f(c)$. By the same token, the **absolute minimum** (or simply minimum) of f is seen to occur at b, with minimum (value) $f(b)$.

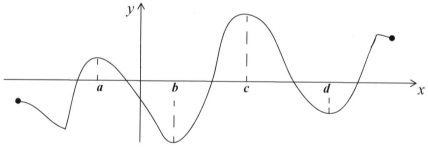

Figure 6.3

Problems in which one is to determine the maximum or minimum value of a function are called **optimization problems**. While in general one needs the calculus to solve such problems, this is not the case with quadratic functions, for, as previously noted, the maximum or minimum is the y-coordinate of the vertex.

EXAMPLE 6.6 Determine the maximum or minimum value of the given function, and indicate whether it is a maximum or a minimum:

(a) $f(x) = 4(x-5)^2 - 7$

(b) $f(x) = -2x^2 - 4x + 1$

SOLUTION:
(a) The graph of $f(x) = 4(x-5)^2 - 7$ opens upward. From the given standard form, we see that the minimum value of -7 occurs at $x = 5$.

(b) The graph of $f(x) = -2x^2 - 4x + 1$ opens downward, with maximum value occurring at $x = -\dfrac{b}{2a} = -\dfrac{-4}{2(-2)} = -1$, and:

Maximum value: $f(-1) = -2(-1)^2 - 4(-1) + 1 = -2 + 4 + 1 = 3$

Answers: (a) 11, maximum (b) $\frac{11}{3}$, minimum.

CHECK YOUR UNDERSTANDING 6.6

Determine the maximum or minimum value of:
(a) $f(x) = -3(x+2)^2 + 11$ (b) $f(x) = 3x^2 - 2x + 4$

The main step in the solution process of an optimization word problem is to express the quantity to be optimized as a function of one variable. To achieve that end, we suggest the following 4-step procedure:

Step 1. See the problem.

Step 2. Express the quantity to be maximized or minimized in terms of any convenient number of variables.

Step 3. If the expression in Step 2 involves more than one variable, use the given information to eliminate all but one variable in that expression.

Step 4. Either "by hand," or using a graphing calculator (for an approximate solution), find where the maximum or minimum of the single variable function obtained in Step 3 occurs, and then use it to determine the specified quantities.

EXAMPLE 6.7
MAXIMUM AREA

A farmer has 1200 feet of fencing with which to enclose a rectangular field. By taking advantage of the straight bank of a river, only three sides of the region will have to be fenced. Find the dimensions of the region of maximum area.

SOLUTION:
Step 1:   SEE THE PROBLEM

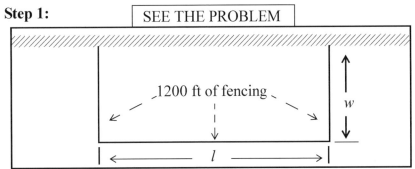

254 Chapter 6 Polynomial Functions

Step 2: Area is to be maximized, and we can easily express it in terms of the two indicated variables:

$$A = lw \qquad (*)$$

Step 3: To eliminate one of the variables (either l or w), we look for a relation involving those variables, and find it in the given condition that 1200 feet of fencing is to be used:

$$l + 2w = 1200 \text{ or: } l = 1200 - 2w \qquad (**)$$

By substituting $1200 - 2w$ for l in $(*)$, we are able to express A as a function of one variable:

$$A = lw = (1200 - 2w)w = -2w^2 + 1200w$$

Step 4: The above is a quadratic function (in the variable w) with parabolic graph opening downward. Applying Theorem 6.1, we find that the maximum value of $A(w)$ occurs at:

$$w = -\frac{b}{2a} = \frac{-1200}{-4} = 300$$

Returning to $(**)$, we obtain the length l:

$$l = 1200 - 2(300) = 600$$

Conclusion: To enclose the greatest area, the farmer should lay out a 300 ft by 600 ft region, with the 600 ft length parallel to the river.

EXAMPLE 6.8
MAXIMUM YIELD

When 20 peach trees are planted per acre, each tree will yield 200 peaches. For every additional tree planted per acre, the yield of each tree diminishes by 5 peaches. How many trees per acre should be planted to maximize yield?

SOLUTION:

Step 1:

SEE THE PROBLEM

Number of trees per acre	Yield per tree
20	200
20 + **1**	200 − **1** · 5
20 + **2**	200 − **2** · 5
20 + **3**	200 − **3** · 5
...	...
20 + x	200 − x · 5

Step 2: Letting x denote the number of trees above 20 to be planted per acre, we express the yield per acre, as a function of x:

number of trees per acre → ← yield per tree

$$Y(x) = (20 + x)(200 - 5x)$$
$$= -5x^2 + 100x + 4000$$

Step 3: Not necessary, as the function to be maximized, $Y(x)$, is already expressed in terms of one variable, x.

Step 4: Applying Theorem 6.1, we find that the maximum value of Y occurs at:

$$x = -\frac{b}{2a} = -\frac{100}{2(-5)} = 10$$

Conclusion: To maximize yield, $20 + x = 20 + 10 = 30$ trees per acre should be planted.

CHECK YOUR UNDERSTANDING 6.7

A rectangular field is to be enclosed on all four sides with a fence. One side of the field borders a road, and the fencing material to be used for that side costs $8 per foot. The fencing material for the remaining sides costs $6 per foot. Find the maximum area that can be enclosed for $2800.

Answer: $\frac{35000}{3}$ ft^2

EXERCISES

Exercises 1–6: Determine the vertex of the parabola.

1. $f(x) = x^2 + 4x + 7$
2. $f(x) = x^2 + 6x + 8$
3. $f(x) = 3x^2 - 12x + 14$
4. $f(x) = -x^2 - 2x + 2$
5. $f(x) = \frac{1}{4}x^2 + \frac{1}{2}x - \frac{11}{4}$
6. $f(x) = 2x^2 + 10x$

Exercises 7–14: Determine the vertex and x- and y-intercepts of the graph of the quadratic function. Sketch the graph, and determine the range of the function.

7. $f(x) = x^2 - 8x + 12$
8. $f(x) = 3(x-4)^2 - 6$
9. $f(x) = -4(x+5)^2 + 16$
10. $f(x) = 4(x-1)^2 - 3$
11. $f(x) = -3x^2 + 6x - 2$
12. $f(x) = 3x^2 - 5x - 1$
13. $f(x) = 2x^2 - 4x + 5$
14. $f(x) = -2x^2 - 3x - 2$

Exercise 15: Find a quadratic function $f(x) = ax^2 + bx + c$ whose graph is a parabola with vertex at $(2, -3)$ and passing through the point $(-1, 1)$.
[Suggestion: Begin with the standard form, $f(x) = a(x - x_0)^2 + y_0$.]

Exercises 16–17: Find the equation of the parabola whose graph is given.
[Suggestion: Begin with the standard form, $y = a(x - x_0)^2 + y_0$.]

16.
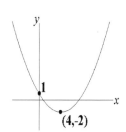

17.
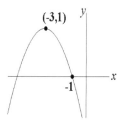

Exercises 18–19: Determine the number of x-intercepts of the graph of f, from the discriminant of the equation $f(x) = 0$.

18. $f(x) = x^2 + 6x + 9$
19. $f(x) = -3x^2 + 2x + 4$

Exercise 20: (**Break-Even Point**) Suppose the daily cost and revenue functions for a certain company are given by:

$$C(x) = \frac{1}{10}x^2 + 40x + 360 \quad \text{and} \quad R(x) = 60x$$

where x denotes the number of units produced and sold. Determine the daily production levels at which break-even points occur.

6.1 Quadratic Functions 257

Exercises GC21–GC23: (**Break-Even Point**) Use a graphing calculator to graph the cost and revenue functions for the production and sale of x units. From the graph, determine the two production levels at which break-even points occur.

21. $C(x) = \dfrac{x^2}{625} + 10x + 746$ and $R(x) = 25x$

22. $C(x) = 10x + 196$ and $R(x) = -\dfrac{x^2}{100} + 20x$

23. $C(x) = \dfrac{x^2}{200} + 10x + 197$ and $R(x) = -\dfrac{x^2}{400} + 20x$

Exercise GC24: (**Break-Even Point**) A company has a monthly fixed cost of $10,000, and a variable cost of production. The variable cost per unit decreases linearly. If the company produces 500 units, the variable cost per unit is $5. If the company produces 1000 units, the variable cost per unit reduces to $4. If the company produces x units per month, it can sell each unit for $\$\left(7 - \dfrac{x}{1000}\right)$.

(a) Determine Cost and Revenue as functions of x.

(b) Use a graphing calculator to find the monthly production levels at which break-even point(s) occur.

Exercises 25–30: Solve the nonlinear system of equations.

25. $\left.\begin{array}{l} 2x^2 + x = 8 + y \\ y = 2x - 7 \end{array}\right\}$
    26. $\left.\begin{array}{l} 2x^2 + x = 8 + y \\ y = 5 \end{array}\right\}$
    27. $\left.\begin{array}{l} y - 2x^2 = -3 \\ y - x^2 = -2 \end{array}\right\}$

28. $\left.\begin{array}{l} x^2 - y = 2 \\ 2x^2 + y = 6x + 7 \end{array}\right\}$
    29. $\left.\begin{array}{l} x^2 + y^2 = 13 \\ x^2 - y = 7 \end{array}\right\}$
    30. $\left.\begin{array}{l} x^2 + y^2 = 9 \\ x - 2y = 3 \end{array}\right\}$

Exercise 31: Find the points of intersection of the two curves

$$(x - 4)^2 + y^2 = 9 \quad \text{and} \quad y = \frac{\sqrt{5}}{2}x - 2\sqrt{5}$$

Exercise 32: (**Classical Problem**) (This problem dates back to the ancient Egyptians, about 2000 B.C.) Divide 100 square measures into two squares such that the side of one of the squares shall be three-fourths the side of the other. [That is, solve the system $x^2 + y^2 = 100$, and $y = \frac{3}{4}x$.]

Exercise 33: (**Field Dimensions**) The area of a rectangular field is 2,275 ft^2 and the perimeter is 200 ft. Find the dimensions of the field.

258   Chapter 6 Polynomial Functions

Exercises 34–38: Solve the system of equations for $0 \le \alpha < 2\pi$ and $0 \le \beta < 2\pi$.

34. $\left. \begin{array}{l} 3 \cot^2 \alpha - 2 \cos \beta = 7 \\ \cos^2 \beta + 4 \cos \beta - \cot^2 \alpha = 2 \end{array} \right\}$

35. $\left. \begin{array}{l} 2 \sin \alpha - 3 \cos \beta = 1 \\ \cos \beta = \dfrac{1}{\sin \alpha} \end{array} \right\}$

36. $\left. \begin{array}{l} 6 \cos \alpha + \tan \beta = 2 \\ 2 \cos^2 \alpha + 5 \cos \alpha + 3 \tan \beta = 13 \end{array} \right\}$

37. $\left. \begin{array}{l} -5 \tan^2 \alpha + 4 = \csc \beta \\ 7 \tan^2 \alpha - 2 \csc \beta = 9 \end{array} \right\}$

38. $\left. \begin{array}{l} 3 \sin^2 \alpha - 4 \cos \beta = -2 \\ 5 \sin^2 \alpha + 2 \cos \beta = 1 \end{array} \right\}$

39. $\left. \begin{array}{l} 2 \sin^2 \alpha - \cos \beta = 2 \\ \sin \alpha + 2 \cos \beta = 1 \end{array} \right\}$

Exercises 40–45: Determine the maximum or minimum value of the quadratic function, and indicate which it is.

40. $f(x) = -4(x+2)^2 + 5$

41. $f(x) = \frac{1}{4}(x-3)^2 + 2$

42. $h(x) = -3x^2 - 1$

43. $g(x) = 5x^2 + 2$

44. $g(x) = 4x^2 + 24x + 38$

45. $h(x) = -x^2 - 3x$

Exercises 46–49: (Optimization - Free Fall) Near the surface of the earth, the height $s(t)$ (in feet) of an object subject only to the force of gravity is given by

$$s(t) = -16t^2 + v_0 t + s_0$$

where t is time in seconds, v_0 is initial velocity and s_0 is the initial height of the object. Velocity is positive when the object is going up, and negative when it is coming down.

46. A ball is thrown directly upward from the roof of a building with a speed of 80 ft/sec. The height of the ball above the ground (not the roof) t seconds later is given by

$$s(t) = -16t^2 + 80t + 225$$

   (a) How tall is the building?

   (b) To what maximum height above the roof will the ball rise?

47. A ball is to be thrown directly upward from the base of a 50 ft building so as to reach the top of the building. Find the minimum velocity with which it must be thrown.

48. A ball is thrown directly upward from the edge of the roof of a 384 foot building at a speed of 100 ft/sec.

   (a) To what maximum height above the roof will the ball rise?

   (b) How long after it was thrown will the ball hit the ground?

49. A faulty proof in a bottle is thrown straight up from the edge of the roof of a 192 ft building at 64 ft/sec. At the same instant, SM-man (Super Math man) tosses a stone upward from the ground along the line of trajectory of the bottle. What should be the initial velocity of the stone, if it is to hit the bottle at roof's height?

Exercise 50: (**Minimize Product**) What two real numbers that differ by 36 have the smallest possible product?

Exercise 51: (**Maximize Area**) What is the largest rectangular area that can be enclosed with 400 feet of fencing?

Exercise 52: (**Maximize Area**) A rectangular field is to be enclosed with a fence. One side of the field borders a road, and the fencing material to be used for that side costs $8 per foot. The fencing material for the remaining sides costs $6 per foot. Find the maximum area that can be enclosed for $1600.

Exercise 53: (**Minimize Cost**) It costs a company $450x$ dollars to produce x units. There are additional expenses given by $2x^2 - 1450x + 65000$. Determine the production level x which minimizes total cost.

Exercise 54: (**Optimize Area**) A 16-inch wire is to be cut into two pieces. One piece will be bent into a square, and the other is to be shaped into a rectangle of width 1 inch.

(a) Find the smallest possible combined area of the two regions.

(b) Find the largest possible combined area.

Exercise 55: (**Maximize Profit**) A company determines that its revenue and cost functions in the production of x units are

$$R(x) = -2x^2 + 800x \text{ and } C(x) = 3x^2 + 270$$

respectively. Find the maximum profit the company can make. [Profit = Revenue − Cost]

Exercise GC56: (**Maximize Profit**) An automobile dealer finds that his profit P (in hundreds of dollars) is a function of the total number x of salespeople in the showrooms, where
$$P(x) = -15.6x^2 + 1743x - 126$$
Use a graphing calculator to view the graph of the profit function and estimate

(a) the number of salespeople that results in maximum profit.

(b) the maximum profit.

Exercise 57: (**Maximize Crop Yield**) The yield of oranges per tree decreases if orange trees are planted too close together. When there are 50 trees per acre, the yield per tree is 300 oranges. The yield decreases by 5 oranges per tree for each additional tree planted per acre. How many trees should be planted per acre to produce the largest crop?

Exercise 58: (**Maximize Revenue**) The Better-Cable Company has 5000 customers paying $40 per month for their service. Assume that for every $1 rate increase, 100 customers will terminate their contract with the company. What should the company charge in order to maximize revenue?

Exercise 59: (**Maximize Revenue**) The Better-Cable Company has 5000 customers paying $40 per month for their service. Assume that every $1 rate reduction in price attracts 500 new customers. What should the company charge in order to maximize revenue?

Exercise GC60: Find a function of x representing the area of a rectangle inscribed in the region bounded below by the x-axis and above by the graph of the function $f(x) = \sin x$ on the interval $[0, \pi]$. (See adjacent figure.) Use a graphing calculator to estimate the value of x which results in maximum area. Estimate the maximum area.

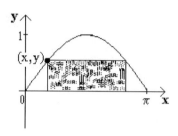

Exercise 61: Find a quadratic function $f(x) = ax^2 + bx + c$ whose graph passes through the points $(1,2)$ and $(-1,8)$. Can there be more than one such function? Justify your answer.

Exercise 62: Does there exist a quadratic function $f(x) = ax^2 + bx + c$ whose graph passes through the four points $(1,2)$, $(-1,1)$, $(0,0)$, and $(2,0)$? Justify your answer. Is it ever possible to find a quadratic function whose graph passes through four given points? Explain.

Exercise 63:

(a) At how many points can a line and a parabola intersect?

(b) At how many points can two parabolas intersect?

(c) At how many points can a parabola and a circle intersect?

(d) Translate your findings in (a)–(c) into statements about systems of equations. (For example, since 2 lines can intersect in no points, one point, or infinitely many points, a system of two linear equations in two unknowns can have no solution, one solution, or infinitely many solutions.)

Exercises 64–67: Show that the function is **not** one-to-one: Exhibit $a \neq b$ such that $f(a) = f(b)$.

64. $f(x) = x^2 + 4x + 6$

65. $f(x) = 3x^2 - 2x + 1$

66. $f(x) = -2x^2 - x - 1$

67. $f(x) = -5x^2 + 5x - 2$

Exercises 68–71: Sketch the graph of the piecewise-defined function, and indicate whether or not the function appears to be one-to-one.

68. $f(x) = \begin{cases} x, & \text{if } x \leq 1 \\ x^2, & \text{if } x > 1 \end{cases}$

69. $g(x) = \begin{cases} -x^2, & \text{if } x < 0 \\ x^2, & \text{if } x \geq 0 \end{cases}$

70. $h(x) = \begin{cases} 3x^2 + 1, \text{ if } x < 0 \\ 2x^2 + 3, \text{ if } x \geq 0 \end{cases}$

71. $k(x) = \begin{cases} x^2 - 1, \text{ if } x < -1 \\ 1 - x^2, \text{ if } x \geq -1 \end{cases}$

Exercises 72–75: Sketch the graph of the piecewise-defined function.

72. $f(x) = \begin{cases} x^2, \text{ if } x < 0 \\ 2x - 1, \text{ if } x \geq 0 \end{cases}$

73. $h(x) = \begin{cases} (x-2)^2 + 1, \text{ if } x < 2 \\ -(x-2)^2, \text{ if } x \geq 2 \end{cases}$

74. $k(x) = \begin{cases} -2(x+1)^2, \text{ if } x \leq 0 \\ 3(x-1)^2, \text{ if } x > 0 \end{cases}$

75. $g(x) = \begin{cases} x^2 - 2x - 3, \text{ if } x \leq 1 \\ x^2 - 7x + 12, \text{ if } 1 < x \leq 2 \\ -2x^2 + 10, \text{ if } x > 2 \end{cases}$

Exercise 76: (**Auto Acceleration**) Starting from rest, an automobile accelerates in such a way that its distance, in feet, from its starting point is given by $s(t) = 3t^2$, where t is measured in seconds. This acceleration is maintained for 6 seconds, after which the car maintains a constant speed of 36 feet/second until it reaches the first curve, some 720 feet from the starting place.

(a) How long will it take the car to reach the curve?

(b) Express the distance of the car from its starting point as a function of time, up to the time the car reaches the curve.

(c) Sketch the graph of the distance function in part (b).

Exercise 77: (**Satellite Distance**) A satellite is propelled directly away from the earth from a space station in orbit 2 miles above the earth (see figure). For the first minute of flight, the distance between the satellite and the space station (in miles) is given by $h(t) = t^2$ (where t is measured in minutes). One minute into the flight, an on-board computer effects a deceleration, and, for the next 3 minutes the distance versus time function is given by $h(t) = t$. At that point, a reverse directional booster is fired, and the satellite achieves orbit, stabilizing at a constant distance from earth.

(a) What is the orbiting distance of the satellite from the earth?

(b) Express the distance of the satellite from the earth as a function of time.

(c) Sketch the graph of the distance function in part (b).

262   Chapter 6 Polynomial Functions

(d) Suppose the deceleration causes the distance from the earth versus time function to be given instead by $h(t) = 0.1t^2 + 2.9$.

   (i) Express the distance of the satellite from the earth as a function of time.

   (ii) Sketch the graph of the distance function in part (i).

Exercise 78:   Write an explanation justifying the following observation:

One can locate points on the graph of the quadratic function $f(x) = ax^2 + bx + c$ by starting at the vertex of the parabola, and then moving over 1 unit (to the right or left) and up if $a > 0$ (or down, if $a < 0$) $|a| \cdot 1^2$ units, moving over 2 units from the vertex, and up if $a > 0$ (or down, if $a < 0$) $|a| \cdot 2^2$ units, etc.

§2. POLYNOMIAL FUNCTIONS AND EQUATIONS

As previously noted, a A **polynomial function of degree n** is a function of the form $p(x) = a_n x^n + a_{n-1} x^{n-1} + \ldots + a_1 x + a_0$, with $a_n \neq 0$. The numbers a_0 through a_n are said to be the **coefficients** of the polynomial, with a_n and a_0 the **leading coefficient** and **constant coefficient**, respectively.

Every polynomial function p gives rise to a polynomial equation $p(x) = 0$. Solving such an equation often hinges on the important fact that a product is zero **if**, and **only if**, one of its factors is zero. Consider the following example:

> It can be shown that every polynomial has a factorization that involves only linear and quadratic factors. However, finding such a factorization can be a difficult, if not impossible, task.

EXAMPLE 6.9 Solve:

(a) $x^3 + x^2 - 6x = 0$

(b) $x^4 = 16$

(c) $4x^4 - 9x^2 + 2 = 0$

(d) $3x^3 + 5x^2 - 6x - 10 = 0$

SOLUTION: (a)

$$x^3 + x^2 - 6x = 0$$

pull out the common factor, x: $\quad x(x^2 + x - 6) = 0$

factor further: $\quad x(x + 3)(x - 2) = 0$

Solution: $\quad x = 0 \text{ or } x = -3 \text{ or } x = 2$

> If $x^n = a$ then, if n is **odd**:
> $$x = a^{\frac{1}{n}}$$
> For example:
> $$x^3 = 8 \Rightarrow x = 8^{1/3} = 2$$
> $$x^3 = -8 \Rightarrow x = (-8)^{1/3} = -2$$
> If n is **even** and $a \geq 0$:
> $$x = \pm a^{\frac{1}{n}}$$
> For example:
> $$x^4 = 16 \Rightarrow x = \pm 16^{1/4} = \pm 2$$
> Note: There are no (real) solutions when n is even and $a < 0$.

(b) One approach:

$$x^4 = 16$$
$$x^4 - 16 = 0$$
$$(x^2 + 4)(x^2 - 4) = 0$$

can't be zero $\rightarrow (x^2 + 4)(x + 2)(x - 2) = 0$

$$x = -2 \text{ or } x = 2$$

also written: $x = \pm 2$

Another approach: $x^4 = 16 \Rightarrow x = \pm 16^{1/4} = \pm 2$

see margin

264   Chapter 6 Polynomial Functions

(c)

Within the equation:	$4x^4 - 9x^2 + 2 = 0$
Lurks a quadratic equation:	$4(x^2)^2 - 9(x^2) + 2 = 0$
Substitute: $Y = x^2$:	$4Y^2 - 9Y + 2 = 0$
And solve for Y:	$(4Y - 1)(Y - 2) = 0$

$$Y = \frac{1}{4} \text{ or } Y = 2$$

Substitute x^2 back for Y: $x^2 = \frac{1}{4}$ or $x^2 = 2$

Solution: $x = \pm\frac{1}{2}$ or $x = \pm\sqrt{2}$

(d) If you look closely at $3x^3 + 5x^2 - 6x - 10$ you might observe that by grouping the first two terms together, and the last two terms together, the common factor $(3x + 5)$ emerges:

$$3x^3 + 5x^2 - 6x - 10 = 0$$
$$x^2(3x + 5) - 2(3x + 5) = 0$$
$$(3x + 5)(x^2 - 2) = 0$$

Solution: $x = -\frac{5}{3}, x = \pm\sqrt{2}$

Answer:

(a) $\pm\sqrt{7}, \ \pm 3, \ -2, \ 7^{\frac{1}{3}}$

(b) $1, \ -3^{\frac{1}{3}}$ (c) -1

CHECK YOUR UNDERSTANDING 6.8

Solve:

(a) $(x^2 - 7)(x^4 - 81)(x^3 + 8)(x^3 - 7) = 0$ (b) $x^6 + 2x^3 = 3$

(c) $x^3 + x^2 = -9x - 9$

Way back on page 12 we stated that:

> A **zero of a polynomial** $p(x)$ is a number which when substituted for the variable x yields zero; that is: the zeros of $p(x)$ are the solutions of the equation $p(x) = 0$.

We then went on to establish the following result:

> c is a zero of a polynomial $p(x)$ if and only if $(x - c)$ is a factor of $p(x)$.

At times, one can spot a zero of a polynomial, as is the case in the following example:

EXAMPLE 6.10   Solve:

$$x^3 - 2x - 1 = 0$$

6.2 Polynomial Functions and Equations **265**

$$
\begin{array}{r}
x^2 - x - 1 \\
x + 1\overline{\smash{\big)}\ x^3 + 0x^2 - 2x - 1} \\
\underline{x^3 + x^2} \\
-x^2 - 2x - 1 \\
\underline{-x^2 - x} \\
-x - 1 \\
\underline{-x - 1} \\
0
\end{array}
$$

SOLUTION: It is easy to see that -1 is a zero of the polynomial $p(x) = x^3 - 2x - 1$ (also called a **root** of the equation $p(x) = 0$). Consequently, $x - (-1) = x + 1$ is a factor, and (see margin):

$$
\begin{array}{r}
x^2 - x - 1 \\
x + 1\overline{\smash{\big)}\ x^3 - 2x - 1}
\end{array}
$$

So:

$$x^3 - 2x - 1 = (x + 1)(x^2 - x - 1) = 0$$

quadratic formula:

$$x + 1 = 0 \qquad x^2 - x - 1 = 0$$

$$x = -1$$

$$x = \frac{1 \pm \sqrt{1^2 - 4(1)(-1)}}{2(1)} = \frac{1 \pm \sqrt{5}}{2}$$

Solution: $-1, \dfrac{1 - \sqrt{5}}{2}, \dfrac{1 + \sqrt{5}}{2}$.

When you are unable to spot a zero of a polynomial, you can turn to the following fact, which you are invited to prove in the exercises:

THEOREM 6.2

THE RATIONAL ZEROS THEOREM

Let

$$p(x) = a_n x^n + a_{n-1} x^{n-1} + \cdots + a_0$$

be a polynomial of degree n with integer coefficients. Each rational zero of $p(x)$ (reduced to lowest terms) is of the form $\dfrac{c}{l}$, where c is a factor of the constant coefficient a_0, and l is a factor of the leading coefficient a_n.

EXAMPLE 6.11 Solve:

$$4x^3 + 24x^2 - 27x + 7 = 0$$

SOLUTION: Theorem 6.2 tells us that these are the possible rational zeros of the polynomial $4x^3 + 24x^2 - 27x + 7$:

$$\frac{c}{l} = \frac{\pm 1 \text{ or } \pm 7}{\pm 1 \text{ or } \pm 2 \text{ or } \pm 4}: \ \pm 1, \pm\frac{1}{2}, \pm\frac{1}{4}, \pm 7, \pm\frac{7}{2}, \pm\frac{7}{4}$$

While neither 1 nor -1 pan out, $\dfrac{1}{2}$ does:

$$4\left(\frac{1}{2}\right)^3 + 24\left(\frac{1}{2}\right)^2 - 27\left(\frac{1}{2}\right) + 7 = \frac{4}{8} + 6 - \frac{27}{2} + 7 = \frac{4 + 48 - 108 + 56}{8} = 0$$

Dividing by the known factor $\left(x - \dfrac{1}{2}\right)$ (steps omitted), we arrive at:

$$
\begin{array}{r}
4x^2 + 26x - 14 \\
x - \frac{1}{2}\overline{\smash{\big)}\ 4x^3 + 24x^2 - 27x + 7}
\end{array}
$$

266 **Chapter 6 Polynomial Functions**

Bringing us to:
$$4x^3 + 24x^2 - 27x + 7 = 0$$

$$\left(x - \frac{1}{2}\right)(4x^2 + 26x - 14) = 0$$

$$\left(x - \frac{1}{2}\right)2(2x^2 + 13x - 7) = 0$$

$$(2x - 1)(2x - 1)(x + 7) = 0$$

Solution: $\frac{1}{2}, 7$

Answers: (a) $-\frac{1}{2}, 1, 5$

(b) $-4, -\frac{1}{2}, 5$ (c) $\pm\frac{1}{2}, \frac{1}{4}$

CHECK YOUR UNDERSTANDING 6.9

Solve: (a) $2x^3 - 11x^2 + 4x + 5 = 0$ (b) $2x^3 - x^2 - 41x - 20 = 0$

(c) $x^3 - \frac{1}{4}x^2 - \frac{1}{4}x + \frac{1}{16} = 0$ (Suggestion: multiply both sides by 16)

EXAMPLE 6.12 Find all $-\frac{\pi}{2} \le x \le \frac{\pi}{2}$ for which:

$$\sec^3 x + \sec^2 x - 6\sec x = 0$$

SOLUTION: Making the substitution $y = \sec x$, we have:
$$y^3 + y^2 - 6y = 0, \text{ where } y = \sec x \quad (*)$$

The above is the equation of Example 6.9(a) (with x replaced by y); which was shown to have solutions $\{0, -3, 2\}$; or, from (*):

:

$\sec x = 0$	$\sec x = -3$	$\sec x = 2$
no solution since $\frac{1}{\cos x}$ cannot equal 0	no solution in $\left[-\frac{\pi}{2}, \frac{\pi}{2}\right]$ since the secant is positive in that interval	$\cos x = \frac{1}{2}$ $\;x = \pm\frac{\pi}{3}$

Conclusion: $\pm\frac{\pi}{3}$ are the solutions of

$$\sec^3 x + \sec^2 x - 6\sec x = 0 \text{ in } \left[-\frac{\pi}{2}, \frac{\pi}{2}\right].$$

Answer: $\frac{\pi}{6}, \frac{5\pi}{6}$

CHECK YOUR UNDERSTANDING 6.10

Solve for $0 \le x \le \pi$:

$$\csc^3 x + \csc^2 x - 6\csc x = 0$$

6.2 Polynomial Functions and Equations   267

APPROXIMATE SOLUTIONS OF POLYNOMIAL EQUATIONS

Students typically have the mistaken impression that exact solutions of equations are always obtainable, and that they turn out to be rather nice numbers, such as 4 or $\frac{1}{2}$. In reality, however, even a simple cubic equation such as $-2x^3 + x + 5 = 0$ cannot be solved without some intricate mathematics. It is in such situations that graphing utilities truly shine, for they can at least supply approximate solutions. This being the case, it would be nice to know, in advance:

1. The number of possible solutions.

And: 2. An interval that contains all the solutions.

As for 1:

THEOREM 6.3
NUMBER OF SOLUTIONS

A polynomial equation of degree n has at most n solutions.

We are using a **proof by contradiction**. We start off in a very contrary fashion, and **assume** that what we are trying to prove is, in fact, false. We then go on to show that the assumption leads to a contradiction. Since a logical argument cannot take you from a true statement to a false statement, the initial **assumption** must itself be false.

PROOF: Suppose that the polynomial equation:

$$p(x) = a_n x^n + a_{n-1} x^{n-1} + \cdots + a_1 x + a_0 = 0, \ a_n \neq 0$$

has more than n solutions, say: $c_1, c_2, ..., c_{n+1}$. Since each c_i is a zero of the polynomial p, $(x - c_i)$ must be a factor of that polynomial (Theorem 1.4, page 12). But this cannot be the case, for

$$(x - c_1)(x - c_2)\cdots(x - c_n)(x - c_{n+1})$$

is already a polynomial of degree $n+1$, and the given polynomial is of degree n.

The following theorem, which you are invited to prove in Exercise 60, addresses the important issue of choosing a window size which will display **all** solutions of a given polynomial equation:

THEOREM 6.4
CONFINEMENT THEOREM

All solutions of the polynomial equation:

$$p(x) = x^n + a_{n-1} x^{n-1} + ... + a_1 x + a_0 = 0$$

are contained in the interval $[-K, K]$, where:

$$K = 1 + \text{largest}\{|a_{n-1}|, |a_{n-2}|, ..., |a_1|, |a_0|\}$$

IN WORDS: K is one more than the largest magnitude of the coefficients of the polynomial (with leading coefficient of 1) in $p(x) = 0$. For example, if $p(x) = x^5 - x^4 + 2x^2 - 3x + 1$, then $K = 1 + 3 = 4$, since -3 is the coefficient of largest magnitude.

It is important to note that the leading coefficient in the polynomial of Theorem 6.4 is **1**: $p(x) = 1x^n + a_{n-1}x^{n-1} + ... a_1 x + a_0$. As is demonstrated in the following example, this restriction can easily be circumvented.

EXAMPLE 6.13 Use a graphing calculator to approximate the solutions of the equation:

$$2x^5 - x^4 + 2x^2 - 4x + 1 = 0$$

SOLUTION: Since the leading coefficient is not 1, we cannot apply Theorem 6.4 directly. We must first divide both sides of the equation by 2 (which does not alter the solution set):

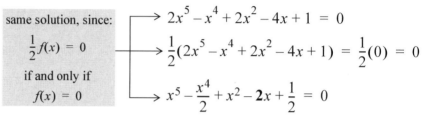

same solution, since:
$$\frac{1}{2}f(x) = 0$$
if and only if
$$f(x) = 0$$

$$\longrightarrow 2x^5 - x^4 + 2x^2 - 4x + 1 = 0$$
$$\longrightarrow \frac{1}{2}(2x^5 - x^4 + 2x^2 - 4x + 1) = \frac{1}{2}(0) = 0$$
$$\longrightarrow x^5 - \frac{x^4}{2} + x^2 - 2x + \frac{1}{2} = 0$$

Confident that all solutions of the given equations are confined to the interval $[-K, K]$, where $K = 1 + 2 = 3$, we set our window size accordingly ([A] of Figure 6.4). We then instructed the unit to position the cursor at each of the three x-intercepts of the function $f(x) = 2x^5 - x^4 + 2x^2 - 4x + 1$, and located the approximate solutions of the given equation: $x \approx -1.274$, $x \approx 0.292$, and $x = 1$.

A TI-92 teaser:

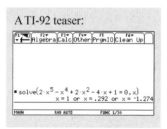

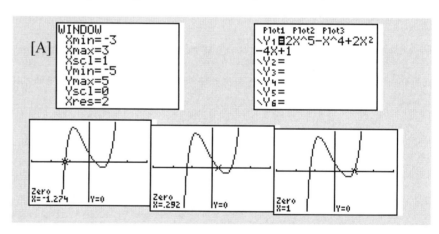

Figure 6.4

CHECK YOUR UNDERSTANDING 6.11

(a) Use Theorem 6.3 to determine the maximum number of solutions of the equation $-2x^3 + x + 5 = 0$.

(b) Use Theorem 6.4 to determine an interval containing all of the solutions of the equation in (a).

(c) Use a graphing calculator to approximate the solutions of the equation in (a), to two-decimal places.

Answers: (a) 3 (b) $\left[-\frac{7}{2}, \frac{7}{2}\right]$
(c) 1.48

| 6.2 Polynomial Functions and Equations | **269** |

ANOTHER OPTIMIZATION PROBLEM

In the next example we will use the four-step procedure for solving maximum/minimum problems that was introduced on page 253:

Step 1. See the problem.

Step 2. Express the quantity to be maximized or minimized in terms of any convenient number of variables.

Step 3. If the expression in Step 2 involves more than one variable, use the given information to eliminate all but one variable in that expression.

Step 4. Either "by hand," or using a graphing calculator (for an approximate solution), find where the maximum or minimum of the single variable function obtained in Step 3 occurs, and then use it to determine the specified quantities.

EXAMPLE 6.14

MAXIMUM VOLUME

A box with a square base is to be constructed. For mailing purposes, the perimeter of the base (the girth of the box), plus the length of the box, cannot exceed 108 in. Find the dimensions of the box of greatest volume.

SOLUTION:

Step 1:

SEE THE PROBLEM

Girth is measured around here ⟶

Step 2: Volume is to be maximized, and we are able to express it in terms of two variables (area of base times height):

$$V = x^2 y \tag{*}$$

Step 3: To eliminate one of the variables in (*), we look for a relationship between x and y, and find it in the condition that *length* + *girth* = 108:

$$y + 4x = 108 \quad \text{or} \quad y = 108 - 4x \tag{**}$$

length ⟶ ↑   ↑ girth

Substituting in (*), we obtain an expression for volume as a function of x:

$$V(x) = x^2(108 - 4x) = -4x^3 + 108x^2$$

270   Chapter 6   Polynomial Functions

Since *y* must be positive, we have:
$$108 - 4x > 0$$
$$-4x > -108$$
$$x < 27$$
x must also be positive. This led us to choose the indicated *x*-range, [0, 30].

Step 4:

 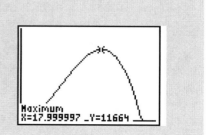

We see that the maximum volume of approximately 11,664 cubic inches occurs at $x \approx 18$. Substituting 18 for *x* in (**) we find the remaining dimension of the box:

$$y = 108 - 4 \cdot 18 = 36$$

Conclusion: The box of maximum volume is (approximately) 18 in. × 18 in. × 36 in. (With the calculus, you will be able to show that our answer is, in fact, exact.)

> A note on window sizing:
> Suppose you instruct your graphing calculator to graph a function *f*, and you see nothing at all. That can be frustrating. To remedy the situation, pick a number, *c*, in the domain of the function, and calculate $f(c)$. Setting the window size to include the point $(c, f(c))$ will surely display part of the graph. You can then, if necessary, modify the window size to reveal the point of interest.

CHECK YOUR UNDERSTANDING 6.12

Best Box Company is to manufacture open-top boxes from 12 in. by 12 in. pieces of cardboard. The construction process consists of two steps: (1) cutting the same size square from each corner of the cardboard, and (2) folding the resulting cross-like configuration into a box. What size square should be cut out, if the resulting box is to have the largest possible volume?

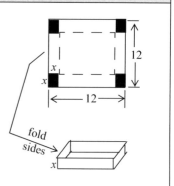

Answer: 2 inch square

6.2 Polynomial Functions and Equations **271**

EXERCISES

Exercises 1–10: Solve the equation, as in Examples 6.9(a),(b), and 6.10.

1. $x^5 = -32$
2. $x^7 + 1 = 0$
3. $x^4 - 1 = 0$

4. $x^6 = 64$
5. $x^3 = 4x$
6. $3x^3 + x^2 = 4x$

7. $2x^4 - 8x^3 + 3x^2 = 0$
8. $3x^4 + 6x^3 - 3x^2 = 0$
9. $9x^3 = 9x^2 - x + 1$

10. $x^4 + 2x^3 = x^2 + 2x$

Exercises 11–18: Make an appropriate substitution and solve [see Example 6.9(c)].

11. $(x^2 - 3x - 2)^2 = 4$
12. $(x^2 + 6x)^2 = 64$

13. $3x^4 + 2x^2 = 8$
14. $8x^6 - 7x^3 - 1 = 0$

15. $(x^2 - 4)^3 - 64 = 0$
16. $(2x - 1)^4 = 81$

17. $3(x^2 - 1)^2 = 10(x^2 - 1) + 8$
18. $(x + 1)^3 + 6 = 2(x + 1)^2 + 5(x + 1)$

Exercises 19–22: Spot a solution, and use it to solve the equation.

19. $x^3 + 5x = 6x^2 - 12$
20. $x^3 - 2x^2 - 5x = -6$

21. $2x^3 - 8x^2 + 3x + 3 = 0$
22. $3x^3 + x^2 = 4x + 2$

Exercises 23–30: Solve the equation, using the Rational Zeros Theorem (Theorem 6.2).

23. $18x^3 - 27x^2 + 13x - 2 = 0$
24. $30x^3 + 13x^2 - 8x - 3 = 0$

25. $9x^3 - 3x^2 - 4x + \frac{4}{3} = 0$
26. $\frac{3}{5}x^4 + x^3 + \frac{1}{5}x^2 + x - \frac{2}{5} = 0$

27. $30x^3 + 7x^2 - 10x = 3$
28. $27x^3 - 45x^2 + 6x = -8$

29. $x^4 - 27x^2 - 14x = -120$
30. $x^4 - 2x^3 - 17x^2 + 18x = -72$

Exercises 31–32: Solve the equation for $0 \leq x < 2\pi$.

31. $\sqrt{3} \cot^4 x + (\sqrt{3} - 1) \cot^3 x - \cot^2 x = 0$
32. $2\cos^4 x - 3\cos^2 x + 1 = 0$

Exercises 33–34: Solve the equation for $0 \leq x \leq \pi$.

33. $\sin^7 x - \sin^3 x = 0$
34. $\tan^5 x - 4\tan^3 x + 3\tan x = 0$

Exercises 35–38: Use the Confinement Theorem (Theorem 6.4), to find an interval containing all of the solutions of the equation. At most how many solutions can there be?

35. $9x^6 - 5x^5 + 12x^3 + 2x = 0$
36. $-\frac{1}{2}x^5 + \frac{2}{3}x^4 - \frac{9}{10}x^2 = 1$

37. $-x^4 + 4x = -3$
38. $6x^3 - 6x^2 + 5x - 3 = 0$

272   Chapter 6 Polynomial Functions

Exercises 39–42: Use the Confinement Theorem (Theorem 6.4), to find an interval containing all the x-intercepts of the graph of the function. At most how many x-intercepts can there be?

39. $f(x) = 5x^4 - 3x^3 + 4x - 1$

40. $g(x) = 8x^5 - 7x^3 + \sqrt{2}\,x^2 - 10x - \frac{1}{2}$

41. $k(x) = -4x^7 + 3x^4 - 12$

42. $h(x) = 3x^8 - 4x^6 + x^3 + 2$

Exercises GC43–GC45: Use a graphing calculator and the Confinement Theorem (Theorem 6.4), to approximate the solutions of the equation.

43. $-x^3 + 12x^2 = 6.4x + 3.8$

44. $5x^3 - 8.3x + 3.8 = 0$

45. $x^4 - 5x^3 = 10 - 7x$

Exercise GC46: (Maximum Profit) By experience, a company has found that its sales increase with more advertising, up to a point. The relationship between the company's profit P (in hundreds of dollars) and their advertising costs x (in thousands of dollars) is given by

$$P(x) = -2x^3 + 194x - 37$$

Use a graphing calculator to approximate

(a) the advertising costs that result in maximum profit.

(b) the maximum profit.

Exercise GC47: (Patient's Temperature) A patient's temperature T (in degrees Fahrenheit) during the first nine hours of a certain illness was given (approximately) by the following function of time t (in hours)

$$T(t) = 98.6 - 0.1(t^3 - 12t^2 + 27t)$$

Use a graphing calculator to approximate

(a) the times when the patient's temperature was $101°$.

(b) the patient's maximum temperature during this period.

(c) the patient's minimum temperature during this period.

Exercise GC48: (Patient's Temperature) A patient's temperature T (in degrees Fahrenheit) during the first four hours of a certain illness was given by

$$T(t) = 98.6 - \frac{1}{2}t^3 + 2t^2$$

Use a graphing calculator to view the graph of the temperature, and estimate the patient's maximum temperature and when it occurred.

Exercise GC49: (Medicine Dosage) The blood level of a certain dosage of medicine, in parts per million, after t hours is given by

$$N(t) = -0.045t^3 + 2.063t + 2$$

(a) Use a graphing calculator to view the graph of $N(t)$ and determine, approximately, the maximum value of the blood level, and when it occurs.

(b) The *minimum effective dosage* of the medication occurs for values of t for which $2 \leq N(t)$. Use the graph to estimate the times when the medication has its *minimum effective dosage*.

6.2 Polynomial Functions and Equations 273

Exercises GC50–GC54: Use the four-step procedure to solve the optimization problem.

50. **(Minimum)** Find two **integers** whose sum is 34, and the sum of the square of one and the fourth power of the other is a minimum.
(Suggestion: Locate the minimum and test the nearest integers to the left and right.)

51. **(Maximum Volume)** An open box whose base is five times as long as it is wide is to be constructed from 250 in.^{2} of cardboard. Estimate the dimensions of the box of maximum volume.

52. **(Maximum Volume)** A closed box with a square base is to be constructed from 100 in.^{2} of cardboard. Estimate the dimensions of the box of maximum volume.

53. **(Minimum Distance)** Find the minimum distance between the point (3,5) and the graph of the function $f(x) = x^2$. [Suggestion: Minimize the square of the distance between the point and a point, (x, x^2), on the graph of f.]

54. **(Maximum Volume)** A cylinder of radius r and height h is inscribed in a right circular cone of height 8 inches and base radius 5 inches. Find the dimensions r and h of the cylinder of maximum volume. (Volume of a cylinder: $V = \pi r^2 h$)

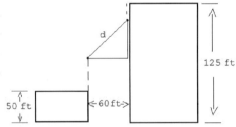

Exercise 55: **(Free Fall)** An object is thrown upward from the edge of the roof of a 50 foot building at a speed of 42 ft/sec, while, at the same time, another is dropped from the edge of the roof of a 125 foot building that is in line with the first building, and 60 feet away (see figure).

(a) Find the closest and furthest distance between the objects (the minimum and maximum values of d in the figure).

(b) Determine at what time the objects are closest, and their positions at that time.

(c) When the objects are closest, is the object that was thrown upward going up or down?

Exercises GC56–GC59: Use a graphing calculator to view the graph of the function, and answer (a)-(d) below.

56. $f(x) = 2x^3 - 5x^2 + 2x - 1$

57. $g(x) = 3x^7 - 6x^2 + 3$

58. $k(x) = 0.01x^2(x-4)^3$

59. $h(x) = 2x^6 + x^5 - x^4 + 2x^3 - x^2 - x + 1$

From the graph, it appears that:

(a) the function has (this many) ___local maxima and they occur at $x \approx$ ___, ...

(b) the local maximum values are $y \approx$ ___, ...

274   Chapter 6 Polynomial Functions

(c) the function has (this many) __local minima and they occur at $x \approx$ ___, ...

(d) the local minimum values are $y \approx$ ___, ...

Exercise GC60:   Use a graphing calculator to view the graph of the function

$$f(x) = 30x^3 - 30x^2 - \frac{x}{50} + \frac{1}{50}$$

(a) From that graph it appears that the function has (this many) ___ zeros, and that they occur at: $x \approx$ ___, ...

(b) Solve the equation $f(x) = 0$ by hand, algebraically (not geometrically), noting that 1 is a zero of the function.

(c) Now find a window that will enable you to spot the three solutions found in (b).

Exercise 61:   Construct a cubic polynomial with three distinct zeros such that your graphing utility will suggest the existence of less than three zeros, independently of the chosen viewing rectangle.
[Suggestion: Consider $f(x) = x(x - a)(x - b)$, with suitable a and b.]

Exercise 62:   Show that the pencil is mightier than the graphing calculator by plotting, on the number line, the three distinct zeros of the polynomial constructed in Exercise 61.

Exercise GC63: (**Rain Forests**) Although efforts to save the world's tropical rain forests have rightly received widespread attention, another type of rain forest is perhaps even more threatened. Now estimated to cover less than half their original area, coastal temperate rain forests are an exceptionally productive and biologically diverse system. They include some of the oldest and most massive tree species in the world, and constitute some of the largest remaining pristine landscapes in the temperate zone.[1] The percentage of protected coastal temperate rain forests in British Columbia is two and a half times the percentage protected in Oregon, and the percentage protected in Tasmania is twelve times greater than that in British Columbia.

Find the percent protected in Oregon, if its sixth power is only 4 percentage points greater than the percent protected in Tasmania, as follows:

(a) Write an equation for the percentage of protected forests in Oregon.

(b) See if you can find any exact solution.

(c) What is the smallest interval that contains all the solutions of the problem?

(d) Use a graphing calculator to see if there are additional solutions, besides any found in (b).

---

[1]For further information, see <u>Conserving the Other Rain Forest</u>, by Derek Denniston, Worldwatch Institute, vital Signs 1994 (pp. 124–125) W.W. Norton & Company, Inc.

6.2 Polynomial Functions and Equations   275

Exercises 64–69: Suppose that c is a solution of the polynomial equation $p(x) = 0$. Discuss in favor or against the given assertion.

64. c is a solution of the equation $\dfrac{p(x)}{q(x)} = 0$ for any polynomial q.

65. c is a solution of the equation $p(x)q(x) = 0$ for any polynomial q.

66. c cannot be a solution of the equation $p(x + 1) = 0$.

67. $c - 1$ is a solution of the equation $p(x + 1) = 0$.

68. $\dfrac{c}{2}$ is a solution of the equation $p(2x) = 0$.

69. $-c$ is a solution of the equation $-p(x) = 0$.

Exercise 70: Prove the Rational Zeros Theorem (Theorem 6.2) in the following steps.

(a) Show that $p\left(\dfrac{c}{l}\right) = 0$ can be written as

$$a_n c^n + a_{n-1} c^{n-1} l + a_{n-2} c^{n-2} l^2 + \ldots + a_1 c l^{n-1} = -a_0 l^n$$

(b) Show that as c divides the left side of the equation in (a), it must also divide the right side, and as it can not divide l^n (why?), c divides a_0.

(c) Rewrite the equation in (a) with only terms containing l on the left side of the equation, and repeat the argument in (b) for l instead of c.

Exercise 71: Extension of the Rational Zeros Theorem (Theorem 6.2) to polynomials with rational coefficients. Suppose $p(x)$ is a polynomial with rational coefficients and $\dfrac{c}{l}$ is a zero of $p(x)$ in lowest terms. Then c is a factor of ka_0 and l is a factor of ka_n, where k is the least common denominator of the coefficients, $a_n, a_{n-1}, \ldots, a_0$, in $p(x)$. Prove this extension in the following steps.

(a) Show that $\dfrac{c}{l}$ is a zero of $kp(x)$.

(b) Show that Theorem 6.2 applies with $kp(x)$ replacing $p(x)$.

Exercise 72: Prove the Confinement Theorem (Theorem 6.4): The zeros of

$$P(x) = x^n + a_{n-1} x^{n-1} + a_{n-2} x^{n-2} + \cdots + a_1 x + a_0$$

all lie in the interval $[-K, K]$, where $K = 1 + \max\{|a_{n-1}|, \ldots, |a_0|\}$.

Establish the following steps to show that if c is a number such that $c > K$, then $P(c) \neq 0$:

(a) Use the reverse triangle inequality ($|a + b| \geq |a| - |b|$), followed by the triangle inequality ($|a + b| \leq |a| + |b|$) to show that

$$|P(c)| \geq |c|^n - |a_{n-1}||c|^{n-1} - |a_{n-2}||c|^{n-2} - \cdots - |a_0|$$

(b) With $|c| = L > K$, show that $|P(c)| \geq L^n - (K - 1)(|L|^{n-1} + L^{n-2} + \cdots + L + 1)$.

(c) Show $|P(c)| \geq L^n - \left(\dfrac{K - 1}{L - 1}\right)(L^n - 1)$.

(d) Show $|P(c)| \geq 1 > 0$.

276   Chapter 6 Polynomial Functions

§3. POLYNOMIAL FUNCTIONS AND INEQUALITIES

Many of the applications that you will see in the calculus require the ability to solve inequalities such as the polynomial inequalities in this section. As you will see later on, the techniques used here generalize to other types of functions as well.

QUADRATIC INEQUALITIES

We can take advantage of the parabolic nature of the graph of quadratic functions to solve any quadratic inequality. Consider the following example.

EXAMPLE 6.15 Solve:

(a) $-3x^2 + 6x - 1 < 0$ (b) $-3x^2 + 6x - 1 \geq 0$

(c) $2x^2 + x + 1 > 0$ (d) $2x^2 + x + 1 \leq 0$

SOLUTION: [For both (a) and (b)] The graph of the function

$$f(x) = -3x^2 + 6x - 1$$

is that of a parabola which opens downward. To determine its x-intercepts, we solve $-3x^2 + 6x - 1 = 0$:

$$x = \frac{-6 \pm \sqrt{6^2 - 4(-3)(-1)}}{2(-3)} = \frac{-6 \pm \sqrt{24}}{-6} = \frac{-6 \pm 2\sqrt{6}}{-6}$$

$$= \frac{-2(3 \pm \sqrt{6})}{-2(3)} = \frac{3 \pm \sqrt{6}}{3}$$

From the graph of $f(x) = -3x^2 + 6x - 1$ in Figure 6.5(a), we see that:

Since

$y = -3x^2 + 6x - 1$, the solution of

$-3x^2 + 6x - 1 < 0$

is the set of x-coordinates of points on the parabola having a negative y-coordinate: the points **below** the x-axis.

(a) Solution of $-3x^2 + 6x - 1 < 0$: $\left(-\infty, \frac{3-\sqrt{6}}{3}\right) \cup \left(\frac{3+\sqrt{6}}{3}, \infty\right)$

where the graph of $y = -3x^2 + 6x - 1$ lies below the x-axis

(b)   Solution of $-3x^2 + 6x - 1 \geq 0$: $\left[\frac{3-\sqrt{6}}{3}, \frac{3+\sqrt{6}}{3}\right]$

where the graph of $y = -3x^2 + 6x - 1$ lies above or on the x-axis

The location of the vertex and the y-axis are of no concern to us here, for we simply want to know where the graph of the quadratic function lies above the x-axis and where it lies below the x-axis.

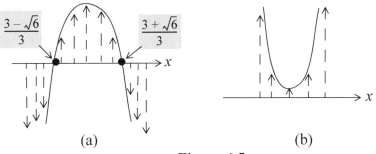

Figure 6.5

For both (c) and (d), The graph of the function
$$f(x) = 2x^2 + x + 1$$
is that of a parabola which opens upward. Since the discriminant of $2x^2 + x + 1$ is negative ($b^2 - 4ac = 1 - 8 = -7$), there are no solutions of $2x^2 + x + 1 = 0$, i.e. **no x-intercepts** of the graph. It follows that the graph must lie entirely above the x-axis [Figure 6.5(b)]. That being the case, we can conclude that:

(c)   Solution of $2x^2 + x + 1 > 0$: $(-\infty, \infty)$ ← where the graph lies **above** the x-axis

(d)   Solution of $2x^2 + x + 1 \le 0$: $\varnothing$ ← the empty set
        where the graph lies **below or on** the x-axis

Answers: (a) $\left[-2, \frac{1}{2}\right]$
(b) $\left(-\infty, \frac{-7-\sqrt{33}}{4}\right) \cup \left(\frac{-7+\sqrt{33}}{4}, \infty\right)$

> **CHECK YOUR UNDERSTANDING 6.13**
>
> Solve:
>   (a) $2x^2 + 3x - 2 \le 0$ (b) $2x^2 + 7x + 2 > 0$

GENERAL POLYNOMIAL INEQUALITIES

The following two observations will enable us to solve polynomial inequalities when the polynomial appears in factored form:

(i) The sign **of a polynomial** can only change from positive to negative or from negative to positive about a zero of the polynomial. (The graph of a polynomial function can only cross the x-axis at a zero of the polynomial.)

(ii) As you know, if c is a zero of a polynomial $p(x)$, then $(x - c)$ is a factor (Theorem 1.4, page 12). When $(x - c)$ occurs an odd number of times in the factorization of $p(x)$ we will say that c is a **odd-zero** of $p(x)$. When $(x - c)$ occurs an even number of times in the factorization of $p(x)$ we will say that c is an **even-zero** of $p(x)$.

For example, 7 is an odd-zero and -5 is an even-zero of:

$$\underbrace{(x-7)^3}_{\text{odd}}\underbrace{(x+5)^2}_{\text{even}}$$

Note that the sign of the above polynomial **will change** about the **odd**-zero 7 [since the sign of $(x-7)^3$ will be positive for $x > 7$ and negative for $x < 7$], and that the sign of the polynomial will **not change** about the **even**-zero -5 [since $(x+5)^2$ is positive on either side of -5]. In general:

> The sign of a polynomial will change as one traverses an odd-zero of the polynomial and will not change as one traverses an even-zero of the polynomial.

EXAMPLE 6.16 Solve:

$$(x+3)^3(x+1)^2(4-x) < 0$$

SOLUTION:

Step 1. Chart the sign of the polynomial as follows:
Locate the zeros of the polynomial on the number line:

Place the letter c above the zeros of factors which are raised to an odd power [the factors $(x+3)^3$ and $(4-x)^1$]; this is to remind us that the sign of the polynomial will **c**hange about the odd-zeros -3 and 4. Place the letter n above the even-zero -1 to remind us that the sign of the polynomial will **n**ot change about that even-zero:

Determine the sign of the polynomial $(x+3)^3(x+1)^2(4-x)$ to the right of its last zero. In this example, for $x > 4$ the product is easily seen to be **negative** [$(4-x)$ is the only negative factor], thus:

> Determining the sign of the polynomial to the left of -3 or within the interval $(-3, -1)$ or $(-1, 4)$ would serve as starting points just as well.

The c above 4 indicates that the sign will change about 4 (from negative to positive):

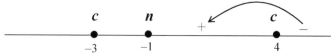

The n above -1 indicates that the sign will not change as you traverse -1 (will remain positive):

6.3 Polynomial Functions and Inequalities 279

Finally, the c above -3 indicates a sign changes:

Here is the end result:

$$\text{SIGN } (x+3)^3(x+1)^2(4-x)$$

Figure 6.6

Step 2. Since we are solving $(x+3)^3(x+1)^2(4-x) < 0$ we read off the intervals where the polynomial is negative (the "$-$" intervals): $(-\infty, -3) \cup (4, \infty)$.

NOTE: The information in Figure 6.6 also enables us to solve the inequalities

$(x+3)^3(x+1)^2(4-x) > 0$: $(-3, -1) \cup (-1, 4)$

$(x+3)^3(x+1)^2(4-x) \geq 0$: $[-3, 4]$

$(x+3)^3(x+1)^2(4-x) \leq 0$: $(-\infty, -3] \cup [4, \infty) \cup \{-1\}$

> One can also determine the SIGN of the expression $(x+3)^3(x+1)^2(4-x)$ by evaluating it at some arbitrarily chosen points within each of the intervals $(-\infty, -3)$, $(-3, -1)$, $(-1, 4)$, and $(4, \infty)$. **If** you calculate correctly, you will arrive at the same sign information depicted in the adjacent figure. (This "clumsier" method is called the *cut-point* method.)

CHECK YOUR UNDERSTANDING 6.14

Solve:

(a) $(x-3)(x+2)(-x+5) < 0$ (b) $(x+1)^2(x+2)^3(x-4)^2 \geq 0$

> Answers:
> (a) $(-2, 3) \cup (5, \infty)$
> (b) $[-2, \infty)$

EXAMPLE 6.17 Solve:
$$(x+3)^2(x^2-8x+15)(3x^2-4x-3) > 0$$

SOLUTION: We begin by factoring:

$$(x+3)^2(x^2-8x+15)(3x^2-4x-3) > 0$$
$$(x+3)^2(x-3)(x-5)(3x^2-4x-3) > 0$$

Step 1. Chart the sign of the polynomial:

The zeros are $x = -3$, $x = 3$, $x = 5$, and $x = \dfrac{2 \pm \sqrt{13}}{3}$ (see Example 1.32, page 56). On page 56 we noted that:

$$3x^2 - 4x - 3 = 3\left(x - \frac{2-\sqrt{13}}{3}\right)\left(x - \frac{2+\sqrt{13}}{3}\right)$$

Of the five zeros, only -3 is an even-zero. So, the sign of the polynomial will **not** change about -3, but will **change** about the rest:

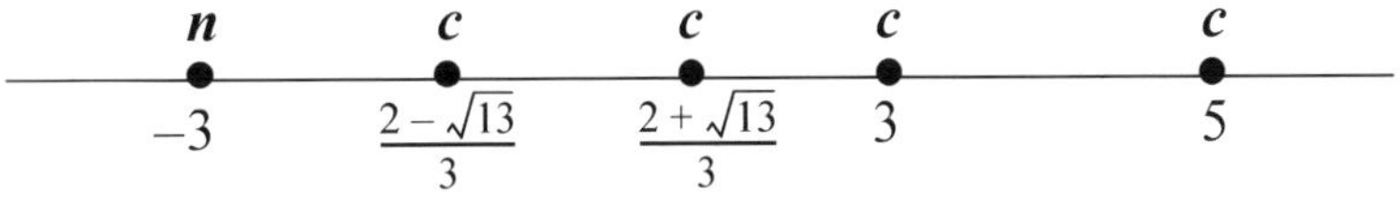

> In general, if a quadratic polynomial has a positive discriminant then it has two distinct zeros, and they are both odd-zeros. If the discriminant is zero, as with:
> $$x^2 - 2x + 1 = (x-1)^2$$
> then that zero is an even-zero. If the discriminant is negative, then there are no (real) zeros.

Determine the sign of the polynomial to the right of its last zero. In this example, for $x > 5$ the product is easily seen to be **positive**:

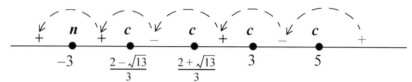

Proceed to the left, changing the sign each time you traverse a *c*, and not changing the sign when traversing an *n*:

Giving us:

$$\text{SIGN } (x+3)^2(x^2 - 8x + 15)(3x^2 - 4x - 3)$$

Step 2. Since we are solving

$$(x+3)^2(x^2 - 8x + 15)(3x^2 - 4x - 3) > 0$$

we read off the intervals where the polynomial is positive (the "+ intervals"):

$$(-\infty, -3) \cup \left(-3, \frac{2-\sqrt{13}}{3}\right) \cup \left(\frac{2+\sqrt{13}}{3}, 3\right) \cup (5, \infty)$$

EXAMPLE 6.18 Solve:
$$(x+3)^2(x^2 - 8x + 15)(3x^2 - 4x + 3) > 0$$

Note that a polynomial that is never zero is always positive or always negative. Just evaluate it at, say, zero to determine which is the case.

SOLUTION: We replaced the factor $(3x^2 - 4x - 3)$ of the previous example which had two zeros with the factor $(3x^2 - 4x + 3)$ which has no zeros: negative discriminant: $b^2 - 4ac = 16 - 36 = -20$. All else remains as in the previous example, leading us to:

$$\text{SIGN } (x+3)^2(x^2 - 8x + 15)(3x^2 - 4x + 3)$$

Solution of: $(x+3)^2(x^2 - 8x + 15)(3x^2 - 4x + 3) > 0$:

$$(-\infty, -3) \cup (-3, 3) \cup (5, \infty)$$

Answer:
$$(-\infty, -2] \cup \left[\frac{1-\sqrt{21}}{2}, 1\right]$$
$$\cup \left[\frac{1+\sqrt{21}}{2}, \infty\right)$$

CHECK YOUR UNDERSTANDING 6.15

Solve:
$$(x^2 + x - 2)(x^2 - x + 5)(-x^2 + x + 5) \leq 0$$

NON-POLYNOMIAL FACTORS

Since the sign of **any** factor (polynomial or not) raised to an **even** power is never negative, the sign of any such factor will **not change** as you traverse its zeros. However, as we observe below, when one traverses a zero of a factor of the form $[\text{trig}(x) - a]^{\text{odd}}$, the sign of that factor may or may not change as you traverse a zero:

For example, $\frac{\pi}{6}$ and $\frac{5\pi}{6}$ are zeros of $(\sin x - \frac{1}{2})$, and the sign of $(\sin x - \frac{1}{2})$ does change about those zeros:

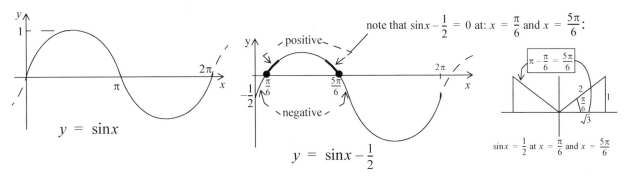

Figure 6.7

The sign of $(\sin x - 1)$ does not change about its zero at $\frac{\pi}{2}$

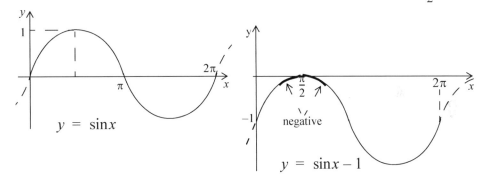

Figure 6.8

As previously noted, the sign of any factor raised to an **even power** will **not** change about its zeros.

Bottom line: Though the sign of any polynomial factor of the form $(x-a)^{\text{odd}}$ will change as you traverse its zero (at a), the same need not hold for non-polynomial factors such as $[\text{trig}(x) - a]^{\text{odd}}$

EXAMPLE 6.19 Find all $0 \leq x < 2\pi$ for which:

(a) $(x-\pi)^3(\sin x - \frac{1}{2}) < 0$

(b) $(x-\pi)^3(\sin x - \frac{1}{2})^2(\sin x - 1) < 0$

282 **Chapter 6 Polynomial Functions**

SOLUTION: (a) The zeros of $(x-\pi)^3(\sin x - \frac{1}{2})$ within the interval $[0, 2\pi)$ occur at π, $\frac{\pi}{6}$, and $\frac{5\pi}{6}$. Since the sign of $(x-\pi)^3$ will change about its odd-zero at π, we position a c above that zero. As noted in Figure 6.7, the sign of $(\sin x - \frac{1}{2})$ will change about the zeros $\frac{\pi}{6}$ and $\frac{5\pi}{6}$, and so we also place a c above those zeros.

To chart the SIGN, we note that $(x-\pi)^3(\sin x - \frac{1}{2})$ is positive at 0 (and immediately to its right), since $(x-\pi)^3$ is negative there, as is $(\sin x - \frac{1}{2})$ [see Figure 6.7]. Walking to the right, and changing the sign as we traverse the c-zeros, brings us to:

$$\begin{array}{ccccccc} & & c & & c & +\ c & \\ - - - - \ | & + & \bullet & - & \bullet & \bullet & - & | - - - - \\ & 0 & \frac{\pi}{6} & & \frac{5\pi}{6} & \pi & & 2\pi \end{array}$$

$$\text{SIGN } (x-\pi)^3(\sin x - \tfrac{1}{2})$$

Conclusion: $(x-\pi)^3(\sin x - \frac{1}{2}) < 0$: $\left(\frac{\pi}{6}, \frac{5\pi}{6}\right) \cup (\pi, 2\pi)$

(b) The zeros of $(x-\pi)^3(\sin x - \frac{1}{2})^2(\sin x - 1)$ in the interval $[0, 2\pi)$ occur at π, $\frac{\pi}{6}$, $\frac{5\pi}{6}$, and $\frac{\pi}{2}$. We again position a c above the odd-zero at π. Since there is now an even exponent associated with $\sin x - \frac{1}{2}$, the sign will not change as one traverses its zeros at $\frac{\pi}{6}$ and $\frac{5\pi}{6}$, and so we position an n above those zeros. As noted in Figure 6.8, the sign of $(\sin x - 1)$ does not change about its zero at $\frac{\pi}{2}$, and we therefore position an n above that zero as well. At 0 (and immediately to its right) $(x-\pi)^3(\sin x - \frac{1}{2})^2(\sin x - 1)$ is positive, as $(x-\pi)^3 < 0$, $(\sin x - \frac{1}{2})^2 > 0$ and $(\sin x - 1) < 0$, bringing us to:

$$\begin{array}{ccccccccc} & & n & & n & & n & c & \\ - - - - \ | & + & \bullet & + & \bullet & + & \bullet & + & \bullet & - & | - - - - \\ & 0 & \frac{\pi}{6} & & \frac{\pi}{2} & & \frac{5\pi}{6} & & \pi & & 2\pi \end{array}$$

$$\text{SIGN } (x-\pi)^3(\sin x - \tfrac{1}{2})^2(\sin x - 1)$$

Conclusion: $(x-\pi)^3(\sin x - \frac{1}{2})^2(\sin x - 1) < 0$: $(\pi, 2\pi)$

Answer: (a) π

(b) $\left[0, \frac{5\pi}{6}\right] \cup \left(\frac{7\pi}{6}, 2\pi\right)$

CHECK YOUR UNDERSTANDING 6.16

Find all $0 \le x < 2\pi$ for which:

(a) $(x+\pi)^2(\cos x + 1) \le 0$

(b) $(x+\pi)^2\left(\cos x + \frac{\sqrt{3}}{2}\right) > 0$

APPROXIMATING SOLUTIONS OF INEQUALITIES

How does one solve a polynomial inequality when the polynomial does not readily factor into linear and/or quadratic factors? With a graphing utility. Consider the following example.

EXAMPLE 6.20 Find, to three decimal places, the approximate solution set of the inequality:

$$2x^5 - x^4 > -2x^2 + 3x + 1$$

SOLUTION: Write the inequality in the form

$$2x^5 - x^4 + 2x^2 - 3x - 1 > 0$$

By the Confinement Theorem, page 267, we know that the x-intercepts of the graph of the function:

$$f(x) = 2x^5 - x^4 + 2x^2 - 3x - 1$$

are all contained in the interval $\left[-\frac{5}{2}, \frac{5}{2}\right]$ (see margin). Then:

Zeros of:
$2x^5 - x^4 + 2x^2 - 3x - 1$
$= 2\left(x^5 - \frac{x^4}{2} - \frac{3}{2}x - \frac{1}{2}\right)$
occur in the interval $[-K, K]$, where:
$K = 1 + \frac{3}{2} = \frac{5}{2}$

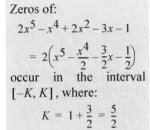

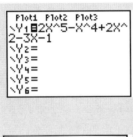

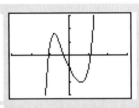
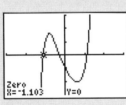
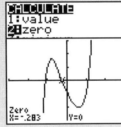
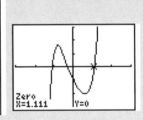

Recalling that a function is positive where its graph lies above the x-axis, we conclude that $(-1.103, -0.283) \cup (1.111, \infty)$ is an approximate solution set of $2x^5 - x^4 > -2x^2 + 3x + 1$.

CHECK YOUR UNDERSTANDING 6.17

Use a graphing utility to find an approximate solution set of:

$$3x^4 - 5x^2 - 4x < x^3 + 5$$

Answer: $\approx (-1.17, 1.87)$

284   Chapter 6   Polynomial Functions

EXERCISES

Exercises 1–7: Solve the quadratic inequality by the parabola method of Example 6.16.

1. $x^2 - 2x + 9 \leq 0$

2. $-2x^2 - 3x + 1 \geq 0$

3. $3x^2 + 4x - 4 \geq 0$

4. $\frac{2}{3}x^2 - 3 < x^2 + x - 5$

5. $x^2 + x - 30 > 0$

6. $2x^2 - x - 5 \leq 0$

7. $-x^2 - 2x + 9 > 0$

Exercises 8–9: Solve the inequality, by first making a substitution.

8. $27x^4 + 15x^2 - 2 \leq 0$

9. $2(1 - x^2)^2 + 5(1 - x^2) - 42 < 0$

Exercises 10–27: Solve the inequality.

10. $(4x - 3)(2x - 1) > 0$

11. $(x + 1)(3x - 1) > 0$

12. $(1 - x)(2x + 3)(x + 2) \leq 0$

13. $(-3x + 2)(5x - 1)(x - 3) > 0$

14. $-4(x - 3)(x + 1) \geq 0$

15. $x^2 - 3x - 10 \leq 0$

16. $x^2 - 2x - 1 > 0$

17. $-2x^2 - 5x + 3 \leq 0$

18. $3(x - 1)^3(x + 3)^3(x - 2)^4 < 0$

19. $2x^2(1 - x)(x + 2)^3(3 - x)^2 \leq 0$

20. $-2x(3 - x)(4 + 3x)(2 - x) > 0$

21. $-\frac{1}{2}(2 - 3x)^4(x - 4)^5(x + 3) < 0$

22. $(x - 1)^4 \leq 0$

23. $3(-x + 2)^2(x - 3) \geq 0$

24. $(2x + 3)^4 > 0$

25. $(x - 1)(2x + 5)^2 > 0$

26. $-(x + 1)^3(5 - x)^2(-2x - 1)^3 \geq 0$

27. $x^4(2x - 1)^3(3x - 1)^2(1 - 4x)^3 > 0$

Exercises 28–38: Put all terms on the left side of the inequality and 0 on the right, factor, and solve the inequality.

28. $x^3 < x$

29. $4x^4 \geq 2x^2$

30. $2x^3 + 3x^2 < 8x - 3$

31. $x^4 > 1$

32. $3x^3 + 2x^2 + 2x \leq 0$

33. $-2x^3 + 3x \leq 1$

34. $2x^3 + 3x + 5 > 0$

35. $(-2x^2 + 3x - 5)(x^3 - 16x) < 0$

36. $(x^2 - 1)^2 \leq x(x^2 - 1)(x + 1)$

37. $(x^4 + x^3 - x^2 - x)(x^4 - 2x^2 + 1) > 0$

38. $(3x^2 - x - 1)(4x^2 + 1)(x^2 - x - 6) \geq 0$

Exercises 39–45: Solve for $0 \leq x < 2\pi$.

39. $\left(\frac{\pi}{2} - x\right)\sin x \geq 0$

40. $\left(x - \frac{\pi}{3}\right)(\tan x - 1)^2 \leq 0$

6.3 Polynomial Functions and Inequalities   **285**

41. $\left(x - \frac{\pi}{6}\right)^3 (\cos x + 1) \cos^2 x \geq 0$ 42. $(2x - \pi)(1 + 2\sin x) < 0$

43. $\tan^2 x \leq 1$ 44. $2\sin x \leq \sin 2x$ 45. $\cos^2 x - \cos x \geq \sin^2 x - 1$

Exercises GC46–GC49: Use a graphing calculator to find an approximate solution set, as in Example 6.21.

46. $0.2x^7 + x^5 - x^4 + 1.3 > 0$ 47. $-3.6x^4 + 6.1x^3 < -22$

48. $-8x^6 + 2x^4 < 3x - 2$ 49. $5x^5 - 8x^2 - 2x + 1 > 0$

Exercise 50: (A Number) The cube of a certain positive integer is less than 14 times its square minus 48 times the integer. Find the integer.

Exercise 51: (Area) A rectangular field is twice as long as it is wide. What range of values can the width of the field have, if the area of the field is between 100 square feet and 150 square feet?

Exercise 52: (Free Fall) A ball is dropped from a height of 75 feet above the ground. When will the ball be at a height strictly between 11 and 71 feet above the ground?

Exercise 53: (Free Fall) A ball is thrown directly upward with a velocity of 32 ft/sec from a height of 148 feet above the ground. When will the ball be at a height between 100 and 160 feet above the ground?

Exercise 54: (Box Construction) A carton is to be constructed by cutting the same size square from each corner of a 5-inch by 4-inch piece of cardboard, and then bending up the sides.

(a) To what interval of values is the side of the square limited?

(b) What range of values can the side of the square have, if the resulting carton is to have a volume of at most 6 cubic inches.

Exercise GC55: (Box Construction) An open box whose base is x by $2x$, for some x, is to be constructed from 32 in.^{2} of cardboard.

(a) To what interval of values is x limited?
    (Suggestion: Consider that the height must be positive.)

(b) Express the volume of the box as a function of x.

(c) Use a graphing calculator to view the volume function in (b), and estimate the possible values x can have, if the volume is at least 10 in.^{3}.

Exercise GC56: (Box Construction) An open box whose base is five times as long as it is wide is to be constructed from 250 in.^{2} of cardboard.

(a) Express the volume of the box as a function of its width.

286   Chapter 6 Polynomial Functions

(b) What is the domain of the volume function in (a)?

(c) Use a graphing calculator to view the volume function in (a), and estimate the possible values of the width of the base of the box for which the volume is greater than 100 in.^{3}

Exercise GC57: (**Box Construction**) A closed box whose base is five times as long as it is wide is to be constructed from 425 in.^{2} of cardboard.

(a) Express the volume of the box as a function of its width.

(b) What is the domain of the volume function in (a)?

(c) Use a graphing calculator to view the volume function in (a), and estimate the possible values of the width of the base of the box for which the volume is less than 150 in.^{3}

Exercise 58: (**Free Fall**) A faulty proof in a bottle is thrown straight up from the edge of a 192 ft building at 64 ft/sec. At the same instant, SM-man (Super Math man) tosses a stone upward from the ground along the line of trajectory of the bottle. Find the range of initial velocities with which the stone must be tossed in order for it to hit the bottle before the bottle hits the ground.

Exercise GC59:   Suppose that an object is moving along a line and that its position $s(t)$ (in feet) with respect to a fixed reference point (the origin on the number line) at time t (in seconds) is given by

$$s(t) = \frac{t^3}{4} - 3t^2 + 5t + 8$$

in the interval of time $0 \leq t \leq 10$. As is customary, position to the right of the origin is said to be positive, and position to the left of the origin is negative.

Use a graphing calculator to view the graph of the position function over the given interval. From that graph, it appears that:

(a) The object is to the right of the origin in the time intervals: _____.

(b) The object is further than 3 feet to the left of the origin in the time intervals: _____.

Exercise 60:   Appealing only to the graph of a quadratic equation, explain why a nonempty solution set of a quadratic inequality consists of one number or is an interval [possibly $(-\infty, \infty)$] or the union of two intervals. Must one of the intervals in the solution set always be an infinite interval? Can there be more than one finite interval in the solution set?

287   Chapter 6 Polynomial Functions

§4   GRAPHING POLYNOMIAL FUNCTIONS

There is no question that graphing calculators can graph most functions better and faster than any of us, but this does not diminish the importance of this section. As you will see, being able to graph a function, requires an understanding of important concepts.

FUNCTIONS OF THE FORM $f(x) = x^n$

As is depicted below, the graphs of polynomial functions of the form $f(x) = x^n$ ($n > 1$, an integer) fall into two categories:

n even	n odd
The graph of every $y = x^{even}$ is similar to those of the functions $y = x^2$, and $y = x^4$ (below). Each such graph passes through the origin, and the points $(-1,1)$ and $(1,1)$.	The graph of every $y = x^{odd}$ is similar to those of the functions $y = x^3$, and $y = x^5$ (below). Each such graph passes through the origin, and the points $(-1,-1)$ and $(1,1)$.
The larger the exponent, the flatter is the graph over $(-1, 1)$ and the steeper outside of $(-1, 1)$.	The larger the exponent, the flatter is the graph over $(-1, 1)$ and the steeper outside of $(-1, 1)$

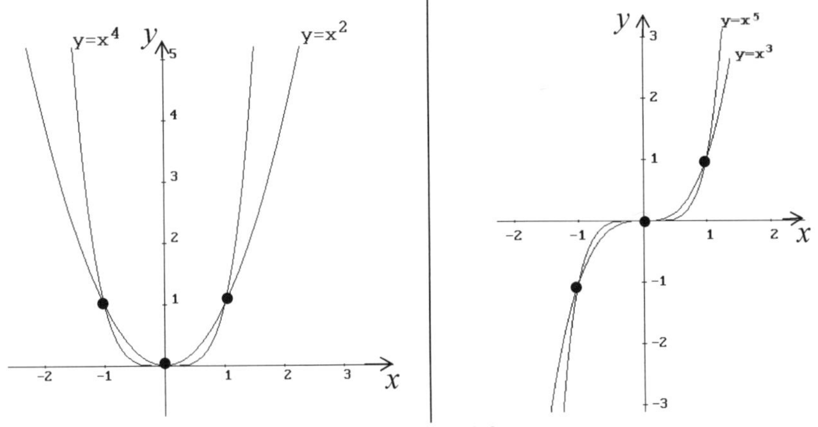

Figure 6.9

BEHAVIOR OF A POLYNOMIAL FUNCTION "FAR" FROM THE ORIGIN

A graphing calculator enables you to view only a portion of the graph of a function. To determine how a graph behaves far from the origin (as x tends to $\pm\infty$, written: $x \to \pm\infty$), use the following fact:

Far away from the origin the graph of the polynomial function:

$$p(x) = a_n x^n + a_{n-1} x^{n-1} + \cdots + a_1 x + a_0$$

resembles, in shape, that of its leading term $g(x) = a_n x^n$

It should be noted that the graph of a polynomial function and that of its leading term need not get arbitrarily close to each other as x tends to $\pm\infty$, but they will have similar shapes.

For example, as $x \to \pm\infty$, the graph of $p(x) = 6x^4 - 3x^3 - 15x^2 - 10$ resembles that of $g(x) = 6x^4$. This makes sense, since as x gets larger and larger in magnitude, the term $6x^4$ becomes more and more dominant.

Since the leading term of a polynomial in factored form is the product of the leading terms of its factors, as $x \to \pm\infty$, the graph of:

$$p(x) = (x^4 + 5)(2x^3 - 3)(3x + 1) = 6x^8 + \ldots$$

resembles that of $g(x) = 6x^8$.

You can see that while the graphs of the functions $p(x) = 6x^4 - 3x^3 - 15x^2 - 10$ and $g(x) = 6x^4$ have different shapes near the origin, their shapes are similar "far away" from the origin.

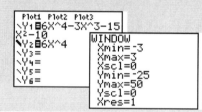
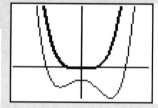

CHECK YOUR UNDERSTANDING 6.18

Determine a function $g(x) = ax^n$, whose graph resembles that of the given polynomial function f as $x \to \pm\infty$. Use a graphing calculator to simultaneously view the graphs of f and g, and check their resemblance far away from the origin.

(a) $f(x) = x^3 - x$ (b) $f(x) = (2x^3 + x)(x^2 - 5x + 1)(x - 1)$

Answers: (a) $g(x) = x^3$
(b) $g(x) = 2x^6$

GRAPHING GENERAL POLYNOMIAL FUNCTIONS

The following procedure will be used to graph polynomial functions that can readily be expressed as a product of linear and quadratic factors.

Step 1. Factor the polynomial.

Step 2. Near the origin:

Determine and plot x- and y-intercepts.
Chart the sign of the function to see where the graph lies above the x-axis and where it lies below the x-axis.

"Near the origin" encompasses all zeros of the polynomial.

289 Chapter 6 Polynomial Functions

Step 3. **Far from the origin:**

Determine the shape of the graph as $x \to \pm\infty$.

Step 4. **Sketch the anticipated graph.**

We say "anticipated," since the calculus is generally needed to "guarantee" the exact shape of the graph.

EXAMPLE 6.22 Sketch the graph of:
$$f(x) = x^4 - x^3 - 6x^2$$

SOLUTION:

Step 1. **Factor:**
$$f(x) = x^4 - x^3 - 6x^2 = x^2(x^2 - x - 6) = x^2(x+2)(x-3)$$

Step 2. **y-intercept:** $f(0) = 0$ [Figure 6.10(a)].

x-intercepts: $f(x) = x^2(x+2)(x-3) = 0$ at 0, -2, and 3 [Figure 6.10(a)].

SIGN $f(x)$: From the sign information [top of Figure 6.10(a)], we see that moving from left to right, the graph crosses from above the x-axis to below the x-axis at -2; touches the origin but is negative on both sides of 0, and then crosses from below the x-axis to above the x-axis at 3 [Figure 6.10(a)].

Step 3. As $x \to \pm\infty$: The farther away from the origin, the more the graph of f resembles that of its leading term $g(x) = x^4$ [Figure 6.10(a)].

Step 4. **Sketch the anticipated graph:** Figure 6.10(b) and the final graph in (c).

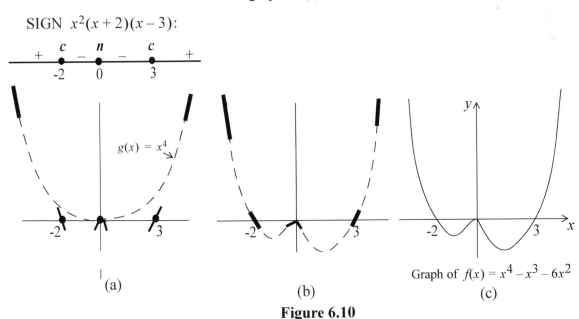

Figure 6.10

6.4 Graphing Polynomial Functions 290

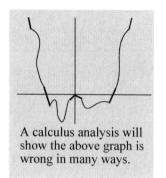

A calculus analysis will show the above graph is wrong in many ways.

Note: Without the calculus, we can not determine where the two local minima depicted in Figure 6.10(c) occur, nor do we know that the local minimum between 0 and 3 is smaller than the local minimum between −2 and 0. Indeed from our restricted analysis we could connect the pieces of Figure 6.10(b) to arrive at other "correct" graphs (see margin). It turns out, however, that the simplest connections are usually the most accurate.

EXAMPLE 6.23 Sketch the graph of:

$$f(x) = x^3 - 3x^2 - 13x + 15$$

SOLUTION:

Step 1. Factor: Using the Rational Zeros Theorem (page 13), we find that $x = 1$ is a zero of $f(x) = x^3 - 3x^2 - 13x + 15$. Theorem 1.4, page 12, tells us that $(x - 1)$ is a factor of $f(x)$, and as you can easily check:

$$\begin{array}{r} x^2 - 2x - 15 \\ x - 1 \overline{\smash{\big)}\, x^3 - 3x^2 - 13x + 15} \end{array}$$

Bringing us to:

$$f(x) = (x-1)(x^2 - 2x - 15) = (x-1)(x-5)(x+3)$$

Step 2. y-intercept: $f(0) = 15$ [Figure 6.11(a)].

x-intercepts: $f(x) = (x-1)(x-5)(x+3) = 0$ at 1, 5, and −3 [Figure 6.11(a)].

SIGN $f(x)$: From the sign information [top of Figure 6.11(a)], we see that moving from left to right, the graph crosses from below the x-axis to above the x-axis at −3, from above the x-axis to below the x-axis at 1, and from below the x-axis to above the x-axis at 5 [Figure 6.11(a)].

Step 3. As $x \to \pm\infty$: The farther away from the origin, the more the graph of f resembles that of its leading term $g(x) = x^3$ [Figure 6.11(a)].

Step 4. Sketch the anticipated graph: Figure 6.11(b) and the final graph in (c).

291   Chapter 6 Polynomial Functions

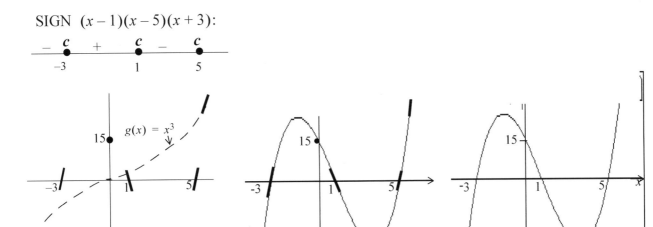

Figure 6.11

CHECK YOUR UNDERSTANDING 6.19

Use the 4-step procedure to sketch the graph of the function:
$$f(x) = -x^4 - 2x^3 + 3x^2$$

Answer: See page C-49.

WHEN YOU ARE NOT ABLE TO FACTOR THE POLYNOMIAL

The following example illustrates how to graph a polynomial function f that does not readily factor.

EXAMPLE 6.24 Sketch the graph of:
$$f(x) = x^3 - 2x^2 - 15x + 3$$

SOLUTION: Since we can not readily factor $x^3 - 2x^2 - 15x + 3$, we first graph the function $h(x) = x^3 - 2x^2 - 15x$ which we can factor, and then raise that graph 3 units to arrive at the graph of f.

Step 1. Factor: $h(x) = x^3 - 2x^2 - 15x = x(x - 5)(x + 3)$

Step 2. y-intercept: $f(0) = 0$ [Figure 6.12(a)].

x-intercepts: $f(x) = x(x - 5)(x + 3) = 0$ at $x = 0, 5,$ and -3 [Figure 6.12(a)].

SIGN $f(x)$: From the sign information [top of Figure 6.12(a)], we see that as you move from left to right, the

6.4 Graphing Polynomial Functions   **292**

graph crosses from below the x-axis to above the x-axis at -3; from above to below at 0; and from below to above at 5 [Figure 6.12(a)].

Step 3. As $x \to \pm\infty$: The farther away from the origin, the more the graph of f resembles that of its leading term $g(x) = x^3$ [Figure 6.12(a)].

Step 4. Sketch the anticipated graph of $h(x) = x^3 - 2x^2 - 15x$ [Figure 6.12(b)].

Step 5. We then obtained the graph of $f(x) = x^3 - 2x^2 - 15x + 3$ by raising that of $h(x)$ three units [see Figure 6.12(c)]. Though we don't know exactly where the graph of f crosses the x-axis, we do have a sense of the general shape of the graph, and additional accuracy could be gained by plotting a few extra points.

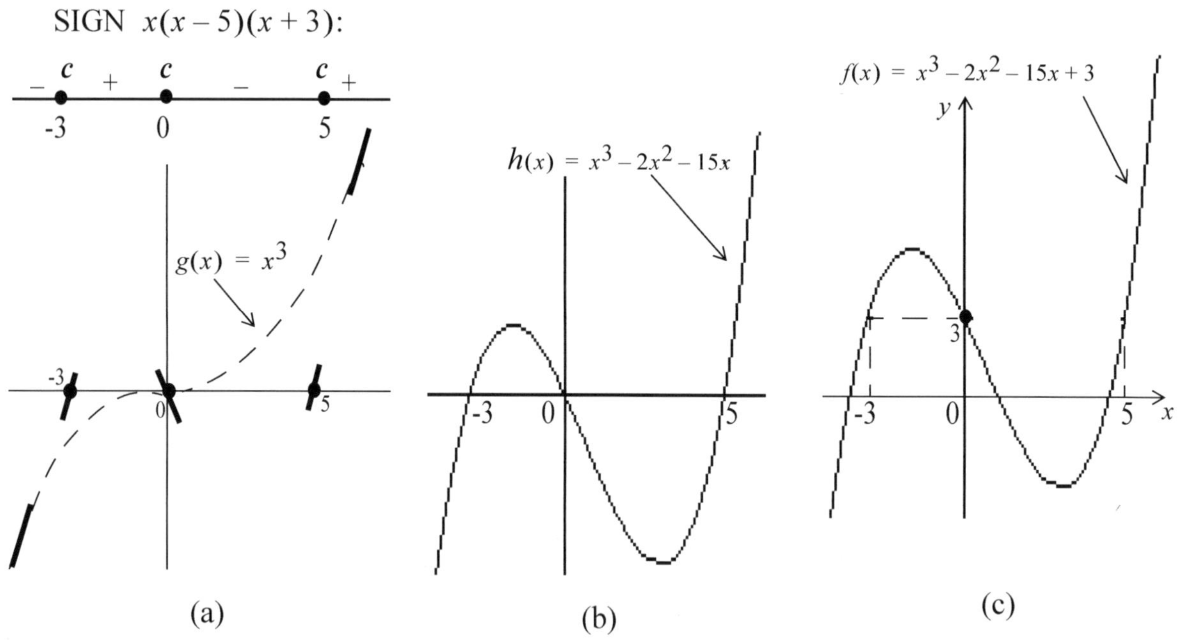

SIGN $x(x-5)(x+3)$:

(a)   (b)   (c)

Figure 6.12

CHECK YOUR UNDERSTANDING 6.20

Follow the above five step procedure to sketch the graph of:
$$f(x) = x^5 - 6x^4 + 8x^3 - 2$$

Answer: See page C-49.

6.4 Graphing Polynomial Functions   293

EXERCISES

Exercises 1–18: Sketch the graph of the function by reflecting, stretching, and shifting the basic curve of the form $y = x^n$, and indicate the x-and y-intercepts on the graph.

1. $f(x) = x^4 + 2$

2. $k(x) = x^3 + 3$

3. $h(x) = x^3 - 3$

4. $k(x) = x^8 + 1$

5. $p(x) = -x^7 + 5$

6. $f(x) = -x^6 - 1$

7. $h(x) = \frac{1}{8}(x - 1)^6$

8. $f(x) = \frac{1}{3}(x + 1)^3$

9. $q(x) = -4(x + 2)^3$

10. $p(x) = -3(x - 4)^2$

11. $q(x) = -(x + 1)^4 + 81$

12. $f(x) = (x + 1)^5 + 1$

13. $h(x) = (x - 2)^4 - 16$

14. $h(x) = -(x - 1)^3 - 8$

15. $k(x) = -5(x + 1)^5 - 5$

16. $f(x) = 2(x - 2)^2 + 2$

17. $h(x) = -\frac{1}{4}(x + 3)^3 + 2$

18. $k(x) = (2x - 1)^2 + 1$

Exercises GC19–GC24: Determine a function $g(x) = ax^n$ whose graph resembles the graph of the given function far away from the origin. Then use a graphing calculator to view simultaneously the graphs of f and g and check their resemblance far away from the origin.

19. $f(x) = x^4 + 3x^2 - 1$

20. $f(x) = 5x^4 + 300x^3$

21. $f(x) = (x - 5)^5(x + 5)^2$

22. $f(x) = x^2(4 - x)^3$

23. $f(x) = (3x - 2)(2x + 20)^3$

24. $f(x) = (2x - 1)^2(3x - 1)^2(2 - x)$

Exercises 25–38: Sketch the graph of the function using the 4-step method.

25. $f(x) = -(3x + 4)(2x - 5)(x - 2)$

26. $k(x) = 4x(4x + 1)(3 - x)$

27. $p(x) = x(2x + 1)^3(4 - x)$

28. $h(x) = -2x(2x + 1)^2(x - 2)$

29. $f(x) = 3(x + 2)^2(x - 2)^3$

30. $p(x) = \frac{1}{2}(x - 4)^2(3x - 1)^5$

31. $f(x) = -2(4x + 3)^2(x - 1)^2(x + 2)^3$

32. $q(x) = 3x^2(3x - 2)^2(x - 1)^4$

33. $f(x) = x^3 - 6x^2 + 11x - 6$

34. $h(x) = -x^3 - x^2 + 2x$

35. $F(x) = x^4 - x^2$

36. $Q(x) = -x^3 + 2x - 1$

37. $f(x) = x^3 - 4x^2 - 3x + 18$

38. $P(x) = -9x^4 - 5x^2 + 4$
(Hint: factor after substituting for x^2)

Exercises 39-42: Sketch the graph of the function, as in Example 6.24.

39. $f(x) = x^3 + 4x^2 - 5x + 1$

40. $f(x) = x^4 - x^2 - 1$

41. $k(x) = x^4 + 3x^3 - 4x + 2$

42. $H(x) = x^5 - 3x^3 - 3$

CHAPTER SUMMARY

POLYNOMIAL FUNCTION	A **polynomial function of degree n** (in the variable x) is a function of the form: $$p(x) = a_n x^n + a_{n-1} x^{n-1} + \ldots + a_1 x + a_0$$ with $a_n \neq 0$. The numbers a_0 through a_n are said to be the **coefficients** of the polynomial, with a_n and a_0 the **leading coefficient** and **constant coefficient**, respectively.
QUADRATIC FUNCTION	A **quadratic function** is a function that can be expressed in the form: $$f(x) = ax^2 + bx + c, \text{ with } a \neq 0$$
GRAPHS OF QUADRATIC FUNCTIONS **VERTEX** **STANDARD FORM**	The graph of a quadratic function $f(x) = ax^2 + bx + c$ is a parabola, which opens upward if $a > 0$, and opens downward if $a < 0$. The vertex of the parabolic graph of the quadratic function $f(x) = ax^2 + bx + c$ occurs at $x = -\dfrac{b}{2a}$. If the vertex is at (x_0, y_0), then: $f(x) = a(x - x_0)^2 + y_0$. 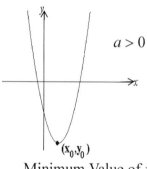 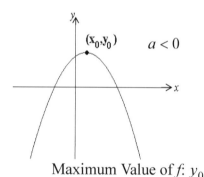 Minimum Value of f: y_0 Maximum Value of f: y_0
Relationship of the graph of the quadratic function: $f(x) = ax^2 + bx + c$ and the solutions of the quadratic equation: $ax^2 + bx + c = 0$	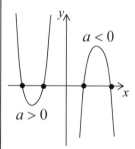 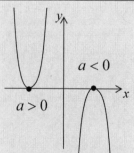 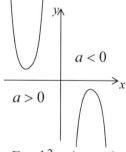 For $b^2 - 4ac > 0$ For $b^2 - 4ac = 0$ For $b^2 - 4ac < 0$ Two solutions: One solution: No solution: Two x-intercepts One x-intercept No x-intercept

OPTIMIZATION PROBLEMS	**Step 1.** See the problem. **Step 2.** Express the quantity to be maximized or minimized in terms of any convenient number of variables. **Step 3.** If the expression in Step 2 involves more than one variable, use the given information to eliminate all but one variable in that expression. **Step 4.** Either "by hand," or using a graphing calculator (for an approximate solution), determine the maximum or minimum value(s) of the function obtained in Step 3.
SOLUTIONS OF $x^n = a$	If $x^n = a$, then: $$x = a^{\frac{1}{n}}, \text{ if } n \text{ is odd, and } x = \pm a^{\frac{1}{n}} \text{ if } n \text{ is even and } a \geq 0.$$ [There are no (real) solutions when n is even and $a < 0$]
Graphing Method	The solutions of $p(x) = 0$ are the x-intercepts of the graph of $y = p(x)$.
Number of Solutions	A polynomial equation of degree n has at most n solutions.
RATIONAL ZEROS THEOREM	Each rational zero of $$p(x) = a_n x^n + a_{n-1} x^{n-1} + \cdots + a_0$$ is of the form $\frac{c}{l}$, where c is a factor of the constant coefficient a_0, and l is a factor of the leading coefficient a_n.
Confinement Theorem	All solutions of the polynomial equation: $$p(x) = x^n + a_{n-1} x^{n-1} + \ldots + a_1 x + a_0 = 0$$ are contained in the interval $[-K, K]$, where: $$K = 1 + \text{largest}\{|a_{n-1}|, |a_{n-2}|, \ldots, |a_1|, |a_0|\}$$
POLYNOMIAL INEQUALITIES	To solve a polynomial inequality $p(x) < 0$ (or ≤ 0, or > 0, or ≥ 0) when $p(x)$ is expressed as a product of linear and quadratic factors: **Step 1.** Chart the sign of the polynomial as follows: Locate the zeros of the polynomial on the number line. Place the letter **c** above the **odd zeros** [zeros of factors which are raised to an **odd** power], and the letter **n** above the **even zeros** [zeros of factors raised to an **even** power]. Determine the sign of the polynomial to the right of its last zero. Proceed to the left, changing the sign when you cross over an odd-zero, and not changing the sign when you cross over an even-zero. **Step 2.** Read off the solution set of the given inequality from the sign information determined in Step 1.

296   Chapter 6 Polynomial Functions

NOTE	Though the sign of any polynomial factor of the form $(x-a)^{\text{odd}}$ will change as you traverse its zero (at a), the same need not hold for non-polynomial factors such as $[\text{trig}(x)-a]^{\text{odd}}$
Graphing Method	The solution set of $p(x) > 0$ is the union of those intervals where the graph of $y = p(x)$ lies **above** the x-axis. The solution set of $p(x) < 0$ is the union of those intervals where the graph of $y = p(x)$ lies **below** the x-axis

GRAPHING POLYNOMIAL FUNCTIONS

Graphs of $f(x) = x^n$ (*n* a positive integer)	
As $x \to \pm\infty$	Far away from the origin the graph of the polynomial function: $$p(x) = a_n x^n + a_{n-1} x^{n-1} + \cdots + a_1 x + a_0$$ resembles, in shape, that of its leading term $g(x) = a_n x^n$
Graphing Procedure	**Step 1. Factor the polynomial function.** **Step 2. Near the origin:** Determine and plot x- and y-intercepts. Chart the sign of the function to see where the graph lies above the x-axis and where it lies below the x-axis. **Step 3. Far from the origin:** Determine the shape of the graph for as $x \to \pm\infty$. **Step 4. Sketch** the anticipated graph.

Chapter Review Exercises   297

CHAPTER REVIEW EXERCISES

Exercises 1–2: Find the vertex of the parabola, and the maximum (or minimum) value of the quadratic function (and state which it is).

1. $g(x) = -2x^2 + 3x - 4$ 　　　　　　　　2. $h(x) = 2(x+3)^2 - 8$

Exercises 3–4: Determine the vertex and x- and y-intercepts of the parabola, and sketch its graph.

3. $k(x) = x^2 - 6x + 5$ 　　　　　　　　4. $F(x) = 4x^2 + 2x - 3$

Exercises 5–7: Solve the system of equations.

5. $\left. \begin{array}{l} y = 3x - 2 \\ y = -(x-2)^2 + 4 \end{array} \right\}$ 　　6. $\left. \begin{array}{l} 2x^2 - 14x = 5 + y \\ x^2 + \; y \; = 0 \end{array} \right\}$ 　　7. $\left. \begin{array}{l} 8\cos^2\alpha - 6\sin\beta = 9 \\ 3\cos^2\alpha + \frac{5}{2}\sin\beta = 1 \end{array} \right\}$

　　　　　　　　　　　　　　　　　　　　　　　　for $0 \le \alpha < 2\pi,\, 0 \le \beta < 2\pi$

Exercises 8–16: Solve the equation.

8. $x^3 - x^2 + x - 1 = 0$ 　　　　　　　　9. $2x^3 + x^2 - 11x + 2 = 0$

10. $10x^3 + 3x^2 = 31x - 6$ 　　　　　　　11. $(x^2 - x - 8)^2 = x^2$

12. $2x^3 + 5x^2 - 5x + 1 = 0$ 　　　　　　13. $x^4 - 2x^3 - 3x^2 + 2x + 2 = 0$

14. $25x^3 = 50$ 　　　　　　　　　　　　15. $16x^3 + 4x^2 - 24x + 9 = 0$

16. $4\sin^3 x - 4\sin^2 x - 3\sin x + 3 = 0$, for x in $\left[-\frac{\pi}{2}, \frac{\pi}{2}\right]$

Exercises 17–20: Make an appropriate substitution and solve.

17. $3x^4 - 2x^2 - 1 = 0$ 　　　　　　　　18. $(x+1)^3 = (x+1)^2 - 2$

19. $2(x^2 + x + 1)^2 - 7(x^2 + x + 1) + 3 = 0$ 　　　20. $x^6 - 2x^3 + 1 = 0$

Exercises 21–35: Solve the inequality.

21. $(5x + 4)(3x + 5) > 0$ 　　　　　　　22. $(x - 3)(2x - 7) < 0$

23. $-8x^2 - 2x + 1 \ge 0$ 　　　　　　　24. $x(x - 2) \le 4$

25. $x(-x + 2) \le -1$ 　　　　　　　　　26. $-x^2 + 2x - 6 > 0$

27. $2x(2x + 1) \ge -1$ 　　　　　　　　28. $x^3 < -1$

29. $x^3 - 4x < x^2 - 4$ 　　　　　　　　30. $3x(x-1)^2 - (2-x)(x-1)^2 < 0$

31. $(3x - 1)^3(2x + 5)^2(x - 1) > 0$ 　　　32. $-x(x+2)^2(2-x)(3-x)^3 \le 0$

33. $(x - 1)(x^2 - 3x - 1) \ge 0$ 　　　　　34. $(2x - 1)^3 \left(x^2 + \frac{1}{2}x + 3\right) \le 0$

35. $\left(x - \frac{\pi}{2}\right)\left(\sin x - \frac{1}{2}\right) > 0$, for x in $[0, 2\pi)$

298   Chapter 6   Polynomial Functions

Exercises GC36–GC41:   Use a graphing calculator to approximate the solution set.

36. $5x^5 - \frac{1}{3}x^3 + 3x - \sqrt{3} = 0$

37. $x^7 - 6x^5 = 2x - \frac{1}{3}$

38. $(2x)^4 - x^3 = 7x + 2$

39. $0.2x^7 + x^5 - x^4 + 1.3 > 0$

40. $-8x^6 + 2x^4 < 3x - 2$

41. $\sqrt{2}x^4 - 5x < x^5 + 17$

Exercises 42–43:   Use the Confinement Theorem (Theorem 6.4), to find an interval containing all the x-intercepts of the graph of the function. At most how many x-intercepts can there be?

42. $g(x) = -3x^5 + x^4 - 5x + 7$

43. $f(x) = 2x^7 + \frac{5}{3}x^5 - 4x + 5$

Exercises 44–48:   Determine the intercepts and sketch the graph by reflecting, stretching and shifting the basic graph.

44. $f(x) = -2(x+1)^5$

45. $g(x) = -3(x+2)^4 - 6$

46. $k(x) = (x-5)^6 + 1$

47. $f(x) = 8(x-1)^3 + 1$

48. $k(x) = -\frac{3}{16}(x-4)^4 + 3$

Exercises GC49–GC52:   Determine a function $g(x) = ax^n$ whose graph resembles the graph of the given function far away from the origin. Then use a graphing calculator to view simultaneously the graphs of f and g and check their resemblance far away from the origin.

49. $f(x) = \frac{1}{2}(2x+3)(6x^2+1)$

50. $f(x) = 8x^2\left(1 - \frac{x}{2}\right)^5$

51. $f(x) = -4x(2x+5)^3(2x-1)^2$

52. $f(x) = 3x^4\left(\frac{2}{3}x+5\right)^2(x^2-6)$

Exercises 53–58:   Find the intercepts and sketch the graph.

53. $f(x) = (x+3)(2x-1)(x-1)$

54. $f(x) = 2x^2(x-2)(3x+1)(4x-5)$

55. $f(x) = -2(x^2+5x+9)(2x+7)$

56. $h(x) = x^3 - 3x + 2$

57. $K(x) = -x^3 + 2x^2 + x - 2$

58. $F(x) = x^4 + 3x^3 + x^2 - 3x - 2$

Exercises 59–60:   Sketch the graph of the function, and indicate its y-intercept.

59. $f(x) = (x-1)(x+2)(x-3) + 2$

60. $f(x) = x^2(2x-1)(3x+7) + 1$

Exercise GC61:   (**Break-Even Point**) A company can produce x units per month at a cost $C(x) = x^2 + 50x + 1000$, where $x \leq 300$. Because of supply and demand, the price per unit decreases linearly. If 200 units are manufactured and sold, each unit will sell for $250, whereas 300 units will sell for $150 each. Determine the revenue function, and then use a graphing calculator to determine, approximately, the monthly production levels at which break-even point(s) occur.

Exercise GC62: (**Break-Even Point**) A company has a fixed monthly cost of $12,000, and a variable cost of production. The variable cost per unit decreases linearly. If the company produces 400 units, the variable cost per unit is $8. If the company produces 700 units, the variable cost per unit reduces to $6.50. If the company produces x units per month, it can sell each unit for $\$\left(10 - \frac{x}{600}\right)$.

(a) Determine Cost and Revenue as functions of x.

(b) Use a graphing calculator to determine, approximately, the monthly production levels at which break-even point(s) occur.

Exercise 63: (**Free Fall**) A stone is dropped from a bridge and hits the water below in three seconds. How high is the bridge above the water?

Exercise 64: (**Free Fall**) Determine the number of seconds required for an object tossed directly upward from a height of 125 ft with a speed of 48 feet per second to reach its maximum height. Find the maximum height.

Exercises 65: (**Free Fall**) The height above the ground of an object fired directly upward is $s(t) = -16t^2 + 16t + 96$, where s is measured in feet and t in seconds. What maximum height is attained by the object? For how many seconds after it was fired does it remain airborne?

Exercise 66: (**Maximum Product**) Find two numbers whose sum is 60 and whose product is as large as possible.

Exercise 67: (**Maximum Difference**) Find the number that exceeds its square by the largest amount.

Exercise 68: (**Profit**) A company determines that its revenue and cost functions in the production of x units are

$$R(x) = -2x^2 + 400x \quad \text{and} \quad C(x) = 5x^2 + 50x + 100$$

Determine the maximum profit the company can make, and how many units should be produced so that this maximum profit is realized.

Exercise 69: (**Maximum Area**) A farmer will spend $600 to enclose a rectangular field along a (straight) river with a fence that costs $1.50 per running foot. No fence is required along the river. The farmer also needs to subdivide the field with two fences perpendicular to the river, and that fencing costs $2.00 per running foot. Determine the dimensions of the field that will maximize area.

Exercise 70: (**Maximum Light**) A Norman window has the shape of a rectangle surmounted by a semicircle. The larger the area of the window, the more light admitted. If the perimeter of the window is to be 50 feet, what should its dimensions be so that it admits the most light?

300   Chapter 6   Polynomial Functions

Exercise 71: (Maximum Area) A rectangular yard will be fenced off and divided into four pens by fences all parallel to one side of the yard. Find the greatest area that can be enclosed with 1000 feet of fencing.

Exercise 72: (Maximize Profit) A 75-unit motel can rent each of its rooms for $25 a night. It costs the motel $7 in room maintenance for each rented unit. Every $1 increase in the rate results in one less rental. Determine the rate which maximizes: (a) Revenue; (b) Profit.

Exercise 73: (Maximize Revenue) A 75-unit motel can rent each of its rooms for $25 a night. Every $1 rate increase results in one less rental. Determine the rate that maximizes revenue.

Exercise 74: (Maximum Revenue) A company that gives tours of the city charges $20 per person if at most 36 people take the tour. For each additional tourist, the charge per person is reduced by $0.25. How many tourists would maximize the company's revenue per tour?

Exercise GC75: (Maximum Profit) A company has found that its profit P (in hundreds of dollars) can be approximated by:

$$P(x) = -\frac{x^3}{5} + 194x - 37$$

where x represents advertising cost (in thousands of dollars), for $x \leq 25$. Use a graphing calculator to determine, approximately, the amount that should be spent on advertising to maximize profit.

Exercise 76: (Square and Rectangle) A 17-inch piece of wire is to e cut into two pieces. One piece will be bent into a square, and the other into a rectangle of length twice its width. How should the wire be cut to obtain the smallest possible combined area of the two regions? How should the wire be cut to obtain the largest possible combined area? (Note that one of the two pieces may be of length 0, i.e. don't cut the wire.)

Exercise GC77: (Profit) The profit, in dollars, for a company is given by

$$P(x) = -0.003x^3 + 0.5x^2 + 17.4x - 1250$$

where x is the number of units produced, with $x \leq 200$. Use a graphing calculator to estimate the production levels, x:

(a) for which profit is positive.

(b) for which profit exceeds $1500.

(c) for which the company loses money.

Cumulative Review Exercises 301

Exercise GC78: Given the parabola $y = x^2 + 2x + 1$:

(a) Express the square of the distance between a point (x, y) on the parabola and the point $(1, 5)$ as a function of x.

(b) Use a graphing calculator to find approximate coordinates of the point on the parabola closest to $(1,5)$. [Suggestion: Minimize the function in (a).]

Exercise GC79: (**Box Inequality**) A closed box whose base is five times as long as it is wide is to be constructed from 250 in.^{2} of cardboard. Use a graphing calculator to view the volume function and estimate the possible values of the width of the base, if the volume is to be at most 100 in.^{3}.

Exercise 80: (**Volume**) The volume of a certain cube is two more than three times the length of its edge, measured in inches. What is the volume of this cube?

Exercise 81: (**Cylinder Inequality**) A storage cylinder whose height is three times its radius is to be constructed so that it will fit into a space 20 ft by 20 ft by 20 ft. What range of values can the height take on so that the cylinder will contain at least 375 ft^3?

Exercise GC82: (**Maximum**) Use a graphing calculator to estimate the maximum value of $x - y$ if $x = 4t^3 - 5t^2 + 3t + 1$ and $y = 5t^3 - 5t^2 - 45t + 1$ and $t > 0$?

CUMULATIVE REVIEW EXERCISES

Exercises 1–2: Simplify.

1. $\left[1 + \dfrac{1}{1 + \frac{1}{a}}\right]\left[1 - \dfrac{1}{1 - \frac{1}{a}}\right]$

2. $\left[\dfrac{(-2)^{-4}(a+b)^{-1}}{\sqrt{2(a+b)}}\right]^{-1}$

Exercise 3: Given $1m = 3.28$ ft, 16 oz $= 1$ lb, and $1g = 0.0353$ oz, convert 10.1 kg^2/cm to lb^2/in.

Exercise 4: Solve the equation $2\sin^2 x + 3\sin x + 1 = 0$, for $0 \le x < 2\pi$.

Exercise 5: Find the domain and range of the function $f(x) = 3\sin 2x + 1$, and sketch the graph.

Exercises 6–11: Solve each of the following.

6. $\left.\begin{array}{l} 5x - 3y = 1 \\ x^2 + \ y = \frac{1}{3} \end{array}\right\}$

7. $16x^4 = 1$

8. $(x^2 - 2x - 4)^2 = 16$

9. $4x^3 + 6x^2 - 4x - 6 = 0$

10. $(1 - 5x)^3 - 2(1 - 5x)^2 = 5(1 - 5x) - 6$

11. $12x^3 - 4x^2 - 5x + 2 = 0$

302   Chapter 6 Polynomial Functions

Exercise 12:   Evaluate each expression.

(a) $\sin\frac{7\pi}{6}$ (b) $\cos^{-1}(-1)$ (c) $\tan^{-1}1$ (d) $\sin\left(\tan^{-1}\frac{4}{3}\right)$ (e) $\cos\left(\cos^{-1}\frac{1}{8}\right)$

Exercises 13–18: Solve the inequality.

13. $x^2 \geq 4$

14. $(2-x)(3x+5)\cos x \geq 0,\quad 0 \leq x < 2\pi$

15. $2x^2 - 3x - 3 \leq 0$

16. $x^4 + x^3 - 5x^2 - 3x + 6 > 0$

17. $(2-x)(x+1)^3(3x+2)^2 > 0$

18. $(x+\pi)^5\left(\cos x - \frac{1}{2}\right)^3 \leq 0,\ $ in $[0, 2\pi)$

Exercises GC19–GC20: Use a graphing calculator to approximate the solution set.

19. $x^3 + 100x^2 - 131x = -23$

20. $x^3 - 3x > -1$

Exercise 21:   Determine the vertex and intercepts of the parabola, $k(x) = -x^2 - 4x - 2$, and sketch its graph.

Exercise 22:   Find the maximum (or minimum) value of the quadratic function $g(x) = 4x^2 - 16x + 3$ (and state which it is).

Exercises 23–25:   Sketch the graph, either starting with the basic curve and stretching/shifting it etc. or using the 4-step graphing method, whichever is appropriate. Indicate the intercepts.

23. $h(x) = 2(x-1)^5 - 2$

24. $p(x) = (2x-1)(x+3)(x-4)$

25. $q(x) = 48x^4 + 12x^3 - 18x^2$

Exercise 26: (**Theater Tickets**) When a theater charges \$10 for admission, 150 people buy tickets. For each \$.50 increase in the price of a ticket, the number of tickets sold decreases by 4. What should the admission charge be to take in the most money?

Exercise 27: (**Free Fall**) A stone is dropped from the roof of a 128 ft building at the same time that one is tossed up directly beneath it from the ground. Find the initial velocity of the stone tossed up from the ground, if the two are to collide at a height of 92 ft? $[s(t) = -16t^2 + v_0 t + s_0]$

Exercise GC28: (**Maximum Volume**) A closed box will be constructed from 300 in.^{2} of material so that its base is $4x$ by $3x$ for some x. Use a graphing calculator to estimate the maximum volume such a box can have.

CHAPTER 7

RATIONAL FUNCTIONS

What you learned about solving polynomial equations and inequalities and graphing polynomial functions in the previous chapter will be put to good use in the first two sections of this chapter where we consider equations, inequalities, and graphs of functions involving rational expressions. In Section 3, we show how rational expressions can be decomposed into a sum of "elementary terms," called partial fractions. This decomposition process will be needed in the calculus when you study techniques of integration.

---

§1. RATIONAL FUNCTIONS, EQUATIONS, AND INEQUALITIES

Just as a polynomial function is defined in terms of a polynomial expression, a rational function is described by a rational expression:

DEFINITION 7.1

RATIONAL FUNCTION

A **rational function** is a function of the form:

$$f(x) = \frac{p(x)}{q(x)}$$

where $p(x)$ and $q(x)$ are polynomials, and $q(x)$ is not the zero polynomial.

Since the domain of a polynomial is all the real numbers, a rational function will only be undefined where its denominator takes on a value of zero. For example, the function:

$$f(x) = \frac{(x-5)(x+2)}{(x+1)(x-3)}$$

has domain: $D_f = \{x \mid x \neq -1, x \neq 3\} = (-\infty, -1) \cup (-1, 3) \cup (3, \infty)$.

CHECK YOUR UNDERSTANDING 7.1

Find the domain of the rational function.

(a) $f(x) = \dfrac{x^2 + 4}{x^2} - 4$

(b) $g(x) = \dfrac{x^2 - 4}{x^2 + x - 6}$

Answers:
(a) $(-\infty, 0) \cup (0, \infty)$ (b) $(-\infty, -3) \cup (-3, 2) \cup (2, \infty)$

304   Chapter 7 Rational Functions

RATIONAL EQUATIONS

Since the value of a quotient is zero if and only if the numerator is zero and the denominator is not zero, we have:

THEOREM 7.1

SOLUTIONS OF A RATIONAL EQUATION

The solutions of the rational equation $\dfrac{p(x)}{q(x)} = 0$ are those values of x for which $p(x) = 0$ and $q(x) \neq 0$.

When applying this theorem, you cannot ignore the condition that $q(x) \neq 0$:

The rational equation: $\dfrac{x^2 + 3x - 10}{x + 5} = 0$ (*)

and the polynomial equation: $x^2 + 3x - 10 = 0$

$(x + 5)(x - 2) = 0$ (**)

do **not** have the same solutions: -5 is a solution of (**), but it is not a solution of (*) (why not?).

A rational equation of the form $\dfrac{p(x)}{q(x)} = \dfrac{p_1(x)}{q_1(x)}$ can be solved by multiplying both sides of the equation by the least common denominator (LCD) of the rational expressions in that equation, and then solving the resulting polynomial equation. It is important to remember, however, that while you can't "lose" a root of an equation by multiplying both sides by any quantity:

> MULTIPLYING BOTH SIDES OF AN EQUATION BY A QUANTITY WHICH CAN BE ZERO MAY INTRODUCE EXTRANEOUS SOLUTIONS. **CHECK YOUR ANSWERS.**

EXAMPLE 7.1   Solve:

$$\frac{2x^2}{x^2 - x - 6} = \frac{2}{x^2 + 2x} - \frac{1}{x}$$

SOLUTION: Factor all expressions:

$$\frac{2x^2}{(x + 2)(x - 3)} = \frac{2}{x(x + 2)} - \frac{1}{x}$$

Clear denominators by multiplying both sides of the equation by $x(x + 2)(x - 3)$, the LCD of the three rational expressions:

$$\frac{2x^2}{(x+2)(x-3)} \cdot x(x+2)(x-3) = \frac{2}{x(x+2)} \cdot x(x+2)(x-3) - \frac{1}{x} \cdot x(x+2)(x-3)$$

$$2x^2(x) = 2(x-3) - 1(x+2)(x-3)$$

$$2x^3 = 2x - 6 - (x^2 - x - 6)$$

$$2x^3 + x^2 - 3x = 0$$

$$x(2x+3)(x-1) = 0$$

$$x = 0, \ x = -\frac{3}{2}, \ x = 1$$

At this point, we see that the only **possible** solutions are 0, $-\frac{3}{2}$, and 1. Any candidate which causes a denominator in the original equation to be zero must be discarded. Discarding 0 (as it renders the denominator of $\frac{1}{x}$ to be zero), we find that $-\frac{3}{2}$ and 1 are the only solutions of the given equation.

EXAMPLE 7.2 Solve:

$$\frac{x+1}{x^2} + \frac{2x^2}{x+1} = 3$$

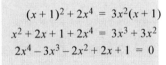

SOLUTION: Multiplying both sides of the equation by $x^2(x+1)$ leads us to:

$$2x^4 - 3x^3 - 2x^2 + 2x + 1 = 0 \quad \text{(see margin)}$$

1 is easily seen to be a zero of $p(x) = 2x^4 - 3x^3 - 2x^2 + 2x + 1$:

$$p(1) = 2 - 3 - 2 + 2 + 1 = 0$$

Theorem 1.4, page 12, tells us that $(x-1)$ is a factor of the polynomial, bringing us to:

$$(x-1)(2x^3 - x^2 - 3x - 1) = 0 \quad \text{(see margin)}$$

Not being able to readily spot a zero of $2x^3 - x^2 - 3x - 1$, we turn to the Rational Zeros Theorem, page 13, which tells us that ± 1 and $\pm\frac{1}{2}$ are the only possible rational zeros of that polynomial. A direct calculation shows that $-\frac{1}{2}$ is a zero. We leave it for you to check that:

$$x + \frac{1}{2} \overline{\smash{\big)}\, 2x^3 - x^2 - 3x - 1} \quad \frac{2x^2 - 2x - 2}{}$$

306   Chapter 7 Rational Functions

$$2x^2 - 2x - 2 = 0$$

$$x^2 - x - 1 = 0$$

$$x = \frac{-b \pm \sqrt{b^2 - 4ac}}{2a}$$

$$= \frac{-(-1) \pm \sqrt{(-1)^2 - 4(1)(-1)}}{2(1)}$$

$$= \frac{1 \pm \sqrt{5}}{2}$$

And so we have:

$$2x^4 - 3x^3 - 2x^2 + 2x + 1 = 0$$

$$(x - 1)\left(x + \frac{1}{2}\right)(2x^2 - 2x - 2) = 0$$

$$x = 1, \quad x = -\frac{1}{2}, \quad \text{and} \quad x = \frac{1 \pm \sqrt{5}}{2} \quad \leftarrow \text{(see margin)}$$

At this point we know that:

$$-\frac{1}{2}, 1, \frac{1 + \sqrt{5}}{2}, \frac{1 - \sqrt{5}}{2}$$

are the only possible solutions of the original equation. Since none of the four renders a denominator of that equation zero, each is a solution.

AN ALTERNATE SOLUTION:

Observing the term $\dfrac{x + 1}{x^2}$ and its reciprocal $\dfrac{x^2}{x + 1}$ in the equation:

$$\frac{x + 1}{x^2} + 2\left(\frac{x^2}{x + 1}\right) = 3 \qquad (*)$$

leads us to make the substitution $Y = \dfrac{x + 1}{x^2}$ in (*):

$$Y + 2\left(\frac{1}{Y}\right) = 3$$

$$Y^2 + 2 = 3Y$$

Solve for Y: $\qquad Y^2 - 3Y + 2 = 0$

$$(Y - 2)(Y - 1) = 0$$

Since $Y = \dfrac{x + 1}{x^2}$:

$Y = 2$

$$\frac{x + 1}{x^2} = 2$$

$$2x^2 - x - 1 = 0$$

$$(2x + 1)(x - 1) = 0$$

$$x = -\frac{1}{2} \text{ or } x = 1$$

$Y = 1$

$$\frac{x + 1}{x^2} = 1$$

$$x^2 - x - 1 = 0$$

$$x = \frac{1 \pm \sqrt{5}}{2}$$

CHECK YOUR UNDERSTANDING 7.2

Solve the given rational equation.

(a) $\dfrac{x - 2}{x^2 - 4} - \dfrac{5}{4} = \dfrac{1}{x - 3}$

(b) $3\left(\dfrac{x}{x^2 + 1}\right) + \left(\dfrac{x^2 + 1}{x}\right) = 4$ Suggestion: Consider the "alternate solution approach" of the previous example.

Answers: (a) -1

(b) $\dfrac{3 \pm \sqrt{5}}{2}$

EXAMPLE 7.3

Find all $0 \leq x < 2\pi$ for which:
$$\frac{-3\sin x + 1}{\sin^2 x} + \frac{5}{\sin x + 2} = 0$$

SOLUTION:

$$\frac{-3\sin x + 1}{\sin^2 x} + \frac{5}{\sin x + 2} = 0$$

$$[-3\sin x + 1][\sin x + 2] + 5\sin^2 x = 0$$

$$-3\sin^2 x - 5\sin x + 2 + 5\sin^2 x = 0$$

$$2\sin^2 x - 5\sin x + 2 = 0$$

$$(2\sin x - 1)(\sin x - 2) = 0$$

$$\sin x = \frac{1}{2} \quad \cancel{\sin x = 2}$$
no solution

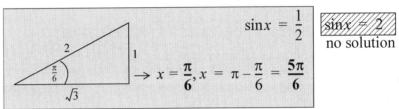

$$\rightarrow x = \frac{\pi}{6}, \ x = \pi - \frac{\pi}{6} = \frac{5\pi}{6}$$

CHECK YOUR UNDERSTANDING 7.3

Find all $0 \leq x < 2\pi$ for which:
$$3\sin x + 1 = \frac{1}{2\sin x - 1}$$

Answer:
$\sin^{-1}\left(\frac{2}{3}\right), \pi - \sin^{-1}\left(\frac{2}{3}\right)$
$\frac{7\pi}{6}, \frac{11\pi}{6}$

APPLICATIONS

EXAMPLE 7.4
MATERIAL COST

A cylindrical drum is to hold 65 cubic feet of chemical waste. Metal for the top of the drum costs $2 per square foot, and $3 per square foot for the bottom. Metal for the side of the drum costs $2.50 per square foot. Find the dimensions that will minimize cost, and the minimal cost.

308   Chapter 7 Rational Functions

SOLUTION: We follow the four-step procedure of page 253.

Step 1:

> Area of a circle of radius r is πr^2, and the circumference is $2\pi r$.

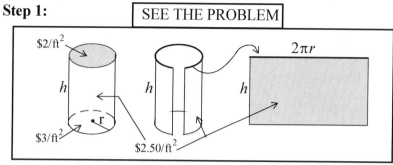

Step 2: Total cost C is to be minimized, and from the above we see that:

$$\$C = \overset{\text{cost of the top}}{\$2.00(\pi r^2)} + \overset{\text{cost of the bottom}}{\$3.00(\pi r^2)} + \overset{\text{cost of the side}}{\$2.50(2\pi rh)}$$

$$C = 5\pi r^2 + 5\pi rh \qquad (*)$$

Step 3: To express C as a function of only one variable, we need to eliminate either r or h. Since the drum is to have a volume of 65 ft^3:

$$\pi r^2 h = 65$$
$$h = \frac{65}{\pi r^2} \qquad (**)$$

Substituting in (*):

$$C = 5\pi r^2 + 5\pi r\left(\frac{65}{\pi r^2}\right) = 5\pi r^2 + \frac{325}{r}$$

Step 4: We turn to a graphing calculator:

> For hints on window sizing, see the marginal note on page 270.

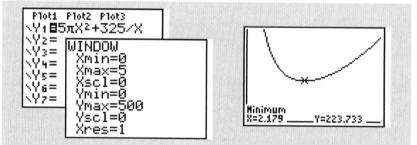

Conclusion: The minimal cost of (approximately) $223.73 occurs when $r \approx 2.18$ feet, and, from (**): $h = \dfrac{65}{\pi(2.18)^2} \approx 4.35$ feet.

EXAMPLE 7.5
POLLUTION COUNT

Two chemical plants are located 12 miles apart. The pollution count from plant A, in parts per million at a distance of x miles from plant A, is given by $\dfrac{4K}{x^2}$ for some constant $K > 0$. The pollution count from the cleaner plant B, at a distance of x miles from plant B, is one quarter that of A. Determine the point along the line joining A and B where the pollution count is minimal.

SOLUTION:

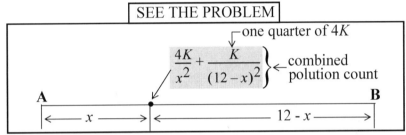

We see the function to be minimized, but what are we going to do with the K? We would have to know its value to determine the values of the pollution count, but by factoring it out:

$$\frac{4K}{x^2} + \frac{K}{(12-x)^2} = K\left(\frac{4}{x^2} + \frac{1}{(12-x)^2}\right)$$

we conclude that the minimum pollution count will occur where the function:

$$P(x) = \frac{4}{x^2} + \frac{1}{(12-x)^2}, \ 0 \le x \le 12$$

is minimal (why?). So:

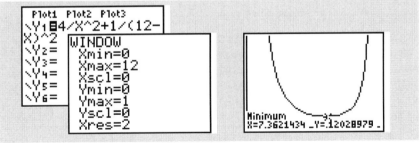

Conclusion: The minimum pollution count along the line between the two plants occurs at a distance of approximately 7.36 miles from plant A.

CHECK YOUR UNDERSTANDING 7.4

You need to design a closed aluminum can to hold 160 cubic inches. Use a graphing calculator to help you determine the dimensions of the can requiring the least amount of material.

Answer: radius ≈ 2.94 in.
height ≈ 5.89 in.

310   Chapter 7 Rational Functions

RATIONAL INEQUALITIES

One can use the SIGN method introduced in the previous chapter to solve rational inequalities where both the numerator and denominator are polynomials expressed in factored form containing only linear and quadratic factors. Consider the following examples.

EXAMPLE 7.6 Solve:

$$\frac{2}{x-2} + \frac{x}{x+1} + 1 \geq 0$$

SOLUTION: Combine terms, and factor:

$$\frac{2(x+1) + x(x-2) + (x-2)(x+1)}{(x-2)(x+1)} \geq 0$$

$$\frac{2x+2 + x^2 - 2x + x^2 - x - 2}{(x-2)(x+1)} \geq 0$$

$$\frac{2x^2 - x}{(x-2)(x+1)} \geq 0$$

$$\frac{x(2x-1)}{(x-2)(x+1)} \geq 0$$

Locate the zeros of either the numerator or denominator on the number line, positioning a *c* above each, as all are odd zeros. Noting that the rational expression is positive to the right of 2, we placed a "+" over the right-most interval, and then moved to the left, changing the sign each time we crossed over an odd zero:

We are adopting the convention of placing a white dot at those points on the number line where the denominator is zero (function not defined), and a black dot where the numerator is zero and the denominator is not zero.

Reading off the intervals with "+" signs, and adding the numbers where the numerator is zero (the black dots), we see that:

$$\frac{2}{x-2} + \frac{x}{x+1} + 1 \geq 0: \quad (-\infty, -1) \cup [0, \tfrac{1}{2}] \cup (2, \infty)$$

EXAMPLE 7.7 Solve:

$$\frac{3}{x} - 2 \leq 5x$$

7.1 Rational Functions, Equations, and Inequalities 311

> Multiplying the **equation**:
> $$\frac{3}{x} - 2 = 5x$$
> by x: $3 - 2x = 5x^2$
> is fine, since multiplying both sides of an **equation** by any nonzero quantity, whether positive or negative, will result in an equivalent equation.

SOLUTION: (**WARNING**) All too often, when confronted with such an inequality, one is tempted to begin by multiplying both sides by x, as one typically does with rational equations (see margin):

$$\frac{3}{x} - 2 \leq 5x$$

WRONG: $\quad 3 - 2x \leq 5x^2$
when $x < 0$

DON'T DO IT! The resulting inequality will **not** be equivalent to the original one when x is a **negative** number (as you know, if you multiply both sides of an inequality by a negative number, you must reverse the inequality sign). Therefore, if you're set on clearing the denominator, then you will have to consider two cases: (1) if $x > 0$, and (2) if $x < 0$. A simpler approach is to bring all terms to the left, and proceed as in the previous example

$$\frac{3}{x} - 2 \leq 5x$$

$$\frac{3}{x} - 2 - 5x \leq 0$$

$$\frac{3 - 2x - 5x^2}{x} \leq 0$$

$$\frac{-5x^2 - 2x + 3}{x} \leq 0$$

$$\frac{(-5x + 3)(x + 1)}{x} \leq 0$$

Noting that each zero is odd, and that the rational expression is negative to the right of the last zero, we have:

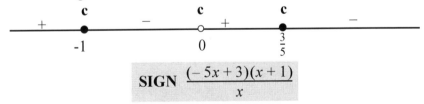

SIGN $\dfrac{(-5x + 3)(x + 1)}{x}$

Reading off the "−" intervals, and where the numerator is zero (the black dots in the SIGN chart), we arrive at the solution set of the given inequality: $[-1, 0) \cup [\frac{3}{5}, \infty)$

CHECK YOUR UNDERSTANDING 7.5

Determine the solution set of the given inequality.

(a) $\dfrac{x + 2}{x^2 - 2x - 3} \geq 0$

(b) $x < \dfrac{1}{3x + 2}$

> Answers:
> (a) $[-2, -1) \cup (3, \infty)$
> (b) $(-\infty, -1) \cup \left(-\frac{2}{3}, \frac{1}{3}\right)$

312   Chapter 7 Rational Functions

EXAMPLE 7.8 Find all $-2\pi < x < 2\pi$ for which:
$$\frac{\sin x}{x} \geq \frac{1}{x \sin x}$$

SOLUTION:
$$\frac{\sin x}{x} \geq \frac{1}{x \sin x}$$

$$\frac{\sin x}{x} - \frac{1}{x \sin x} \geq 0$$

$$\frac{\sin^2 x - 1}{x \sin x} \geq 0$$

$\sin^2 x + \cos^2 x = 1$: $\quad \dfrac{-\cos^2 x}{x \sin x} \geq 0 \quad$ or: $\quad \dfrac{\cos^2 x}{x \sin x} \leq 0$

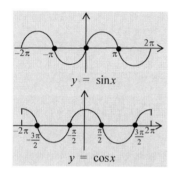
$y = \sin x$

$y = \cos x$

Within the interval $-2\pi < x < 2\pi$, the zeros of $x \sin x$ and of $\cos^2 x$, respectively are: $0, \pm\pi, \pm\dfrac{3\pi}{2}$ and $\pm\dfrac{\pi}{2}$ (see graphs of the sine and cosine function in the margin). The sine function changes signs when traversing its zeros at $\pm\pi$ and 0. But the x in the denominator of $\dfrac{\cos^2 x}{x \sin x}$, also changes its sign about 0, and those two changes of sign at 0 will "cancel each other out."

In expressions such as $x^2 \sin x$ or $x \sin^4 x$, where one of the factors changes sign about 0 and the other does not, the change of sign will not "cancel out."

In general:
Just as the product of an odd (even) number of negatives is negative (positive), so then does the product of an odd (even) number of changes of sign about a point result in a change (no change) of sign about that point.

(Think of the expression $x \cdot x = x^2$: Though each of the two x's do change their sign at 0, the product does not. Were there three such x's, namely $x \cdot x \cdot x = x^3$, then the net effect would be a change of sign at 0.)

Due to the even power, no change of sign will occur when traversing the zeros of $\cos^2 x$ at $\pm\dfrac{3\pi}{2}$ and $\pm\dfrac{\pi}{2}$.

Immediately to the right of 0, each of the terms in the expression $\dfrac{\cos^2 x}{x \sin x}$ is positive, and so therefore is the expression itself. This gets us started with a $+$ to the right of zero. From there, we just moved along the line, changing signs when traversing a "*c*" and not changing when traversing an *n*:

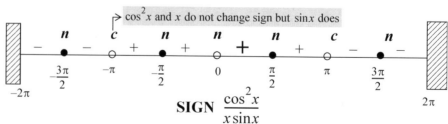

Having observed that $\dfrac{\sin x}{x} \ge \dfrac{1}{x \sin x}$ is equivalent to $\dfrac{\cos^2 x}{x \sin x} \le 0$, we read off our solution set from the above chart, being careful not to include the 0's of the denominator:

$$(-2\pi, -\pi) \cup (\pi, 2\pi) \cup \left\{ \pm\frac{\pi}{2} \right\}$$

Answer:

$$\left[-2\pi, -\frac{3\pi}{2}\right) \cup \left(\frac{\pi}{2}, \frac{\pi}{2}\right) \cup \left(\frac{3\pi}{2}, 2\pi\right]$$

CHECK YOUR UNDERSTANDING 7.6

Find all $-2\pi \le x \le 2\pi$ for which:

$$\frac{\sin x}{\cos x} \le \sec x$$

APPROXIMATING SOLUTIONS OF RATIONAL EQUATIONS AND INEQUALITIES

Since any solution of the rational equation $\dfrac{p(x)}{q(x)} = 0$ must be a solution of the polynomial equation $p(x) = 0$, The Confinement Theorem of page 267 extends to rational equations:

Theorem 6.3 of page 267 also extends to rational equations: The rational equation $\dfrac{p(x)}{q(x)} = 0$ can have at most n solutions, where n is the degree of $p(x)$.

THEOREM 7.2
CONFINEMENT THEOREM

All solutions of the rational equation:

$$\frac{p(x)}{q(x)} = \frac{x^n + a_{n-1}x^{n-1} + \cdots + a_1 x + a_0}{q(x)} = 0$$

are contained in the interval $[-K, K]$, where:

$$K = 1 + \text{largest}\{|a_{n-1}|, |a_{n-2}|, \ldots, |a_1|, |a_0|\}$$

As it was with polynomial equations, the above theorem comes in handy when using a graphing calculator to find approximate solutions of rational equations and inequalities. Consider the following example.

EXAMPLE 7.9 Use a graphing calculator to find, to two decimal places, the approximate solutions of:

$$\frac{2x}{x+3} - \frac{1}{x^2+2} + \frac{x-1}{x^3} = 0$$

and $$\frac{2x}{x+3} - \frac{1}{x^2+2} + \frac{x-1}{x^3} > 0$$

314 Chapter 7 Rational Functions

SOLUTION: We must first determine a viewing rectangle which will display all of the sign changes of the function:

$$f(x) = \frac{2x}{x+3} - \frac{1}{x^2+2} + \frac{x-1}{x^3}$$

details omitted: $= \dfrac{2x^6 + 4x^4 - x^3 - x^2 + 4x - 6}{x^3(x^2+2)(x+3)}$

Such sign changes can only occur where the graph crosses the x-axis (numerator is zero), or where the function is not defined (denominator is zero). Looking at the factored form of the denominator, we see that it is zero only at 0 and at −3. Dividing the numerator by 2 and then applying the Confinement Theorem, we find that the zeros of the numerator are contained in the interval [−4, 4] (see margin). Then:

$2x^6 + 4x^4 - x^3 - x^2 + 4x - 6 = 0$

$x^6 + 2x^4 - \dfrac{x^3}{2} - \dfrac{x^2}{2} + 2x - 3 = 0$

largest coefficient

$K = 1 + 3 = 4$

That vertical line occurring at $x = -3$ is not part of the graph; it can't be since the function is not defined at that point. It represents a vertical asymptote, a concept discussed in detail in the next section. Another vertical asymptote occurs at $x = 0$ (denominator is zero at that point). You can't see that vertical asymptote, since it coincides with the y-axis.

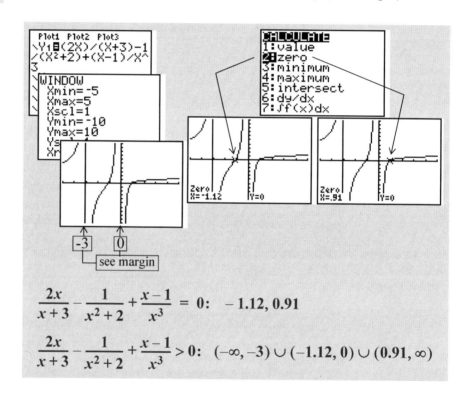

$\dfrac{2x}{x+3} - \dfrac{1}{x^2+2} + \dfrac{x-1}{x^3} = 0:\ \ -1.12, 0.91$

$\dfrac{2x}{x+3} - \dfrac{1}{x^2+2} + \dfrac{x-1}{x^3} > 0:\ \ (-\infty, -3) \cup (-1.12, 0) \cup (0.91, \infty)$

CHECK YOUR UNDERSTANDING 7.7

Use a graphing calculator to approximate the solution set of:

(a) $\dfrac{x}{x^2-3} = \dfrac{x-1}{3x^4+1}$ (b) $\dfrac{x}{x^2-3} < \dfrac{x-1}{3x^4+1}$

Answer: (a) 0.64
(b) $(-\infty, -\sqrt{3}) \cup (0.64, \sqrt{3})$

7.1 Rational Functions, Equations, and Inequalities 315

EXERCISES

Exercises 1–8: Find the domain of the function.

1. $k(x) = \dfrac{3x(x^3 - 4)}{x^2 - x - 2}$

2. $g(x) = \dfrac{x^3 - 1}{(x^2 - 4)(x^3 - x^2)}$

3. $f(x) = \dfrac{x + 1}{(3x - 4)^2(2x + 3)(x^2 + 4)} - 1$

4. $h(x) = \dfrac{4}{x} - \dfrac{2}{x + 3} + \dfrac{4x - 1}{4x - 1}$

5. $f(x) = \dfrac{8x^2 - 3}{10x^3 - 31x^2 + 15x}$

6. $g(x) = \dfrac{(3x - 1)^2}{(6x^2 + 1)(6x + 1)}$

7. $k(x) = \dfrac{2}{x^2 + 2x + 3} - \dfrac{3x}{3x + 1} + x - 2$

8. $h(x) = \left(\dfrac{x}{2} - \dfrac{2}{x}\right)^3$

Exercises 9–18: Solve the equation.

9. $\dfrac{-x + 5}{2} = \dfrac{x^2 + 5x}{x}$

10. $\dfrac{x + 2}{3} = \dfrac{1}{x + 4}$

11. $x = \dfrac{2}{x} + \dfrac{3}{2x} + 1$

12. $\dfrac{x + 3}{2x + 4} = \dfrac{4x}{x^2 - x - 6}$

13. $\dfrac{2x}{x - 1} - \dfrac{8}{3x + 1} = 2$

14. $\dfrac{x - 3}{x + 1} = \dfrac{x + 3}{3} + \dfrac{3x - 3}{x^2 - 1}$

15. $\dfrac{x - 5}{8x^2 + 8x} = \dfrac{3}{x^2 + 6x + 5}$

16. $\dfrac{1}{x^2 - 1} + \dfrac{1}{x - 1} = \dfrac{5}{x + 1}$

17. $\dfrac{x - 2}{x + 4} = \dfrac{x + 2}{x - 4} + \dfrac{4}{x^2 - 16}$

18. $\dfrac{4x}{x^2 - 2x - 3} - \dfrac{3x - 1}{x^2 + 4x + 3} - \dfrac{3}{x + 1} = 0$

Exercises 19–22: Make an appropriate substitution and solve the equation.

19. $\dfrac{x^2 + 1}{(x^2 + 1) + 2} = \dfrac{1}{x^2 + 1}$

20. $\dfrac{2x - 1}{x^2} + \dfrac{x^2}{2x - 1} = -2$

21. $\dfrac{(x - 5)^2}{3} = \dfrac{2}{(x - 5)^2 + 1}$

22. $\dfrac{2}{2x + 1} = \dfrac{3}{1 + (2x + 1)} + 3$

Exercises 23–28: Solve for x in $[0, 2\pi)$. (You may want to first substitute t for the trigonometric function in the equation, then solve for t, and substitute back.)

23. $\dfrac{2\sin x + 1}{\sin x + 1} - \dfrac{3\sin x}{\sin x - 1} = -1$

24. $\dfrac{1}{\cos x - 3} - \dfrac{\cos x - 2}{\cos^2 x - 4} = -\dfrac{5}{4}$

25. $\dfrac{19(\sin x + 3)}{\sin^2 + 4} - \dfrac{1}{\sin x + 1} = -\dfrac{2}{\sin x - 1}$

26. $\dfrac{\cos x}{20(\cos^2 x + 1)} = \dfrac{\cos^3 x}{21(2 - \cos^2 x)}$

27. $\dfrac{\tan x - 2}{\tan x} = \dfrac{\tan x + 1}{\tan x - 1} + 3$

28. $\dfrac{2}{\csc x} - \dfrac{1}{\csc x - 2} = 3$

316 Chapter 7 Rational Functions

Exercises GC29–GC34: Use a graphing calculator and the four-step procedure to find an approximate solution to the optimization problem.

29. (**Minimum Area**) A poster is to surround 45 in.^{2} of printing material with a top and bottom margin of 4 in. and side margins of 3 in. Find the outside dimensions of the poster that will require the minimum amount of paper.

30. (**Smallest Perimeter**) A rectangle has an area of 500 in.^{2}. What are the dimensions of the rectangle of smallest perimeter?

31. (**Lightest Box**) A closed box with a square base has a volume of 200 in^3. The base of the box weighs four times as much per square inch as each of the sides and twice as much per square inch as the top. Determine the dimensions of the lightest box possible.

32. (**Minimum Cost**) A cylindrical drum is to hold 300 ft^3. The cost of the top of the drum is $6.00 per square foot, and the cost of the bottom is $10.50 per square foot. The side costs $8.00 per square foot. Find the minimum cost of materials.

33. (**Minimum Cost**) A fenced-in rectangular garden is divided into 3 areas by two fences running parallel to one side of the rectangle. The two fences cost $6 per running foot, while the outside fencing only costs $4 per running foot. If the garden is to have an area of 8,000 ft^2, find the dimensions of the garden that minimizes the total cost of fencing.

34. (**Minimum Area**) The positive x- and y-axes and any line of negative slope m through the point $(1,2)$ determine a right triangle. Express the area of the triangle as a function of the slope m of the line, and then find the approximate slope of the line for which this triangle has minimum area. What is the minimum area, approximately? What is the approximate equation of the line giving minimum area?

Exercises 35–46: Solve the inequality.

35. $\dfrac{2x-1}{x+3} < 0$

36. $\dfrac{2x^2+x-3}{x+2} \le 0$

37. $\dfrac{4-x^2}{3x-2} \ge 0$

38. $\dfrac{x^2-4x+4}{x-5} < 0$

39. $\dfrac{x}{1-x} < 2$

40. $3x+2 < \dfrac{1}{x}$

41. $x > \dfrac{1}{3x+2}$

42. $\dfrac{1}{x} > \dfrac{2x-1}{x}$

43. $\dfrac{1}{2x-3} \le x-2$

44. $\dfrac{1}{x^2} + \dfrac{3}{x} \le \dfrac{2}{x^3}$

45. $\dfrac{3x}{3x+1} - 1 \le \dfrac{1}{x+5}$

46. $\dfrac{-x}{x+1} < \dfrac{1}{x-1} - \dfrac{1}{x^2-1}$

Exercises 47–49: Solve the inequality in the interval indicated.

47. $x \tan x \le -\dfrac{x}{\tan x}$, $[0, 2\pi)$

48. $\dfrac{\cos x}{x^2} < \dfrac{1}{x^2 \cos x}$, $[\pi, \pi]$

49. $x \sec \ge \dfrac{x}{\sec x}$, $[-2\pi, 2\pi]$

7.1 Rational Functions, Equations, and Inequalities   317

Exercises GC50–GC57: Use a graphing calculator as in Example 7.9 to approximate to two decimal places, the solutions of the equation and of the inequality indicated in parenthesis.

50. $\dfrac{5x^5 - 3x - \sqrt{2}}{x - 100} - x^2 - 1 = 0;\ \ (<0)$ 51. $2x^4 + 5x - \dfrac{1}{x^2 - 1} = 0;\ \ (>0)$

52. $\dfrac{9}{6x - 11} + \dfrac{4x^5}{x^6 - 1} - 1 = 0;\ \ (>0)$ 53. $\dfrac{-4x^2 + 1}{x + 4} + \dfrac{2}{x + 3} + 1 = 0;\ \ (<0)$

54. $\dfrac{3}{7x^2} - \dfrac{4}{(x^4 + 1)^2} = 0;\ \ (>0)$ 55. $\dfrac{4}{x^2} + \dfrac{7x^3 - 2x + 2}{x - 1} = 0;\ \ (<0)$

56. $\dfrac{x^5}{2x - 3} - \dfrac{2}{x^3} + x = 0;\ \ (>0)$ 57. $-\dfrac{2}{x} + \dfrac{3}{x + 1} - \dfrac{2}{x^2 + 1} = 0;\ \ (<0)$

Exercise GC58: (Acceleration) The acceleration A, in mi/hr^2 of an experimental car as a function of fuel consumption, C, in gallons per hour is approximated by the function

$$A = \frac{2C^3 - 3C + 1}{7C + 4}$$

with the approximation being acceptable only when acceleration falls between 5 mi/hr^2 and 11 mi/hr^2. Use a graphing calculator to approximate the level of fuel consumption for which the acceleration A is acceptable.

Exercise GC59: (Profit) The Ajax Company's profit P (in thousands of dollars) from the sale of x hundred units is given by the function

$$P = \frac{190x - 1166}{2x^2 - 24x + 77}$$

where $0 \le x \le 15$. Use a graphing calculator to find the required production for:

(a) profit to be maximum.   (b) profit to be positive.   (c) profit to exceed \$12,000.

Exercises 60–61: Express P as a function only of t.

60. $P = \dfrac{st}{2u}, \quad s = \dfrac{3}{u} - \dfrac{1}{t}, \quad u = 5t - 2$ 61. $P = \dfrac{3u}{4s - 3}, \quad s = \dfrac{6t}{2u - 1}, \quad u = \dfrac{2}{t}$

Exercises 62–63: Without expressing P as a function only of t, find the value of P when $t = 1$.

62. $P = \dfrac{2s - u}{s + u}, \quad s = 4 - 3t, \quad u = \dfrac{t}{t + 1}$ 63. $P = \dfrac{u}{9s^2}, \quad s = \dfrac{4 - tu}{3u}, \quad u = 3t - \dfrac{4}{t}$

Exercises 64–69: Suppose c is a solution of the rational equation $\dfrac{p(x)}{q(x)} = 0$, $q(x)$ is not the zero polynomial, and $k(x)$ is a polynomial. Indicate True or False. Justify your answer.

64. c is a solution of the equation $\dfrac{k(x)p(x)}{q(x)} = 0$.

318 **Chapter 7 Rational Functions**

65. $-c$ is a solution of the equation $\dfrac{p(x)}{k(x)q(x)} = 0$.

66. c is a solution of the equation $\dfrac{p(x)}{k(x)q(x)} = 0$.

67. $c - 1$ is a solution of the equation $\dfrac{p(x+1)}{q(x+1)} = 0$.

68. $c - 1$ is a solution of the equation $\dfrac{p(x+1)}{q(x^2-1)} = 0$.

69. $c - 1$ is a solution of the equation $\dfrac{p(x+1)}{q(x)} = 0$.

Exercises 70–71: (Identities) An equation which is satisfied for all values of the variable for which all expressions are defined is said to be an identity. To see that

$$\frac{2x+1}{x-1} + 1 = \frac{3x}{x-1}$$

is an identity, multiply both sides by $x - 1$:

$$2x + 1 + x - 1 = 3x$$
$$3x = 3x$$

which holds for all x. (Note that the original equation is undefined for $x = 1$.)

Show that the equation is an identity.

70. $\dfrac{3x+1}{x+2} = \dfrac{3x}{x-1} - \dfrac{8x+1}{x^2+x-2}$ 　　71. $2x + \dfrac{1}{x} = \dfrac{4x+1}{x} + \dfrac{2x^2-2x-4}{x+1}$

Exercises 72–73: (Contradictions) An equation that leads to an absurdity is called a contradiction. To see that

$$\frac{2x+1}{x-1} + 1 = \frac{x}{x-1} + 2$$

is a contradiction, multiply both sides by $x - 1$:

$$2x + 1 + x - 1 = x + 2x - 2$$
$$3x = 3x - 2$$
$$0 = -2$$

Show that the equation is a contradiction.

72. $\dfrac{2x}{x-3} - \dfrac{1}{3-x} = \dfrac{2x}{x-3}$ 　　73. $\dfrac{5x+1}{x^2-1} + \dfrac{4}{x-1} = \dfrac{9}{x+1}$

7.1 Rational Functions, Equations, and Inequalities **319**

Exercises 74–76: (**Variation**) A quantity y is said to *vary inversely* with another quantity x, if there exists a nonzero constant k such that $y = \frac{k}{x}$. (The expression "y is inversely proportional to x" is also used.)

74. A quantity Q varies inversely with x. What will be the value of Q when $x = 15$, if $Q = 4$ when $x = 30$?

75. (Combined Variation) A quantity Q is said to vary directly with x and inversely with y if there is a nonzero constant k such that $Q = \frac{kx}{y}$. The volume V of a mass of gas varies directly with the temperature T and inversely with the pressure P. If $V = 250$ in.^{3} when $T = 300°\,$F and $P = 25$ lb/in.^{2}, what is the volume when $T = 200°\,$F and $P = 20$ lb/in.^{2}?

76. Above and near the earth's surface, the weight of an object varies inversely as the square of its distance from the center of the earth. If a man weighs 200 pounds at sea level (approximately 4000 miles from the center of the earth), what will he weigh at a point in space that is 100 miles above sea level?

320   Chapter 7 Rational Functions

§2. GRAPHING RATIONAL FUNCTIONS

In many ways, the procedure for sketching the graph of a rational function (which readily factors) is quite similar to that for polynomial functions.

Since the leading term of a polynomial function dominates its behavior far away from the origin, to anticipate the behavior of the graph of a rational function as $x \to \pm\infty$ you will want to consider the leading terms of both the numerator and denominator of that function:

Far away from the origin the graph of the rational function:

$$f(x) = \frac{a_n x^n + a_{n-1} x^{n-1} + \cdots + a_1 x + a_0}{b_m x^m + b_{m-1} x^{m-1} + \cdots + b_0}$$

resembles, in shape, that of:

$$g(x) = \frac{a_n x^n}{b_m x^m}$$

For example, far from the origin (as $x \to \pm\infty$), the graphs of:

$$f(x) = \frac{3x^5 - 2x^3 - 3x}{2x^2 + x - 5} \text{ and } g(x) = \frac{3x^5}{2x^2} = \frac{3}{2}x^3$$

have similar shapes.

HORIZONTAL ASYMPTOTE

When the degree of the numerator of a rational function f is less than or equal to the degree of the denominator, the graph will approach a horizontal line, called a **horizontal asymptote** for the graph of f.

For example, as $x \to \pm\infty$, the graph of $f(x) = \dfrac{2x^3 + 3x^2 - 5}{5x^3 + x}$ approaches the horizontal line $y = \dfrac{2x^3}{5x^3} = \dfrac{2}{5}$. Consequently $y = \dfrac{2}{5}$ is a horizontal asymptote for the graph of f.

The graph of $h(x) = \dfrac{2x^3 + 3x^2 - 5}{5x^4 + x}$ approaches that of the function $y = \dfrac{2x^3}{5x^4} = \dfrac{2}{5x}$, which approaches 0 as $x \to \pm\infty$. Consequently, the x-axis ($y = 0$) is a horizontal asymptote for the graph of h.

OBLIQUE ASYMPTOTE

When the degree of the numerator of a rational function f is one more than the degree of the denominator, the graph will approach an oblique line, called an **oblique asymptote** for the graph of f.

7.2 Graphing Rational Functions 321

For example, the graph of $f(x) = \dfrac{2x^2}{x+2}$ approaches a line of slope $\dfrac{2x^2}{x} = 2$ as $x \to \pm\infty$. To be more specific, from:

$$f(x) = \frac{2x^2}{x+2} = 2x - 4 + \frac{8}{x+2} \quad \text{(see margin)}$$

we see that the graph of f will get arbitrarily close to the line $y = 2x - 4$ as $x \to \pm\infty$, and this is because $\dfrac{8}{x+2}$ approaches zero as x tends to $\pm\infty$.

```
            2x - 4
       ┌─────────────
x + 2  │ 2x²
         2x² + 4x
         ─────────
              -4x
              -4x - 8
              ───────
                  8
```

CHECK YOUR UNDERSTANDING 7.8

If the graph of the given function has a horizontal or an oblique asymptote, find the equation of that asymptote. If not, then find the function whose graph resembles that of the given function as $x \to \pm\infty$.

(a) $f(x) = \dfrac{3x^4 - 4x^2 + 1}{x^4 - 5x^2 + 4}$ (b) $f(x) = \dfrac{4x^3 - 2x^2 - 4}{3x^4 + 1}$ (c) $f(x) = \dfrac{6x^3 - 9x^2 - 4x}{3x^2 - 1}$

Answers:
(a) $y = 3$ (b) $y = 0$
(c) $y = 2x - 3$

VERTICAL ASYMPTOTES

The main difference between graphing polynomial functions and rational functions is that vertical asymptotes might come into play when graphing a rational function; where:

> A **vertical asymptote** for the graph of a function f is a vertical line about which the graph tends to either plus or minus infinity.

Consider, for example, the function:

$$f(x) = \frac{2x - 1}{(x - 2)(x + 1)}$$

This function is not defined at $x = 2$ and at $x = -1$ (why not?). As x gets closer and closer to 2, the numerator, $2x - 1$ approaches the value $2 \cdot 2 - 1 = 3$, while the denominator $(x - 2)(x + 1)$ approaches zero. The magnitude of the quotient $\dfrac{2x - 1}{(x - 2)(x + 1)}$ must therefore get larger and larger, tending to plus or minus infinity, depending on the sign of f. From the sign information:

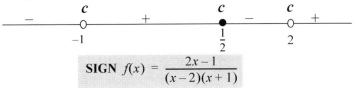

SIGN $f(x) = \dfrac{2x - 1}{(x - 2)(x + 1)}$

we conclude that:

As x approaches 2 from the left, the values of $f(x)$, being negative, tend to $-\infty$ (see margin).

As x approaches 2 from the right, the values of $f(x)$, being positive, tend to ∞ (see margin).

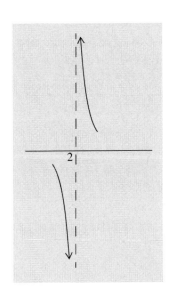

322   Chapter 7 Rational Functions

A vertical asymptote need not occur at a point at which both the numerator and denominator of the rational expression are zero (see Exercises 39-41).

Answer: Same as at $x = 2$

Another vertical asymptote occurs at $x = -1$, and you are asked for a similar analysis in the following Check Your Understanding.

CHECK YOUR UNDERSTANDING 7.9

The nature of the graph of the function $f(x) = \dfrac{2x - 1}{(x - 2)(x + 1)}$ about the vertical asymptote at $x = 2$ is depicted in the margin of the previous page. Do the same for the vertical asymptote at $x = -1$.

OBTAINING THE GRAPH

We return to the 4-step procedure for graphing polynomial functions, modifying Step 2 to accommodate vertical asymptotes:

Here, "near the origin" encompasses all zeros of the numerator or denominator.

Step 2.   **Near the origin:**
Determine and plot x- and y-intercepts.
Sketch the dotted lines for the vertical asymptotes.
Chart the sign of the function to determine the nature of the graph about the x-intercepts, and the direction of the graph near its vertical asymptotes.

EXAMPLE 7.10   Sketch the graph of the function:

$$f(x) = \frac{4x - 7}{2x + 5}$$

SOLUTION:

Step 1. Factor: Already in factored form.

Step 2. y-intercept: $y = f(0) = -\dfrac{7}{5}$ [Figure 7.1(a)].

x-intercepts: $f(x) = 0$ at $x = \dfrac{7}{4}$ [Figure 7.1(a)].

Vertical Asymptotes: The line $x = -\dfrac{5}{2}$ [Figure 7.1(a)].

SIGN $f(x)$: From the sign information at the top of Figure 7.1(a), we conclude that the graph goes from below the x-axis to above the x-axis as you move from left to right across the x-intercept at $x = \dfrac{7}{4}$ [Figure 7.1(a)]. Since the function is positive to the left of the vertical asymptote at $x = -\dfrac{5}{2}$, the graph must tend to $+\infty$ as x approaches $-\dfrac{5}{2}$ from the left. Since the function is negative just to the right of $x = -\dfrac{5}{2}$, the graph tends to $-\infty$ as x approaches $-\dfrac{5}{2}$ from the right.

Step 3. As $x \to \pm\infty$: The graph of $f(x) = \dfrac{4x-7}{2x+5}$ approaches the horizontal asymptote $y = \dfrac{4x}{2x} = 2$ [Figure 7.1(a)].

$$\begin{array}{r}2\\2x+5\overline{\smash{\big)}4x-7}\\4x+10\\\hline-17\end{array}$$

Additional information can be derived by observing that:
$$f(x) = \frac{4x-7}{2x+5} = 2 - \frac{17}{2x+5} \quad \text{(see margin)}$$
From the above form, we can see that the graph will approach the line $y = 2$ from below as $x \to \infty$ (for $\dfrac{17}{2x+5}$ will be positive), and from above $x \to -\infty$ (for $\dfrac{17}{2x+5}$ will be negative).

Step 4. Sketch the anticipated graph: Figure 7.1(b) and the final graph in (c).

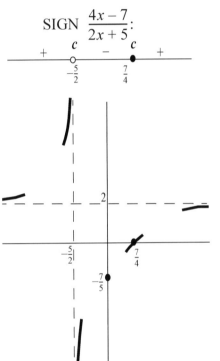
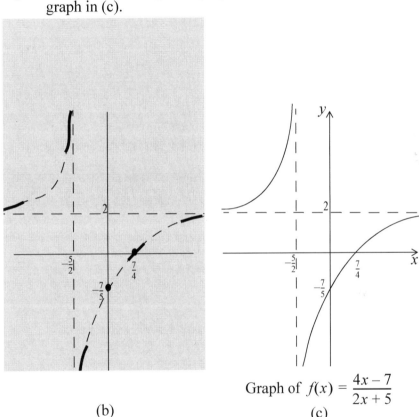

(a)   (b)   Graph of $f(x) = \dfrac{4x-7}{2x+5}$ (c)

Figure 7.1

EXAMPLE 7.11

Sketch the graph of the function:
$$f(x) = \frac{x}{x^2 + 4x - 5}$$

SOLUTION:

Step 1. Factor: $f(x) = \dfrac{x}{(x-1)(x+5)}$

Step 2. y-intercept: $y = f(0) = \dfrac{0}{-5} = 0$ [Figure 7.2(a)].

x-intercepts: $f(x) = 0$: at $x = 0$ [Figure 7.2(a)].

Vertical Asymptotes: The lines $x = 1$ and $x = -5$ [Figure 7.2(a)].

SIGN $f(x)$: From the sign information at the top of Figure 7.2(a), we conclude that the graph goes from above the x-axis to below the x-axis, as you move, from left to right, across the x-intercept at $x = 0$ [Figure 7.2(a)]. Since the function is negative just to the left of the vertical asymptote at $x = -5$, the graph must tend to $-\infty$ as x approaches -5 from the left. Since the function is positive just to the right of -5, the graph tends to $+\infty$ as x approaches -5 from the right. Similarly, you can see that the graph tends to $-\infty$ as x approaches 1 from the left, and to $+\infty$ as x approaches 1 from the right [Figure 7.2(a)].

Step 3. As $x \to \pm\infty$: The graph resembles that of $g(x) = \dfrac{x}{x^2} = \dfrac{1}{x}$ which tends to 0 as x tends to $\pm\infty$. This tells us that the line $y = 0$ (the x-axis) is a horizontal asymptote for the graph [Figure 7.2(a)]. From the sign information we conclude that the graph approaches the x-axis from below as $x \to -\infty$, and from above the x-axis as $x \to \infty$.

Step 4. Sketch the anticipated graph: Figure 7.2(b) and the final graph in (c).

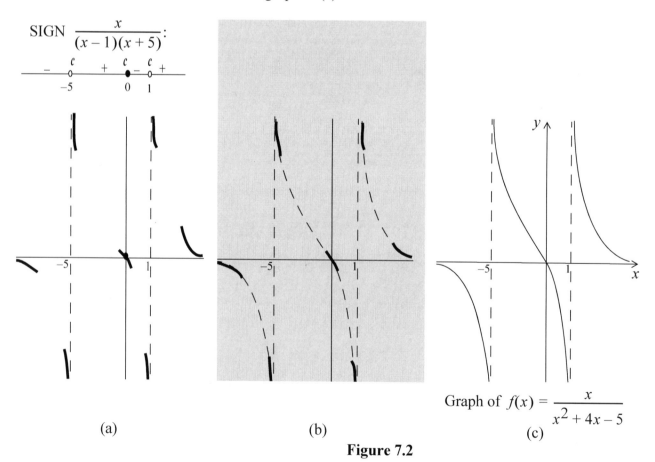

Graph of $f(x) = \dfrac{x}{x^2 + 4x - 5}$

(a)　　(b)　　(c)

Figure 7.2

7.2 Graphing Rational Functions   **325**

EXAMPLE 7.12   Sketch the graph of the function:

$$f(x) = \frac{x^4-16}{8x^2-8}$$

SOLUTION:

Step 1. Factor: $f(x) = \dfrac{(x^2+4)(x^2-4)}{8(x^2-1)} = \dfrac{(x^2+4)(x+2)(x-2)}{8(x+1)(x-1)}$

Step 2. y-intercept: $y = f(0) = \dfrac{-16}{-8} = 2$ [Figure 7.3(a)].

x-intercepts: $f(x) = 0$: at $x = -2$ and $x = 2$ [Figure 7.3(a)].

Vertical Asymptotes: The lines $x = -1$ and $x = 1$ [Figure 7.3(a)].

SIGN $f(x)$: From the sign information at the top of Figure 7.3(a), we conclude that the graph goes from above to below the x-axis, as you move from left to right across the x-intercept at $x = -2$; and from below to above as you move across the intercept at $x = 2$ [Figure 7.3(a)]. Since the function is negative immediately to the left of the vertical asymptote at $x = -1$, the graph must tend to $-\infty$ as x approaches -1 from the left. Since the function is positive just to the right of -1, the graph tends to $+\infty$ as x approaches -1 from the right. Similarly, since the function values are positive just to the left of 1 and negative immediately to its right, we know that the graph tends to $+\infty$ as x approaches 1 from the left, and to $-\infty$ as x approaches from the right [Figure 7.3(a)].

Step 3. As $x \to \pm\infty$: The graph resembles that of

$$g(x) = \frac{x^4}{8x^2} = \frac{1}{8}x^2 \text{ [Figure 7.3(a)].}$$

Step 4. Sketch the anticipated graph: Figure 7.3(b) and the final graph in (c).

Being an even function, the graph of:

$$f(x) = \frac{x^4-16}{8x^2-8}$$

must be symmetrical about the y-axis (think about it).

326   Chapter 7 Rational Functions

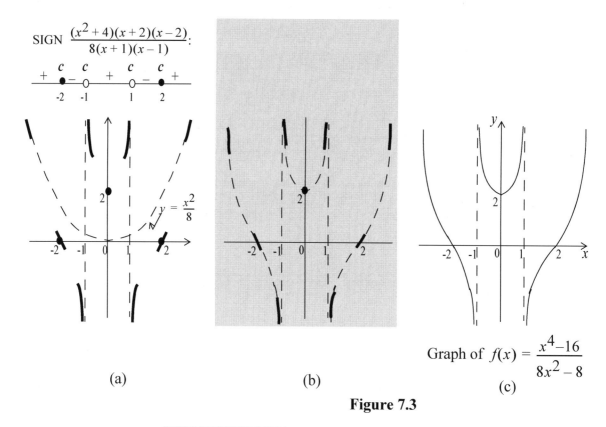

(a)   (b)   Graph of $f(x) = \dfrac{x^4 - 16}{8x^2 - 8}$

(c)

Figure 7.3

EXAMPLE 7.13   Sketch the graph of the function:
$$f(x) = \frac{2x^2}{x+1}$$

SOLUTION:

Step 1. **Factor:** The function is already in factored form.

Step 2. **y-intercept:** $y = f(0) = \dfrac{0}{1} = 0$ [Figure 7.4(a)].

x-intercepts: Where $f(x) = 0$: $x = 0$ [Figure 7.4(a)].

Vertical Asymptotes: The line $x = -1$ [Figure 7.4(a)].

SIGN $f(x)$: From the sign information at the top of Figure 7.4(a), we conclude that the graph lies above the x-axis on both sides of its x-intercept at $x = 0$ [Figure 7.4(a)]. Since the function is negative to the left of the vertical asymptote at $x = -1$, the graph must tend to $-\infty$ as x approaches -1 from the left [Figure 7.4(a)]. By the same token, since the function is positive to the right of -1, the graph tends to $+\infty$ as x approaches -1 from the right.

7.2 Graphing Rational Functions

Step 3. As $x \to \pm\infty$: The graph of $f(x) = \dfrac{2x^2}{x+1}$ will resemble that of a line of slope 2 ($\dfrac{2x^2}{x} = 2x$) [Figure 7.4(a)].

Additional information can be derived by observing that:
$$f(x) = 2x - 4 + \dfrac{8}{x+2} \quad \text{(see margin)}$$

From the above form, we can see that the graph will approach the oblique asymptote $y = 2x - 4$ from above as $x \to \infty$ (for $\dfrac{8}{x+2}$ will be positive), and from below as $x \to -\infty$ (for $\dfrac{8}{x+2}$ will be negative).

Step 4. Sketch the anticipated graph: Figure 7.4(b) and the final graph in (c).

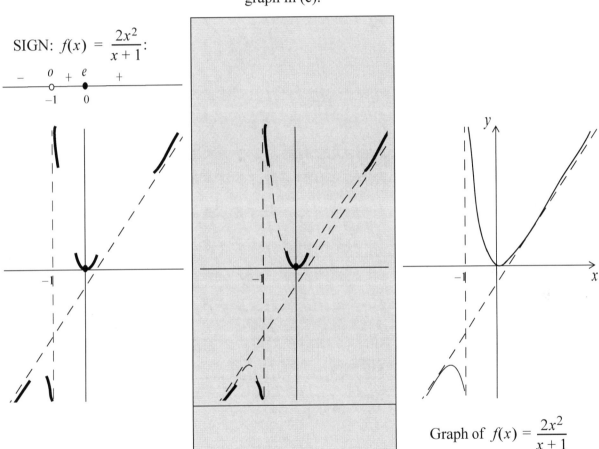

Figure 7.4

CHECK YOUR UNDERSTANDING 7.10

Use the four-step graphing procedure to sketch the graph of the given function.

(a) $f(x) = \dfrac{x-1}{x+1}$ (b) $f(x) = \dfrac{x^7}{x^4 - 16}$

Answers:
(a) See page C-53.
(b) See page C-54.

328 Chapter 7 Rational Functions

EXAMPLE 7.14 Sketch the graph of the function:
$$f(x) = \frac{\cos^2 x}{x \sin x}$$
over the interval $[-2\pi, 2\pi]$

SOLUTION: Most of the work was done on page 312, where we determined the SIGN of $\frac{\cos^2 x}{x \sin x}$. The rest, follows suit:

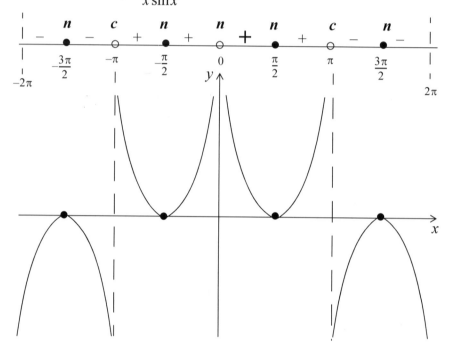

The graph's symmetry is due to the fact that $f(x) = \frac{\cos^2 x}{x \sin x}$ is an even function (see page 92). While the graph may suggest that $f(x) = \frac{\cos^2 x}{x \sin x}$ is periodic, that is not the case (note the non-periodic factor x in the denominator).

CHECK YOUR UNDERSTANDING 7.11

Sketch the graph of $f(x) = \dfrac{\cos x \sin x}{x^2 - 1}$ over the interval $[-\pi, \pi]$.

Answer: See page C-55.

7.2 Graphing Rational Functions   **329**

EXERCISES

Exercises 1–8: Determine a function $g(x)$ whose graph the given function resembles as $x \to \pm\infty$.

1. $f(x) = \dfrac{-9x^4 + 3x^3 + 2x + 7}{3x^4 - 2x^2 - 9}$

2. $f(x) = \dfrac{\frac{2}{3}x^5 - x^3 + 5x + 8}{4x^6 - x + 12}$

3. $f(x) = \dfrac{16x^8 + \frac{1}{2}x^5 - 3x}{12x^3 - 12x^2 - 6x + 11}$

4. $f(x) = \dfrac{20x^4 - 2x^3 + 4x^2 - 13x - 8}{4x^3 - 2x^2 + x - 3}$

5. $f(x) = \dfrac{5 - \frac{1}{x}}{10x^2 + 3}$

6. $f(x) = \dfrac{7x^3 + 5x^2 - 4x - 2}{2 - 3x}$

7. $f(x) = \dfrac{-6x^3 + 3x^2 + 2}{3x^2 - 1}$

8. $f(x) = \dfrac{(2x + 1)(4x - 1)^2}{(3x^2 + 4)(8x - 5)}$

Exercises GC1–GC8: Use a graphing calculator to check your result in Exercises 1-8, by simultaneously viewing the graphs of f and g far away from the origin.

Exercises 9–14: Determine the asymptotes (vertical, and horizontal or oblique) , if any.

9. $h(x) = \dfrac{(3x + 1)(2x - 1)}{(4x - 1)^2}$

10. $f(x) = \dfrac{2x^3}{3 - x}$

11. $h(x) = -\dfrac{3x^2 + 2}{x - 1}$

12. $k(x) = \dfrac{2(x + 1)^3(2x + 1)^2}{(1 - x)^4(5x - 3)}$

13. $h(x) = \dfrac{x^3 + x^2}{x^3 - x^2 - 4x + 4}$

14. $f(x) = \dfrac{4x + 3}{x^4 + 2x^3 - 3x^2}$

Exercises 15–30: Sketch the graph of the function, using the 4-step method.

15. $f(x) = \dfrac{2}{x + 3}$

16. $h(x) = \dfrac{x - 1}{x}$

17. $k(x) = \dfrac{6 - x}{x + 1}$

18. $h(x) = \dfrac{16 - 9x^2}{2x + 1}$

19. $f(x) = \dfrac{3x^2 - 9}{x + 2}$

20. $f(x) = \dfrac{x + 1}{x^2 - 2x + 1}$

21. $f(x) = \dfrac{x^2 - x - 2}{2x^2 + 5x - 3}$

22. $k(x) = \dfrac{3}{x + 5} + \dfrac{4}{x - 2}$

23. $h(x) = \dfrac{2x^2 - 3x - 9}{x^2 - 6x + 9}$

24. $f(x) = \dfrac{x^2 + 5x + 6}{2x^2 + 3x - 2}$

25. $f(x) = \dfrac{(2 - x)^3(4x - 3)^2}{2(x - 1)^3}$

26. $f(x) = \dfrac{3x^5}{(x - 5)^2}$

330   Chapter 7 Rational Functions

27. $h(x) = \dfrac{x \sin^2 x}{\cos x}$, on $[-2\pi, 2\pi]$

28. $k(x) = \dfrac{\cos^2 x}{x^2 \sin x}$, on $[-2\pi, 2\pi]$

29. $f(x) = \dfrac{\cos x}{x \sin x}$, on $[-2\pi, 2\pi]$

30. $h(x) = \dfrac{x \sin x}{\cos^2 x}$, on $[-2\pi, 2\pi]$

Exercises GC31–GC34: Use a graphing calculator to view the graph of the denominator of the function, and estimate where the vertical asymptotes of the function occur.

31. $f(x) = \dfrac{x^5 - 6}{x^4 + x - 7}$

32. $k(x) = \dfrac{4x^2 + 6x - 15}{8x^6 - 7x^2 + 5x + 5}$

33. $h(x) = \dfrac{5x^3 - 9x + 3}{2x^5 + x^3 - 2x^2 - 4}$

34. $f(x) = \dfrac{(x - 1)^4}{6x^5 - 3x^4 - 9x^3 + 1}$

Exercises 35–38: Sketch the graph of the function, using the 4-step process. You will have to take some liberties in labeling the units along the x-axis in order to be able to show all points of interest near the origin. [Then try to check your graph with a graphing calculator (not so easy to do).]

35. $f(x) = \dfrac{(x - 1)(x - 500)}{(x + 1)(x + 1000)}$

36. $f(x) = \dfrac{(x - 1)(x - 500)}{(x + 1)^2(x + 100)}$

37. $f(x) = \dfrac{x^2(x - 1)(x - 500)}{(x + 1)(x + 1000)}$

38. $f(x) = \dfrac{(x - 1)^4(x - 500)}{(x + 1)(x + 1000)}$

Exercises 39–41: (Theory) Determine the domain of the function, and sketch its graph. Does the graph have any vertical asymptotes?

39. $f(x) = \dfrac{x^2}{x}$

40. $f(x) = \dfrac{x}{x^2}$

41. $f(x) = \dfrac{x^2}{x^2}$

Exercises 42–45: Answer True or False. If true, sketch the graph of a function satisfying the given condition. If false, explain why.

42. The graph of a rational function can have more than one vertical asymptote.

43. The graph of a rational function can intersect any of its vertical asymptotes.

44. The graph of a rational function can have more than one horizontal asymptote.

45. The graph of a rational function can intersect its horizontal asymptote.

§3. PARTIAL FRACTIONS

There is no question that you can perform the sum:

$$\frac{2}{x+3} + \frac{1}{x-2}$$

You find the least common denominator, and go on from there:

$$\frac{2}{x+3} + \frac{1}{x-2} = \frac{2x-4+x+3}{(x+3)(x-2)} = \frac{3x-1}{(x+3)(x-2)}$$

In the calculus, you will have to reverse this process: Start with the rational expression:

$$\frac{3x-1}{(x+3)(x-2)}$$

and decompose it into a sum of simpler rational expressions, called partial fractions:

$$\frac{3x-1}{(x+3)(x-2)} = \frac{2}{x+3} + \frac{1}{x-2}$$

The purpose of this section is to illustrate a method by which such a decomposition can be achieved.

We begin by noting that any rational expression can be represented as the sum of a polynomial and rational expressions of the form:

$$\frac{A}{(ax+b)^n} \quad \text{or} \quad \frac{Ax+B}{(ax^2+bx+c)^n} \tag{*}$$

where n is a positive integer, a, b, c, and A and B denote constants, and $ax^2 + bx + c$ is irreducible (has no linear factors; i.e. $b^2 - 4ac < 0$).

PARTIAL FRACTION DECOMPOSITION

Such a representation is said to be the **partial fraction decomposition** of the rational expression.

We first consider rational expressions such as:

$$\frac{-4x+9}{2x^2+5x-3} \quad \text{and} \quad \frac{2x^2+x+3}{(x+1)(x^2+x+4)}$$

in which the **degree of the denominator exceeds that of the numerator**. Here, the decomposition process consists of two stages. In the first stage, we obtain the general form of the decomposition, by writing the given rational expression as a sum of partial fractions of type (*), with unspecified values for the constants A and B appearing in the numerators of the rational expressions $\frac{A}{(ax+b)^n}$ and $\frac{Ax+B}{(ax^2+bx+c)^n}$. In the second stage, the values of those constants are determined.

Later, we will consider rational expressions in which the degree of the denominator is less than or equal to that of the numerator. In this case, the partial fraction decomposition will also include a polynomial term.

332   **Chapter 7 Rational Functions**

OBTAINING THE GENERAL FORM

The first step toward obtaining a partial fraction decomposition of a rational expression in which the degree of the numerator is less than that of the denominator is to factor its denominator into a product of powers of distinct linear factors, $ax + b$, and powers of irreducible quadratic polynomials, $ax^2 + bx + c$. There are four types of these factors:

Case 1. A non-repeated linear factor [such as, $(2x - 1)$]

Case 2. A repeated linear factor [such as, $(2x - 1)^2$]

Case 3. A non-repeated irreducible quadratic factor [such as, $(x^2 + x + 1)$ which has a negative discriminant: $b^2 - 4ac = 1 - 4 = -3$, and is therefore irreducible.]

Case 4. A repeated irreducible quadratic factor [such as $(x^2 + x + 1)^3$]

We then write the given expression as the most general sum possible of rational expressions of the form:

$$\frac{A}{(ax + b)^n} \quad \text{or} \quad \frac{Ax + B}{(ax^2 + bx + c)^n}$$

whose least common denominator is the denominator of the given rational expression. The table in Figure 7.5 reveals exactly what terms are to be included in the general sum, for each factor in the denominator of the given expression:

Case	Factor in denominator	General terms in partial fraction decomposition
1.	$ax + b$	$\dfrac{A}{ax + b}$
2.	$(ax + b)^2$	$\dfrac{A}{ax + b} + \dfrac{B}{(ax + b)^2}$
	$(ax + b)^3$	$\dfrac{A}{ax + b} + \dfrac{B}{(ax + b)^2} + \dfrac{C}{(ax + b)^3}$
$\vdots$	$\vdots$	$\vdots$
3.	$ax^2 + bx + c$	$\dfrac{Ax + B}{ax^2 + bx + c}$
4.	$(ax^2 + bx + c)^2$	$\dfrac{Ax + B}{ax^2 + bx + c} + \dfrac{Cx + D}{(ax^2 + bx + c)^2}$
	$(ax^2 + bx + c)^3$	$\dfrac{Ax + B}{ax^2 + bx + c} + \dfrac{Cx + D}{(ax^2 + bx + c)^2} + \dfrac{Ex + F}{(ax^2 + bx + c)^3}$
$\vdots$	$\vdots$	$\vdots$

Figure 7.5

EXAMPLE 7.15

NON-REPEATED FACTORS

Find the general form of the partial fraction decomposition of:

$$\frac{2x^2 + 3x - 4}{(x-1)(x^2 + 2x + 5)}$$

SOLUTION: In accordance with the table in Figure 7.5, we need two partial fractions: $\dfrac{A}{x-1}$ to accommodate the $(x-1)$ linear factor in the denominator, and $\dfrac{Bx + C}{x^2 + 2x + 5}$ for that irreducible $(x^2 + 2x + 5)$ factor:

For $x^2 + 2x + 5$:

$b^2 - 4ac = 4 - 20 < 0$
(irreducible)

$$\frac{2x^2 + 3x - 4}{(x-1)(x^2 + 2x + 5)} = \frac{A}{x-1} + \frac{Bx + C}{x^2 + 2x + 5}$$

The next example is quite involved. Still, by following its development you can be assured that you can find the general partial fraction decomposition of any rational expression in factored form (with degree of numerator less than that of the denominator).

EXAMPLE 7.16

SOME REPEATED FACTORS

Find the general form of the partial fraction decomposition of:

$$\frac{x+3}{(5x+2)(x-1)^3(x^2+x+4)(3x^2+1)^2}$$

For $x^2 + x + 4$:

$b^2 - 4ac = 1 - 16 < 0$

For $3x^2 + 1$:

$b^2 - 4ac = 0 - 12 < 0$
(both irreducible)

SOLUTION: From Figure 7.5:

$$(5x + 2) \xrightarrow{\text{gives rise to}} \frac{A}{5x + 2}$$

$$(x - 1)^3 \longrightarrow \frac{B}{x-1} + \frac{C}{(x-1)^2} + \frac{D}{(x-1)^3}$$

$$(x^2 + x + 4) \longrightarrow \frac{Ex + F}{x^2 + x + 4}$$

$$(3x^2 + 1)^2 \longrightarrow \frac{Gx + H}{3x^2 + 1} + \frac{Ix + J}{(3x^2 + 1)^2}$$

Note that the numerator has no influence whatsoever on the general decomposition of the given rational expression. It will come into play when determining the values of the constants A, B,... appearing in that decomposition.

Putting it all together we have:

$$\frac{x+3}{(5x+2)(x-1)^3(x^2+x+4)(3x^2+1)^2} = \frac{A}{5x+2} + \frac{B}{(x-1)} + \frac{C}{(x-1)^2} + \frac{D}{(x-1)^3}$$

$$+ \frac{Ex+F}{x^2+x+4} + \frac{Gx+H}{3x^2+1} + \frac{Ix+J}{(3x^2+1)^2}$$

334   Chapter 7 Rational Functions

Answer:

$$\frac{A}{x-3} + \frac{Bx+C}{x^2+x+5} + \frac{D}{2x+1}$$
$$+ \frac{E}{(2x+1)^2} + \frac{Fx+G}{x^2+5}$$
$$+ \frac{Hx+I}{(x^2+5)^2}$$

CHECK YOUR UNDERSTANDING 7.12

Find the general form of the partial fraction decomposition of the given rational expression.

$$\frac{x-4}{(x-3)(x^2+x+5)(2x+1)^2(x^2+5)^2}$$

PARTIAL FRACTION DECOMPOSITION

The following examples illustrate techniques for determining the values of the unknown constants in the general form of the partial fraction decomposition of a rational expression.

EXAMPLE 7.17   Find the partial fraction decomposition of:

$$\frac{-4x+9}{2x^2+5x-3}$$

SOLUTION: Factor the denominator, and then write the general form of the decomposition for the given rational expression:

$$\frac{-4x+9}{2x^2+5x-3} = \frac{-4x+9}{(2x-1)(x+3)} = \frac{A}{2x-1} + \frac{B}{x+3}$$

Clear denominators by multiplying both sides of the equation:

$$\frac{-4x+9}{2x^2+5x-3} = \frac{A}{2x-1} + \frac{B}{x+3}$$

by $(2x-1)(x+3)$:

$$-4x+9 = A(x+3) + B(2x-1) \qquad (*)$$

One can always substitute two **arbitrary** distinct values of x in (*) to arrive at a system of two equations in the two unknowns A and B, and then proceed to solve the resulting system. We offer two simpler methods for your consideration:

Evaluation Method

Simply choose x so that a linear factor becomes 0.

By setting x equal to **–3** in (*), the term $A(x+3)$ in (*) will drop out, and this will enable us to easily find the value of B:

$$-4(-3)+9 = A(-3+3) + B[2(-3)-1]$$
$$21 = -7B$$
$$\boxed{B = -3}$$

By the same token, setting $x = \frac{1}{2}$ in (*) makes $B(2x-1)$ zero, and A can then be found:

$$\boxed{\text{Equating Coefficients Method}}$$

$$-4(\tfrac{1}{2}) + 9 = A(\tfrac{1}{2} + 3) + B[2(\tfrac{1}{2}) - 1]$$

$$7 = \tfrac{7}{2}A$$

$$\boxed{A = 2}$$

Expand the right side of the equation in (*), and collect like terms:

$$-4x + 9 = A(x + 3) + B(2x - 1) = Ax + 3A + 2Bx - B$$

$$-4x + 9 = (A + 2B)x + (3A - B)$$

Given that two polynomials are equal if and only if corresponding coefficients are identical, leads us to a system of two equations in two unknowns:

$$\left. \begin{array}{l} \text{equating coefficients of } x: \quad A + 2B = -4 \\ \text{equating constant coefficients:} \quad 3A - B = 9 \end{array} \right\} \xrightarrow{\text{solving}} \begin{array}{l} A + 2B = -4 \\ \underline{6A - 2B = 18} \\ 7A \qquad\quad = 14 \\ \qquad\qquad\boxed{A = 2} \\ 2 + 2B = -4 \\ \qquad\qquad\boxed{B = -3} \end{array}$$

Both methods yielded the same decomposition:

$$\frac{-4x + 9}{2x^2 + 5x - 3} = \frac{2}{2x - 1} - \frac{3}{x + 3}$$

> You can also visually check the decomposition by viewing the graphs of both sides of the equation.

Checking our answer:

$$\frac{2}{2x - 1} - \frac{3}{x + 3} = \frac{2(x + 3) - 3(2x - 1)}{(2x - 1)(x + 3)} = \frac{-4x + 9}{2x^2 + 5x - 3}$$

> When the denominator factors into a product of distinct linear factors, the Evaluation Method is simpler than the Equating Coefficients method. In any other situation, the Equating Coefficients method alone or in combination with the Evaluation Method is best.

EXAMPLE 7.18 Find the partial fraction decomposition of:

$$\frac{2x^2 + x + 3}{(x + 1)(x^2 + x + 4)}$$

SOLUTION: General decomposition:

$$\frac{2x^2 + x + 3}{(x + 1)(x^2 + x + 4)} = \frac{A}{x + 1} + \frac{Bx + C}{x^2 + x + 4} \qquad (*)$$

336   Chapter 7 Rational Functions

Multiply both sides by $(x + 1)(x^2 + x + 4)$ to clear denominators:

$$2x^2 + x + 3 = A(x^2 + x + 4) + (Bx + C)(x + 1) \qquad (**)$$

Because there is only one value of x $(x = -1)$ which causes all but one term on the right side of $(**)$ to be zero, we will blend the Evaluation method with the Equating Coefficients method to arrive at the values of A, B, and C:

$$x = -1: \quad 2(-1)^2 + (-1) + 3 = A[(-1)^2 + (-1) + 4] + \boxed{0}$$

$$4 = A(4)$$

$$\boxed{A = 1}$$

With only two coefficients left to be determined, B and C, we need only consider the coefficients of any two of x^2, x, and the constant term in the equating coefficients method. We chose x^2 and the constant term:

Without going through the trouble of expanding the right side of $(**)$, you can probably see that its x^2-coefficient is $A + B$, and that its constant coefficient is $4A + C$, bringing us to:

If you find it necessary:

$$A(x^2 + x + 4) + (Bx + C)(x + 1)$$
$$= Ax^2 + \quad Ax + 4A$$
$$+ Bx^2 + (B + C)x + C$$

equating x^2 coefficients: $2 = A + B = 1 + B$

$$\boxed{B = 1}$$

equating constant coefficients: $3 = 4A + C = 4 + C$

$$\boxed{C = -1}$$

Substituting these values into $(*)$, we arrive at the partial fraction decomposition:

$$\frac{2x^2 + x + 3}{(x + 1)(x^2 + x + 4)} = \frac{1}{x + 1} + \frac{x - 1}{x^2 + x + 4}$$

CHECK YOUR UNDERSTANDING 7.13

Find the partial fraction decomposition of:

$$\frac{-x^2 - 4x - 3}{x^3 + x^2 + x}$$

Answer: $-\dfrac{3}{x} + \dfrac{2x - 1}{x^2 + x + 1}$

7.3 Partial Fractions   **337**

> **GENERAL RATIONAL EXPRESSION**

How does one go about obtaining the partial fraction decomposition of a rational expression such as:

$$\frac{2x^4 + 3x^3 - 8x^2 - x + 9}{2x^2 + 5x - 3}$$

where the degree of the numerator is **not** less than that of the denominator? One first divides the denominator into the numerator (as described on page 9):

$$
\begin{array}{r}
x^2 - x \\
2x^2 + 5x - 3 \,\overline{\big)\, 2x^4 + 3x^3 - 8x^2 - x + 9} \\
2x^4 + 5x^3 - 3x^2 \\
\hline
\text{subtract:} \quad -2x^3 - 5x^2 - x + 9 \\
-2x^3 - 5x^2 + 3x \\
\hline
-4x + 9
\end{array}
$$

Thus (see the Division Algorithm Theorem of page 10):

$$\frac{2x^4 + 3x^3 - 8x^2 - x + 9}{2x^2 + 5x - 3} = \underset{\text{quotient}}{x^2 - x} + \underset{\text{remainder}}{\frac{-4x + 9}{2x^2 + 5x - 3}}$$

The next step is to determine the partial fraction decomposition of:

$$\frac{-4x + 9}{2x^2 + 5x - 3}$$

which we already did in Example 7.17:

$$\frac{-4x + 9}{2x^2 + 5x - 3} = \frac{2}{2x - 1} - \frac{3}{x + 3}$$

Putting the two steps together, we arrive at the partial fraction decomposition

$$\frac{2x^4 + 3x^3 - 8x^2 - x + 9}{2x^2 + 5x - 3} = x^2 - x + \frac{2}{2x - 1} - \frac{3}{x + 3}$$

> ## CHECK YOUR UNDERSTANDING 7.14
>
> Find the partial fraction decomposition of the rational expression:
>
> $$\frac{x^4 - x^3 + x^2 + 2x - 10}{x^2 - x - 2}$$

Answer:

$$x^2 + 3 + \frac{2}{x - 2} + \frac{3}{x + 1}$$

338   Chapter 7 Rational Functions

EXERCISES

Exercises 1–12: Find the general form (only) of the partial fraction decomposition.

1. $\dfrac{4x - 1}{(x - 3)(2x - 5)}$

2. $\dfrac{x - 1}{(4x - 7)^4}$

3. $\dfrac{4x^3 - 2x + 1}{(x + 1)^3(x - 2)^2}$

4. $\dfrac{x^2 - 1}{(3x - 2)(4x^2 + 5)}$

5. $\dfrac{3x^3 - 2x^2 + x + 1}{(x^2 + 1)(x^2 + 3x - 4)(x^2 + 2x + 3)}$

6. $\dfrac{1}{(x^2 + 1)^3(x^2 + x + 1)^2(x - 2)^4}$

7. $\dfrac{x^4 + 5}{(x^4 - 1)^2}$

8. $\dfrac{2x^2 - 7}{2x^5 + 7x^4 - 4x^3}$

9. $\dfrac{3x}{(x^2 - 2x + 1)^2(x^2 + 1)}$

10. $\dfrac{2x^3 + 9x^2 + 6x}{(x + 1)^2}$

11. $\dfrac{3x^3 - 8}{(x - 3)(x + 1)^2}$

12. $\dfrac{x^5 + 6x^4}{(x + 4)(x^2 + 4)}$

Exercises 13–30: Find the partial fraction decomposition.

13. $\dfrac{6x + 2}{x^2 - x - 6}$

14. $\dfrac{12x - 10}{8x^2 + 10x - 3}$

15. $\dfrac{6x^2 - 5x + 1}{(x^2 - 1)(x - 2)}$

16. $\dfrac{4x^2 + 21x + 25}{(x + 3)^3}$

17. $\dfrac{7x^2 - 20x - 63}{(x + 1)(x - 5)^2}$

18. $\dfrac{5x^4 - 3x^3 + 2x^2 + 3x - 3}{x^5 - x^4}$

19. $\dfrac{x^2 - 3x - 2}{x^3 + x^2 + x}$

20. $\dfrac{5x^2 - 3x + 7}{x^3 - x^2 + 2x - 2}$

21. $\dfrac{3x^3 + 3x^2 + 7x + 11}{(x^2 + 3)(x + 1)^2}$

22. $\dfrac{9x^3 + 7x^2 + 36x - 20}{(x^2 + 4)(x^2 + 2x + 4)}$

23. $\dfrac{4x^3 + x^2 + 3x - 3}{(x^2 + x + 1)^2}$

24. $\dfrac{4x^3 - 4x^2 + 8x - 7}{(2x^2 + 3)(x^2 + 2)}$

25. $\dfrac{6x^3 + x^2 + 4x + 2}{(x^2 + 1)^2}$

26. $\dfrac{12x^4 + 8x^3 + 5x^2 + 6x + 15}{6x^2 - 5x + 1}$

27. $\dfrac{5x^3 - 3x^2 + 1}{x^2 + 2x + 1}$

28. $\dfrac{12x^4 + 3x^3 + 8x^2 - x - 1}{4x^2 + x}$

29. $\dfrac{-3x^4 + 16x^3 + 22x^2 + 6x - 15}{(3x - 1)(x^2 + 1)}$

30. $\dfrac{-12x^3 + 8x^2 - 2x - 6}{6x^2 - x - 1}$

CHAPTER SUMMARY

RATIONAL FUNCTION	A **rational function** is a function of the form: $$f(x) = \frac{p(x)}{q(x)}$$ where $p(x)$ and $q(x) \neq 0$ are polynomials, and $q(x)$ is not the zero polynomial.
Solutions of a Rational Equation	The solutions of the rational equation $\frac{p(x)}{q(x)} = 0$ are those values of x for which $p(x) = 0$ and $q(x) \neq 0$. NOTE: Multiplying both sides of a rational equation by a quantity which can be zero may introduce extraneous solutions. **Check your answers**.
Confinement Theorem	All solutions of the rational equation: $$\frac{p(x)}{q(x)} = \frac{x^n + a_{n-1}x^{n-1} + \cdots + a_1 x + a_0}{q(x)} = 0$$ are contained in the interval $[-K, K]$, where: $$K = 1 + \text{largest}\{|a_{n-1}|, |a_{n-2}|, \ldots, |a_1|, |a_0|\}$$
Solving Rational Inequalities	To solve an inequality containing rational expressions: 1. Express it in the form $\frac{p(x)}{q(x)} < 0$ (or > 0, or ≤ 0, or ≥ 0). 2. Factor $p(x)$ and $q(x)$. 3. Chart the sign of $\frac{p(x)}{q(x)}$, and then read off the solution set.

GRAPHING RATIONAL FUNCTIONS

Vertical Asymptote	Vertical line at $x = x_0$ which the graph of $f(x) = \frac{p(x)}{q(x)}$ approaches as x approaches x_0. The values of x_0 are the zeros of $q(x)$, when $\frac{p(x)}{q(x)}$ is in lowest terms.

340 **Chapter 7 Rational Functions**

As $x \to \pm\infty$	Far away from the origin the graph of the rational function: $$f(x) = \frac{a_n x^n + a_{n-1} x^{n-1} + \cdots + a_1 x + a_0}{b_m x^m + b_{m-1} x^{m-1} + \cdots + b_0}$$ resembles, in shape, that of: $$g(x) = \frac{a_n x^n}{b_m x^m}$$
Horizontal Asymptote	If $n = m$, then the graph of f approaches the horizontal line $y = \dfrac{a_n}{b_n}$ as x tends to $\pm\infty$. If $n < m$, then the graph approaches the x-axis.
Oblique Asymptote	If $n = m + 1$, then the graph of f will approach the oblique line $y = \dfrac{a_n}{b_m}x + d$ as x tends to $\pm\infty$. The value of d can be determined by dividing the denominator of the rational function f into its numerator, as is illustrated in Example 7.13, page 326.
Graphing Procedure	**Step 1. Factor the rational function.** **Step 2. Nature of the graph "near the origin:"** Determine and plot x- and y-intercepts. Sketch the dotted lines for the vertical asymptotes. Chart the sign of the function to determine the nature of the graph about the x-intercepts, and the direction of the graph near its vertical asymptotes. **Step 3. Nature of the graph "far from the origin:"** Determine the behavior of the graph as $x \to \pm\infty$. **Step 4. Sketch the anticipated graph of the function.**

PARTIAL FRACTIONS	Any rational expression can be decomposed into the sum of a polynomial (if the degree of the numerator is greater than or equal to that of the denominator), and rational expressions of the form:
	$$\frac{A}{(ax+b)^n} \quad \text{or} \quad \frac{Ax+B}{(ax^2+bx+c)^n}$$
	To arrive at the partial fraction decomposition of a rational expression $\frac{p(x)}{q(x)}$ with degree $p(x)$ less than degree $q(x)$:
	1. Factor $q(x)$ into a product of powers of distinct linear and irreducible quadratic factors.
	2. Each of the resulting factors in the denominator generates terms in the general decomposition as follows: $$(ax+b)^n \text{ generates } \frac{A_1}{ax+b} + \ldots + \frac{A_n}{(ax+b)^n}$$ $$(ax^2+bx+c)^n \text{ generates } \frac{A_1x+B_1}{ax^2+bx+c} + \ldots + \frac{A_nx+B_n}{(ax^2+bx+c)^n}$$ (see Figure 7.5, page 332 for more details)
	3. You can combine the Evaluation Method and the Equating Coefficients Method to determine the values of the constants A_i and B_i in step 2.

		A COMMON PITFALL	
WRONG:	$\frac{3}{x-2} < \frac{1}{x}$ $3x < x - 2$	**Because:**	If you multiply both sides of $\frac{3}{x-2} < \frac{1}{x}$ by $x(x-2)$, you may be multiplying by a **negative quantity**. You should begin with: $$\frac{3}{x-2} < \frac{1}{x}$$ $$\frac{3}{x-2} - \frac{1}{x} < 0$$ and go on from there.

342   Chapter 7   Rational Functions

CHAPTER REVIEW EXERCISES

Exercises 1–4: Find the domain of the function.

1. $f(x) = \dfrac{4x - 3}{2x + 1}$

2. $g(x) = \dfrac{x^3 + 8}{x^2 - 9}$

3. $h(x) = \dfrac{x^4 - 5x^3}{x^3 + x^2}$

4. $k(x) = \dfrac{3}{x - 2} - \dfrac{x - 2}{3} + \dfrac{1}{x}$

Exercises 5–11: Solve the equation.

5. $\dfrac{x + 1}{2} = \dfrac{3}{x - 1}$

6. $\dfrac{x^2}{x + 1} + \dfrac{3}{x - 1} = \dfrac{x^3 + x^2 + 4x + 4}{x^2 - 1}$

7. $\dfrac{1}{x^2} + \dfrac{3}{x} = \dfrac{2}{x^3}$

8. $\dfrac{3x}{3x + 1} - 1 = \dfrac{1}{x + 5}$

9. $3\left[\dfrac{x}{(x + 1)^2}\right] - \dfrac{1}{2}\left[\dfrac{(x + 1)^2}{x}\right] = \dfrac{5}{2}$

10. $\dfrac{-x}{x + 1} = \dfrac{1}{x - 1} - \dfrac{1}{x^2 - 1}$

11. $\cos^2 x = \dfrac{5\cos x + 2}{2\cos x - 1}$, in $[0, 2\pi)$

Exercise 12: (**River Current**) In a certain river, a boat can travel 25 miles upstream in the same time that it would take to travel 50 miles downstream. What is the rate of the current, if in still water the speed of the boat is 15 mi/hr?

Exercises GC13–GC14: Use a graphing calculator to find approximate solutions.

13. $\dfrac{x}{x^3 + 3} = 3 + \dfrac{x - 1}{x + 4}$

14. $\dfrac{3x^3 + x - \pi}{x + 1} - \dfrac{x}{x^2 + 1} = 2$

Exercises GC15–GC18: Use a graphing calculator and the four-step procedure to find an approximate solution to the optimization problem.

15. (**Minimum Cost**) A farmer wants to enclose a 3,000 ft^2 rectangular field along a (straight) river with a fence that costs \$1.50 per running foot. No fence is required along the river. The farmer also needs to subdivide the field with two fences perpendicular to the river, and that fencing costs \$2.00 per running foot. Determine the dimensions of the field that will minimize cost.

16. (**Traffic Accidents**) The number, N, of accidents per 100 million vehicle driving miles is approximated by the function

$$N(s) = 1.05s^3 - 21.9s^2 + 156s - 327, \quad \text{for } 6 \le s \le 11$$

where s is the number of miles per hour a vehicle is being driven above or below the prevailing traffic speed. Use a graphing calculator to determine, to one decimal place, the speeds at which 55 accidents per 100 million vehicle driving miles occur, at a prevailing traffic speed of 60mph. (For example: If $s = 7$, then speeds are $60 - 7 = 53$ mph or $60 + 7 = 67$ mph.)

Chapter Review Exercises   343

17. (**Profit**) The weekly cost function (in dollars) for a company is given by

$$C(x) = 14x - \frac{15x^2}{x + 200} + 2500, \quad \text{for} \quad 0 \le x \le 300$$

where x is the number of units produced.

(a) Use a graphing calculator to find the production levels at which weekly cost exceeds \$3,500.

(b) The revenue function (in dollars) is given by $R(x) = 35x$. Use a graphing calculator to find the production level at which the break-even point occurs.

(c) Use a graphing calculator to determine weekly production levels for which weekly profit exceeds \$1,000.

18. (**Demand and Profit**) The monthly number of Vision TVs sold a Happy's Appliances is given by

$$N(p) = 200 - \frac{p^2}{p + 10}$$

where p is the price per unit (in dollars) and $100 \le p \le 200$. Happy's total cost per unit is \$95.

(a) How many units will be sold at \$100 per unit?

(b) What is Happy's monthly Vision profit, if each unit is priced at \$100?

(c) What is Happy's monthly Vision profit, if each unit is priced at \$150?

(d) Use a graphing calculator to determine the price which will result in maximum profit.

Exercises 19–26: Determine the solution set of the inequality.

19. $\dfrac{3}{x - 4} < 7$

20. $\dfrac{4x - 3}{2 - 3x} \le 1$

21. $\dfrac{5x + 2}{x^2 - x} \ge 0$

22. $\dfrac{4x^2(3 - 7x)}{(x - 4)^3(x + 2)^4(3x + 2)} \ge 0$

23. $\dfrac{3x - 5}{x^2 + x} > \dfrac{4}{x}$

24. $x + 1 < \dfrac{1}{x - 2} + 2x$

25. $\dfrac{1}{x - 2} + 2 \le \dfrac{x + 1}{x}$

26. $x^3 \sin x > \dfrac{x^3}{\sin x}$, in $\left[-\dfrac{3\pi}{2}, \dfrac{3\pi}{2}\right]$

Exercises GC27–GC28: Use a graphing calculator to find an approximate solution set of the inequality.

27. $\dfrac{5x^6 - 3x^3 + 2x - 6}{5x^2 + 7} > 0$

28. $\dfrac{x^6}{(4 - x^2)^3} - \dfrac{3}{x} < 0$

344   Chapter 7 Rational Functions

Exercises GC29–GC30: Use a graphing calculator to find an approximate solution.

29. (**Average Cost**) The cost (in dollars) of producing x units is

$$C(x) = \frac{100x}{1 + 0.005x} + 500$$

The average cost per unit function is $\frac{C(x)}{x}$. Determine the number of units that must be produced before the average cost per unit is less than \$140.

30. (**Gas consumption**) Gas consumption, G, in miles per gallon, for a certain automobile driven at speeds, s, between 35 mph and 70 mph is given by

$$G(s) = \frac{40}{3 - \dfrac{50s^2}{s^3 + 10}}$$

What range of speed will result in gas consumption between 19 and 22 miles per gallon?

Exercises 31–32: Determine a function $g(x)$ whose graph the given function resembles as $x \to \pm\infty$.

31. $f(x) = \dfrac{23x^5 - 17x^4 + 5x^3 - 1}{8x^6 - 5x^3 - 2x + 2}$

32. $f(x) = \dfrac{9x^4 - 3x^3 + 2x - 5}{6x^4 + 3x^3 + 6x - 3}$

Exercises GC31–GC32: Use a graphing calculator to check your result in Exercises 31–32, by simultaneously viewing the graphs of f and g far away from the origin.

Exercises 33–36: Determine the asymptotes (vertical, and horizontal or oblique), if any.

33. $f(x) = \dfrac{3x - 1}{5x + 4}$

34. $f(x) = \dfrac{9}{x^2 + 4x + 4}$

35. $f(x) = \dfrac{4x^7 + 2x^6 + 3}{x^6 - x^4}$

36. $f(x) = \dfrac{3x^8}{(x - 1)^3(x - 3)^2(x^3 + 1)}$

Exercises 37–42: Sketch the graph of the function, using the 4-step method.

37. $f(x) = \dfrac{3x + 2}{x - 1}$

38. $g(x) = \dfrac{x^2 - 5x + 6}{4x^3 - 4x}$

39. $h(x) = \dfrac{1}{x} + \dfrac{2}{x + 3}$

40. $k(x) = \dfrac{2(2 - x)(4x + 3)}{(x - 1)(x + 4)(x - 5)}$

41. $f(x) = \dfrac{x^2 \sin x}{\cos^2 x}$, in $[-2\pi, 2\pi]$

42. $g(x) = \dfrac{\cos x}{x \sin^2 x}$, in $[-2\pi, 2\pi]$

Cumulative Review Exercises   **345**

Exercises 43–48: Find the general form (only) of the partial fraction decomposition.

43. $\dfrac{3x+7}{x(2x-3)(5x-2)}$

44. $\dfrac{x+4}{(x-1)^2}$

45. $\dfrac{4-x}{(3x^2+1)(3x-2)^3}$

46. $\dfrac{6x^5+9x^4-1}{2x^4+7x^3-4x^2}$

47. $\dfrac{x+2}{(x^2+2x+1)^2(x^2+x+3)}$

48. $\dfrac{3x^3-4x^2+23x-8}{(x-3)(x^2+4)^2}$

Exercises 49–52: Find the partial fraction decomposition.

49. $\dfrac{-9x^2+2x+67}{(3x-1)(x+3)^2}$

50. $\dfrac{-2x^4-4x^2-x+16}{x(x^2+4)^2}$

51. $\dfrac{x+19}{4(x^2-2x-15)}$

52. $\dfrac{6x^3+5x^2+2x-3}{4x^2+2x-6}$

CUMULATIVE REVIEW EXERCISES

Exercises 1–2: Simplify.

1. $\dfrac{\left(\frac{1}{2}\right)^{-2}(x-a)^3}{(-2a+2x)^4}$

2. $-x(4x+3)^{-2}-2x^{-1}(4x+3)^{-1}$

Exercises 3–5: If $f(x)=\sin^2 x$ and $g(x)=x-\frac{\pi}{2}$, determine

3. $(g\circ f)\left(\frac{\pi}{4}\right)$

4. $(f\circ f)(x)$

5. $(f\circ g)\left(\frac{\pi}{3}\right)$

Exercise 6: Solve the triangle with $a=c=7$, and $b=5$.

Exercise 7: Sketch the graph of $f(x)=-(x-3)^4+1$, starting with the basic curve and stretching/shifting it etc., and determine the intercepts.

Exercise 8: Determine the vertex and intercepts of the parabola $2y-x^2=8x-4$ and sketch the graph.

Exercise 9: Evaluate each expression.

(a) $\sin\dfrac{3\pi}{2}$

(b) $\cos 0$

(c) $\tan\dfrac{\pi}{6}$

Exercise 10: Given $\sin x=\frac{5}{13}$, determine all possible values of

(a) $\sin 2x$

(b) $\cos 2x$

346   Chapter 7   Rational Functions

Exercises 11–22: Solve each of the following.

11. $\left.\begin{array}{r} x^2 + x - 3y = 9 \\ 3x - 2y = -4 \end{array}\right\}$

12. $(3x - 8)^3 = 8$

13. $\dfrac{2x - 1}{x + 2} = 3$

14. $\dfrac{30}{x^2 - 9} - \dfrac{5}{x - 3} + 2 = 0$

15. $\sin^3 x - \sin x = 0$

16. $\dfrac{x^2 - 1}{x^2 - x - 2} = \dfrac{x^2 - x}{x + 4}$

17. $\dfrac{\tan^2 x}{\tan^2 x + 2} - \dfrac{\tan^2 x - 2}{\tan^2 x} = \dfrac{4}{15}$, in $[0, 2\pi)$

18. $x(2x - 4) \le 2x + 1$

19. $1 < x^2$

20. $\dfrac{x - 2}{x + 1} \ge \dfrac{x - 1}{x + 2}$

21. $\dfrac{(2x - 1)(x + 2)}{x + 1} \ge 0$

22. $\dfrac{4x + 3}{x - 4} < 3$

Exercises 23–24: Sketch the graph, using the 4-step graphing method.

23. $f(x) = \dfrac{x + 2}{1 - x}$

24. $k(x) = \dfrac{4x + 1}{x^2 + x - 6}$

Exercise 25:   Determine the domain of $f(x) = \left(\dfrac{2x - 5}{2x^2 - 9x + 4}\right)^{\frac{1}{4}}$

Exercises 26–27: Find the partial fraction decomposition.

26. $\dfrac{5x^2 - 6x - 13}{x^3 - x^2 - 8x + 12}$

27. $\dfrac{4x^4 - x^3 + 14x^2 - 4x + 6}{x^3 + x}$

Exercise GC28:   (**Maximum Area**) A rectangular flag has a blue border and a yellow center. The border is 10 in. wide along the sides and 8 in. wide along the top and bottom. The total area is 40 ft². Use a graphing calculator to find the largest possible area of the yellow center.

Exercise GC29:   (**Minimum Material**) A closed box is to have a base that is twice as long as it is wide, and a volume of 1500 in.³. Use a graphing calculator to find the approximate dimensions of the box requiring the least amount of material in its construction.

CHAPTER 8
ABSOLUTE VALUE AND RADICAL FUNCTIONS

In this brief chapter we consider functions, equations, and inequalities where the variable appears in absolute values, or a radical, or is raised to a rational exponent.

§1. ABSOLUTE VALUE FUNCTIONS, EQUATIONS, AND INEQUALITIES

We begin by recalling the definition of the absolute value function and its graph (page 89):

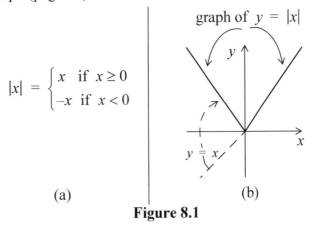

$$|x| = \begin{cases} x & \text{if } x \geq 0 \\ -x & \text{if } x < 0 \end{cases}$$

(a)   (b)

Figure 8.1

GRAPH OF $y = |g(x)|$

As is illustrated in Figure 8.1(b), the graph of the function $f(x) = |x|$ can be obtained by reflecting the "negative" part of the graph of the function $g(x) = x$ about the x-axis. In general, to graph $f(x) = |g(x)|$:

Step 1. Sketch the graph of $g(x)$.

Step 2. Reflect the part of the graph of g that lies below the x-axis about the x-axis.

EXAMPLE 8.1 Sketch the graph of the function:
$$f(x) = |-3x^2 + 6x - 1|$$

SOLUTION: The first step, that of graphing the function:
$$g(x) = -3x^2 + 6x - 1$$

was accomplished in Example 6.1, page 247 [see Figure 8.2(a) below]. The second step is reflected in Figure 8.2(b)

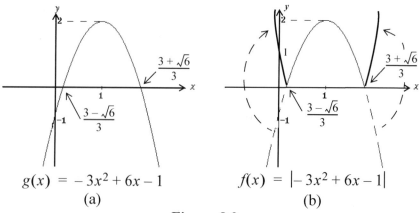

$g(x) = -3x^2 + 6x - 1$
(a)

$f(x) = |-3x^2 + 6x - 1|$
(b)

Figure 8.2

GRAPHING CALCULATOR GLIMPSE 8.1

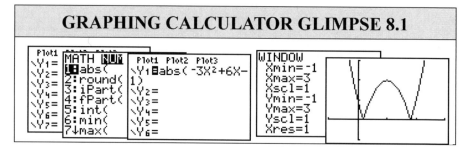

CHECK YOUR UNDERSTANDING 8.1

Sketch the graph of the function:
$$f(x) = |x^4 - x^3 - 6x^2|$$
Suggestion: Consider Example 6.22, page 289.

Answer: See page C-57.

PROPERTIES OF ABSOLUTE VALUES

The following properties are direct consequences of the definition of the absolute value function and its geometrical interpretation:

$|a|$ denotes the distance between a and the 0 on the number line.

THEOREM 8.1 For any numbers a and b:

(i) $|a| \geq 0$, with equality holding if and only if $a = 0$

(ii) $|a| = |-a|$

(iii) $|ab| = |a||b|$ and, if $b \neq 0$, $\left|\dfrac{a}{b}\right| = \dfrac{|a|}{|b|}$

(iv) $|a|^2 = a^2$

(v) $\sqrt{a^2} = |a|$

(vi) **(Triangle Inequality)** $|a + b| \leq |a| + |b|$

Property (iv) appeared as Theorem 2.1, page 109.

Property (i) simply states that distance is a nonnegative quantity, and only the number 0 is zero units from the origin.

Property (ii) says that a number a and its negative, $-a$, are the same distance from the origin.

Property (iii) tells us that:

> **The absolute value of a product equals**
> **the product of the absolute values.**
>
> and
>
> **The absolute value of a quotient equals**
> **the quotient of the absolute values.**

To see that the product property (iii) holds, note that the only possible difference between ab and $|a||b|$ is that ab might be the negative of $|a||b|$, and taking the absolute value of ab will erase any such discrepancy. For example: $|(-2)(4)| = |-2||4| = 8$. Similarly, we have: $\left|\dfrac{-5}{7}\right| = \dfrac{|-5|}{|7|} = \dfrac{5}{7}$

Property (iv) is just as transparent: even if $|a|$ and a differ by a negative sign, their squares will certainly be the same.

Property (v) follows from Property (iv): $\sqrt{a^2} = \sqrt{|a|^2} = |a|$. For example:

$$\cancel{\sqrt{(-5)^2} \neq -5} \quad \text{rather:} \quad \sqrt{(-5)^2} = \sqrt{25} = 5$$

(Equality does hold if and only if the number is nonnegative.)

Property (vi) says that the absolute value of a sum is at most the sum of the absolute values. After all, if the numbers a and b are opposite in sign, some "additive cancellation" will occur in the expression $a + b$, whereas no such cancellation is possible in $|a| + |b|$, since neither $|a|$ nor $|b|$ is negative. For example: $|4 + (-2)| = 2$, while $|4| + |-2| = 6$.

Answer: See page C-57.

CHECK YOUR UNDERSTANDING 8.2
Verify each of the properties in Theorem 8.1, for $a = -5$ and $b = 3$.

350   Chapter 8 Absolute Value and Radical Functions

ABSOLUTE VALUE EQUATIONS

Recalling, once more, that $|x|$ represents the distance between x and 0 on the number line, we see that:

> The equation $|x| = a$ has no solution if $a < 0$ (why not?).

THEOREM 8.2
ABSOLUTE VALUE EQUATIONS

(a) For any $a \geq 0$:

$$|x| = a \text{ if and only if } x = a \text{ or } x = -a$$

(b) For any a and b:

$$|a| = |b| \text{ if and only if } a = b \text{ or } a = -b$$

"PATTERNS" FROM THE ABOVE THEOREM:

(a)

$$|expression| = a \text{ if and only if } expression = a \text{ or } expression = -a$$

(b)

$$|expression_1| = |expression_2|$$

if and only if

$$expression_1 = expression_2 \text{ or } expression_1 = -\,expression_2$$

> Recalling that $|x - 2|$ represents the distance between x and 2 on the number line, we can also see the solution of the equation $|x - 2| = 4$:
>
> ```
> |← 4 →|← 4 →|
> -2     2     6
> ```

EXAMPLE 8.2 Solve:

$$|x - 2| = 4$$

SOLUTION: From Theorem 8.2(a):

$$|x - 2| = 4$$

$$
\begin{array}{ccc}
x - 2 = 4 & \textbf{OR} & x - 2 = -4 \\
x = 6 & & x = -2
\end{array}
$$

EXAMPLE 8.3 Solve:

$$|x^2 - 3x - 3| = 1$$

SOLUTION: From Theorem 8.2(a):

$$|x^2 - 3x - 3| = 1$$

$$
\begin{array}{ccc}
x^2 - 3x - 3 = 1 & \textbf{OR} & x^2 - 3x - 3 = -1 \\
x^2 - 3x - 4 = 0 & & x^2 - 3x - 2 = 0 \\
(x - 4)(x + 1) = 0 & & x = \dfrac{3 \pm \sqrt{9 + 8}}{2} = \dfrac{3 \pm \sqrt{17}}{2} \\
x = 4 \text{ or } x = -1 & &
\end{array}
$$

8.1 Absolute Value Functions, Equations, and Inequalities 351

EXAMPLE 8.4 Solve:
$$|2x - 3| = |3x + 1|$$

SOLUTION: From Theorem 8.2(b):

$$|2x - 3| = |3x + 1|$$

$2x - 3 = 3x + 1$ **OR** $2x - 3 = -(3x + 1)$
$x = -4$ $\qquad\qquad 2x - 3 = -3x - 1$
$\qquad\qquad\qquad\qquad x = \frac{2}{5}$

EXAMPLE 8.5 Solve:
$$|2x + 5| = -3x$$

SOLUTION: It is certainly true that the absolute value function cannot assume negative values. But that $-3x$ on the right side of the equation is not negative when $x < 0$, and the absolute value equation can (and does) have a solution:

$$|2x + 5| = -3x$$

$2x + 5 = -3x$ **OR** $2x + 5 = -(-3x)$
$x = -1$ $\qquad\qquad\qquad x = 5$

Checking the two values in the given equation, we find that -1 is the only solution.

CHECK YOUR UNDERSTANDING 8.3

Answers: (a) $4, -\frac{2}{3}$
(b) $3, -2, 2, -1$ (c) $-\frac{1}{9}, 5$

Solve:
(a) $|3x - 5| = 7$ (b) $|x^2 - x - 4| = 2$ (c) $|4x + 3| = |2 - 5x|$

EXAMPLE 8.6 Solve, for $0 \leq x < 2\pi$:
$$|\cos^2 x - \sin x| = 1$$

SOLUTION:

$\cos^2 x - \sin x = 1$ **OR** $\cos^2 x - \sin x = -1$

From Example 4.19, page 211: $\qquad\qquad (1 - \sin^2 x) - \sin x = -1$

$x = 0, \pi, \frac{3\pi}{2}$ $\qquad\qquad \sin^2 x + \sin x - 2 = 0$

$\qquad\qquad\qquad\qquad (\sin x + 2)(\sin x - 1) = 0$

$\qquad\qquad\qquad\qquad$ $\boxed{\sin x = -2}$ or $\sin x = 1$
$\qquad\qquad\qquad\qquad$ no solution

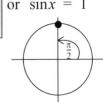

Answer:

$\tan^{-1}\left(\frac{1}{2}\right), \pi - \tan^{-1}\left(\frac{1}{2}\right)$

$\pi + \tan^{-1}\left(\frac{1}{2}\right), 2\pi - \tan^{-1}\left(\frac{1}{2}\right)$

$\frac{\pi}{4}, \frac{3\pi}{4}, \frac{5\pi}{4}, \frac{7\pi}{4}$

CHECK YOUR UNDERSTANDING 8.4

Solve, for $0 \leq x < 2\pi$:
$$|2\tan x - \cot x| = 1$$
Suggestion: Consider Example 4.25, page 214.

ABSOLUTE VALUE INEQUALITIES

Let $a > 0$. Since $|x|$ represents the distance between x and 0 on the number line, to say that $|x| < a$ is to say that x is contained in the interval $(-a, a)$ [see Figure 8.3(a)]. By the same token, to say that $|x| > a$ is to say that x is outside the interval $[-a, a]$ [see Figure 8.3(b)].

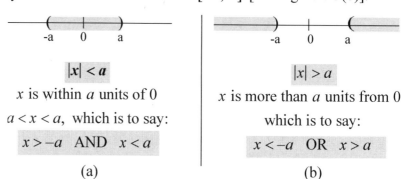

Figure 8.3

Summarizing these observations, we have:

THEOREM 8.3
ABSOLUTE VALUE INEQUALITIES

For any $a > 0$:

$|x| < a$ if and only if $-a < x < a$, that is:
           if and only if $x > -a$ **AND** $x < a$

$|x| > a$ if and only if $x < -a$ **OR** $x > a$

EXAMPLE 8.7
Solve:
$$|2x - 7| < 5$$

SOLUTION: With $2x - 7$ replacing x, and 5 replacing a in $|x| < a$ of Theorem 8.3 we have:

$$|2x - 7| < 5$$
$$-5 < 2x - 7 < 5$$
adding 7: $2 < 2x < 12$
dividing by 2: $1 < x < 6$

We conclude that the interval $(1, 6)$ is the solution set of $|2x - 7| < 5$.

8.1 Absolute Value Functions, Equations, and Inequalities 353

EXAMPLE 8.8 Solve:
$$|-4x + 3| \geq 7$$

SOLUTION: With $-4x + 3$ replacing x, and 7 replacing a $|x| > a$ in Theorem 8.3, and including the equal sign, we have:

$$|-4x + 3| \geq 7$$

$$
\begin{array}{ccc}
-4x + 3 \leq -7 & \textbf{OR} & -4x + 3 \geq 7 \\
-4x \leq -10 & & -4x \geq 4 \\
x \geq \dfrac{5}{2} & & x \leq -1
\end{array}
$$

Conclusion: $(-\infty, -1\,] \cup [\frac{5}{2}, \infty)$ is the solution set of $|-4x + 3| \geq 7$.

In Exercise 100, you are asked to verify that Theorem 8.3 remains valid when the constant a is replaced by a function of x. That result is used in the following example.

EXAMPLE 8.9 Solve:
$$|2x - 3| < -5x + 1$$

SOLUTION:

$$|2x - 3| < -5x + 1$$

$$
\begin{array}{ccc}
2x - 3 > -(-5x + 1) & \textbf{AND} & 2x - 3 < -5x + 1 \\
2x - 3 > 5x - 1 & & 7x < 4 \\
-3x > 2 & & x < \dfrac{4}{7} \\
x < -\dfrac{2}{3} & &
\end{array}
$$

We have shown that the solution set consists of those numbers that are **both** less than $-\frac{2}{3}$ **and** less than $\frac{4}{7}$, which is to say: $(-\infty, -\frac{2}{3})$.

Answers: (a) $\left[\frac{2}{5}, \frac{6}{5}\right]$

(b) $(-\infty, -1) \cup (4, \infty)$

CHECK YOUR UNDERSTANDING 8.5
Solve:
(a) $\|5x - 4\| \leq 2$ (b) $\|2x - 3\| > 5$

354   Chapter 8   Absolute Value and Radical Functions

EXERCISES

Exercises 1–10:   Sketch the graph of the function, indicating the intercepts.

1. $f(x) = |(2x - 1)(x + 3)|$

2. $K(x) = |2x^2 - x - 15|$

3. $f(x) = |-(x + 1)^5 + 32|$

4. $k(x) = |3x^5 - 3|$

5. $f(x) = |(x - 1)^4 - 4|$

6. $k(x) = |-3(x + 2)(x + 1)(2x - 1)|$

7. $h(x) = |x^3 + x^2 - 4x - 4|$

8. $h(x) = |-x^3 - x^2 + 2x + 8|$

9. $g(x) = |\sin x|$

10. $f(x) = |\cos(2x + \pi)|$

Exercises 11–13:   For each function, sketch the graphs of

$$y = f(x), \qquad y = |f(x)|, \qquad \text{and} \qquad y = f(|x|)$$

Does it appear that   (a) $f(x) = |f(x)|$,   or (b) $f(x) = f(|x|)$,   or (c) $|f(x)| = f(|x|)$?

11. $f(x) = 4x - 1$

12. $f(x) = x^3$

13. $f(x) = 3x^2 - 12$

Exercises GC14–GC17:   Use a graphing calculator to graph the function, as in Graphing Calculator Glimpse 8.1, and approximate the x-intercepts.

14. $f(x) = |2x^2 - 4x + 1|$

15. $k(x) = |2x^3 + 6x^2 - 1|$

16. $H(x) = |-3x^5 - 4x^4 + 6x^2|$

17. $h(x) = |-x^4 + 3x + 3|$

Exercises 18–28:   Indicate whether the statement is true or false. Assume no denominator is zero. Justify your answer.

18. $(-x - y)^2 = |x + y|^2$

19. $|x|^3 = x^3$

20. $|x|^6 = x^6$

21. $\left|\dfrac{2x + y}{z}\right| = \dfrac{|2x| + |y|}{|z|}$

22. $|x + y|^2 = (|x| + |y|)^2$

23. $\left|\dfrac{x}{y + z}\right| = \dfrac{|x|}{|y| + |z|}$

24. $|x| = x$

25. If $x < 0$ then $\sqrt{x^2} = -x$.

26. $\sqrt{x^2 + y^2} = |x| + |y|$

27. $\sqrt{(x + y)^2} = |x + y|$

28. $\sqrt{x^4} = x^2$

Exercises 29–40:   Solve the equation.

29. $|x - 5| = 2$

30. $|2x - 3| = 0$

31. $|-3x - 7| = 4$

32. $\left|2x - \dfrac{3}{2}\right| = \dfrac{1}{4}$

33. $|x^2 - 3| = 1$

34. $|x + 2||x - 1| = 10$

35. $|x + 1| = |2x - 1|$

36. $|3x^2 - 4x| = |x^2 + x - 2|$

37. $\dfrac{|x - 2|}{|x - 1|} = -2$

38. $\dfrac{|x - 1|}{|x + 2|} = 1$

39. $|4x - 2| = 3|x - 1|$

40. $|x^2 - 2x| = 3x - 6$

8.1 Absolute Value Functions, Equations, and Inequalities   **355**

Exercises 41–44: Solve the equation for x in $[0, 2\pi)$.

41. $\left|\sin^2 x - \cos x\right| = 1$

42. $\left|\csc x - 2\sin x\right| = 1$

43. $\left|\sec^2 x + \tan^2 x\right| = 3$

44. $\left|\sec x - \cos x\right| = 0$

Exercises 45–52: Without writing anything down, use the concept of distance on the number line to determine the solution set.

45. $|x - 1| \le 0$

46. $|x - 1| > 1$

47. $|x - 5| > 0$

48. $|x - 5| < 2$

49. $|x + 3| > |x - 3|$

50. $|x - 5| < |x - 4|$

51. $|x + 3| \le |x - 2|$

52. $|x + 2| \ge |x - 3|$

Exercises 53–70: Solve the inequality.

53. $|x - 3| < 5$

54. $|x - 3| > 5$

55. $|2x - 5| \le 2$

56. $|-2x - 1| \ge 3$

57. $|-x + 6| > 0$

58. $|4x - 5| < -2$

59. $|-2x + 1| \ge 5$

60. $|2 - 3x| \le 3$

61. $|3x - 7| \le 1$

62. $\left|7x - \frac{1}{2}\right| > 4$

63. $\left|\frac{2x - 3}{3}\right| \ge \frac{1}{6}$

64. $\left|\frac{x - 1}{x + 3}\right| \le 2$

65. $\dfrac{|2x - 3|}{|x + 4|} > 1$

66. $|3x + 2| \ge x - 5$

67. $|x - 4| < 2x + 1$

68. $|\cos x| < 1$

69. $|\sin x| > 0$

70. $|\tan x| \le 1$, in $\left(-\frac{\pi}{2}, \frac{\pi}{2}\right)$

Exercise GC71: (**Temperature**)

(a) Suppose a liquid is being heated in such a way that its temperature T in degrees Fahrenheit as a function of time t in seconds is given by $T = 3t^2 - t + 25$. Use a graphing calculator to approximate the time interval in which the temperature of the liquid is within 5 degrees Fahrenheit of its boiling point, given that the boiling point of the liquid is 73 degrees Celsius, and degrees Fahrenheit (F) vs degrees (C) is given by the formula $F = \frac{9}{5}C + 32$.

(b) In addition to the liquid in (a), another liquid is being heated in such a way that its Fahrenheit temperature versus time equation is $T = 4t^2 + 15$. Use a graphing calculator to approximate the interval of time in which the two liquids are within 3 degrees Fahrenheit of each other.

Exercise 72: (**Distance**) Anna is no farther than 4 feet from Jared, and Nina is at least 5 feet from Anna. What, if anything, can be said about the distance between Nina and Jared?

Exercises 73–75: If p is within 2.3 units of q, and r is within 1.6 units of p, what is the

73. largest possible distance between q and r?

74. smallest possible distance between q and r?

75. largest possible distance between $-q$ and $-r$?

Exercise GC76: (Minimum Distance) Let $f(x) = 5x^3 - 2x$ and $g(x) = 3x^4 + 2$. Use absolute values and a graphing calculator to estimate where the smallest vertical distance between the graphs of f and g occurs.

Exercises 77–80: Write the function without absolute values, as a piece-wise defined function.

77. $f(x) = |7x - 4|$ 78. $f(x) = |x + 5|$ 79. $f(x) = 2|x| - 1$ 80. $f(x) = |x^2 + 2x|$

Exercise 81: The graph of a function f appears on the right. Sketch the graph of the function:

$$g(x) = -3|f(x - 2)| + 1$$

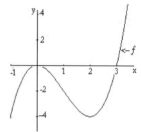

Exercises GC82–GC83: (Maximum/Minimum) Use a graphing calculator to view the graph of the function and estimate where the function has a local maximum or a local minimum.

82. $f(x) = |x^3 - 3x^2 - x - 6|$

83. $g(x) = |x|^3 - 3|x|^2 - |x| - 6$

Exercises GC84–GC87: Use a graphing calculator to determine, approximately, the solution set of the equation.

84. $|x^2 - 3x + 5| = x^3$

85. $4|x^3 - x - 5| = |2x - 1|$

86. $\left|\dfrac{2x^4 - 5x + 1}{x + 7}\right| = 2 + |x|$

87. $|x| - |3x + 2| = -3$

Exercises GC88–GC89: Use a graphing calculator to estimate the solution of the inequality.

88. $|3x^3 + x - 1| < x^2 - 2$

89. $|x^2 - 2| < 3x^3 + x - 1$

Exercise 90: Without a consideration of graphs, determine the solution set of

$$|x^3 - 3x^2 - x - 6| < |x|^3 - 3|x|^2 - |x| - 6$$

and discuss the reasoning process leading to your conclusion.

Exercise 91: Show that for any $x, y,$ and z, $|xyz| = |x||y||z|$.

8.1 Absolute Value Functions, Equations, and Inequalities **357**

Exercise 92: Establish the Triangle Inequality, $|a + b| \leq |a| + |b|$, as follows:

(a) Show that $a \leq |a|$.
[Suggestion: Consider the case $a \geq 0$ and then the case $a < 0$.]

(b) Show that $ab \leq |a||b|$.

(c) Show that $(a + b)^2 \leq (|a| + |b|)^2$.

[Suggestion: Expand $(a + b)^2$ and apply part (b) of this exercise.]

(d) Use Property (iv): $A^2 = |A|^2$, and the fact that if A and B are nonnegative then $A^2 < B^2$ if and only if $A < B$.

Exercise 93: Show that for any two numbers a and b, $|a - b| \geq |\,|a| - |b|\,|$.
[Suggestion: Apply the Triangle Inequality to $|(a - b) + b|$ to show that $|a - b| \geq |a| - |b|$. Then show $|a - b| \geq -(|a| - |b|)$.]

Exercise 94: Using the Triangle Inequality, show that for any two numbers a and b, $|a - b| \leq |a| + |b|$.

Exercises 95–96: Solve the given inequality.
[Suggestion: Consider two cases: When the right-hand side of the inequality equals zero and when it is not zero. In the latter case, divide both sides by the right-hand side and apply Theorem 8.3.]

95. $|x^3 - 2| < |x^3 + 1|$

96. $|x^2 + 3x - 1| \leq |3x + 1|$

Exercises 97–98: Solve the given inequality. (See previous exercise.)

97. $\left|\frac{1}{2}x - 3\right| > |2x|$

98. $|x + 1| > \dfrac{1}{|x - 1|}$

Exercise 99: Show that the inequalities in Theorem 8.3, hold for any "a", positive, negative, or zero.
[Suggestion: For the first inequality consider, separately, the solution set of $|x| < a$ and that of $x > -a$ and $x < a$. Similarly for the second inequality.]

Exercise 100: Generalize Theorem 8.3: Show that the "a" can be replaced by a function, $g(x)$, of x.
[Suggestion: Show that $|f(x)| < g(x)$ if and only if $-g(x) < f(x) < g(x)$, as follows: First show that any x for which $g(x) \leq 0$ is not in the solution set of $|f(x)| < g(x)$. Then show that for any x for which $g(x) > 0$, x is in the solution set of $|f(x)| < g(x)$ if and only if $-g(x) < f(x) < g(x)$. The case $|f(x)| > g(x)$ follows along similar lines.]

358   Chapter 8 Absolute Value and Radical Functions

§2. RADICAL EQUATIONS AND INEQUALITIES

In this section we will be solving equations and inequalities in which the variable appears under a radical sign or to a fractional exponent.

EQUATIONS INVOLVING ONLY ROOTS

We begin by noting that raising both sides of an equation to an odd power yields an equivalent equation. However:

> Raising both sides of an equation to an **even** power can introduce extraneous solutions.
> ### CHECK YOUR ANSWERS

A case in point:

$$2 \neq -2$$

but: $(2)^2 = (-2)^2$

No such problem can occur when raising both sides of an equation to an odd power.

EXAMPLE 8.10   Solve:

$$\sqrt{2x + 13} + 1 = x$$

SOLUTION:

$$\sqrt{2x + 13} + 1 = x$$

Isolate the radical: $\sqrt{2x + 13} = x - 1$

Square both sides: $(\sqrt{2x + 13})^2 = (x - 1)^2$

And solve for x: $2x + 13 = x^2 - 2x + 1$

$$x^2 - 4x - 12 = 0$$

$$(x - 6)(x + 2) = 0$$

$$x = 6 \text{ or } x = -2$$

The above argument shows that 6 and −2 are the only **possible** solutions. Below, we find that while 6 is a solution, −2 is not:

For $x = 6$:	For $x = -2$:
$\sqrt{2 \cdot 6 + 13} + 1 \overset{?}{=} 6$	$\sqrt{2 \cdot (-2) + 13} + 1 \overset{?}{=} -2$
$\sqrt{25} + 1 \overset{?}{=} 6$ — yes	$\sqrt{9} + 1 \overset{?}{=} -2$ — no

Can you SEE why the equation:

$$\sqrt{2x + 1} + 2\sqrt{x} = -1$$

can have **no** solution?

EXAMPLE 8.11   Solve:

$$\sqrt{2x + 1} - 2\sqrt{x} = -1$$

8.2 Radical Equations and Inequalities 359

SOLUTION:

> No matter how you rearrange the terms of the equation you cannot eliminate both radicals by squaring once.

$$\sqrt{2x+1} - 2\sqrt{x} = -1$$

$$\sqrt{2x+1} = 2\sqrt{x} - 1$$

Square both sides: $\quad 2x + 1 = 4x - 4\sqrt{x} + 1$

$$-2x = -4\sqrt{x}$$

$$x = 2\sqrt{x}$$

Square again: $\quad x^2 = 4x$

$$x^2 - 4x = 0$$

$$x(x-4) = 0$$

$$x = 0 \text{ or } x = 4$$

For $x = 0$:	For $x = 4$:
$\sqrt{2\cdot 0 + 1} - 2\sqrt{0} \stackrel{?}{=} -1$	$\sqrt{2\cdot 4 + 1} - 2\sqrt{4} \stackrel{?}{=} -1$
$\sqrt{1} \stackrel{?}{=} -1$ — no	$\sqrt{9} - 4 \stackrel{?}{=} -1$ — yes

We see that the given equation has only one solution: $x = \mathbf{4}$.

EXAMPLE 8.12   Solve:

$$(x+7)^{\frac{1}{3}} = x + 1$$

SOLUTION:

Cubing both sides of: $\quad (x+7)^{\frac{1}{3}} = x + 1$

we find: $\quad x + 7 = (x+1)^3$

$$x + 7 = x^3 + 3x^2 + 3x + 1$$

Observe that 1 is a zero: $\quad x^3 + 3x^2 + 2x - 6 = 0$

See margin: $\quad (x-1)(x^2 + 4x + 6) = 0$

$$x = 1 \text{ or } (x^2 + 4x + 6) = 0$$

no solution since
$$b^2 - 4ac = 16 - 4(1)(6) < 0$$

> $$\begin{array}{r} x^2 + 4x + 6 \\ x-1\overline{\smash{\big)}\,x^3 + 3x^2 + 2x - 6} \\ \underline{x^3 - x^2} \\ 4x^2 + 2x - 6 \\ \underline{4x^2 - 4x} \\ 6x - 6 \\ \underline{6x - 6} \\ 0 \end{array}$$

We find that 1 is the only solution of the given equation.

CHECK YOUR UNDERSTANDING 8.6

> Answers:
>
> (a) 6   (b) $\dfrac{-1 \pm \sqrt{33}}{2}$

Solve:

(a) $\sqrt{x-2} + \sqrt{x+3} = 5$ (b) $(x^2 + 2x - 1)^{\frac{1}{5}} = (x+7)^{\frac{1}{5}}$

Chapter 8 Absolute Value and Radical Functions

EXAMPLE 8.13 Solve for $0 \leq x < 2\pi$:
$$\sqrt{\sin x + 1} = \sin x + \cos x$$

SOLUTION:
$$\sqrt{\sin x + 1} = \sin x + \cos x$$
$$\sin x + 1 = \sin^2 x + 2\sin x \cos x + \cos^2 x$$
$$\sin x + 1 = 1 + 2\sin x \cos x$$
$$\sin x - 2\sin x \cos x = 0$$
$$\sin x (1 - 2\cos x) = 0$$
$$\sin x = 0 \quad \text{or} \quad \cos x = \frac{1}{2}$$
$$x = 0, \pi \quad \text{or} \quad x = \frac{\pi}{3}, \frac{5\pi}{3}$$

A direct substitutuion in the original equation reveals that of the above four candidates, only $x = 0$ is a solution of the given equation.

EXAMPLE 8.14 Solve for $0 \leq x < 2\pi$:
$$\sin x + \cos x = 1$$

SOLUTION: It appears than no initial trigonometric step will improve the situation. Hoping for the best, we decided to square both sides of the equation, realizing that in the process **extraneous roots might be introduced**:
$$\sin x + \cos x = 1$$
$$(\sin x + \cos x)^2 = 1^2$$
$$\sin^2 x + 2\sin x \cos x + \cos^2 x = 1$$
$$(\sin^2 x + \cos^2 x) + 2\sin x \cos x = 1$$
$$1 + 2\sin x \cos x = 1$$
$$2\sin x \cos x = 0$$

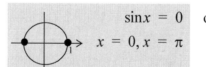

$$\sin x = 0 \quad \text{or} \quad \cos x = 0$$
$$x = 0, x = \pi \quad \quad x = \frac{\pi}{2}, x = \frac{3\pi}{2}$$

Direct substitution into the original equation shows that of the four candidates, only $x = 0$ and $x = \frac{\pi}{2}$ are solutions.

8.2 Radical Equations and Inequalities   361

CHECK YOUR UNDERSTANDING 8.7

Solve, for $0 \le x < 2\pi$:

Answer: (a) $\frac{\pi}{6}, \frac{5\pi}{6}$ (b) $\frac{5\pi}{3}$

    (a) $\sqrt{4\sin x - 1} = 2\sin x$ (b) $\cot x = \csc x + \dfrac{1}{\sqrt{3}}$

EQUATIONS INVOLVING ROOTS AND POWERS

If $x^{\frac{m}{n}} = 0$, then $x = 0$.

We now turn our attention to equations of the form $x^{\frac{m}{n}} = c \ne 0$ where the rational number $\frac{m}{n}$ is reduced to lowest term. Here are four such equations:

$$x^{\frac{2}{3}} = -4, \quad x^{\frac{3}{2}} = -8, \quad x^{\frac{2}{3}} = 4, \quad \text{and} \quad x^{\frac{3}{2}} = 8$$

Since $x^{\frac{2}{3}} = \left(x^{\frac{1}{3}}\right)^2$ can not be negative (note the **2**-exponent), the equation $x^{\frac{2}{3}} = -4$ has no solution. Also, since $x^{\frac{3}{2}} = (\sqrt{x})^3$, and since $\sqrt{x}$ is never negative, the equation $x^{\frac{3}{2}} = -8$ also has no solution. In general:

> Let the rational number $\frac{m}{n}$ be reduced to lowest terms.
>
> If m or n is **even** and $c < 0$, then $x^{\frac{m}{n}} = c$ has **no** solution.

Taking odd powers or roots of both sides of an equation will not alter the solution set of that equation. Consequently, when confronted with an equation of the form $x^{\frac{m}{n}} = c$ where either m or n is odd, first simplify the equation by taking an odd root or an odd power, whichever is appropriate. Consider the following example:

EXAMPLE 8.15 Solve:

        (a) $x^{\frac{2}{3}} = 4$ (b) $x^{\frac{3}{2}} = 8$

SOLUTION: (a) $x^{\frac{2}{3}} = 4$ (b) $x^{\frac{3}{2}} = 8$

$$\left(x^{\frac{2}{3}}\right)^3 = 4^3 \qquad\qquad \left(x^{\frac{3}{2}}\right)^{\frac{1}{3}} = 8^{1/3}$$

$$x^2 = 64 \qquad\qquad\qquad x^{\frac{1}{2}} = 2$$

$$x = \pm 8 \qquad\qquad\qquad\quad x = 4$$

362 **Chapter 8 Absolute Value and Radical Functions**

In general:	For example:
For m even, n odd, and $c \geq 0$: If $x^{\frac{m}{n}} = c$, then $x = \pm c^{\frac{n}{m}}$	If $x^{\frac{4}{3}} = 16$ then $x = \pm 16^{\frac{3}{4}} = \pm \left(16^{\frac{1}{4}}\right)^3 = \pm 2^3 = \pm 8$
For m odd, n even, and $c \geq 0$: If $x^{\frac{m}{n}} = c$, then $x = c^{\frac{n}{m}}$	If $x^{\frac{3}{4}} = 8$ then $x = 8^{\frac{4}{3}} = \left(8^{\frac{1}{3}}\right)^4 = 2^4 = 16$
For m and n odd, if $x^{\frac{m}{n}} = c$, then $x = c^{\frac{n}{m}}$, for any c.	

CHECK YOUR UNDERSTANDING 8.8

Answers: (a) 624   (b) $\pm\frac{1}{8}$

Solve:

(a) $(x+1)^{\frac{3}{4}} = 125$ (b) $x^{\frac{2}{3}} = \frac{1}{4}$

INEQUALITIES INVOLVING RATIONAL EXPONENTS

In the calculus, you may encounter inequalities of the form:

$$x^{\frac{1}{3}}(x+2)^{\frac{2}{5}} > 0$$

In such cases, you will find the following theorem useful:

You are invited to prove this theorem in the exercises.

THEOREM 8.4 Let $\frac{m}{n}$ be a positive rational number in lowest terms.

If m is odd, then c is an odd zero of $(x-c)^{\frac{m}{n}}$

If m is even then c is an even zero of $(x-c)^{\frac{m}{n}}$

For example, since 0 is an odd zero of $x^{\frac{1}{3}}(x+2)^{\frac{2}{5}}$, the sign of the expression will change about 0. It will not change, however, about the even zero -2:

8.2 Radical Equations and Inequalities

EXAMPLE 8.16 Solve:
$$5x^{\frac{2}{3}} - x^{\frac{5}{3}} > 0$$

SOLUTION: The first step is to factor by pulling out the common factor x raised to the **smallest** exponent, namely $x^{\frac{2}{3}}$:

$$5x^{\frac{2}{3}} - x^{\frac{5}{3}} > 0$$
$$x^{\frac{2}{3}}(5 - x) > 0$$

Then:

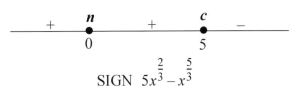

SIGN $5x^{\frac{2}{3}} - x^{\frac{5}{3}}$

Solution: $(-\infty, 0) \cup (0, 5)$

EXAMPLE 8.17 Solve:
$$\frac{(x-5)^2(x-4)^{\frac{1}{3}}}{(x+3)^{\frac{2}{5}}} \le 0$$

SOLUTION: Noting that 4 is an odd zero, that 5 is an even zero, and that -3 is an even zero (in the denominator), we have:

Solution: $(-\infty, -3) \cup (-3, 4] \cup \{5\}$

CHECK YOUR UNDERSTANDING 8.9

Solve:

(a) $x^{\frac{4}{5}}(x-1)^{\frac{1}{3}}(-x+3)^3 < 0$ (b) $x^{\frac{3}{5}}(x-1)^{\frac{2}{3}} \ge 0$

Answers:
(a) $(-\infty, 0) \cup (0, 1) \cup (3, \infty)$
(b) $[0, \infty)$

Particular care has to be taken when the denominator of a rational exponent is even, for you cannot take an even root of a negative quantity. Consider the following example.

EXAMPLE 8.18 Solve:
$$(x+2)^2(x-5)^3(x-1)^{\frac{3}{2}} \le 0$$

364 Chapter 8 Absolute Value and Radical Functions

SOLUTION: The expression $(x-1)^{\frac{3}{2}}$ is **not defined** when
$x - 1 < 0$
$x < 1$:

```
~~~~~~~~~~~~~~~~~~~~~~~~~~~~~~●————————
              not defined     1
```

We then superimpose the additional relevant zero information for
$(x+2)^2(x-5)^3(x-1)^{\frac{3}{2}}$:

```
~~~~~~~~~~~~●~~~~~~●————————●————————
  not defined      1    −   5    +
            −2                c
```

Solution: $[1, 5]$

CHECK YOUR UNDERSTANDING 8.10

Solve:
$$(-x+3)(x+2)^{\frac{5}{2}}(x+1)^{\frac{2}{3}} \le 0$$

Answer: $[3, \infty) \cup \{-1, -2\}$

APPLICATIONS

The next two examples afford you additional opportunities to hone your translating skills from words to equations.

EXAMPLE 8.19

A RATE PROBLEM

A man in a boat 3 miles from a straight coastline wants to get to a dock that is 10 miles down the coast by rowing to some point P on the coast and running the rest of the way. The journey is to consist of two linear paths, one from the boat to P, and the other from P to the dock. Assuming that the man rowed at a rate of 4 miles per hour, and ran at a rate of 5 miles per hour, determine the distance rowed, given that it took him 2 hours and 27 minutes to reach the dock.

SOLUTION:

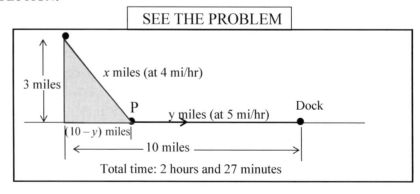

We want to find x. Knowing that it took 2 hours and 27 minutes for the journey, leads us to the equation:

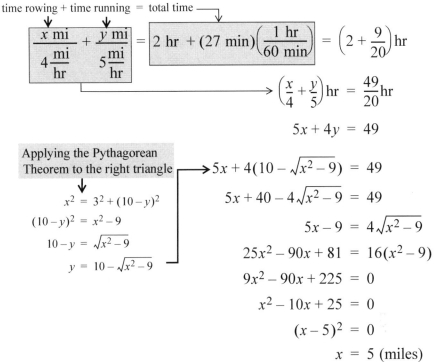

$$5x + 4y = 49$$
$$5x + 4(10 - \sqrt{x^2 - 9}) = 49$$
$$5x + 40 - 4\sqrt{x^2 - 9} = 49$$
$$5x - 9 = 4\sqrt{x^2 - 9}$$
$$25x^2 - 90x + 81 = 16(x^2 - 9)$$
$$9x^2 - 90x + 225 = 0$$
$$x^2 - 10x + 25 = 0$$
$$(x - 5)^2 = 0$$
$$x = 5 \text{ (miles)}$$

We leave it for you to verify that it will indeed take the man 2 hours and 27 minutes to row the 5 miles to the shore and run the $y = 10 - \sqrt{5^2 - 9} = 6$ miles to the dock.

EXAMPLE 8.20

MINIMUM LABOR COST

Point A is at ground level. Point B is 100 feet horizontally from A, and 35 feet below ground level. The first 15 feet below ground level is soil, after which there is shale. Two pipes are to connect the two points A and B; one in the soil region and the other in the shale region. Approximate the minimum labor cost of the project, if it costs $76 per foot to lay piping in the soil layer, and $245 per foot to lay piping in the shale layer.

SOLUTION: We follow the four-step optimization procedure of page 253:

Step 1:

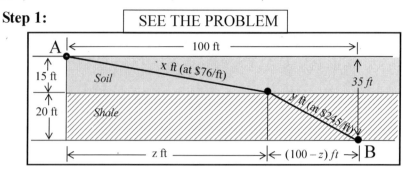

SEE THE PROBLEM

Step 2: Labor cost is to be minimized, and we are able to easily express it in terms of two variables:

Labor Cost = Cost to lay pipe in soil + Cost to lay pipe in shale:

$$C = \left(76\frac{\$}{ft}\right)(x \text{ ft}) + \left(245\frac{\$}{ft}\right)(y \text{ ft}) = 76x + 245y \text{ (dollars)} \quad (*)$$

Step 3: There is no apparent relationship between the variables x and y. However, employing the Pythagorean Theorem, we can express both x and y in terms of the variable z:

$$x^2 = 15^2 + z^2 \quad \text{or:} \quad x = \sqrt{15^2 + z^2}$$

$$y^2 = 20^2 + (100-z)^2 \quad \text{or:} \quad y = \sqrt{20^2 + (100-z)^2}$$

From (*):

$$C = 76\sqrt{15^2 + z^2} + 245\sqrt{20^2 + (100-z)^2}$$

Step 4: Use a graphing calculator to approximate the minimum value of C:

From Step 1, we know that $0 \leq z \leq 100$. Wanting to find any one point on the graph of C, we chose to find the value of C at 50:

76√(15²+50²)+245
√(20²+50²)
               17160.97

This told us that the point (50, 17161) lies on the graph, and that we should go with a tall window.

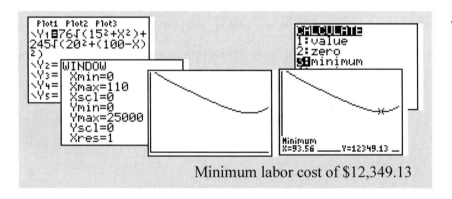

Minimum labor cost of $12,349.13

CHECK YOUR UNDERSTANDING 8.11

Suppose the man in Example 8.19 can row at a rate of 3 mph and run at 4 mph. Use a graphing calculator to find the minimum time to get to the dock.

Answer: 3 hrs. 10 min.

8.2 Radical Equations and Inequalities **367**

EXERCISES

Exercises 1–14: Solve the equation.

1. $\sqrt{x} - 2 = 4$

2. $\sqrt{x-2} = 2$

3. $2 - \sqrt{x+1} = -4$

4. $1 - 2\sqrt{3x+5} = 0$

5. $\sqrt{x+1} = 2x$

6. $\sqrt{x+2} + 4 - x = 0$

7. $\sqrt{x+1} = \sqrt{2x-2}$

8. $\sqrt{x-1} + 1 = \sqrt{x}$

9. $\sqrt{x-3} - \sqrt{2x+1} = 2$

10. $\sqrt{x-2} + \sqrt{x+5} = 7$

11. $2\sqrt{x-1} - \sqrt{2x-1} = 1$

12. $(x+4)^{-\frac{1}{3}} = -\frac{1}{2}$

13. $(2x-5)^{\frac{1}{3}} = x - 2$

14. $(19x+11)^{\frac{1}{3}} = x - 1$

Exercises 15–25: Solve the equation for x in $[0, 2\pi)$, unless otherwise indicated.

15. $\sqrt{1 - \cos x} = \cos x - \sin x$

16. $\sqrt{4\sqrt{3}\sin x - 3} = 2\sin x$

17. $\sqrt{4 - 5\cos x} = 2\sin x$

18. $\sqrt{\sin x + 2\cos x} = 1$

19. $\tan x = \sec x + 1$

20. $\cos x - \sin x = \sqrt{2}$

21. $\sin x - \cos x = -1$,

22. $\cot x - \csc x = 1$,

23. $\tan x = \sec x - 2$

24. $\tan 2x + \sqrt{3} = \sec 2x$

25. $\csc 4x = 1 - \cot 4x$ in $[0, \pi]$

Exercise 26: Determine x such that the distance between $(x, 2)$ and $(1, 1)$ equals the distance between $(1, 2)$ and $(1, 4)$.

Exercise 27: Determine y such that the distance between $(1, y)$ and $(2, 2)$ equals the distance between $(1, y)$ and $(-1, 1)$.

Exercises 28–36: Solve the equation.

28. $x^{\frac{4}{5}} = 81$

29. $x^{\frac{4}{5}} = -81$

30. $x^{\frac{3}{5}} = \frac{1}{8}$

31. $(x-2)^{\frac{3}{2}} = 27$

32. $(x+1)^{\frac{2}{3}} = 4$

33. $(3x-1)^{\frac{5}{4}} = 32$

34. $(2x+5)^{\frac{9}{2}} = -1$

35. $(4x+3)^{\frac{4}{3}} = 1$

36. $(6x-3)^{\frac{5}{3}} = -243$

Exercises 37–40: Solve the equation, by first making a substitution.

37. $\sqrt{x+1} - \dfrac{3}{\sqrt{x+1}} = 2$

38. $\dfrac{\sqrt{x-3}}{4} + \dfrac{4}{\sqrt{x-3}} = 2$

39. $\dfrac{12}{\sqrt{2-x}} - 2 = 2 - x$

40. $(x+2)^{\frac{2}{3}} - 5(x+2)^{\frac{1}{3}} = \dfrac{4}{(x+2)^{\frac{1}{3}}} - 8$

368   Chapter 8   Absolute Value and Radical Functions

Exercises 41–46: Without attempting to solve directly, describe the solution set or explain why the equation has no (real) solutions.

41. $\sqrt{-3x-1} = -2$ 42. $(-x)^{\frac{3}{2}} = \sqrt{x-1}$ 43. $\sqrt{-x} = \sqrt{x}$

44. $\sqrt{-x-1} = -x^2-1$ 45. $x^{-\frac{3}{2}} = -x$ 46. $x^{\frac{2}{3}} = x^2+3$

Exercises 47–54: Solve the inequality.

47. $(x-1)^{\frac{1}{5}}(x-3)^{\frac{4}{3}} \le 0$

48. $(1-x)^{\frac{3}{5}}(x-3)^3 > 0$

49. $x^{\frac{7}{2}} - 3x^{\frac{5}{2}} \ge 0$

50. $x^{\frac{4}{3}} - 9x^{\frac{1}{3}} > 0$

51. $x(x+1)^2(x-4)^{\frac{2}{3}} < 0$

52. $(x-1)^{\frac{3}{2}}(x+2)^5(2-x)^{\frac{1}{3}} \ge 0$

53. $\dfrac{(x+3)^{\frac{5}{2}}(x+4)^2}{(x-1)^{\frac{1}{3}}} \le 0$

54. $\dfrac{x^{\frac{3}{4}}(x-4)^{\frac{3}{5}}}{(-x+2)^2} \ge 0$

Exercises 55–57: Without attempting to solve directly, describe the solution set or explain why the inequality has no (real) solutions.

55. $x^{\frac{2}{3}} + 1 \ge 0$ 56. $x^{\frac{3}{2}} + 1 \ge 0$ 57. $\sqrt{x} + x < 0$

Exercise GC58: After arriving at an appropriate function, use a graphing calculator to help you find (approximately) all pairs of numbers whose sum is three and such that adding two to one of the two numbers and then raising that sum to the power $\frac{2}{3}$ yields the other number.

Exercises GC59–GC62: Use a graphing calculator to find an approximate solution of the optimization problem.

59. **(Drilling a Tunnel)** A tunnel is to be drilled to connect point A to point B in the adjacent figure (where the units are feet). Determine the path (find x) of minimum cost, if it costs $11 per foot to drill through the sand layer, and $15 per foot to drill in the shale layer.

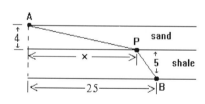

60. **(Area)** What is the largest possible area of a right triangle whose hypotenuse is 10 cm?

61. **(Lake-to-Shore)** Karen is in a canoe on a lake one kilometer from the closest point Q of a straight shoreline. She wants to get to a point R that is 10 kilometers along the shore from Q by paddling to P between Q and R and then walking. She can paddle at 3 km/hr and walk at 5 km/hr. Where should P be so that she gets to R as quickly as possible?

62. (**Fermat's principle**) Fermat's principle in optics states that light always travels from one point to another along the quickest route.[1] Consequently, since light travels at a constant speed in any given medium, within that medium a light ray will travel along a straight line. However, since light travels at different speeds through different mediums, when a ray moves across a medium, it changes its direction.

What is the minimum time required for light to travel from point A to point B in the figure below?

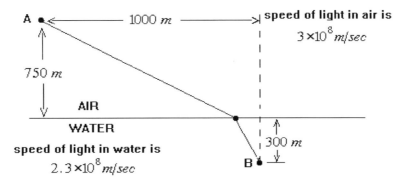

Exercise GC63: An airline flies between Newark and Boston. The monthly cost (in millions of dollars) to the airline is

$$C(x) = \sqrt{1 + 0.3x}$$

where x is measured in thousands of passengers. Use a graphing calculator to graph the cost function and estimate the range of values for the number of passengers that flew during a month when the total cost was between 1.5 and 2 million dollars.

Exercise 64: Prove Theorem 8.4.

---

[1] Pierre De Fermat (1601-1665) is one of the greatest giants of mathematics. He is best known for his famous "Fermat's Last Theorem", which states that the equation $x^n + y^n = z^n$, where n is an integer greater than 2, has no solution in positive integers. Many great mathematicians, and others, have attempted, without success, to prove this evasive theorem. In 1993, Andrew Wiles, was able to present a proof incorporating some recently developed theory in elliptical geometry. Nonetheless, this famous problem will not be laid to rest until a proof emerges (if one exists) which uses only those tools available to Fermat some three hundred years ago.

Chapter Summary

Absolute Value Function	$$|x| = \begin{cases} x & \text{if } x \geq 0 \\ -x & \text{if } x < 0 \end{cases}$$ 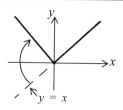
Graphing an Absolute Value Function	To graph $y = |f(x)|$, first graph $y = f(x)$ and then reflect the negative portion of that graph about the x-axis. For example: 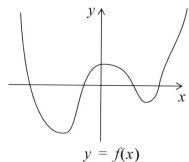 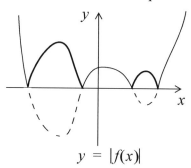
Properties of Absolute Values	For any numbers a and b: (i) $|a| \geq 0$, with equality holding if and only if $a = 0$ (ii) $|a| = |-a|$ (iii) $|ab| = |a||b|$ and, if $b \neq 0$, $\left|\dfrac{a}{b}\right| = \dfrac{|a|}{|b|}$ (iv) $|a|^2 = a^2$ (v) $\sqrt{a^2} = |a|$ (vi) **(Triangle Inequality)** $|a + b| \leq |a| + |b|$
Absolute Value Equations	For any $a \geq 0$: $$|x| = a \text{ if and only if } x = a \text{ or } x = -a$$ For any a and b: $$|a| = |b| \text{ if and only if } a = b \text{ or } a = -b$$

Absolute Value Inequalities	$|x| < a$ x is within a units of 0 $-a < x < a,$ which is to say: $x > -a$ AND $x < a$	$|x| > a$ x is more than a units from 0 $x < -a$ OR $x > a$

Radical Equations

Involving only Roots

Raising both sides of the equation to an **even** power can introduce extraneous solutions

CHECK YOUR ANSWERS

Involving Roots and Powers

Let the rational number $\dfrac{m}{n}$ be reduced to lowest terms.

If m or n *is* **even** and $c < 0$, then $x^{\frac{m}{n}} = c$ has **no** solution.

For m even, n odd, and $c \geq 0$: If $x^{\frac{m}{n}} = c$, then $x = \pm c^{\frac{n}{m}}$

For m odd, n even, and $c \geq 0$: If $x^{\frac{m}{n}} = c$, then $x = c^{\frac{n}{m}}$

For m and n odd, if $x^{\frac{m}{n}} = c$, then $x = c^{\frac{n}{m}}$, for any c.

Inequalities containing factors of the form $(x - c)^{\frac{m}{n}}$

Let $\dfrac{m}{n}$ be a positive rational number in lowest terms.

If m is odd, then c is an odd zero of $(x - c)^{\frac{m}{n}}$

If m is even, then c is an even zero of $(x - c)^{\frac{m}{n}}$

372   Chapter 8   Absolute Value and Radical Functions

C H A P T E R   R E V I E W   E X E R C I S E S

Exercises 1–36: Solve each of the following.

1. $|3x + 4| = 5$

2. $|\sqrt{3}x - 1| = 4$

3. $|5 + 2x| = |3 - x|$

4. $|-2||3x - 5| = |-4||x|$

5. $\left|\dfrac{5x + 1}{2x + 3}\right| = 2$

6. $|x||1 - x| = |4x - 2|$

7. $|4x + 2| \leq 1$

8. $|x - 5| > 3$

9. $\left|\dfrac{x + 1}{2}\right| \geq \dfrac{3}{4}$

10. $\dfrac{|3x - 1|}{|x + 4|} < 2$

11. $\left|\dfrac{2}{x}\right| > 3$

12. $|x + 2| \leq 3x - 1$

13. $|2\sin^2 x + \sin x| = 1$, in $[0, 2\pi)$

14. $\sqrt{2x} = 3x$

15. $2 + 3\sqrt{x - 3} = 4$

16. $\sqrt{x + 2} - \dfrac{12}{\sqrt{x + 2}} = 1$

17. $\sqrt{x - 2} = \sqrt{x} - 2$

18. $\sqrt{15x + 4} = \sqrt{5x + 5} + 1$

19. $x^{\frac{1}{3}} = -4$

20. $x^{\frac{1}{4}} = 2$

21. $(2 - 3x)^{\frac{1}{5}} = -2$

22. $\sqrt{2}\sin x = -\cos x$, in $[0, 2\pi)$

23. $\sqrt{2}\sin x + 3 = 4\sin x$, in $[0, 2\pi)$

24. $\tan x + \sec x = 1$, in $[0, 2\pi)$

25. $\tan x = \sec x + \dfrac{1}{\sqrt{3}}$, in $[0, 2\pi)$

26. $3\cos x + \sin x = 1$, in $[0, 2\pi)$

27. $x^{\frac{3}{2}} = -8$

28. $x^{\frac{4}{3}} = 100$

29. $x^{\frac{2}{3}} = 5$

30. $x^{\frac{20}{17}} = 1$

31. $x^{\frac{17}{20}} = 1$

32. $(3x + 5)^{\frac{3}{4}} = 125$

33. $\dfrac{(x + 2)^{\frac{3}{4}}}{x^2 - 2x} \geq 0$

34. $x(x + 3)^{\frac{2}{5}}(x - 2)^{\frac{3}{7}} > 0$

35. $(x + 5)^{\frac{2}{3}}(x - 1)^{\frac{1}{5}} > 0$

36. $(x + 1)^{\frac{3}{2}}(x - 2)^{\frac{3}{5}}(x - 3)^4 \leq 0$

Exercises 37–38: Sketch the graph of the function, indicating the x- and y-intercepts.

37. $f(x) = |20x^3 - 17x^2 - 10x|$

38. $f(x) = |(x + 2)(3x - 4)|$

Exercise GC39:   (Minimum Cost) Points A and B are directly across from each other on opposite shores of a straight river 3 km wide. Point C is on the same shore as B but 2 km down the river from B. A telephone company wishes to lay a cable from A to C, by laying it underwater from A to a point D between B and C, and then on land from D to C. It costs $300 per kilometer to lay the cable underwater, and $100 per kilometer to lay the cable on land. Use a graphing calculator to estimate where D should be to minimize cost.

Exercises GC40:   Use a graphing calculator to graph the function $f(x) = |2x^3 - 4x - 1|$ and approximate the intercepts.

CUMULATIVE REVIEW EXERCISES

Exercise 1: Simplify $x^{\frac{4}{3}}(x-3)^{-1} - \dfrac{x^{-\frac{2}{3}}(x^2 - 3x + 9)}{x-3}$.

Exercise 2: For $f(x) = 3x - 2$ and $g(x) = 2x^2 - 5x + 1$, determine each of the following.

(a) $g(x+h)$ (b) $(f \circ f)(x)$ (c) $(g \circ f)(x)$ (d) $g\left(\dfrac{1}{x}\right)$

Exercise 3: Solve the equation $2\cos^2 x = \cos x$ for x in $[0, 2\pi)$.

Exercise 4: Express $3\cos 2x - 2\sin^2 2x$ in terms only of $\sin x$.

Exercise 5: Find the partial fraction decomposition of $\dfrac{4x^3 + 11x - 3}{x^4 + x^2 - 2}$.

Exercises 6–20: Solve each of the following.

6. $\dfrac{-2x-1}{x-5} \leq \dfrac{1}{2}$

7. $\left|\dfrac{4x-1}{2x+3}\right| = 3$

8. $|x^2 + x - 5| = 1$

9. $|\cos^2 x - 2\sin x| = 2$, in $[0, 2\pi)$

10. $|6 - 5x| \geq 3$

11. $|x^2 - 2x - 4| < 1$

12. $(x+2)^{\frac{1}{3}} = -3$

13. $5x + 2 = \dfrac{2}{\sqrt{5x+2}} - 1$

14. $6 - \sqrt{5x-1} = -\sqrt{2x+5}$

15. $\sec x = \tan x + \dfrac{1}{\sqrt{3}}$, in $[0, 2\pi)$

16. $x^{\frac{2}{5}} = 2$

17. $(2x-1)^{\frac{3}{2}} = 27$

18. $\sqrt{x} \leq 3$

19. $x^{\frac{5}{3}} + 4x^{\frac{2}{3}} \leq 0$

20. $(x-6)^{\frac{5}{7}}(x-3)^{\frac{1}{4}}(x+1)^3 \leq 0$

Exercises 21–24: Determine the intercepts, and where appropriate, the vertical and horizontal asymptotes. Sketch the graph of the function.

21. $F(x) = |10x^2 + 7x - 12|$

22. $G(x) = |(x+1)^3 - 8|$

23. $H(x) = \dfrac{x(2x-1)(3-x)}{3x+5}$

24. $K(x) = -\tan x$, $-\dfrac{\pi}{2} < x < \dfrac{\pi}{2}$

Exercise GC25: (**Maximum Area**)
A rectangle has its upper two vertices on the parabola $y = 15 - 2x^2$ and its lower two vertices on the line $y = -1$ (see figure). Express the area of the rectangle as a function of one variable, and then use a graphing calculator to estimate the largest area the rectangle can have.

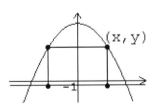

Exercise GC26: (**Minimum Distance**) For time $t \geq 0$, bug A has coordinates (t, t^2) in the plane, while bug B has coordinates $(5t - 5, 5t)$. Use a graphing calculator to determine the minimum distance between the bugs.

374

CHAPTER 9

EXPONENTIAL AND LOGARITHMIC FUNCTIONS

Exponential and logarithmic functions have important applications in many fields. In particular, exponential functions are used to describe biological growth and radioactive decay, while measurements of sound intensity and chemical acidity involve logarithmic functions.

Exponential functions are introduced in Section 1, and their inverse, the logarithmic functions, in Section2. Methods for solving exponential and logarithmic equations and inequalities are discussed in Section 3. Exponential growth and decay problems are considered in Section 4, and additional applications of exponential and logarithmic functions surface in Section 5.

§1. EXPONENTIAL FUNCTIONS

As is indicated in the experimental data in Figure 9.1, a culture of E.coli bacteria increased from 120 million at 1:00 p.m. to 3,932,160 million at 6:00 p.m.

Time	Population (in Millions)	Increase in Population (in Millions)
1:00 p.m.	120	
2:00 p.m.	960	840
3:00 p.m.	7,680	6,720
4:00 p.m.	61,440	53,760
5:00 p.m.	491,520	430,080
6:00 p.m.	3,932,160	3,440,640

Figure 9.1

While it is evident that the bacteria are increasing at a rapid rate, just how many bacteria will be present at 7:00 p.m. is not evident. However, dividing each hour's population by the previous hour's population we find that:

$$\frac{\text{population at 2:00}}{\text{population at 1:00}} = \frac{960 \times 10^6}{120 \times 10^6} = 8$$

$$\frac{\text{population at 3:00}}{\text{population at 2:00}} = \frac{7,680 \times 10^6}{960 \times 10^6} = 8$$

$$\frac{\text{population at 4:00}}{\text{population at 3:00}} = \frac{61,440 \times 10^6}{7,680 \times 10^6} = 8$$

and so on

376 **Chapter 9 Exponential and Logarithmic Functions**

Noting that each hour's population is 8 times the previous hour's population, we can easily calculate the number of bacteria that will be present at 7:00 p.m.:

$$\text{population at 7:00 p.m.} = (\text{population at 6:00 p.m.}) \cdot 8$$

$$= (3{,}932{,}160 \times 10^6) \cdot 8 = 31{,}457{,}280 \times 10^6$$

We can now make an important addition to the table of Figure 9.1, a "last" row which gives an expression for the population n hours after 1:00 p.m.:

Time	No. of Hours After 1:00 P.M	Population (in Millions)
1:00 p.m.	0	$120 = \boxed{120 \cdot 8^0}$
2:00 p.m.	1	$960 = (120 \cdot 8^0) \cdot 8 = \boxed{120 \cdot 8^1}$
3:00 p.m.	2	$7{,}680 = (120 \cdot 8^1) \cdot 8 = \boxed{120 \cdot 8^2}$
4:00 p.m.	3	$61{,}440 = (120 \cdot 8^2) \cdot 8 = \boxed{120 \cdot 8^3}$
.	.	.
.	.	.
.	.	.
.	n	$(120 \cdot 8^{n-1}) \cdot 8 = \boxed{120 \cdot 8^n}$.

In particular, the number of bacteria (in millions) present at 11:00 p.m., which is 10 hours after 1:00 p.m., can now easily be calculated:

$$\text{population at 11:00 p.m.} = 120 \cdot 8^{10} \approx 1.29 \times 10^{11}$$

The function $P(x) = 120 \cdot 8^x$ giving the population of bacteria x hours after 1:00 p.m., involves the so-called **exponential** function $f(x) = 8^x$. In general:

We note that the exponential properties of Theorem 1.1, page 4 extend to exponential functions. In particular:

$$b^s b^t = b^{s+t}$$

$$\frac{b^s}{b^t} = b^{s-t}$$

$$(b^s)^t = b^{st}$$

DEFINITION 9.1

EXPONENTIAL FUNCTION

Let b be a positive number other than 1. The function f given by:

$$f(x) = b^x$$

is said to be the **exponential function with base b**.

Why can't the base be negative? Because an expression such as $(-4)^x$ is not defined (in the real number system) for values of x such as $\frac{1}{2}$ or $-\frac{5}{8}$ (even roots). Why can't the base be 1? Because the function $f(x) = 1^x = 1$ is a constant function. Why can't the base be 0? Because the function $f(x) = 0^x = 0$ is also a constant function [with domain $(-\infty, 0) \cup (0, \infty)$, since 0^0 is not defined].

9.1 Exponential Functions   377

GRAPHS OF EXPONENTIAL FUNCTIONS

After calculating several points on the graphs of the functions $f(x) = 4^x$ and $g(x) = \left(\frac{1}{4}\right)^x$ (Figure 9.2), we were able to sketch their graphs (Figure 9.3).

x	$y = 4^x$	Point on graph		x	$y = \left(\frac{1}{4}\right)^x$	Point on graph
-1	$y = 4^{-1} = \frac{1}{4}$	$\left(-1, \frac{1}{4}\right)$		-1	$y = \left(\frac{1}{4}\right)^{-1} = 4$	$(-1, 4)$
$-\frac{1}{2}$	$y = 4^{-\frac{1}{2}} = \frac{1}{2}$	$\left(-\frac{1}{2}, \frac{1}{2}\right)$		$-\frac{1}{2}$	$y = \left(\frac{1}{4}\right)^{-1/2} = 2$	$\left(-\frac{1}{2}, 2\right)$
0	$y = 4^0 = 1$	$(0, 1)$		0	$y = \left(\frac{1}{4}\right)^0 = 1$	$(0, 1)$
$\frac{1}{2}$	$y = 4^{\frac{1}{2}} = 2$	$\left(\frac{1}{2}, 2\right)$		$\frac{1}{2}$	$y = \left(\frac{1}{4}\right)^{1/2} = \frac{1}{2}$	$\left(\frac{1}{2}, \frac{1}{2}\right)$
1	$y = 4^1 = 4$	$(1, 4)$		1	$y = \left(\frac{1}{4}\right)^1 = \frac{1}{4}$	$\left(1, \frac{1}{4}\right)$
$\frac{3}{2}$	$y = 4^{\frac{3}{2}} = 8$	$\left(\frac{3}{2}, 8\right)$		$\frac{3}{2}$	$y = \left(\frac{1}{4}\right)^{3/2} = \frac{1}{8}$	$\left(\frac{3}{2}, \frac{1}{8}\right)$

Figure 9.2

We must point out that we have not defined expressions such as $4^{\sqrt{2}}$ and $\left(\frac{1}{4}\right)^{\pi}$ in which the exponent is not a rational number. This is done in the calculus. We will accept the graphs of Figure 9.3 each of which has $(-\infty, \infty)$ as its domain.

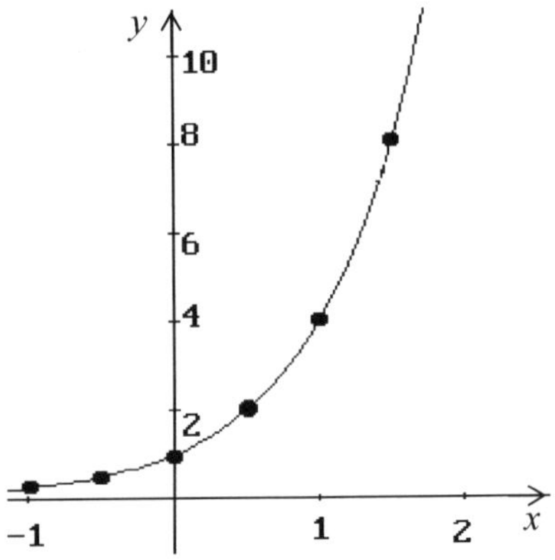

In Exercise 49 you are asked to show that the graph of $f(x) = \left(\frac{1}{b}\right)^x$ is the reflection of the graph of $g(x) = b^x$ about the y-axis.

Graph of $f(x) = 4^x$ Graph of $f(x) = \left(\frac{1}{4}\right)^x$

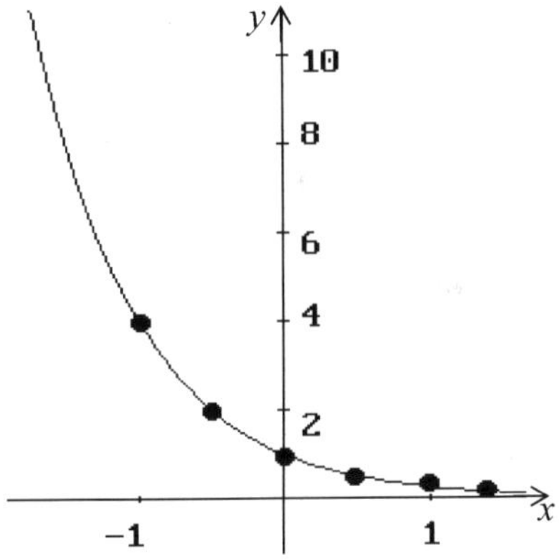

Figure 9.3

The graphs of several exponential functions appear in Figure 9.4. Since $b^0 = 1$ for every base b, all pass through the point $(0, 1)$. The exponential functions of base greater than 1 are **increasing functions** (graphs climb as you move to the right) [see Figure 9.4(a)], and exponential functions of base less than 1 are **decreasing functions** (graphs fall as you move to the right) [see Figure 9.4(b)].

378   Chapter 9 Exponential and Logarithmic Functions

Note also that the x-axis is a horizontal asymptote for the graph of every exponential function:

When $b > 1$, the graph approaches the x-axis as $x \to -\infty$.

When $0 < b < 1$, the graph approaches the x-axis as $x \to +\infty$.

A particularly important exponential function is that of base e, where $e \approx 2.7183$:

Like the number π, the number e, named after Leonhard Euler, arises naturally in mathematics, economics, and the sciences.

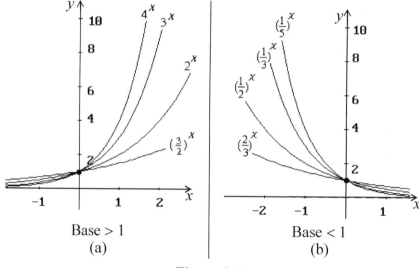

Figure 9.4

As is indicated by the graphs of Figure 9.4, exponential functions have domain $(-\infty, \infty)$ (they are defined everywhere), and range $(0, \infty)$ (they take on all positive values).

CHECK YOUR UNDERSTANDING 9.1

Sketch the graph of the given function, indicating its asymptote and y-intercept.

Answer: See page C-61

(a) $f(x) = \left(\dfrac{8}{3}\right)^x$ (b) $g(x) = \left(\dfrac{3}{8}\right)^x$

EXPONENTIAL EQUATIONS: COMMON BASE

The following result, which is evident from the graphs of Figure 9.4, can be used to solve exponential equations involving a common base:

THEOREM 9.1 Every exponential function is one-to-one:

$$b^s = b^t \text{ if and only if } s = t$$

More generally

$$b^{\text{expression}_1} = b^{\text{expression}_2}$$

if and only if

$$\text{expression}_1 = \text{expression}_2$$

9.1 Exponential Functions 379

EXAMPLE 9.1 Solve:

$$3^{5x+1} = 9^{3x-5}$$

SOLUTION:

$$3^{5x+1} = \mathbf{9}^{3x-5}$$

Common base: $3^{5x+1} = (\mathbf{3^2})^{3x-5}$

$(b^s)^t = b^{st}$: $3^{5x+1} = 3^{2(3x-5)}$

$$3^{5x+1} = 3^{6x-10}$$

Theorem 9.1: $5x+1 = 6x-10$

$$-x = -11$$

$$\boxed{x = 11}$$

EXAMPLE 9.2 Solve:

$$\left(\frac{1}{8}\right)^{x^2+1} = \left(\frac{1}{2}\right)^{4x+2}$$

SOLUTION:

$$\left(\frac{1}{8}\right)^{x^2+1} = \left(\frac{1}{2}\right)^{4x+2}$$

Common base: $\left[\left(\frac{1}{2}\right)^3\right]^{x^2+1} = \left(\frac{1}{2}\right)^{4x+2}$

$(b^s)^t = b^{st}$: $\left(\frac{1}{2}\right)^{3x^2+3} = \left(\frac{1}{2}\right)^{4x+2}$

Theorem 9.1: $3x^2+3 = 4x+2$

$$3x^2-4x+1 = 0$$

$$(3x-1)(x-1) = 0$$

$$\boxed{x = \frac{1}{3} \ \text{or} \ x = 1}$$

CHECK YOUR UNDERSTANDING 9.2

Solve:

$$3^{2x^2-1} = \left(\frac{1}{27}\right)^x$$

Answer: $\dfrac{-3 \pm \sqrt{17}}{4}$

EXAMPLE 9.3 Find all $0 \le x < 2\pi$ which satisfy the equation:

$$2^{5\sin x-1} = 4^{3\sin^2 x}$$

380 Chapter 9 Exponential and Logarithmic Functions

SOLUTION:

$$2^{5\sin x - 1} = 4^{3\sin^2 x}$$
$$2^{5\sin x - 1} = (2^2)^{3\sin^2 x}$$
$$2^{5\sin x - 1} = 2^{6\sin^2 x}$$
$$5\sin x - 1 = 6\sin^2 x$$
$$6\sin^2 x - 5\sin x + 1 = 0$$

Let $y = \sin x$: $\quad\longrightarrow\quad 6y^2 - 5y + 1 = 0$
$$(2y - 1)(3y - 1) = 0$$
$$y = \frac{1}{2} \text{ or } y = \frac{1}{3}$$
$$\longrightarrow \sin x = \frac{1}{2} \text{ or } \sin x = \frac{1}{3}$$

OR

$\sin x = \frac{1}{2}$: $\qquad\qquad\qquad\qquad \sin x = \frac{1}{3}$:

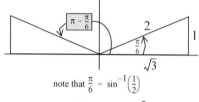 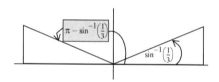

$$x = \frac{\pi}{6} \text{ or } x = \frac{5\pi}{6} \qquad x = \sin^{-1}\frac{1}{3} \text{ or } x = \pi - \sin^{-1}\frac{1}{3}$$

Answer:
$$\frac{3\pi}{4}, \frac{7\pi}{4}, \tan^{-1}\left(\frac{1}{2}\right),$$
$$\pi + \tan^{-1}\left(\frac{1}{2}\right)$$

CHECK YOUR UNDERSTANDING 9.3

Find all $0 \le x < 2\pi$ which satisfy the equation:

$$9^{\tan x} = \frac{3^{\cot x}}{3}$$

EXAMPLE 9.4 Solve:
$$2^{2x+1} - 5 \cdot 2^x - 12 = 0$$

SOLUTION: To solve this equation observe that:

$$2^{2x+1} = 2^{2x} \cdot 2^1 = (2^x)^2 \cdot 2 = 2(2^x)^2$$

We then have:

It is relatively easy to see that:
$$2(2^x)^2 = 2^{2x+1}$$
But here, you have to go the other way!

9.1 Exponential Functions 381

$$2^{2x+1} - 5 \cdot 2^x - 12 = 0$$

$$2(2^x)^2 - 5(2^x) - 12 = 0$$

Make the substitution $Y = 2^x$:
$$2Y^2 - 5Y - 12 = 0$$

$$(2Y+3)(Y-4) = 0$$

$$Y = -\frac{3}{2} \quad \text{or} \quad Y = 4$$

substitute back 2^x for Y:

no solution since exponential functions are always positive $\rightarrow$ $\cancel{2^x = -\dfrac{3}{2}}$ or $2^x = 4 = 2^2$

Theorem 9.1: $\boxed{x = 2}$

CHECK YOUR UNDERSTANDING 9.4

Solve:
$$3^{2x-1} + 3^{x-1} = \frac{2}{3}$$

Answer: $x = 0$

EXPONENTIAL INEQUALITIES: COMMON BASE

Since the function $f(x) = b^x$ is an increasing function if $b > 1$ and is a decreasing function if $0 < b < 1$ (see Figure 9.4), we have:

More generally,

If $b > 1$:
$$b^{\text{expression}_1} < b^{\text{expression}_2}$$
if and only if
$$\text{expression}_1 < \text{expression}_2$$

If $0 < b < 1$:
$$b^{\text{expression}_1} < b^{\text{expression}_2}$$
if and only if
$$\text{expression}_1 > \text{expression}_2$$

THEOREM 9.2 (a) If $b > 1$:
$$b^s < b^t \text{ if and only if } s < t$$

(b) If $0 < b < 1$:
$$b^s < b^t \text{ if and only if } s > t$$

EXAMPLE 9.5 Solve:
$$2^{x-1} > 8^{x+2}$$

SOLUTION:

$$2^{x-1} > 8^{x+2}$$

$$2^{x-1} > (2^3)^{x+2}$$

$(b^s)^t = b^{st}$:
$$2^{x-1} > 2^{3x+6}$$

Theorem 9.2(a):
$$x - 1 > 3x + 6$$

$$-2x > 7$$

Dividing by the negative quantity -2 reverses the inequality symbol:
$$x < -\frac{7}{2}$$

Solution set: $\left(-\infty, -\dfrac{7}{2}\right)$

382 Chapter 9 Exponential and Logarithmic Functions

You can also rewrite the inequality in the form:

$$(3^{-1})^{2x+3} > 3^{-2}$$

and go on from there:

$$(3^{-1})^{2x+3} < (3^{-1})^2$$

$$3^{-2x-3} < 3^{-2}$$

Theorem 9.2(a): $-2x - 3 < -2$

$$-2x < 1$$

Dividing by a negative quantity: $x > -\dfrac{1}{2}$

EXAMPLE 9.6 Solve:

$$\left(\frac{1}{3}\right)^{2x+3} < \frac{1}{9}$$

SOLUTION:

$$\left(\frac{1}{3}\right)^{2x+3} < \frac{1}{9}$$

$$\left(\frac{1}{3}\right)^{2x+3} < \left(\frac{1}{3}\right)^2$$

Theorem 9.2(b): $\quad 2x + 3 > 2$

$$2x > -1$$

$$x > -\frac{1}{2} \quad \text{or:} \left(-\frac{1}{2}, \infty\right)$$

Answers: (a) $(1, \infty)$

(b) $\left(-\infty, \frac{2}{3}\right)$

CHECK YOUR UNDERSTANDING 9.5

Solve:

(a) $4^{x+1} > 8 \cdot 2^x$ (b) $\left(\dfrac{2}{3}\right)^{5x-2} > \left(\dfrac{2}{3}\right)^{2x}$

9.1 Exponential Functions **383**

EXERCISES

Exercises 1–8: Find the y-intercept and the asymptote, and sketch the graph of the function. Determine the range from the graph. [Hint: Stretch/shrink, reflect, and/or shift the basic graph of b^x.]

1. $f(x) = 3 \cdot 2^x$

2. $f(x) = -4^x + 2$

3. $f(x) = -2e^x + 1$

4. $f(x) = e^x - 3$

5. $f(x) = 2^{x-2} - 2$

6. $f(x) = \left(\frac{9}{5}\right)^{x+1}$

7. $f(x) = \frac{1}{3}\left(\frac{1}{2}\right)^x + 3$

8. $f(x) = \left(\frac{2}{3}\right)^{-x+2} - 1$

Exercises 9–26: Solve the equation. If there are trigonometric functions of x in the equation, then x is in $[0, 2\pi)$.

9. $2^x = 8^{2x+1}$

10. $3^{2x-3} = \left(\frac{1}{3}\right)^{x^2}$

11. $\left(\frac{1}{2}\right)^{x-2} = \left(\frac{1}{4}\right)^{x+2}$

12. $3^{x^2-16} = 1$

13. $\left(\frac{1}{10^x}\right)\left(\frac{1}{100}\right)^{2x} = 10^{2x^2-3}$

14. $2^{-3} \cdot 3^{x-1} = \left(\frac{3}{8}\right)\left(\frac{1}{27}\right)^x$

15. $\dfrac{3^{-x}}{9} = 27^{1+x}$

16. $2^{|x-1|} = \dfrac{1}{16}$

17. $4^{3x-1} = \dfrac{8}{2^x}$

18. $3^{\sqrt{x+2}} = 81$

19. $\dfrac{1}{2^{\tan x+1}} = \dfrac{8^{\tan^2 x}}{8}$

20. $3 \cdot 9^{\sin x} = 27^{\csc x}$

21. $2^{\sec x} = \dfrac{4^{\cos x}}{2}$

22. $4^{6\cot x+3} = \dfrac{8^{\cot^4 x}}{8^{7-4\cot x}}$

23. $3^{2x+1} + 5 \cdot 3^x - 2 = 0$

24. $2 \cdot 4^{2x+1} - 17 \cdot 4^x + 2 = 0$

25. $2^{2x+3} - 10 \cdot 2^{x+1} + 8 = 0$

26. $5^{2x} - 26 \cdot 5^x + 5^2 = 0$

Exercises 27–30: Without attempting to solve the equation, explain why it cannot have a solution.

27. $3^x = -5^x$

28. $3^x + 1 = 1 - x^2$

29. $3^x + 1 = \dfrac{1}{x^2+1}$

30. $-3^x = \sqrt{x-2}$

Exercises 31–42: Determine the solution set of the inequality.

31. $2^x < 4^{2x}$

32. $\left(\frac{1}{2}\right)^x > \left(\frac{1}{2}\right)^{2x-1}$

33. $\left(\frac{1}{4}\right)^{2x-1} \geq 16$

34. $\left(\frac{3}{2}\right)^{2x^2-x} \leq 1$

35. $25^{x^2} < \dfrac{5^3}{5^{5x}}$

36. $2^{x+1} \geq -1$

37. $\dfrac{3^{x-1}(2x+1)}{x-5} > 0$

38. $\dfrac{9}{3^{1-x}} > 9^{2x}$

39. $2^{|-2x+3|} \leq 4$

40. $x^2 e^x + x e^x > 0$

41. $\dfrac{2^{2x-1}}{2x^2+3x+1} \geq 0$

42. $\dfrac{2e^{-x}}{x^2-7} \leq 0$

384   Chapter 9 Exponential and Logarithmic Functions

Exercises 43–45: Use a graphing calculator to approximate the solution set. [Suggestion: Graph each side of the equation, and find the points of intersection.]

43. $2^{-3x} = x - 1$ 44. $3 \cdot 2^{x-1} = 3x + 1$ 45. $3^{-x} = 5^{-x^2}$

Exercise GC46:   Experimentation has shown that for altitudes (heights above sea level) less than 80 kilometers, the atmospheric pressure in grams per square centimeter at an altitude of h kilometers is given by

$$P(h) = 1035\, e^{-0.12h}$$

(a) Use a calculator to estimate the atmospheric pressure at an altitude of 50 kilometers.

(b) Use a graphing calculator to view the graph of $P(h)$ and estimate the height at which the atmospheric pressure is 500 grams per square centimeter.

(c) What is the atmospheric pressure at sea level in pounds per square inch?
[1g = 0.0353 oz, 1m = 3.28 ft]

Exercises 47–48:   This section began with some experimental data about a culture of E.coli bacteria, which led us to the formula for the population (in millions) of bacteria x hours after 1:00 P.M.:

$$P(x) = 120 \cdot 8^x$$

Being that the world is not the ideal environment of a laboratory, the data in the following tables will not be as "uniform" as that in the bacteria table. Nonetheless, by analyzing it as we did the bacteria data, you will be able to arrive at a formula similar to that for $P(x)$ above.

47. All radioisotopes have a characteristic rate of decay. From a consideration of the following data pertaining to a sample of the radioisotope Tritium, find a formula that will give the number of grams of the substance x years after 1950. Use your formula to predict how much of the substance was present in 1900, and how much will be present in the year 2050.

Year	Grams of Tritium Sample
1950	300
1960	172
1970	99
1980	57

48. From the data in the following table, find a formula for world population, and then use your formula to predict the world population in 1995. In 2010.

Year	1980	1981	1982	1983	1984	1985	1986	1987	1988
Population (in Billions)	4.457	4.534	4.614	4.695	4.775	4.856	4.942	5.029	5.117

9.1 Exponential Functions **385**

Exercise 49: Show that for every base b, the graph of $f(x) = \left(\frac{1}{b}\right)^x$ is the reflection of the graph of $g(x) = b^x$ about the y-axis.

Exercises GC50–GC53: Use a graphing calculator to view the graph of the function, and estimate the local maximum and minimum values and the intercepts.

50. $f(x) = x^2 \cdot 2^{-x}$

51. $k(x) = \left(\frac{3}{4}\right)^{2x^2 - 3x + 1}$

52. $g(x) = (3 - x)10^{0.85x}$

53. $h(x) = -xe^{-x}$

Exercise GC54:

(a) Use a graphing calculator to view the graph of the function $f(x) = 2x \cdot 5^{2x-1}$. Does there appear to be a local maximum or a local minimum? If so, locate it?

(b) Repeat part (a), but now with a viewing rectangle of xmin $= -3$, xmax $= 1$, ymin $= -0.07$, and ymax $= 0.1$.

Exercises 55–56: Solve the system of equations.

55. $\left.\begin{array}{l} 2^{x+y} = 4^x 8^y \\ 3^{x-y} = 9^x \end{array}\right\}$

56. $\left.\begin{array}{l} 2^{3x-4} = 4^{2y} \\ 3x + 2y = 6 \end{array}\right\}$

Exercises 57–71: Indicate True or False. Assume that a and b are any two bases. Justify your answer.

57. If $f(x) = e^{2x}$, and $g(x) = 2e^x$ then $f(x) = g(x)$ for all x.

58. If $c \neq 1$, and $f(x) = b^{cx}$ and $g(x) = c \cdot b^x$, then $f(x) = g(x)$ for all x.

59. The graph of the function $y = b^x + 2$ does not lie below the graph of the function $y = \left(\frac{1}{b}\right)^x + 2$ anywhere.

60. The graph of the function $y = b^x + 2$ does not lie below the graph of the function $y = \left(\frac{1}{b}\right)^x - 2$ anywhere.

61. For any real number $c > 0$, the graph of the function $y = b^x + c$ does not lie below the graph of the function $y = b^x - c$ anywhere.

62. The functions $y = e^x$ and $y = e^x + 10$ have the same domain and range.

63. The functions $y = e^x$ and $y = 10e^x$ have the same domain and range.

64. If b^x is an increasing function, then b must be greater than 1.

65. If b^x is an increasing function, then b^{-x} is a decreasing function.

66. The graphs of the functions ca^x and db^x will always intersect, where c and d are constants.

386   Chapter 9 Exponential and Logarithmic Functions

67. The graphs of the functions ba^x and ab^x will always intersect.

68. If $a \neq b$, then the equation $a^x = b^x$ has no solution.

69. If $a \neq b$, then the equation $a^x = b^x$ has $x = 0$ as its only solution.

70. If $a < b$, then $a^x < b^x$ for all x.

71. If $a^x < b^x$ on some interval I, then $a < b$.

§2. LOGARITHMIC FUNCTIONS

For any positive number b other than one, the exponential function b^x is one-to-one and therefore has an inverse. That inverse function is called the **logarithmic function of base b**, and is denoted by the symbol $\log_b x$ (read: *log base b of x*):

Being the inverse of b^x, the domain of $\log_b x$ is the range of b^x: $(0, \infty)$; and the range of $\log_b x$ is the domain of b^x: $(-\infty, \infty)$.

DEFINITION 9.2

LOGARITHMIC FUNCTION

For any positive number b, other than 1, and any $x > 0$:

$$\log_b x = y \text{ if and only if } b^y = x$$

IN WORDS:	To say that "log base b of x equals y" is the same as saying that "b to the y equals x."

In still other words:

$\log_b x$ is the exponent that b must be raised to in order to get x.

For example:

Logarithmic Form		**Exponential Form**
$\log_3 9 = 2$	since	$3^2 = 9$
$\log_4(\frac{1}{16}) = -2$	since	$4^{-2} = \frac{1}{16}$
$\log_{\frac{1}{5}} 125 = -3$	since	$(\frac{1}{5})^{-3} = 125$

Two logarithmic functions deserve special mention:

the **common logarithm:** $\log_{10} x$, or simply $\log x$

and

$e \approx 2.71828$

the **natural logarithm**: $\log_e x$, or simply $\ln x$

Both the common logarithmic function and the natural logarithmic function are accommodated by calculators. For example, using your calculator you will find that:

$$\log 22 \approx 1.3424 \quad (\text{i.e. } 10^{1.3424} \approx 22)$$

and

$$\ln 22 \approx 3.0910 \quad (\text{i.e. } e^{3.0910} \approx 22)$$

388 **Chapter 9 Exponential and Logarithmic Functions**

CHECK YOUR UNDERSTANDING 9.6

(a) Evaluate, without a calculator.

    (i) $\log_2 16$ (ii) $\log_{\frac{1}{3}} 9$ (iii) $\log_5 1$

(b) Express $\log_3\left(\dfrac{1}{27}\right) = -3$ in exponential form.

(c) Express $\left(\dfrac{1}{2}\right)^4 = \dfrac{1}{16}$ in logarithmic form.

(d) Use a calculator to find a two decimal place approximation for:

    (i) $\dfrac{\ln 37.3}{\ln 1.8 + \ln 13.2}$ (ii) $\dfrac{\log 50}{5 - \log 12}$

Answers: (a-i) 4   (a-ii) -2
(a-iii) 0   (b) $3^{-3} = \dfrac{1}{27}$
(c) $\log_{\frac{1}{2}} \dfrac{1}{16} = 4$
(d-i) ≈ 1.14 (d-ii) ≈ 0.43

The following properties are direct consequences of the fact that logarithmic and exponential functions of the same base are inverses of each other: Applying one, and then the other, brings you back to where you started:

From (i):
$$\log_b 1 = \log_b b^0 = 0$$
$$\log_b b = \log_b b^1 = 1$$

In particular:
$$\ln 1 = \log 1 = 0,$$
$$\ln e = 1 \text{ and } \log 10 = 1$$

From (ii):
$$e^{\ln x} = x \text{ and } 10^{\log x} = x$$

THEOREM 9.3

INVERSE PROPERTIES

(i) For any x: $\log_b b^x = x$

(ii) For any $x > 0$: $b^{\log_b x} = x$

                   Since $\log_b x$ is only defined for $x > 0$

EXAMPLE 9.7   Determine the exact value of:

$$\frac{(2^{\log_2 9})(\log_4 16)\left(\log \dfrac{1}{10}\right)}{(e^{\ln 4})(\ln e^2)(10^{\log 2})}$$

SOLUTION: The simplification reduces to six little problems:

By Theorem 9.3 (i):	By Theorem 9.3 (ii):
$\log_4 16 = \log_4 4^2 = 2$	$2^{\log_2 9} = 9$
$\log\dfrac{1}{10} = \log 10^{-1} = -1$	$e^{\ln 4} = 4$
$\ln e^2 = 2$	$10^{\log 2} = 2$

And so we have:

$$\frac{(2^{\log_2 9})(\log_4 16)\left(\log \dfrac{1}{10}\right)}{(e^{\ln 4})(\ln e^2)(10^{\log 2})} = \frac{(9)(2)(-1)}{(4)(2)(2)} = -\frac{9}{8}$$

9.2 Logarithmic Functions

CHECK YOUR UNDERSTANDING 9.7

Evaluate:

(a) $15\log_3 9 + \log_2 2^{\sqrt{5}}$ 　　(b) $\ln e^{3^{\log_3 5}}$ 　　(c) $\dfrac{\ln e - \ln\sqrt{e}}{3 - \log 100}$

Answers: (a) $30 + \sqrt{5}$　(b) 5　(c) $\dfrac{1}{2}$

GRAPHS OF LOGARITHMIC FUNCTIONS

Since $\log_b x$ is the inverse function of the function b^x, we know that we can obtain the graph of $y = \log_b x$ by reflecting the graph of $y = b^x$ about the line $y = x$ (see Theorem 2.3, page 109). This technique is used in Figure 9.5 to arrive at the graphs of $y = \log_4 x$ and $y = \log_{\frac{1}{4}} x$.

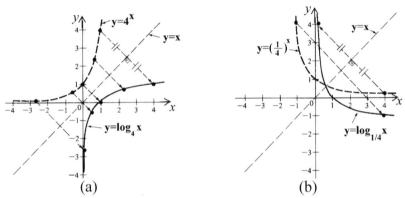

Figure 9.5

Note that the reflected image of the horizontal asymptote (the x-axis) for exponential functions, becomes the **vertical asymptote** (the y-axis) for logarithmic functions.

Like the function $\log_4 x$, all logarithmic functions of base $b > 1$ are increasing functions, and the closer the base is to 1, the more rapidly the function increases [see Figure 9.6(a)]. Like the function $\log_{\frac{1}{4}} x$, the logarithmic functions of base b with $0 < b < 1$ are decreasing functions, which decrease more rapidly the closer the base is to 1 [see Figure 9.6(b)].

Note also that since the point $(0, 1)$ lies on the graph of every exponential function (as $b^0 = 1$), the point $(1, 0)$ lies on the graph of every logarithmic function ($\log_b 1 = 0$).

390 Chapter 9 Exponential and Logarithmic Functions

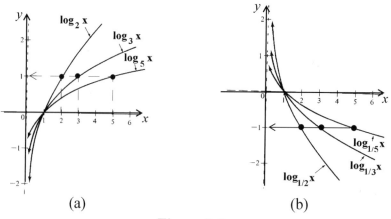

Figure 9.6

It is important to remember that, since the domain of $\log_b x$ is $(0, \infty)$:

For $\log_b(\text{expression})$ to be defined, **expression** must be **positive**.

EXAMPLE 9.8 Find the domain of the function:
$$f(x) = \ln(x^2 - x - 2)$$

SOLUTION: The problem reduces to solving the inequality:
$$x^2 - x - 2 > 0$$
$$(x-2)(x+1) > 0:$$

SIGN $(x-2)(x+1)$ $\rightarrow D_f = (-\infty, -1) \cup (2, \infty)$

Note that some negative numbers are in the domain of: $f(x) = \ln(x^2 - x - 2)$. That's fine: because logarithms are only defined on positive quantities, it is the **expression** in:

$\ln(\text{expression})$

that must be positive.

CHECK YOUR UNDERSTANDING 9.8
Find the domain of the function:
$$f(x) = \frac{\ln(x-3)}{x-5}$$

Answer: $(3, 5) \cup (5, \infty)$

| **LOGARITHMIC EQUATIONS: COMMON BASE** |

Since logarithmic functions are one-to-one, we have:

THEOREM 9.4 For $s, t > 0$:

$$\log_b s = \log_b t \quad \text{if and only if} \quad s = t$$

More generally:

For **expression$_1$** and **expression$_2$** positive:

$$\log_b(\text{expression}_1) = \log_b(\text{expression}_2)$$

if and only if

$$\text{expression}_1 = \text{expression}_2$$

EXAMPLE 9.9 Solve:

$$\ln(x^2 - x) = \ln(-6x + 6)$$

SOLUTION:

$$\ln(x^2 - x) = \ln(-6x + 6)$$

Theorem 9.4:
$$x^2 - x = -6x + 6$$

$$x^2 + 5x - 6 = 0$$

$$(x + 6)(x - 1) = 0$$

$$x = -6 \quad \text{or} \quad x = 1$$

We cannot conclude that since -6 is negative and 1 is positive, that -6 is not a solution and that 1 is a solution of the given equation. Indeed, it's the other way around, for $\log_b(\text{expression})$ is defined when **expression** > 0, and:

For $x = -6$ both $x^2 - x$ and $-6x + 6$ are **positive**

while for $x = 1$, $x^2 - x = 0$ [Also: $-6(1) + 6 = 0$]

Conclusion: -6 is the only solution of $\ln(x^2 - x) = \ln(-6x + 6)$.

Please remember:

When solving logarithmic equations,
CHECK YOUR ANSWER
by verifying that all logarithms in the original equation are defined.

EXAMPLE 9.10 Solve for $0 \leq x < 2\pi$:

$$\ln(2\sin x) = \ln(\csc x + 1)$$

SOLUTION:

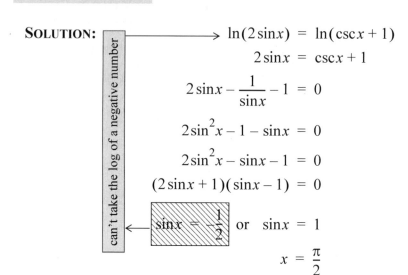

$$\ln(2\sin x) = \ln(\csc x + 1)$$
$$2\sin x = \csc x + 1$$
$$2\sin x - \frac{1}{\sin x} - 1 = 0$$
$$2\sin^2 x - 1 - \sin x = 0$$
$$2\sin^2 x - \sin x - 1 = 0$$
$$(2\sin x + 1)(\sin x - 1) = 0$$
$$\sin x = -\frac{1}{2} \quad \text{or} \quad \sin x = 1$$
$$x = \frac{\pi}{2}$$

Since both $2\sin x$ and $\csc x + 1$ are positive for $x = \frac{\pi}{2}$, we conclude that $\frac{\pi}{2}$ is a solution of the given inequality.

CHECK YOUR UNDERSTANDING 9.9

(a) Solve: $\ln(x^2 - 15) = \ln(-2x)$

(b) Solve for $0 \leq x < 2\pi$: $\log(\cos 2x + 2\cos x) = \log 3$

Answer: (a) -5 (b) 0

LOGARITHMIC INEQUALITIES: COMMON BASE

Since the function $\log_b x$ is an increasing function if $b > 1$ and is a decreasing function if $0 < b < 1$ (see Figure 9.6), we have:

THEOREM 9.5 (a) If $b > 1$:

$$\log_b s < \log_b t \text{ if and only if } s < t$$

(b) If $0 < b < 1$:

$$\log_b s < \log_b t \text{ if and only if } s > t$$

In general, if $b > 1$:
$\log_b \text{expression}_1 < \log_b \text{expression}_2$
if and only if
$\text{expression}_1 < \text{expression}_2$

If $0 < b < 1$:
$\log_b \text{expression}_1 < \log_b \text{expression}_2$
if and only if
$\text{expression}_1 > \text{expression}_2$

EXAMPLE 9.11 Solve: $\ln(2x - 7) < \ln(5x + 2)$

SOLUTION:

$$\ln(2x - 7) < \ln(5x + 2)$$

Theorem 9.5(a): $\quad 2x - 7 < 5x + 2$

$$-3x < 9$$

Dividing by a negative quantity: $\quad x > -3 \quad$ (*)

At this point, we know that the solution set of the given inequality is contained in the interval $(-3, \infty)$. Keeping in mind that the two expressions $2x - 7$ and $5x + 2$ must both be positive, we turn to the two inequalities:

$2x - 7 > 0$	$5x + 2 > 0$
$2x > 7$	$5x > -2$
$x > \dfrac{7}{2}$ (**)	$x > -\dfrac{2}{5}$ (***)

Conclusion: The solution set of the given inequality consists of all numbers satisfying (*), (**), and (***); namely: $\left(\dfrac{7}{2}, \infty\right)$.

CHECK YOUR UNDERSTANDING 9.10

Solve:

$$\log_{\frac{1}{2}}(3x - 1) > \log_{\frac{1}{2}}(2x + 5)$$

Answer: $\left(\dfrac{1}{3}, 6\right)$

PROPERTIES OF LOGARITHMS

A number of useful results which follow directly from properties of exponents, are tabulated in the following theorem.

The restrictions on variables that you see in Theorem 9.6 stem from the fact that the domain of the logarithmic function is the set of **positive** numbers.

THEOREM 9.6 For any base b, and any r, s, and t with $s > 0, t > 0$:

(i) $\quad \log_b(st) = \log_b s + \log_b t$

(ii) $\quad \log_b\left(\dfrac{s}{t}\right) = \log_b s - \log_b t$

(iii) $\quad \log_b s^r = r \log_b s$

PROOF: To establish these properties, let:

$$m = \log_b s \quad \text{and} \quad n = \log_b t.$$

Changing to exponential form: $s = b^m$ and $t = b^n$.

394 Chapter 9 Exponential and Logarithmic Functions

(i) Since $st = b^m b^n = b^{m+n}$:

$$\log_b st = \log_b b^m b^n = \log_b b^{m+n} \overset{\text{Theorem 9.6(i)}}{=} m + n = \log_b s + \log_b t$$

(ii) Since $\dfrac{s}{t} = \dfrac{b^m}{b^n} = b^{m-n}$:

$$\log_b \frac{s}{t} = \log_b \frac{b^m}{b^n} = \log_b b^{m-n} \overset{\text{Theorem 9.6(ii)}}{=} m - n = \log_b s - \log_b t$$

(iii) Since $s^r = (b^m)^r = b^{mr}$:

$$\log_b s^r = \log_b (b^m)^r = \log_b b^{mr} = \log_b b^{rm} \overset{\text{Theorem 9.6(iii)}}{=} rm = r \log_b s$$

Simply stated, these properties (i) - (iii) say that:

> The log of a product is the sum of the logs.
> The log of a quotient is the difference of the logs.
> The log of a power is the power times the log.

We must warn you that there is no result concerning "the log of a sum" or "the log of a difference." In particular:

$\log_b(s + t)$ does NOT equal:

$\log_b s + \log_b t$

EXAMPLE 9.12

Express:

$$\frac{1}{2} \ln 25 + \ln 7 - 2 \ln 2$$

as a single logarithm.

SOLUTION:

$r \ln s = \ln s^r$:
$$\frac{1}{2} \ln 25 + \ln 7 - 2 \ln 2 = \ln 25^{\frac{1}{2}} + \ln 7 - \ln 2^2$$

$$= \ln 5 + \ln 7 - \ln 4$$

$\ln s + \ln t = \ln st$:
$$= \ln(5 \cdot 7) - \ln 4 = \ln 35 - \ln 4$$

$\ln s - \ln t = \ln \dfrac{s}{t}$:
$$= \ln \frac{35}{4}$$

EXAMPLE 9.13

Assuming that both A and B are positive, expand:

$$\log_3 \sqrt{\frac{A^4}{2B}}$$

into a sum and difference of logarithms.

SOLUTION:

$$\log_3 \sqrt{\frac{A^4}{2B}} = \log_3 \left(\frac{A^4}{2B}\right)^{1/2}$$

$\log_b s^r = r\log_b s$:
$$= \frac{1}{2}\log_3\left(\frac{A^4}{2B}\right)$$

$\log_b\left(\frac{s}{t}\right) = \log_b s - \log_b t$:
$$= \frac{1}{2}[\log_3 A^4 - \log_3(2B)]$$

$\log_b s^r = r\log_b s$:
$$= \frac{1}{2}[4\log_3 A - \log_3(2B)]$$

$\log_b(st) = \log_b s + \log_b t$:
$$= \frac{1}{2}[4\log_3 A - (\log_3 2 + \log_3 B)]$$

$$= 2\log_3 A - \frac{1}{2}\log_3 2 - \frac{1}{2}\log_3 B$$

CHECK YOUR UNDERSTANDING 9.11

(a) Express $2\log(x+1) - 3\log x + \log 12$ as a single logarithm.

(b) Given that $\ln 3 = A$ and $\ln 5 = B$, express $\ln 45$ in terms of A and B.
[Suggestion: Express 45 as a product involving 3 and 5.]

(c) Express $\dfrac{\ln \sin x - \ln \cos x}{\ln(e^{\sin^2 x} e^{\cos^2 x})}$ as the logarithm of a single trigonometric function.

Answers: (a) $\log\dfrac{12(x+1)^2}{x^3}$

(b) $2A+B$ (c) $\ln\tan x$

CHANGE OF BASE FORMULA

The following theorem, enables you to express any logarithmic function in terms of any other logarithmic function:

THEOREM 9.7

CHANGE OF BASE FORMULA

For any bases a and b, and any $x > 0$:

$$\log_b x = \frac{\log_a x}{\log_a b}$$

PROOF:

Begin with: $\qquad b^{\log_b x} = x$

Apply $\log_a x$ to both sides: $\qquad \log_a b^{\log_b x} = \log_a x$

Apply $\log_a x^r = r\log_a x$ (with $x = b$ and $r = \log_b x$): $\quad (\log_b x)(\log_a b) = \log_a x$

Divide both sides by $\log_a b$: $\qquad \log_b x = \dfrac{\log_a x}{\log_a b}$

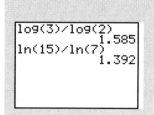

You can now use your calculator to approximate values of $\log_b x$ for any base b. All you have to do is to convert to either common logarithms, $\log x$, or natural logarithms, $\ln x$. For example:

$$\log_2 3 = \frac{\log 3}{\log 2} \approx 1.585 \quad \text{and} \quad \log_7 15 = \frac{\ln 15}{\ln 7} \approx 1.392$$

In science courses, one often has to convert from natural logs to common logs, and vice versa. Using The Change of Base formula, we have:

$$\log x = \frac{\ln x}{\ln 10} \approx 0.4343 \ln x \quad \text{and} \quad \ln x = \frac{\log x}{\log e} \approx 2.3026 \log x$$

CHECK YOUR UNDERSTANDING 9.12

Use a calculator to approximate, to two decimal places, the value of:

(a) $\log_5 32$

(b) $\log_{\frac{1}{2}} 17$

Answers: (a) ≈ 2.15
(b) ≈ -4.09

9.2 Logarithmic Functions   397

E X E R C I S E S

Exercises 1–6: Evaluate each of the following.

1. $\log_2 \frac{1}{8}$ 2. $\log_3 1$ 3. $\log_{\frac{1}{3}} 3$ 4. $\log_4 64$ 5. $\log_{16} 4$ 6. $\log_{\frac{1}{2}} 32$

Exercises 7–10: Express in exponential form.

7. $\log_{64} 4 = \frac{1}{3}$ 8. $\log_4 16 = 2$ 9. $\log_{\frac{1}{2}} 1 = 0$ 10. $\log_7 7 = 1$

Exercises 11–14: Express in logarithmic form.

11. $5^3 = 125$ 12. $12^0 = 1$ 13. $\left(\frac{1}{2}\right)^2 = \frac{1}{4}$ 14. $\left(\frac{1}{4}\right)^{-1} = 4$

Exercises 15–18: Use a calculator to approximate the value to two decimal places.

15. $\dfrac{\ln 4.8 + \log 4.8}{\log e}$ 16. $\dfrac{\ln 3.8 + \ln 1.9}{2\ln 4.12}$ 17. $\dfrac{2\ln 4 - \log 2}{\log 8}$ 18. $\dfrac{\log 4 - 3\log 5}{2 - \ln 9}$

Exercises 19–27: Evaluate.

19. $\log_5 5^4$

20. $\log_2 2^3 - (\log_2 2)^3$

21. $\log_2 2^{\log_{\frac{1}{2}} 4}$

22. $3\log_5 \frac{1}{25}$

23. $\log_2(4^2 \cdot 8^3)$

24. $\log_b b^2$

25. $\log_b b^{\log_a a^2}$

26. $\dfrac{\ln e^5 + 2\log_3 3^{-2}}{\log_5 5^4 + \frac{1}{2}\log \frac{1}{100}}$

27. $\dfrac{e^{\ln 6} - \log 1000}{3^{\log_3 5} - 3\log 1}$

Exercises 28–35: Determine the domain of the function.

28. $f(x) = 6 - 3\log_2(x + 5)$

29. $f(x) = \log_{\frac{2}{3}}\left(\dfrac{4 - x}{x^2 - 9}\right)$

30. $f(x) = \log_7(x^2 + x - 6)$

31. $f(x) = \ln(-x)$

32. $f(x) = \log_{\frac{1}{2}}\left(\dfrac{x - 1}{3x + 5}\right)$

33. $f(x) = \log(6x^3 + 23x^2 + 7x)$

34. $f(x) = \dfrac{10^{x-3}\log(x^2 - 1)}{x + 2}$

35. $f(x) = \dfrac{1}{\log_3 x} + \dfrac{1}{2 - x}$

Exercises 36–45: Find the domain and sketch the graph of the function. Indicate the y-intercept (if any) and the vertical asymptote.
[Hint: Stretch/shrink, reflect, and/or shift the basic graph of $\log_b x$.]

36. $f(x) = 3\log_3 x$

37. $f(x) = \frac{1}{2}\log_{\frac{1}{4}} x$

38. $f(x) = -\log_{\frac{1}{3}} x$

39. $f(x) = \ln x - 1$

40. $f(x) = \log(x + 3)$

41. $f(x) = \ln(x - 1)$

42. $f(x) = -\ln x$

43. $f(x) = \log_{\frac{2}{3}}(x - 3) + 2$

44. $f(x) = \log_2 x + 1$

45. $f(x) = 4 - 2\log(x + 1)$

398   Chapter 9 Exponential and Logarithmic Functions

Exercises 46–57: Solve the equation.

46. $\ln(3x^2 + x) = \ln(4 - 3x)$

47. $\log_3(6x + 7) = \log_3(2 - x^2)$

48. $\ln(x^3 - x) = \ln 3x$

49. $\log(5x^2 - 4) = \log(-19x)$

50. $\log_{\frac{1}{2}}(x^2 - 3x) = \log_{\frac{1}{2}}(-3x + 4)$

51. $\log 3^{2x} = \log \frac{1}{27}$

52. $\log_2(2x - 1) = \log_2(3x - 5)$

53. $\log_3(x - 5) = \log_3(-x^2 + 1)$

54. $\ln(2\cos x) = \ln(\sec x - 1)$
    in $[0, 2\pi)$

55. $\log(\sin x \cos x) = \log \frac{1}{2}$
    in $[0, 2\pi)$

56. $\log_5(\cos 2x + 5) = \log_5(6 + \sin x)$
    in $[0, 2\pi)$

57. $\log_2(6 \sin x + 17) = \log_2(6 - 4\csc x)$
    in $[0, 2\pi)$

Exercises 58–61: Solve the inequality.

58. $\log(4x + 5) < \log(x + 8)$

59. $\ln(x - 2) > \ln(3x - 10)$

60. $\log_{\frac{2}{5}}(3x + 2) > \log_{\frac{2}{5}}(x + 4)$

61. $\log_{\frac{1}{3}}(2x - 1) < \log_{\frac{1}{3}}(x - 1)$

Exercises 62–65: Express as a single logarithm. Assume all expressions are defined.

62. $2\log x + \log y$

63. $3\ln x^2 - 2\ln x + 3\ln 10$

64. $-\log_2 3 + 4\log_2(x + 1)$

65. $\log_3(2x - 5) + \frac{1}{2}\log_3 14 - 2\log_3(x + 4)^2$

Exercises 66–70: Expand the single logarithm to a sum and/or difference of logarithms. Assume each variable represents a positive number.

66. $\log_3 \dfrac{27x}{y^2}$

67. $\log_2 \left(\dfrac{xy}{z^2}\right)^4$

68. $\ln \dfrac{4x^2 z}{(ey)^3}$

69. $\log \left[\sqrt{\dfrac{x}{y}}\,(z^2)\right]$

70. $\log_3 \left(\dfrac{3^{2x} z^3}{27}\right)^{\frac{1}{3}}$

Exercises 71–74: Given $\log_b 3 = A$, $\log_b 4 = B$ and $\log_b 5 = C$, express the given logarithm in terms of A, B, and C.

71. $\log_b 36$

72. $\log_b \sqrt{15}$

73. $\log_b \dfrac{45}{16}$

74. $\log_b \dfrac{80}{b^2}$

Exercises 75–76: Evaluate.

75. $\dfrac{\ln e^3 - e^{-2\ln 4}}{\ln \sqrt{e}}$

76. $\dfrac{\ln \sqrt{e} + e^{3\ln 2}}{e^{2\ln 3} - e^{-\ln 2}}$

Exercises 77–78: Simplify fully.

77. $e^{\ln \tan x + \ln \cot x}$

78. $\ln \left[\dfrac{e^{\cos^4 x}}{e^{\sin^4 x}}\right]$

9.2 Logarithmic Functions   **399**

Exercises 79–81: Express in terms of the logarithm of a single trigonometric function.

79. $\left[\ln\left(\dfrac{e^{\sec^2 x}}{e^{\tan^2 x}}\right)\right]\left[\ln\sin x + \ln\cos x - \ln\frac{1}{2}\right]$

80. $(2\ln\cos x - \ln\sin^2 x)\cdot 2(\ln\cos x - \ln\sin x)$ 81. $\dfrac{\ln[e^{-\cot^2 x}\,e^{\csc^2 x}]}{\ln\sec x - \ln\csc x}$

Exercises 82–85: Express

82. $\log_3 17$ in terms of common logarithms.

83. $\log_2 \frac{3}{5}$ in terms of natural logarithms.

84. $\ln 15$ in terms of common logarithms.

85. $\log_{\frac{1}{2}} 3 + 4\log_{\frac{1}{2}} x^4$ as a single natural logarithmic function.

Exercises 86–93: Use the Change of Base formula to rewrite the expression, and then evaluate it. $\left[\text{For example, } \log_4 8 = \dfrac{\log_2 8}{\log_2 4} = \dfrac{3}{2}.\right]$

86. $\log_{16} 64$

87. $\log_{27} 9$

88. $\log_8\left(\frac{1}{16}\right)^{-5}$

89. $4^{\log_2 5}$

90. $\log_{b^2} b$

91. $\log_{\frac{1}{2}} x + \log_2 x$

92. $2^{-\log_4 9}$

93. $\log_3 5 \cdot \log_5 3$

Exercises 94–97: Use a calculator to approximate the value to two decimal places.

94. $2 - 3\log_3 20$

95. $\dfrac{\log_4 6}{\log_4 9}$

96. $(\log 5 - \ln 3)(1 + \log_{\frac{1}{2}} 7)$

97. $\dfrac{\log_2 6}{2\log_3 5}$

Exercises 98–101: Determine the base of the logarithmic function $f(x) = \log_b x$ whose graph contains the given point.
(Suggestion: Obtain an equation in logarithmic form, and then change to exponential form.)

98. $\left(8, \frac{1}{2}\right)$

99. $\left(\frac{1}{16}, 4\right)$

100. $(16, 4)$

101. $(27, -3)$

Exercise 102: Show that for any bases a and b,

$$\log_b a = \frac{1}{\log_a b}$$

Exercises 103–119: Indicate True or False. Assume that all expressions are defined. Justify your answer.

103. For every negative number x, $\ln(-x) = -\ln x$.

104. For some number x, $\ln(-x) = -\ln x$.

105. For every base b, the domain of $f(x) = b^{\ln x}$ is the set of all real numbers.

106. For every base b, the domain of $f(x) = \ln b^x$ is the set of all real numbers.

400   Chapter 9 Exponential and Logarithmic Functions

107. The function $f(x) = \ln|x|$ is an even function.

108. The functions $f(x) = \ln x$ and $g(x) = \ln x + 10$ have the same domain and range.

109. The functions $f(x) = \ln x$ and $g(x) = \ln(x + 10)$ have the same domain and range.

110. $e^{\log_5 x} = \dfrac{x}{\ln 5}$

111. $10^{\log x - \log y} = x - y$

112. $\log_b xy = (\log_b x)(\log_b y)$

113. $\log_b \dfrac{x}{y} = \dfrac{\log_b x}{\log_b y}$

114. $\log_b(x^2 y^3) = 2\log_b x + 3\log_b y$

115. $\log_5 x = -\log_{\frac{1}{5}} x$

116. $\ln(x + y) = \ln x + \ln y$

117. $\log_b xy^3 = 3\log_b x^{\frac{1}{3}} y$

118. $\ln(xy) = (\ln x)(\ln y)$

119. $\ln \dfrac{1}{x} = -\ln x$

9.3 Additional Exponential and Logarithmic Equations and Inequalitites   401

§3. ADDITIONAL EXPONENTIAL AND LOGARITHMIC EQUATIONS AND INEQUALITIES

The logarithmic properties of the previous section enable us to solve a wide range of exponential and logarithmic equations and inequalities.

EXPONENTIAL EQUATIONS AND INEQUALITIES

Given an exponential equation or inequality in which both sides cannot readily be expressed as powers of the same base, applying a logarithmic function may lead to solutions. Typically, one uses the natural logarithm in the process.

EXAMPLE 9.14 Solve the exponential equation:
$$2^{2x-1} = 3^{3x-7}$$

SOLUTION:

$$2^{2x-1} = 3^{3x-7}$$

To bring the variable exponents "down," take the natural log of both sides:
$$\ln(2^{2x-1}) = \ln(3^{3x-7})$$

$\ln s^r = r \ln s$:
$$(2x-1)\ln 2 = (3x-7)\ln 3$$
$$2x\ln 2 - \ln 2 = 3x\ln 3 - 7\ln 3$$
$$2x\ln 2 - 3x\ln 3 = \ln 2 - 7\ln 3$$
$$(2\ln 2 - 3\ln 3)x = \ln 2 - 7\ln 3$$
$$x = \frac{\ln 2 - 7\ln 3}{2\ln 2 - 3\ln 3}$$

Or, more compactly
$$x = \frac{\ln 2 - 7\ln 3}{2\ln 2 - 3\ln 3} = \frac{\ln \frac{2}{3^7}}{\ln \frac{4}{27}}$$

EXAMPLE 9.15 Solve the exponential equation:
$$3 \cdot 5^{x+1} = 2^{4x-1}$$

SOLUTION:

$$3 \cdot 5^{x+1} = 2^{4x-1}$$
$$\ln(3 \cdot 5^{x+1}) = \ln(2^{4x-1})$$
$$\ln 3 + \ln 5^{x+1} = (4x-1)\ln 2$$
$$\ln 3 + (x+1)\ln 5 = 4x\ln 2 - \ln 2$$
$$\ln 3 + x\ln 5 + \ln 5 = 4x\ln 2 - \ln 2$$
$$(\ln 5 - 4\ln 2)x = -\ln 2 - \ln 3 - \ln 5$$
$$x = \frac{-\ln 2 - \ln 3 - \ln 5}{\ln 5 - 4\ln 2} = \frac{\ln 2 + \ln 3 + \ln 5}{4\ln 2 - \ln 5}$$

Or, more compactly:
$$x = \frac{\ln 2 + \ln 3 + \ln 5}{4\ln 2 - \ln 5} = \frac{\ln 30}{\ln \frac{16}{5}}$$

Answers: (a) $\dfrac{\ln 450}{\ln \frac{8}{15}}$ (b) $\dfrac{\ln \frac{16}{21}}{\ln \frac{3}{2}}$

CHECK YOUR UNDERSTANDING 9.13

Solve:

(a) $2^{3x-1} = 15^{x+2}$ (b) $7 \cdot 3^{x+1} = 4 \cdot 2^{x+2}$

402 Chapter 9 Exponential and Logarithmic Functions

EXAMPLE 9.16 Find all solutions of the equation:
$$\cos(e^{x-3}) = 1$$

SOLUTION:

$$\cos(e^{x-3}) = 1$$

let $y = e^{x-3}$; (note: $y > 0$)

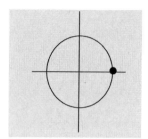

$$\cos y = 1$$
$$y = 2n\pi \text{ for any positive integer } n \text{ (see margin)}$$
$$e^{x-3} = 2n\pi$$
$$x - 3 = \ln 2n\pi$$
$$x = 3 + \ln 2n\pi \text{ (for any positive integer } n)$$

Answer: $\frac{1}{2}\left[-1 + \ln\left(\frac{\pi}{2} + 2n\pi\right)\right]$ for any nonnegative integer n.

CHECK YOUR UNDERSTANDING 9.14

Solve:
$$\sin(e^{2x+1}) = 1$$

EXAMPLE 9.17 Solve the exponential inequality:
$$2^{3x+1} > 5$$

SOLUTION:

Since both sides are positive, and since the logarithmic function is an increasing function, applying it to both sides of the inequality retains the direction of the inequality sign:

$$2^{3x+1} > 5$$
$$\ln(2^{3x+1}) > \ln 5$$

$\ln s^r = r \ln x$: $(3x+1)\ln 2 > \ln 5$
$$3x \ln 2 + \ln 2 > \ln 5$$
$$3x \ln 2 > \ln 5 - \ln 2$$
$$x > \frac{\ln 5 - \ln 2}{3 \ln 2}$$

Solution set: $\left(\frac{\ln 5 - \ln 2}{3 \ln 2}, \infty\right) = \left(\frac{\ln \frac{5}{2}}{\ln 8}, \infty\right)$

Answers: (a) $\left(-\infty, \frac{\ln 90}{\ln \frac{5}{3}}\right]$

(b) $\left(-\infty, \frac{\ln \frac{3}{5}}{\ln 15}\right)$

CHECK YOUR UNDERSTANDING 9.15

Solve:
(a) $2 \cdot 3^{x+2} \geq 5^{x-1}$ (b) $3^{x-1} < \frac{1}{5^{x+1}}$

LOGARITHMIC EQUATIONS

Unlike exponential functions which are defined everywhere, logarithmic functions are only defined for positive numbers. Consequently:

When solving logarithmic equations,
CHECK YOUR ANSWER
by verifying that all logarithms in the original equation are defined.

9.3 Additional Exponential and Logarithmic Equations and Inequalitites   **403**

EXAMPLE 9.18   Solve:

$$\log_3(x-1) + \log_3(2x+3) = 1$$

SOLUTION:

$$\log_3(x-1) + \log_3(2x+3) = 1$$

$\log_b s + \log_b t = \log_b st$:

$$\log_3[(x-1)(2x+3)] = 1$$

Apply the inverse function of $\log_3 x$,
the exponential function of base 3, to both sides:

$$3^{\log_3[(x-1)(2x+3)]} = 3^1$$

$b^{\log_b x} = x$:

$$(x-1)(2x+3) = 3$$

$$2x^2 + x - 3 = 3$$

$$2x^2 + x - 6 = 0$$

$$(2x-3)(x+2) = 0$$

$$x = \frac{3}{2} \text{ or } x = -2$$

We have two solution candidates, and now challenge each of them, to make sure that each logarithm in the original equation is defined:

Since $x - 1$ is negative when $x = -2$, -2 is **not** a solution of the equation (you can only take logs of positive numbers). Since $x - 1$ and $2x + 3$ are positive for $x = \frac{3}{2}$, $\frac{3}{2}$ **is** a solution of the equation.

EXAMPLE 9.19   Solve:

$$2\ln 2x - \ln(x+1) = \ln x$$

SOLUTION:

$$2\ln 2x - \ln(x+1) = \ln x$$

$r \ln s = \ln s^r$:

$$\ln(2x)^2 - \ln(x+1) = \ln x$$

$\ln s - \ln t = \ln\left(\frac{s}{t}\right)$:

$$\ln \frac{4x^2}{x+1} = \ln x$$

one-to-one property of logarithmic functions:

$$\frac{4x^2}{x+1} = x$$

$$4x^2 = x^2 + x$$

$$3x^2 - x = 0$$

$$x(3x-1) = 0$$

$$x = 0 \text{ or } x = \frac{1}{3}$$

Now check that each original logarithm is defined:
Since you can only take logs of positive numbers, and since $2x$ is zero when $x = 0$, 0 is **not** a solution of the equation. Since $2x$, $x + 1$, and x are all positive for $x = \frac{1}{3}$, $\frac{1}{3}$ **is** a solution of the equation.

CHECK YOUR UNDERSTANDING 9.16

Solve:

(a) $\log_{\frac{1}{3}}\left(\frac{1}{x}\right) = 3$ (b) $\log_8(x-6) = 2 - \log_8(x+6)$

Answers: (a) 27   (b) 10

EXAMPLE 9.20 Find all $-\frac{\pi}{2} < x < \frac{\pi}{2}$ which satisfy the equation:

$$2\ln(2\tan x) - \ln(\tan x + 1) = \ln(\tan x)$$

SOLUTION:

$$2\ln(2\tan x) - \ln(\tan x + 1) = \ln(\tan x)$$

let $y = \tan x$: $\rightarrow 2\ln 2y - \ln(y+1) = \ln y$

from Example 9.19: $y = \frac{1}{3}$

$\rightarrow \tan x = \frac{1}{3}$

$x = \tan^{-1}\left(\frac{1}{3}\right)$

EXAMPLE 9.21 Find all solutions of the equation:

$$2\sin(\ln x) = 1$$

SOLUTION:

$$2\sin(\ln x) = 1$$

$$\sin(\ln x) = \frac{1}{2}$$

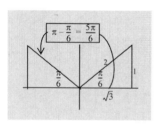

let $y = \ln x$: $\rightarrow \sin y = \frac{1}{2}$

$y = \frac{\pi}{6} + 2n\pi$ and $y = \frac{5\pi}{6} + 2n\pi$

For any integer n (see margin)

$\rightarrow \ln x = \frac{\pi}{6} + 2n\pi$ and $\ln x = \frac{5\pi}{6} + 2n\pi$

$x = e^{\frac{\pi}{6} + 2n\pi}$ and $x = e^{\frac{5\pi}{6} + 2n\pi}$, n any integer

CHECK YOUR UNDERSTANDING 9.17

(a) Solve, for $-\frac{\pi}{2} < x < \frac{\pi}{2}$:
$$\ln(\tan x) + \ln(1 + 2\tan x) = \ln 3$$

(b) Find all solutions of:
$$\cos(\ln x) = -\frac{1}{2}$$

Answers: (a) $\frac{\pi}{4}$

(b) $e^{\frac{2\pi}{3} + 2n\pi}, e^{\frac{4\pi}{3} + 2n\pi}$, for any integer n.

LOGARITHMIC INEQUALITIES

Solving logarithmic inequalities is similar to solving square root inequalities, one should first determine where the functions are defined. Consider the following examples.

EXAMPLE 9.22 Solve:
$$\log_2(x^2 - 1) \le 3$$

SOLUTION: Since logarithms are only defined on positive quantities, we **must have**:
$$x^2 - 1 > 0$$
$$(x+1)(x-1) > 0$$
$$x < -1 \quad \text{or} \quad x > 1$$

At this point, we know that any solution to the given inequality must be contained in $(-\infty, -1) \cup (1, \infty)$:

Since the function 2^x is an increasing function [Figure 9.4(a), page 378], applying that function to both sides of the given inequality will **not** change the sense of that inequality:
$$\log_2(x^2 - 1) \le 3$$
$$2^{\log_2(x^2-1)} \le 2^3$$

Theorem 9.3(ii), page 388: $\quad x^2 - 1 \le 8$
$$x^2 - 9 \le 0$$
$$(x+3)(x-3) \le 0$$
$$-3 \le x \le 3$$

Taking into account the fact that solutions must be contained in the interval $(-\infty, -1) \cup (1, \infty)$, we conclude that $[-3, -1) \cup (1, 3]$ is the solution set of the given inequality:

406   Chapter 9 Exponential and Logarithmic Functions

EXAMPLE 9.23 Solve:

$$\log_{\frac{1}{2}}(-3x+1) \geq 2$$

SOLUTION: Since logarithms are only defined on positive quantities, we **must have**:

$$-3x + 1 > 0$$
$$-3x > -1$$
$$x < \frac{1}{3}$$

Since the function $\left(\frac{1}{2}\right)^x$ is a decreasing function (Figure 9.4(b), page 378), applying that function to both sides of the given inequality will **reverse** the sense of that inequality:

$$\log_{\frac{1}{2}}(-3x+1) \geq 2$$

$$\left(\frac{1}{2}\right)^{\log_{\frac{1}{2}}(-3x+1)} \leq \left(\frac{1}{2}\right)^2$$

Theorem 9.3(ii), page 386:

$$-3x + 1 \leq \frac{1}{4}$$

$$-3x \leq -\frac{3}{4}$$

$$x \geq \frac{1}{4}$$

Taking into account the fact that x must be less than $\frac{1}{3}$, we conclude that $\left[\frac{1}{4}, \frac{1}{3}\right)$ is the solution set of the given inequality.

CHECK YOUR UNDERSTANDING 9.18

Answers: (a) $\left(\frac{19}{54}, \frac{1}{2}\right)$

(b) $(2, 4]$

Solve:

(a) $\log_{\frac{2}{3}}(1 - 2x) > 3$ 　　(b) $\log_3(3x - 6) \leq \log_3(x + 2)$

APPROXIMATING SOLUTIONS

You have seen how graphing calculators can be used to find approximate solutions of equations and inequalities that can not be solved by hand, and realize that determining the proper window size is an important part of the solution process. Knowing the general nature of the functions involved can be a great help.

EXAMPLE 9.24 Use a graphing calculator to approximate the solutions of:

$$\ln(x + 500) = -(x + 475)^2 + 10$$

SOLUTION: You know that the graph of the function on the left side of the equation is simply the natural logarithmic function shifted 500 units to the left, and you can sketch in that function by hand (see Figure 9.7). You also know that the graph of the function on the right side of the equal sign is that of the parabola $-x^2$, but shifted 475 units to the left and 10 units up, and you can sketch that graph as well (see Figure 9.7).

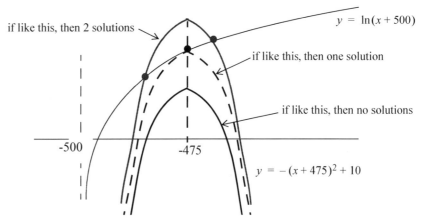

Figure 9.7

You may not know if there are solutions to the equation, but you do know where to go looking for them; and, indeed, we find that there are two solutions: $x \approx -477.63$ and $x \approx -472.42$.

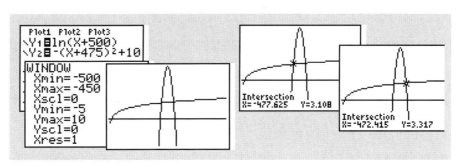

CHECK YOUR UNDERSTANDING 9.19

Approximate the solution set of:
$$-\log(x - 450) < -(x - 475)^2 - 5$$

Answer:
$(450, 473.09) \cup (476.89, \infty)$

408   **Chapter 9 Exponential and Logarithmic Functions**

EXERCISES

Exercises 1–15: Solve the exponential equation, expressing your answer in terms of natural logarithms.

1. $3^{3x-5} = 12$

2. $3^{2x} = 4$

3. $\left(\frac{1}{2}\right)^x = 5^{x-2}$

4. $\left(\frac{1}{2}\right)^{x^2} = 3$

5. $4^{2x-1} = 3^{5x+3}$

6. $5^x = 3^{\sqrt{2x}}$

7. $4^{x+2} = 2 \cdot 3^{x-1}$

8. $7 \cdot 3^{x-4} = 6 \cdot 4^{x+1}$

9. $\dfrac{3^{x+2}}{5^x} = 4^{x-1}$

10. $15 \cdot 4^{2x} + 4^{x+1} - 3 = 0$

11. $6 \cdot 2^{2x+2} - 13 \cdot 2^{x+1} + 5 = 0$

12. $\sin\left(e^{2x+4}\right) = -1$

13. $\sec\left(e^{4x+1}\right) = -2$

14. $\cot\left(e^{-5x}\right) = -1$

15. $\cos\left(e^{2-3x}\right) = -\dfrac{\sqrt{3}}{2}$

Exercises 16–17: Find the x-intercepts of the graph of $f(x)$.

16. $f(x) = -5 \cdot 2^{x+3} + 6$

17. $f(x) = 3e^{2x} - 1$

Exercises 18–19: Sketch the graph of $f(x)$, determining the x- and y-intercepts and the asymptote.

18. $f(x) = -3 \cdot 10^{x-2} + 1$

19. $f(x) = 2 \cdot 3^{x+2} - 6$

Exercises 20–25: Solve the inequality, expressing your answer in terms of natural logs.

20. $2^{x-5} \geq 3$

21. $2^{3x-5} < 5$

22. $\left(\frac{1}{2}\right)^{4x-1} > 6$

23. $\left(\frac{1}{2}\right)^x < 5^{x-2}$

24. $2^{x+1} < 3 \cdot 4^{2x-1}$

25. $5 \cdot 3^{x-4} \geq 3 \cdot 4^{x-1}$

Exercises 26–49: Solve the logarithmic equation.

26. $\log_x 4 = 2$

27. $\ln x = 8$

28. $\log 0.001 = x$

29. $\log_x \dfrac{1}{27} = -\dfrac{3}{2}$

30. $\log_{\sqrt{2}} 8 = x$

31. $\log_3(x - 1) = 3$

32. $2 \log_4(3x + 1) = 1$

33. $\log_2(\log_2 x) = 1$

34. $\log_2 x - 4 = \log_2 5$

35. $2 \log_3 x = 3 - \log_3 2$

36. $\log(3x + 1) = 1 - \log x$

37. $\log_2 x + \log_2(x - 6) = 4$

38. $\ln(2x + 5) - \ln(2x - 1) = -\ln 3$

39. $\log_3(x + 3) - \log_3(x - 1) = 2$

40. $\log_4 x + \log_4(2x - 3) = \dfrac{1}{2}$

41. $\log(x - 2) + \log(x + 13) = 2$

42. $\log_{\frac{1}{3}}(x^2 - 16) - \log_{\frac{1}{3}}(x - 4) = -2$

43. $\log_4(4x + 3) + \log_4(2 - 4x) = 1$

44. $2 \log_2(3x + 8) - \log_2(3x + 7) = 2$

45. $\log_5(x^2 + 2x) - \log_5(x^2 - x) = 1$

9.3 Additional Exponential and Logarithmic Equations and Inequalities 409

46. $\log_{\frac{1}{2}}(x-1) - 2\log_{\frac{1}{2}}(x+1) = 0$

47. $3\log_5(x-1) = 2\log_5(x-2) + \log_5 x$

48. $[\log_2(x-5)]^2 - \log_2(x-5) = 12$

49. $3[\ln(x+3)]^2 - \ln(x+3)^2 - 1 = 0$

Exercises 50–53: Solve the equation for $-\frac{\pi}{2} < x < \frac{\pi}{2}$.

50. $2\ln(3\sin x) - \ln(\sin x + 2) = \ln \sin x$

51. $\ln(4\sin x + 1) + \ln(4\sin x - 1) = \ln 8$

52. $2\ln(3\tan x) = \ln(9\tan x + 4)$

53. $\ln 3 - \ln(8\cos x + 2) = \ln \cos x$

Exercises 54–57: Find all solutions of the equation.

54. $\sec(\ln x) = -\dfrac{2}{\sqrt{3}}$

55. $\tan[\ln(x-1)] = \sqrt{3}$

56. $\cos[\ln(3x-2)] = \dfrac{\sqrt{3}}{2}$

57. $\csc(\ln x) = 4$

Exercises 58–59: Find the x-intercepts of the graph of $f(x)$.

58. $f(x) = 2\log_3(x+1) + 4$

59. $f(x) = 4\ln(x-2) + 3$

Exercises 60–61: Sketch the graph of $f(x)$, determining the x- and y-intercepts and the asymptote.

60. $f(x) = -3\log_2(x-1) + 3$

61. $f(x) = 2\log_4(x+2) - 4$

Exercises 62–71: Solve the inequality.

62. $\log x > 0$

63. $\ln x < 4$

64. $\log_{\frac{3}{4}} x < 0$

65. $\log(x-3) \geq 2$

66. $\log_{\frac{1}{2}}(x+5) \geq 0$

67. $\log_2(x-3) < \log_2(3x+1)$

68. $\log_{\frac{1}{2}}(x-1) + \log_{\frac{1}{2}}(x+1) \geq 1$

69. $\log_{\frac{1}{2}}(x-3) < \log_{\frac{1}{2}}(3x+1)$

70. $\dfrac{\ln x}{\sqrt{x}} \geq 0$

71. $x(x-3)\log_2 x > 0$

Exercises GC72–GC75: Find any viewing rectangle that shows part of the graph of the function.

72. $f(x) = -254\ln(x+375) + 74$

73. $g(x) = \ln(20x - 1000) - 600$

74. $h(x) = \dfrac{\ln(x-100) - 200}{x^2 - 999}$

75. $k(x) = 3^{x+20}\ln(x-5) + 50$

Exercises GC76–GC79: Use a graphing calculator, as in Example 9.24, to approximate the solution set of the equation or inequality.

76. $300\ln(x-750) = 500$

77. $\ln(x-300) = (x-350)^2 - 50$

78. $\ln(x+100) > x - 500$

79. $(x-25)\ln(x-25) < 50$

410   Chapter 9   Exponential and Logarithmic Functions

Exercises 80–87: Solve for x, and then approximate the value (to 1 decimal place) with a calculator.

80. $\log x = 2.34$

81. $10^x = 2$

82. $\ln x = 4$

83. $e^x = 7.3$

84. $(2.93)^{3x} = 14.7$

85. $\log_{\sqrt{2}} (x - 1) = \pi$

86. $\ln x^3 = \log_2 9$

87. $e^{3x-2} = \log_3 12$

Exercises GC88–GC98: Use a graphing calculator to approximate the solution sets of the equation and related inequality.

88. $2^{x+4} = 3^{x+2}$; $\quad 2^{x+4} < 3^{x+2}$

89. $3^x = 2^x + 4$; $\quad 3^x > 2^x + 4$

90. $2^{x-4} = x^2 - 2$; $\quad 2^{x-4} < x^2 - 2$

91. $2^{3x+1} = 4 - 5^{x-1}$; $\quad 2^{3x+1} > 4 - 5^{x-1}$

92. $x^3 + x - 1 = 2^{x+1} - 5$; $\quad x^3 + x - 1 > 2^{x+1} - 5$

93. $-5 \cdot 2^x = -4 \cdot 3^x$; $\quad -5 \cdot 2^x < -4 \cdot 3^x$

94. $\log_3(x - 1) = 2^{x-5}$; $\quad \log_3(x - 1) < 2^{x-5}$

95. $\log_3(x + 1) = x - 3$; $\quad \log_3(x + 1) > x - 3$

96. $\log x = 3^{-(x+1)}$; $\quad \log x > 3^{-(x+1)}$

97. $\log(x + 5) = -3^{x+1}$; $\quad \log(x + 5) < -3^{x+1}$

98. $\ln(x + 4) = 3x^2 - 10$; $\quad \ln(x + 4) > 3x^2 - 10$

Exercise 99: Find the exact value of b for the graph of the function $f(x) = 2^{bx-1} - 13$ to have an x-intercept at 5. Use a graphing utility to check your answer.

Exercises 100–101: Solve the system of equations.

100. $\left.\begin{array}{l} \log_2 x^2 - \log_2(x - y) = 2 \\ y + \frac{1}{3}x^2 - x = -\frac{3}{2} \end{array}\right\}$

101. $\left.\begin{array}{l} \log_2(x + 6) + \log_2(x - y) = 3 \\ 2x - y = 3 \end{array}\right\}$

Exercise 102: Find the values of a and b for $\{(x = 1, y = 2)\}$ to be the solution set of the system.

$$\left.\begin{array}{l} \log_2(ax + b) = y \\ \log_2 ax + 8 = \log_2 by + 2^y \end{array}\right\}$$

Exercises 103–106: Find the value of a for which the equation has $x = 1$ as a solution.

103. $\ln(ax + 3) = x - 5$

104. $\ln(ax + 3) = \ln 3x + \ln a$

105. $[\ln(ax + 3)]^2 - 9 = 0$

106. $a \ln(x + 3) = a^{3x}$

§4. EXPONENTIAL GROWTH AND DECAY

The behavior of the E.coli bacteria discussed in Section 9.1, page 375, illustrates a fact of nature. As it is with all living organisms **in an ideal environment**, the rate of change of the number of bacteria present in a particular bacterial culture increases at a rate proportional to the number currently present in that culture:

The larger the population, the faster it grows.

Radioactive substances behave in a similar manner, except that the amount of the substance decreases at a rate proportional to the amount present:

The larger the amount, the faster it decreases.

The following result, which is established in the calculus, reveals the exponential nature of the growth or decline of such substances.

THEOREM 9.8

EXPONENTIAL GROWTH/DECAY FORMULA

If the rate of change of the amount of a substance is proportional to the amount present, then the amount present t units of time before $(t < 0)$, or after $(t > 0)$, an established initial time $(t = 0)$ is given by the formula:

$$A(t) = A_0(e^k)^t$$

where A_0 is the initial amount present, and k is a constant that depends on the particular substance.

Note that at $t = 0$,
$$A(t) = A_0:$$
$$A(0) = A_0(e^k)^0 = A_0$$

From our knowledge of exponential functions, we know that the amount $A(t) = A_0(e^k)^t$ increases with time if $k > 0$ (since $e^k > 1$), and decreases with time if $k < 0$ (since $e^k < 1$). The former case is referred to as **exponential growth**, and the latter **exponential decay**.

The constant k represents the rate of exponential growth or decay of the particular substance. It may often be read from tables.

The quantity $b = e^k$ is the base of the exponential growth or decay, and one may write: $A(t) = A_0 b^t$.

POPULATION GROWTH

In an ideal environment, the population of living organisms (humans, rabbits, bacteria, etc.) increases exponentially. Consider for example, the bacteria cell process depicted in Figure 9.8. In a fixed period of time (called the **doubling time** of the organism), one cell divides into two; then, in the same period of time, the two become four—the four become eight—and so on:

412   Chapter 9 Exponential and Logarithmic Functions

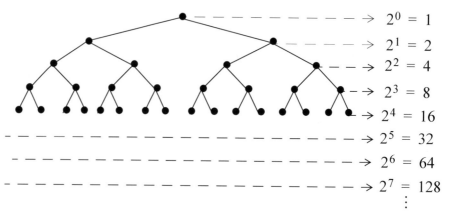

Figure 9.8

EXAMPLE 9.25 The doubling time of E.coli bacteria is 20 min-
E.COLI BACTERIA utes. If a culture of the bacteria contains one
million cells, determine how long it will take
before the culture increases to 9 million cells.

SOLUTION: Figure 9.9 reflects the fact that every 20 minutes the bacteria doubles. From that figure we can safely conclude that it will take a little longer than 60 minutes for there to be 9 million cells.

20 min.→	20 min.→	20 min →	20 min →	20 min →
1 million	2 million	4 million	9×10^6 8 million	16 million

Figure 9.9

To more accurately determine the time required, we turn to the formula of Theorem 9.8, where $A(t)$ now denotes the number of cells (in millions) present at time t (in minutes):

From Theorem 9.8: $\quad A(t) = A_0(e^k)^t$

Since $A(t) = 2A_0$ when $t = 20$: $\quad 2A_0 = A_0(e^k)^{20}$

$$2 = (e^k)^{20}$$

Taking the positive sqaure root: $\quad e^k = 2^{\frac{1}{20}}$
(since $e^k > 0$)

The exponential growth formula for E.coli bacteria: $\quad A(t) = A_0\left(2^{\frac{1}{20}}\right)^t$

Or: $\quad A(t) = A_0 \cdot 2^{\frac{t}{20}}$

Since there are 1 (million) initially, $A_0 = 1$:	$A(t) = 1 \cdot 2^{\frac{t}{20}}$
To determine how long it will take for there to be 9 million cells, set $A(t) = 9$ and solve for t:	$9 = 1 \cdot 2^{\frac{t}{20}}$
	$2^{\frac{t}{20}} = 9$
	$\ln 2^{\frac{t}{20}} = \ln 9$
$\ln s^r = r \ln s$:	$\frac{t}{20} \ln 2 = \ln 9$
	$t = \frac{20 \ln 9}{\ln 2} \approx 63$

We conclude that it will take approximately 63 minutes for the culture to increase from 1 million cells to 9 million cells. In other words, in an ideal situation, E.coli bacteria will increase by a factor of 9 approximately every 63 minutes.

GRAPHING CALCULATOR GLIMPSE 9.1

The function $A(t)$ of Example 9.25 tells the whole E.coli story, and its graph would be one way of witnessing its nature. Another way is to look at a tabulation of some of its function values. A consideration of the function values reveals that the number of bacteria reaches the 9 million mark somewhere between 60 and 70 minutes (left—where the values of the variable start at 40 and are incremented by ten). Refining the process, we instructed the unit to start the variable at 60 and increment it by 1 (right), and see that:

$A(63) = 8.88\text{E}6 = 8.88 \times 10^6 \approx 9$ million (as in Example 9.25)

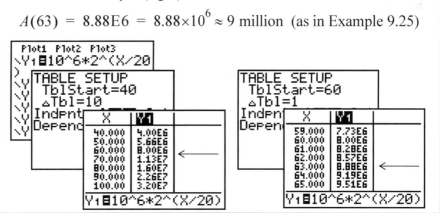

In the previous example, we saw that a doubling time of 20 minutes led to the exponential growth formula:

$$A(t) = A_0 \cdot 2^{\frac{t}{20}}$$

414   Chapter 9 Exponential and Logarithmic Functions

In general:

THEOREM 9.9

DOUBLING TIME

In an exponential growth situation, if the doubling time of a substance is D, then the amount of substance present at time t is given by:

$$A(t) = A_0 \cdot 2^{\frac{t}{D}}$$

where A_0 denotes the initial amount of the substance.

EXAMPLE 9.26

WORLD POPULATION

The world population was 5.28 billion in 1990, and 6.37 billion in 2004. Assuming that, at any given time, the population increases at a rate proportional to the population at that time, determine:

(i) The population in the year 2010.

(ii) The population in the year 1985.

(iii)The year in which the population will reach 7 billion.

SOLUTION: Establishing 1990 as time $t = 0$, from Theorem 9.8, the population (in billions) is given by:

$$A(t) = 5.28(e^k)^t \qquad (*)$$

To find the value of e^k, use the fact that in 2004 ($t = 2004 - 1990 = \mathbf{14}$), $A(t) = \mathbf{6.37}$:

$$A(t) = 5.28(e^k)^t$$

$$\mathbf{6.37} = 5.28(e^k)^{\mathbf{14}}$$

$$(e^k)^{14} = \frac{6.37}{5.28}$$

$$e^k = \left(\frac{6.37}{5.28}\right)^{\frac{1}{14}}$$

Substituting this value for e^k into $(*)$:

$$A(t) = 5.28\left[\left(\frac{6.37}{5.28}\right)^{\frac{1}{14}}\right]^t = 5.28\left(\frac{6.37}{5.28}\right)^{\frac{t}{14}} \qquad (**)$$

We are now able to answer (i)-(iii).

9.4 Exponential Growth and Decay 415

(i) To approximate the population in the year 2010, substitute $2010 - 1990 = 20$ for t in (**):

$$A(20) = 5.28\left(\frac{6.37}{5.28}\right)^{\frac{20}{14}} \approx 6.90$$

which tells us that the world population in 2010 is projected to be about 6.90 billion.

(ii) To approximate the population in the year 1985, we substitute $1985 - 1990 = -5$ for t in (**):

The actual population in 1985 was 4.85 billion. Unlike with bacteria and rabbits, the value of k in the exponential growth formula changes with time, as humans definition of an "ideal environment" is subject to change.

$$A(-5) = 5.28\left(\frac{6.37}{5.28}\right)^{\frac{-5}{14}} \approx 4.94$$

which tells us that the population in 1985 was approximately 4.94 billion.

(iii) To find the year in which the population will reach 7 billion, set $A(t) = 7.00$ in (**) and solve for t:

$$7.00 = 5.28\left(\frac{6.37}{5.28}\right)^{\frac{t}{14}}$$

$$\left(\frac{6.37}{5.28}\right)^{\frac{t}{14}} = \frac{7.00}{5.28}$$

$$\ln\left(\frac{6.37}{5.28}\right)^{\frac{t}{14}} = \ln\left(\frac{7.00}{5.28}\right)$$

$\ln s^r = r \ln s$: $\quad \dfrac{t}{14}\ln\left(\dfrac{6.37}{5.28}\right) = \ln\left(\dfrac{7.00}{5.28}\right)$

$$t = \frac{14\ln\left(\dfrac{7.00}{5.28}\right)}{\ln\left(\dfrac{6.37}{5.28}\right)} \approx 21$$

This extrapolation operates under the assumption that the current "growth trend" will continue to the year 2010—a very iffy proposition, at best.

We conclude that the world population will reach 7 billion in the year $1990 + 21 = 2011$.

CHECK YOUR UNDERSTANDING 9.20

Answer: ≈ 70.75 years

The population of a town grows at a rate proportional to its population. The initial population of 500 increased by 15% in 9 years. How long will it take for the population to triple?

RADIOACTIVE DECAY

By emitting alpha and beta particles and gamma rays, the radioactive mass of a substance decreases with time, and at a rate proportional to the amount present. Consequently, Theorem 9.8 applies.

The half-life of radioactive substances varies greatly. While the half-life of uranium^{235} is 713 million years, that of polonium^{212} is less than a millionth of a second.

A radioactive substance is represented in Figure 9.10. In a fixed period of time (called the **half-life** of the substance), one gram of the radioactive substance decreases to one-half gram; then, in the same period of time, only one-quarter gram remains radioactive, then one-eighth, and so on:

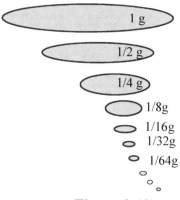

Figure 9.10

CARBON-14 DATING

Organic substances contain both carbon-14 and non-radioactive carbon in known proportions. A living organism absorbs no more carbon when it dies. The carbon-14 decays, thus changing the proportions of the two kinds of carbon in the organism. By comparing the present proportion of carbon-14 with the assumed original proportion, one can determine how much of the original carbon-14 is present, and therefore how long the organism has been dead; hence how old it is. The next example illustrates this method, called **carbon-14 dating**.

EXAMPLE 9.27

AGE OF SKELETON

A skeleton is found to contain one-sixth of its original amount of carbon-14. How old is the skeleton, given that carbon-14 has a half-life of 5730 years?

SOLUTION: Figure 9.11 reflects the fact that every 5730 years after death, the skeleton's carbon-14 is reduced by a factor of $\frac{1}{2}$. From that figure, since $\frac{1}{8} < \frac{1}{6} < \frac{1}{4}$, we can safely conclude that the age of the skeleton is somewhere between $2(5730) = 11,460$ and $3(5730) = 17,190$ years.

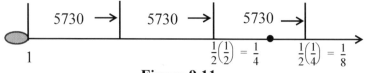

Figure 9.11

To more accurately determine the time, we turn to Theorem 9.8, where $A(t)$ now denotes the amount of carbon-14 present at time t (in years), with $t = 0$ at the time of death:

$A(t) = \dfrac{A_0}{2}$ when $t = 5730$:

$$A(t) = A_0(e^k)^t$$

$$\frac{1}{2}A_0 = A_0(e^k)^{5730}$$

Solve for e^k:

$$\frac{1}{2} = (e^k)^{5730}$$

$$e^k = \left(\frac{1}{2}\right)^{\frac{1}{5730}} = (2^{-1})^{\frac{1}{5730}} = 2^{-\frac{1}{5730}}$$

This brings us to the exponential decay formula for carbon-14:

$$A(t) = A_0 \cdot 2^{-\frac{t}{5730}} \qquad (*)$$

In the present problem, we are told that the skeleton contains one-sixth of its original amount of carbon-14, that is: $A(t) = \dfrac{A_0}{6}$. To find the skeleton's age, we substitute $\dfrac{A_0}{6}$ for $A(t)$ in $(*)$, and solve for t:

Divide both sides by A_0:

$$\frac{A_0}{6} = A_0 \cdot 2^{-\frac{t}{5730}}$$

$$2^{-\frac{t}{5730}} = \frac{1}{6}$$

$$\ln\left(2^{-\frac{t}{5730}}\right) = \ln\left(\frac{1}{6}\right) = \ln(6^{-1})$$

$\ln s^r = r \ln s$:

$$-\frac{t}{5730}\ln 2 = -\ln 6$$

$$t = \frac{5730\ln 6}{\ln 2} \approx 14,812$$

We conclude that the skeleton is approximately 14,812 years old.

In the last example, we saw that a half-life of 5730 years led to the exponential decay formula:

$$A(t) = A_0 \cdot 2^{-\frac{t}{5730}}$$

In general:

418   **Chapter 9 Exponential and Logarithmic Functions**

THEOREM 9.10

HALF-LIFE

In an exponential decay situation, if the half-life of a substance is H, then the amount of substance present at time t is given by:

$$A(t) = A_0 \cdot 2^{-\frac{t}{H}}$$

where A_0 denotes the initial amount present.

EXAMPLE 9.28

RADIOACTIVE SUBSTANCE

A certain radioactive substance loses $\frac{1}{3}$ of its original mass in four days. How long will it take for the substance to decay to $\frac{1}{10}$ of its original mass?

SOLUTION: We are told that after four days, $\frac{1}{3}$ of the substance decays, or equivalently, that the amount present 4 days later is $\frac{2}{3}A_0$. Turning to the exponential formula of Theorem 9.8:

$$A(t) = A_0(e^k)^t \qquad\qquad (*)$$

we set $t = \mathbf{4}$ and $A(t) = \frac{2}{3}\mathbf{A_0}$, and solve for e^k:

$$\frac{2}{3}\mathbf{A_0} = A_0(e^k)^{\mathbf{4}}$$

$$\frac{2}{3} = (e^k)^4$$

$$e^k = \left(\frac{2}{3}\right)^{\frac{1}{4}}$$

Substituting this value of e^k back into $(*)$ yields the exponential decay formula for the radioactive substance under consideration:

$$A(t) = A_0\left(\frac{2}{3}\right)^{\frac{t}{4}}$$

To determine how long it will take for the substance to decay to $\frac{1}{10}$ of its original mass, we substitute $\frac{A_0}{10}$ for $A(t)$ in the above equation, and solve for t:

$$\frac{A_0}{10} = A_0\left(\frac{2}{3}\right)^{\frac{t}{4}}$$

$$\frac{1}{10} = \left(\frac{2}{3}\right)^{\frac{t}{4}}$$

$$\ln\frac{1}{10} = \ln\left(\frac{2}{3}\right)^{\frac{t}{4}}$$

$\ln s^r = r\ln s$: $\quad \ln 10^{-1} = \ln\left(\frac{2}{3}\right)^{\frac{t}{4}}$

$$-\ln 10 = \frac{t}{4}\ln\frac{2}{3}$$

$$t = \frac{-4\ln 10}{\ln\frac{2}{3}} \approx 22.7$$

We conclude that it will take approximately 22.7 days for the substance to decay to $\frac{1}{10}$ of its original mass.

CHECK YOUR UNDERSTANDING 9.21

A certain radioactive substance decays exponentially in accordance with the formula:

$$A(t) = A_0 e^{-\frac{t}{4}}$$

where $A(t)$ is the number of grams present after t years.

(a) How many grams of the substance will there be in 2007, given that 35 grams were present in 2004?

(b) What is the half-life of the substance?

Answers: (a) ≈ 16.53 g
(b) $\ln 16 \approx 2.77$ years

420   Chapter 9 Exponential and Logarithmic Functions

EXERCISES

Exercises GC1–GC2: (Tabulating Values) By tabulating function values, as in Graphing Calculator Glimpse 9.1, approximate the value of t for the given value of $A(t)$.

1. $A(t) = 150 \, e^{\frac{t}{3}}$; when $A(t) = 200$ 2. $A(t) = 70 \, e^{-\frac{t}{165}}$; when $A(t) = 50$

Exercise 3: (Bacterial Growth) Suppose that a certain type of bacteria is subject to the following law of exponential growth: $A(t) = A_0 \, e^{0.1t}$, where t is measured in minutes.

(a) How long will it take for the bacteria to double?

(b) If 2000 bacteria were present at 1 P.M., how many will be present at 1:25 P.M.?

(c) If 2000 bacteria are present at 1 P.M., how many were present at 12:50 P.M.?

Exercise 4: (World Population) The world population in 2002 was approximately 6.2 billion, and in 2008 it was about 6.7 billion. Assuming an exponential rate of growth,

(a) calculate the world population for last year.

(b) Refer to the internet to find your percentage error in (a), where percentage error is 100 times the error divided by the exact value.

(c) Use the population in 2002 and that of last year to arrive at a new formula, and use it to predict the population in the year 2020.

Exercise 5: (U.S. Population) The population of the United States in 2000 was approximately 281.4 million, and in 2010 it was about 308.7 million. Assuming an exponential growth rate, determine

(a) the population in the year 2017.

(b) the population in the year 1980.

(c) the year in which the population reaches 320 million.

(d) the year in which the population will be twice what it was in 1975.

Exercise 6: (Ant Population) On April 15th there were 1500 ants in a particular ant colony. On April 20th there were 2500. Assuming that the ants increase at a rate proportional to their number, determine, approximately, how many ants there are on

(a) April 30.   (b) April 10.

Exercise 7: (Radioactive Decay) A radioactive substance has a half-life of 74 years.

(a) How long will it take for the substance to decay to one-fourth of its initial mass?

(b) If 300 grams are present initially, how many grams will remain after 35 years?

9.4 Exponential Growth and Decay **421**

Exercise 8: (**Radioactive Decay**) A certain radioactive substance decays exponentially in accordance with the formula $A(t) = A_0 \cdot 5^{-\frac{1}{4}t}$, where t is measured in years. Given that 35 grams of the substance were present on Jan. 1, 1998, determine:

(a) The number of grams that will be present on Jan. 1, 2015.

(b) The number of grams that were present on July 1, 1990.

(c) The half-life of this substance.

Exercise 9: (**Strontium-90 Decay**) The artificial isotope Strontium-90 is a fission product recovered from nuclear reactors and released in quantity by nuclear weapons. Because of its chemical similarity to calcium, long term exposures to the isotope result in substantial substitution of Strontium-90 for calcium in animals and humans resulting in a destruction of the blood-cell-forming bone marrow. How long will it take for the isotope to be reduced by 95%, if it has a half-life of 28 years?

Exercise 10: (**Strontium-90 Decay**) Along with Strontium-90, the isotope iodine-131 is another dangerous residual of nuclear reactors. The half-life of iodine-131 is only 8 days. How long must waste material that contains 50 times the acceptable disposal level of iodine-131 be stored before it can be properly disposed?

Exercise 11: (**Piltdown Skull**) The American chemist Willard Frank Libby, was awarded the 1960 Nobel prize for his development of radioactive dating. His theory helped to expose the greatest scientific hoax ever perpetrated. Early in the twentieth century part of a human skull and an apelike jawbone were found in a gravel pit in Piltdown, Sussex, England, giving birth to the so-called Piltdown Man, and to a fierce evolution controversy which lasted for many years.

It was determined that the Piltdown skull fragment contains 93% of its original carbon-14. How old was the skull at that time? (See Example 9.27.) (Incidentally, the jaw was found to be that of a rather recently deceased orangutan.)

Exercise 12: (**Dead Sea Scrolls**) Approximately 20% of the original carbon-14 remains in the Dead Sea Scrolls. How old are they? (See Example 9.27.)

Exercise 13: (**Newton's Law of Cooling**) Newton's Law of Cooling asserts that a body loses heat at a rate directly proportional to the difference in temperature between the body and its surroundings. If $T(t)$ is the temperature of the body at time t, then for some constant $k > 0$ which depends on the particular situation,

$$T(t) = T_m + (T_0 - T_m)e^{-kt}$$

where $T_0 = T(0)$ is the initial temperature of the body, and T_m is the constant temperature of the surroundings.

Suppose it took two minutes for the temperature of a cup of coffee to decrease from 175°F to 150°F in an 80°F room. How much more time is needed for the temperature of the coffee to drop to 125°F?

422   **Chapter 9 Exponential and Logarithmic Functions**

---

§5. ADDITIONAL APPLICATIONS

You have already seen how exponential and logarithmic functions are instrumental in solving exponential growth and decay problems. In this section, you will find other applications of these important functions: in the field of finance (continuously compounded interest), in physics (intensity of sound), and in geology (earthquake magnitude). Additional applications appear in the exercises.

COMPOUND INTEREST

Compound Interest is interest that is paid on the original principal and on all interest earned. While an annual interest rate r is typically quoted, the interest itself is often calculated and added to the principal on a more frequent basis. In such a case, the interest is said to be **compounded**, and the **interest rate per period** is the annual interest rate divided by the number of periods per year:

INTEREST RATE PER PERIOD

interest rate per period: $i = \dfrac{r}{n}$ $\leftarrow$ annual interest rate

$\leftarrow$ number of times compounded during the year

In particular:

Interest compounded	Length of each interest period	Number of interest periods per year	Interest rate per period (r denotes annual interest rate)
Annually	One year	$n = 1$	$i = r$
Semiannually	Six months	$n = 2$	$i = \dfrac{r}{2}$
Quarterly	Three months	$n = 4$	$i = \dfrac{r}{4}$
Monthly	One month	$n = 12$	$i = \dfrac{r}{12}$

Suppose an amount A_0 (the principal) is invested at an annual interest rate of 5% compounded annually for 3 years. Letting $A(n)$ denote the amount accumulated in the account at the end of n years, we find that:

$$A(1) = A_0 + .05A_0 = A_0(1 + .05)$$

$$\begin{aligned} A(2) &= A(1) + .05A(1) \\ &= [A_0(1 + .05)] + .05[A_0(1 + .05)] \\ &= [A_0(1 + .05)](1 + .05) = A_0(1 + .05)^2 \end{aligned}$$

9.5 Additional Applications   423

$$A(3) = A(2) + .05A(2)$$
$$= [A_0(1 + .05)^2] + .05[A_0(1 + .05)^2]$$
$$= [A_0(1 + .05)^2](1 + .05) = A_0(1 + .05)^3$$

Generalizing the above pattern, we conclude that:

THEOREM 9.11

FUTURE VALUE AT COMPOUND INTEREST

If an amount A_0 (the principal) is invested at an interest rate of i **per interest period**, then the amount $A(n)$ accumulated after n **interest periods** is given by:

$$A(n) = A_0(1 + i)^n$$

EXAMPLE 9.29

INVESTMENT

You invest \$10,000 at an annual interest rate of 5%. Determine the value of the account after 18 months, if interest is compounded:

(a) Quarterly   (b) Monthly

SOLUTION: The solution of both parts of the problem appear below. While the initial amount and annual interest rate are $A_0 = 10,000$ and 5% in both cases, the interest **per period** with quarterly compounding is $\frac{.05}{4}$, and it is $\frac{.05}{12}$ with monthly compounding. Since the exponent n in the formula represents the number of interest periods, in the case of quarterly compounding $n = 6$ (18 months is 6 quarters), while in the monthly compounding situation $n = 18$:

Compounded Quarterly:	Compounded Monthly:
$A(6) = \$10,000\left(1 + \frac{.05}{4}\right)^6$	$A(18) = \$10,000\left(1 + \frac{.05}{12}\right)^{18}$
$= \$10,773.83$	$= \$10,777.16$

EXAMPLE 9.30

INTEREST RATE

At what annual interest rate must capital be invested in order for it to double in 10 years, if interest is compounded:

(a) Annually   (b) Monthly

424   Chapter 9 Exponential and Logarithmic Functions

SOLUTION: (a) Since interest is compounded annually, i in:

$$A(n) = A_0(1+i)^n$$

denotes the (unknown) annual interest rate. Setting $n = 10$ (the number of interest periods in 10 years), and $A(10) = 2A_0$ (double the initial amount A_0) in the equation, we have:

$$2A_0 = A_0(1+i)^{10}$$
$$2 = (1+i)^{10}$$
$$1+i = 2^{\frac{1}{10}}$$
$$i = 2^{\frac{1}{10}} - 1 \approx .0718$$

We conclude that a principal invested at approximately 7.18% compounded annually will double in 10 years.

(b) There are $n = 10 \cdot 12 = 120$ interest periods in ten years when interest is compounded monthly. Moreover, the interest **per period** is $i = \dfrac{r}{12}$ where r denotes the annual interest rate. Substituting 120 for n, $2A_0$ for $A(120)$, and $\dfrac{r}{12}$ for i in $A(n) = A_0(1+i)^n$, we can solve for the annual interest rate r:

$$2A_0 = A_0\left(1 + \frac{r}{12}\right)^{120}$$
$$2 = \left(1 + \frac{r}{12}\right)^{120}$$
$$1 + \frac{r}{12} = 2^{\frac{1}{120}}$$
$$r = 12 \cdot 2^{\frac{1}{120}} - 12 \approx .0695$$

We conclude that a principal invested at approximately 6.95% compounded monthly will double in 10 years.

CHECK YOUR UNDERSTANDING 9.22

At what annual interest rate must capital be invested so that it will double in 10 years, if interest is compounded daily? (Assume 365 days in a year.)

Answer: $\approx 6.93\%$

INTEREST COMPOUNDED CONTINUOUSLY

We already know that if money is invested at an annual interest rate r that is compounded m times a year, then the interest rate per interest period is $i = \dfrac{r}{m}$. Since there are mt interest periods at the end of t years, we can use Theorem 9.11 to obtain a formula for the amount $A(t)$ accumulated after t years, or, equivalently, after mt interest periods:

THEOREM 9.12

ANNUAL FUTURE VALUE AT COMPOUND INTEREST

If an amount A_0 is invested at an annual interest rate of r compounded m times a year, then the amount $A(t)$ accumulated after t years is given by:

$$A(t) = A_0\left(1 + \frac{r}{m}\right)^{mt}$$

What happens to the future value $A(t)$ as m increases from 1 (compounded annually), to 2 (compounded semiannually), to 4, to 12, to 365, and so on? For a clue, consider Figure 9.12 which displays the amount accumulated at the end of one year ($t = 1$) when one dollar ($A_0 = 1$) is invested at an annual interest rate of 100% ($r = 1.00$), compounded m times a year:

Compounding	m	$A(1) = \left(1 + \frac{1}{m}\right)^m$
Annually	1	$\left(1 + \frac{1}{1}\right)^1 = 2$
Semiannually	2	$\left(1 + \frac{1}{2}\right)^2 = 2.25$
Quarterly	4	$\left(1 + \frac{1}{4}\right)^4 \approx 2.44141$
Monthly	12	$\left(1 + \frac{1}{12}\right)^{12} \approx 2.61304$
Daily	365	$\left(1 + \frac{1}{365}\right)^{365} \approx 2.71457$
Every Hour	8,760	$\left(1 + \frac{1}{8760}\right)^{8760} \approx 2.71813$
Every Minute	525,600	$\left(1 + \frac{1}{525600}\right)^{525600} \approx 2.71828$

a difference of \$0.25

a difference of \$0.19

essentially no difference

Figure 9.12

Surprisingly, the amount accumulated approaches the number $e \approx 2.71828$. Indeed, in the calculus one shows that the larger m becomes, the closer the value of $\left(1 + \dfrac{1}{m}\right)^m$ gets to e. One can then show that as m gets larger and larger, $\left(1 + \dfrac{r}{m}\right)^m$ approaches e^r, and the resulting compounding is said to be "continuous." This leads to the following modification of Theorem 9.12:

426   **Chapter 9 Exponential and Logarithmic Functions**

Does this formula look familiar? It is the exponential growth formula of Theorem 9.8, page 411. Indeed, when money is compounded continuously, its rate of growth is proportional to the amount present.

THEOREM 9.13

CONTINUOUS COMPOUNDING

If an amount A_0 is invested at an annual interest rate of r, and the interest is compounded continuously, then the amount $A(t)$ accumulated after t years is given by:

$$A(t) = A_0 e^{rt}$$

EXAMPLE 9.31

INTEREST RATE

Determine the annual interest rate r required for capital to double in 10 years, when interest is compounded continuously.

SOLUTION: We substitute $t = 10$ and $A(t) = 2A_0$ in the formula of Theorem 9.13, and solve for r:

$$2A_0 = A_0 e^{10r}$$

$$2 = e^{10r}$$

$$\ln 2 = 10r \ln e = 10r$$

$$r = \frac{1}{10}\ln 2 \approx .0693$$

In Example 9.30 we found that when interest is compounded monthly, an interest rate of about 6.95% is required for the principal to double in 10 years.

We find that an investment at approximately 6.93% compounded continuously doubles every 10 years.

EXAMPLE 9.32

INVESTMENT

A thousand dollars is invested at an annual rate of 8% compounded semiannually, and nine hundred dollars is invested at the same rate compounded continuously. How long will it take before the two accounts show the same balance?

SOLUTION: From Theorem 9.12 we know that after t years, the $1000 invested at 8% compounded semiannually will accumulate to the following dollar amount:

$$A_s(t) = 1000\left(1 + \frac{.08}{2}\right)^{2t}$$

Theorem 9.13 tells us that after t years, the $900 invested at 8% compounded continuously will have grown to the following dollar amount:

$$A_c(t) = 900e^{.08t}$$

To determine when the two accounts show the same balance, we set $A_s = A_c$ and solve for t:

$$1000\left(1 + \frac{.08}{2}\right)^{2t} = 900e^{.08t}$$

$$10(1.04)^{2t} = 9e^{.08t}$$

$$\ln[10(1.04)^{2t}] = \ln(9e^{.08t})$$

$\ln st = \ln s + \ln t:$ $\ln 10 + \ln(1.04)^{2t} = \ln 9 + \ln e^{.08t}$

$\ln s^r = r\ln s: \rightarrow \ln 10 + 2t\ln 1.04 = \ln 9 + .08t \leftarrow$ $: \ln e^x = x$

$$2t\ln 1.04 - .08t = \ln 9 - \ln 10$$

$$t(2\ln 1.04 - .08) = \ln\frac{9}{10} \longleftarrow \quad : \ln s - \ln t = \ln\frac{s}{t}$$

$$t = \frac{\ln\dfrac{9}{10}}{2\ln 1.04 - .08} \approx 67.60$$

We see that it will take approximately sixty-seven and a half years for the two accounts to be equal.

CHECK YOUR UNDERSTANDING 9.23

How much should be invested at an annual rate of 4% compounded continuously in order to have a total of $10,000 at the end of 5 years?

Answer: $8,187.31

SOUND INTENSITY

Alexander Graham Bell, Scottish-born inventor, best known for his invention of the telephone (1847-1922)

James Watt, Scottish inventor (1736-1819).

The intensity level L (in bels) of sound is defined in terms of the common logarithm (base 10) of the intensity I (i.e. energy density) of a sound-wave when it hits your eardrums. It is measured in bels:

$$L = \log\frac{I}{I_0}$$

where I is measured in Watts per square meter, and I_0 is the constant intensity of 10^{-12} Watts per square meter (roughly the intensity of the faintest audible sound).

In this logarithmic scale, when the energy of a sound is $I = 10I_0$, its intensity level, L, is 1 bel. When $I = 100I_0$, its intensity level is 2 bels, and so on. Every time the energy density increases by a factor of 10, the intensity level increases by one bel.

Conversion bridge:

10 db $= 1$ bel

Because the bel is a large unit, it is customary to express the intensity level in decibels [db]:

428   Chapter 9 Exponential and Logarithmic Functions

DEFINITION 9.3

INTENSITY LEVEL OF SOUND

The intensity level of sound, in decibels, is given by:

$$L = 10 \log \frac{I}{I_0}$$

where I is the sound intensity in Watts per square meter, and $I_0 = 10^{-12} \frac{\text{Watts}}{m^2}$.

EXAMPLE 9.33

WHISPERS

It is known that the sound intensity due to independent sources is the sum of the individual intensities. Given that the intensity level of the average whisper is 20 db, how many students would have to be whispering simultaneously in order to produce an intensity level of 60 db, which approximates the intensity level of ordinary conversation?

SOLUTION: Let I_w denote the intensity of a whisper, and I_c the intensity required to produce a conversational intensity level of 60 db. Using Definition 9.3, we express both I_w and I_c as multiples of I_0:

$$10 \log \frac{I_w}{I_0} = 20 \qquad\qquad 10 \log \frac{I_c}{I_0} = 60$$

$$\log \frac{I_w}{I_0} = 2 \qquad\qquad \log \frac{I_c}{I_0} = 6$$

Raise 10 to both sides:

$$10^{\log \frac{I_w}{I_0}} = 10^2 \qquad\qquad 10^{\log \frac{I_c}{I_0}} = 10^6$$

$$\frac{I_w}{I_0} = 10^2 \qquad\qquad \frac{I_c}{I_0} = 10^6$$

$$I_w = 10^2 I_0 \qquad\qquad I_c = 10^6 I_0$$

Since the combined sound intensity of n whispering students equals $n I_w = n \cdot 10^2 I_0$, we set this number equal to $I_c = 10^6 I_0$ and solve for the value of n which will produce an intensity level of 60 db:

$$n \cdot 10^2 I_0 = 10^6 I_0$$

$$n = \frac{10^6}{10^2} = 10{,}000$$

We find that it would take 10,000 whispering students to create the intensity level of ordinary conversation.

9.5 Additional Applications   429

RICHTER SCALE

Like the intensity level L of sound, the intensity of an earthquake, as measured by the Richter magnitude scale, is also defined in terms of the common logarithm:

The Richter magnitude scale was developed in 1935 by the American Physicist Charles F. Richter (1900-1985).

DEFINITION 9.4

EARTHQUAKE MAGNITUDE

The magnitude R (on the Richter scale) of an earthquake of intensity I is given by:

$$R = \log \frac{I}{I_0}$$

where I_0 is a "minimum" intensity used for comparison.

EXAMPLE 9.34

TWO EARTHQUAKES

The 1985 earthquake in Mexico City measured 8.1 on the Richter scale, while the 1989 California earthquake measured 7.0. How much more intense was the Mexico City earthquake?

SOLUTION: We apply Definition 9.4 to express both the intensity of the Mexico earthquake, I_M, and that of the California earthquake, I_C, as multiples of I_0:

$$\log \frac{I_M}{I_0} = 8.1 \qquad\qquad \log \frac{I_C}{I_0} = 7.0$$

Raise 10 to both sides:

$$10^{\log \frac{I_M}{I_0}} = 10^{8.1} \qquad\qquad 10^{\log \frac{I_C}{I_0}} = 10^{7.0}$$

$$\frac{I_M}{I_0} = 10^{8.1} \qquad\qquad \frac{I_C}{I_0} = 10^{7.0}$$

$$I_M = 10^{8.1} I_0 \qquad\qquad I_C = 10^{7.0} I_0$$

From the quotient:

$$\frac{I_M}{I_C} = \frac{10^{8.1} I_0}{10^{7.0} I_0} = 10^{1.1} \approx 12.6$$

we see that the intensity of the Mexico City earthquake was more than twelve times that of the California earthquake.

CHECK YOUR UNDERSTANDING 9.24

On August 16, 1999, an earthquake measuring 7.4 on the Richter scale struck Turkey. The following day, an earthquake measuring 5.0 occurred in California. How much more intense was the earthquake in Turkey?

Answer: Approximately 251 times more intense.

430   Chapter 9 Exponential and Logarithmic Functions

EXERCISES

Exercise 1: (**Compound Interest**) Suppose that $2,000 is invested in a savings account at an annual interest rate of 6%, compounded quarterly. Determine the amount accumulated after 3 years and 3 months.

Exercise 2: (**Compound Interest**) Suppose $5,000 is invested in a savings account at 7% annual interest. Determine the amount accumulated after 20 years, if interest is compounded

  (a) annually.   (b) quarterly.   (c) monthly.   (d) continuously.

Exercise 3:   (**Compound Interest**) Suppose $10,000 is invested in a savings account at $6\frac{1}{2}\%$ annual interest. Determine the amount accumulated after 10 years, if interest is compounded quarterly. How much accumulates if the amount is compounded monthly?

Exercise 4: (**Compound Interest**) How long will it take money invested at an annual interest rate of 10%, compounded quarterly, to triple?

Exercise 5: (**Compound Interest**) How long will it take a principal of $2,000 to grow to $2,200, if it is invested at an annual interest rate of 6% compounded

  (a) semi-annually?   (b) continuously?

Exercise 6: (**Compound Interest**) What annual interest rate compounded continuously is equivalent to 5% compounded semi-annually?

Exercises 7–9: (**Mortgage Payments**) If an amount B is borrowed at an interest rate of i per interest period, to be repaid by paying an amount P at the end of each period for n periods, then P is given by

$$P = B \cdot \frac{i}{1 - (1+i)^{-n}}$$

  7. What are the monthly payments on a 20-year mortgage of $400,000 at an annual rate of 7%?

  8. What is the largest 15-year mortgage at an annual rate of 8% that one can carry with payments of at most $700 per month?

  9. What is the largest 30-year mortgage at an annual rate of 8% that one can carry with payments of at most $700 per month?

Exercise 10: (**Sound Intensity**) Find the intensity of the given sounds.

  (a) Heavy city traffic at 90 db.

  (b) Dripping faucet at 30 db.

  (c) Rustle of leaves at 10 db.

Exercise 11: (**Sound Intensity**) What is the difference in the intensity level of two sounds, if the intensity of one sound is 70 times that of the other?

Exercise 12: (**Earthquake Magnitude**) If an earthquake has an intensity which is 300 times the intensity of a smaller earthquake, how much larger would its Richter scale measurement be?

Exercise 13: (**Earthquake Magnitude**) The 1906 San Francisco earthquake registered 8.4 on the Richter scale. Compare its intensity with the 1971 Los Angeles earthquake that measured 6.7 on the Richter scale.

Exercise 14: (**Earthquake Magnitude**) Letting $I_0 = 1$, find the Richter magnitude of an earthquake which has an intensity that is 300 times larger than that of the 1971 Los Angeles earthquake. How many times larger or smaller would that earthquake be than the 1906 San Francisco earthquake (see Exercise 13)?

Exercises 15–16: (**Learning curves**) are graphs of exponential functions of the form $L(t) = a(1 - b^t)$ where $a > 0$ and $0 < b < 1$. As you can see from the adjacent figure, while initially rapid, the learning process levels off with time.

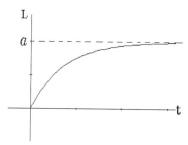

15. Practicing one hour a day, it took Dave 9 days to learn to type 30 words per minute. How many days of practice will he need in order to get his speed up to 60 words per minute, assuming that an average experienced typist can type 73 words per minute?

16. (a) Find the learning curve formula for Mary's riveting abilities, if it took her 5 days before she could do 27 rivets per hour, given that the average experienced riveter can do 43 rivets per hour.

    (b) In how many more working days will she be able to do 30 rivets per hour?

    (c) How long will it take before she can be expected to do 40 rivets per hour?

Exercise 17: (**Chemistry - pH**) The pH (hydrogen potential) of a solution is a measure of the acidity of the solution: $pH = -\log[H^+]$, where $[H^+]$ is the hydrogen ion concentration in moles per liter. pH values vary from 0 (very acidic) to 14 (very basic, alkaline). Pure water has a pH of 7.0 and is neutral, neither acidic nor alkaline.

(a) Find the pH, if the $[H^+]$ value
    (i) of sea water is 6.31×10^{-9} (ii) of wine is 0.000316   (iii) of blood is 3.98×10^{-8}

(b) Find the $[H^+]$ value, given the pH value
    (i) of milk is 6.6   (ii) of lemon juice is 2.3   (iii) of toothpaste is 9.9

Exercise 18: **(Chemistry - pH)**, Referring to the previous exercise, determine how much greater the hydrogen ion concentration of lemon juice (pH = 2.3) is than that

(a) of tomato juice (pH = 4.1). (b) of pure water (pH = 7.0).

Exercises 19–21: (Annuities) If at the beginning of each time period an amount P is deposited in an account at an interest rate of i per time period, then immediately after the nth deposit the worth of the account W is given by

$$W = P \cdot \frac{(1+i)^n - 1}{i}$$

19. Determine the worth of an account after 10 years, if $50 is deposited in the account every month, at an annual interest rate of 6% compounded monthly.

20. Determine the worth of an account after 10 years, if $60 is deposited in the account every month, at an annual interest rate of 6% compounded monthly.

21. How much must be invested monthly at an annual interest rate of 10% compounded monthly, for an account to be worth $100,000 at the end of 20 years?

Exercises 22–24: (Logistic curves) (also called **sigmoidal curves**) are graphs of functions of the form

$$f(t) = \frac{B}{1 + A \cdot b^t}$$

where A, B, and b are positive constants, with $0 < b < 1$. If you graph the curves in these exercises, you will find that they have an initial rapid growth pattern, followed by a declining rate of increase, as is indicated in the adjacent figure.

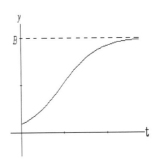

22. The total number of people infected to date by a certain disease t weeks after its outbreak is given by the logistic formula

$$I(t) = \frac{790}{1 + 11\left(\frac{1}{3}\right)^t}$$

(a) How many people were infected when the disease first broke out?

(b) How many are infected by the end of the first week?

(c) How many are infected by the end of the second week?

(d) Explain why the graph of $I(t)$ must approach the horizontal line $y = 790$ from below, as $t \to \infty$. Interpret this feature in the context of this problem.

9.5 Additional Applications   **433**

23. Five hundred pheasants were introduced at a game farm, and their population increased in accordance with the logistic formula

$$P(t) = \frac{1500}{1 + 2\left(\frac{1}{5}\right)^t}$$

where t is measured in quarters (three-month periods).

(a) Determine the pheasant population at the end of the first year.

(b) Explain why the graph of $P(t)$ must approach the horizontal line $y = 1500$ from below, as $t \to \infty$. Interpret this feature in the context of this problem.

(c) Determine the number of quarters necessary for the flock to grow to 90% of the maximum number possible.

24. A number of trout were seeded at a fish hatchery. Assuming that the number increased in accordance with a logistic formula

$$F(t) = \frac{B}{1 + 11b^t}$$

where t is measured in months,

(a) how many trout were seeded initially?

(b) Find the logistic formula, if 1150 trout were present at the end of five months.

CHAPTER SUMMARY

EXPONENTIAL FUNCTIONS	All exponential functions $f(x) = b^x$ of base $b > 1$ are increasing functions, while those of base b with $0 < b < 1$ are decreasing functions. In either case, $D_f = (-\infty, \infty)$ and $R_f = (0, \infty)$. The x-axis is a horizontal asymptote of the graph of f. 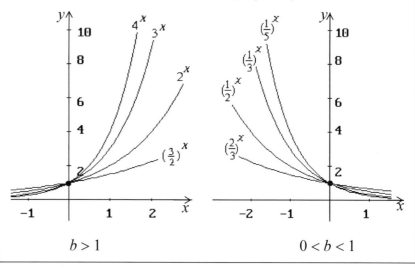	
One-to-One	Every exponential function is one-to-one: $$b^s = b^t \quad \text{if and only if} \quad s = t$$	
LOGARITHMIC FUNCTIONS	$\log_b x$ is the inverse of the exponential function b^x. All logarithmic functions, $f(x) = \log_b x$ of base $b > 1$ are increasing functions, while those of base b with $0 < b < 1$ are decreasing functions. In either case, $D_f = (0, \infty)$ and $R_f = (-\infty, \infty)$. The y-axis is a vertical asymptote of the graph of f. 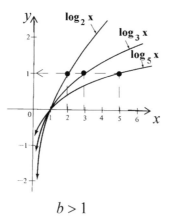 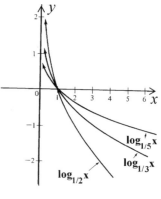	

WARNING	You can only take logs of <u>positive</u> numbers.
One-to-One	Every logarithmic function is one-to-one: For s and t positive: $\log_b s = \log_b t$ if and only if $s = t$
Inverse Properties	For any base b: $\log_b b^x = x$ for all x $b^{\log_b x} = x$ for all $x > 0$
Logarithmic Properties *log of product equals sum of logs* *log of quotient equals difference of logs* *log of a power is the power times the log*	For any base b, any r, s, and t with $s > 0, t > 0$: (i) $\log_b (st) = \log_b s + \log_b t$ (ii) $\log_b \left(\dfrac{s}{t}\right) = \log_b s - \log_b t$ (iii) $\log_b s^r = r\log_b s$
WARNING	None of the above deals with "logs of sums" or "logs of differences."
Change of base formula	For any positive numbers a and b other than 1, and any $x > 0$: $$\log_b x = \frac{\log_a x}{\log_a b}$$
Solving Exponential Equations	With a common base: Since exponential functions are one-to-one, we have: $$b^{\textbf{expression}_1} = b^{\textbf{expression}_2}$$ if and only if $$\textbf{expression}_1 = \textbf{expression}_2$$ With different bases: Applying the natural logarithmic function to both sides of the equation, and using the property $\ln s^r = r\ln s$, we have: $$a^{\textbf{expression}_1} = b^{\textbf{expression}_2}$$ if and only if $$(\textbf{expression}_1)(\ln a) = (\textbf{expression}_2)(\ln b)$$

436 Chapter 9 Exponential and Logarithmic Functions

Solving Exponential Inequalities	If $b > 1$: $b^{\text{expression}_1} < b^{\text{expression}_2}$ if and only if $\text{expression}_1 < \text{expression}_2$ If $0 < b < 1$: $b^{\text{expression}_1} < b^{\text{expression}_2}$ if and only if $\text{expression}_1 > \text{expression}_2$ (note that the inequality symbol is reversed)
Solving Logarithmic Equations	Since logarithmic functions are one-to-one, if: $\log_b(\text{expression}_1) = \log_b(\text{expression}_2)$ then: $\text{expression}_1 = \text{expression}_2$ Since $b^{\log_b x} = x$, if: $\log_b(\text{expression}) = c$ then: $\text{expression} = b^c$ **WARNING**: Since you can only take logs of positive numbers, only potential solutions for which the expression(s) are **positive** are actual solutions.
Solving Logarithmic Inequalities	First determine where the logarithmic functions are defined. Then: If $b > 1$: $\log_b(\text{expression}_1) < \log_b(\text{expression}_2)$ if and only if $\text{expression}_1 < \text{expression}_2$ If $0 < b < 1$: $\log_b(\text{expression}_1) < \log_b(\text{expression}_2)$ if and only if $\text{expression}_1 > \text{expression}_2$ (note that the inequality symbol is reversed)

APPLICATIONS	

EXPONENTIAL GROWTH AND DECAY	Amount at time t → Initial amount
	
	$$A(t) = A_0(e^k)^t$$
	
	     if k>0 A constant if k<0
	
	     Exponential Growth Exponential Decay
	
	The constant k is dictated by the substance under consideration.

Doubling Time	If D denotes the time required for an amount of substance to double in an exponential growth situation, then the amount at time t is given by: $$A(t) = A_0 \cdot 2^{\frac{t}{D}}$$
Half-Life	If H denotes the amount of time required for an amount of substance to halve in an exponential decay situation, then the amount at time t is given by: $$A(t) = A_0 \cdot 2^{-\frac{t}{H}}$$
COMPOUND INTEREST	
Future Value at Compound Interest	If an amount A_0 (the principal) is invested at an interest rate of i **per interest period**, then the amount $A(n)$ accumulated after **n interest periods** is given by: $$A(n) = A_0(1 + i)^n$$
Annual Future Value at Compound Interest	If an amount A_0 is invested at an annual interest rate of r compounded m times a year, then the amount $A(t)$ accumulated after t years is given by: $$A(t) = A_0\left(1 + \frac{r}{m}\right)^{mt}$$
Continuous Compounding	If an amount A_0 is invested at an annual interest rate of r, and the interest is compounded continuously, then the amount $A(t)$ accumulated after t years is given by: $$A(t) = A_0 e^{rt}$$

438   Chapter 9 Exponential and Logarithmic Functions

SOUND INTENSITY	The intensity level of sound, in decibels, is given by: $$L = 10 \log \frac{I}{I_0}$$ where I is the sound intensity in Watts per square meter, and $$I_0 = 10^{-12} \frac{\text{Watts}}{m^2}.$$
EARTHQUAKE MAGNITUDE	The magnitude R (on the Richter scale) of an earthquake of intensity I is given by: $$R = \log \frac{I}{I_0}$$ where I_0 is a "minimum" intensity used for comparison.

Chapter Review Exercises   **439**

CHAPTER REVIEW EXERCISES

Exercises 1–4: Sketch the graph of the function, indicating the domain and range, x- and y-intercepts, and vertical and horizontal asymptotes (if any).

1. $f(x) = 2^{x+2} - 1$

2. $f(x) = -\left(\frac{1}{2}\right)^x + 8$

3. $f(x) = -2\log_2(x-3) + 4$

4. $f(x) = \log_{\frac{1}{2}}(x+3) - 2$

Exercises 5–6: Express in logarithmic form.

5. $3^4 = 81$

6. $6^{-2} = \frac{1}{36}$

Exercises 7–8: Express in exponential form.

7. $\log_2 \frac{1}{4} = -2$

8. $\log_2 32 = 5$

Exercises 9–12: Determine the exact value of the expression (without a calculator).

9. $\log_3 9$

10. $\log_9 3$

11. $\log_{\frac{1}{2}} 8$

12. $\dfrac{2\ln e^3 - e^{3\ln 4}}{10\log\frac{1}{10} + \log 1^8}$

Exercises 13–14: Express as a single logarithm. Assume all expressions are defined.

13. $2\log_4 5 - \log_4 x + \log_4 3$

14. $3\log(x+1) - \log(x-1)$

Exercises 15–16: Expand the single logarithm to a sum and/or difference of logarithms. Assume each variable represents a positive number.

15. $\ln\sqrt{x^3\sqrt{y}}$

16. $\log_3 \dfrac{(xy)^2(x+z)}{27w}$

Exercises 17–20: Use a calculator to approximate the given value.

17. $\dfrac{\log e}{\ln 10} - \log_2 5$

18. $\log_3 2$

19. $10^e - e^e$

20. $e^4 - \frac{1}{e}$

Exercises 21–39: Solve the equation.

21. $9^{\sqrt{x^2-9}} = 3$

22. $4^{-3x} = \dfrac{16^{3x+2}}{2^{x-1}}$

23. $\left(\frac{1}{4}\right)^{4-x} = \left(\frac{1}{2}\right)^{3x+1}$

24. $\left(\frac{1}{9}\right)^{2x+3} = \left(\frac{1}{27}\right)^{x+1}$

25. $3^{-4\sin^2 x} = \frac{1}{27}, \ 0 \le x < 2\pi$

26. $2 \cdot 4^{2x+2} - 18 \cdot 4^x + 1 = 0$

27. $2^{6x-4} = 3 \cdot 5^{2-x}$

28. $e^{x+2} = 7$

29. $\tan(e^{6x-1}) = 1$

30. $\log x^4 = \log x^2$

31. $\ln(2x^2 - 9) = \ln(3x)$

32. $\log(4\sin^2 x - 1) = \log(4\cos x + 4)$

33. $\ln(3\sec^2 x) = \ln(\log 10000)$

34. $\ln x = -3$

35. $\log(2x-5) + \log(x-3) = 1$

36. $\log_5(x^2 + 8x + 13) = \log_5(x+4) - \log_5 2$

37. $[\log_4(x+1)]^2 - 2\log_4(x+1) = 3$

38. $2\log_3(x-4) - \log_3 x = 2$

39. $[\ln\cot x - \ln\cos x]\ln(e^{\cos 2x}e^{2\sin^2 x}) = \ln 2$

440   Chapter 9   Exponential and Logarithmic Functions

Exercises 40–50: Solve the inequality.

40. $\left(\frac{1}{3}\right)^{2x^2-x} \le 0$

41. $-3^x < \dfrac{9}{-9^x}$

42. $\left(\frac{1}{3}\right)^{\sqrt{x+1}} > \left(\frac{1}{9}\right)^2$

43. $3^{x-1} \le 2$

44. $5^x > 4^{3x+2}$

45. $2^{1-2x} \ge \dfrac{4}{3^{4x}}$

46. $\ln(6x+1) > \ln(9x-2)$

47. $\log_{\frac{1}{2}}(3x+7) < \log_{\frac{1}{2}}(2x-9)$

48. $\log x < 2$

49. $\log_3(x+2) \ge 0$

50. $\log_{\frac{1}{4}}(3x-2) > 1$

Exercises GC51–GC55: Use a graphing calculator to solve, approximately.

51. $-3x^2 + 2x + 1 = -15e^{-0.3x}$

52. $\ln(x+3) = \dfrac{1}{2x^2}$

53. $\log(x-250) = (x-300)^2 - 5$

54. $\ln(x+5) < \dfrac{5}{x^3}$

55. $x - 4 > -100 \cdot 6^{-x}$

Exercise 56: (Population) The population of a certain small town is given approximately by $P(t) = 3000\, e^{0.046t}$ where t is the time in years, with $t = 0$ denoting 1996. Find the population in 1994 and 2004.

Exercise 57: (Radon) There is growing concern about radon as a health hazard, because it seeps into houses through cracks in walls and floors. Radon tests are now routinely performed in order to measure the amount of radon $^{222}_{86}\text{Rn}$ present.

Suppose that 3.5×10^7 radon atoms are trapped in a basement at the time the basement is sealed against further entry of the gas. How many radon atoms remain after 25 days, if the half-life of radon is 3.83 days?

Exercise 58: (Half-Life) A radioactive isotope has a half-life of 25 years. If there are 250 mg present initially, then the amount $A(t)$ remaining after t years is given by

$$A(t) = 250 \cdot 2^{-\frac{t}{25}}$$

(a) How much remains after 25 years? 50 years? 55 years?
(b) Sketch the graph of $A(t)$ from $t = 0$ to $t = 100$ with a graphing calculator and check your results in (a).

Exercise 59: (Half-Life) Radioactive lead-212 has a half-life of 11 days. How long will it take a 10 gram sample to decay to 7 grams?

Exercise 60: (Noise Levels) The following table lists some noise levels.

	Intensity I (Watts/m²)	Intensity level L (db)
Threshold of hearing	10^{-12}	0
Normal conversation	?	60
Live rock concert	1.0	?
Threshold of pain	?	130

(a) Fill in the missing values in the table.

(b) By what minimal factor must the sound intensity at a live rock concert be reduced so as to be at most ten times the intensity of normal conversation?

(c) By what factor will the intensity level at a live rock concert be reduced in (b)?

(d) How many people at a rock concert would have to hold a normal conversation simultaneously in order to produce a sound intensity level equal to that of the concert?

Exercise 61: (**Compound Interest**) How long will it take an investment of $15,000 at an annual interest rate of $5\frac{1}{4}\%$ to grow to $25,000, if interest is compounded

   (a) annually   (b) quarterly   (c) monthly   (d) continuously

Exercise 62: (**Earthquake Magnitude**) Compare the intensity levels of the following three earthquakes, and determine how much greater was the level of the Japan quake than the Alaska quake, the Alaska quake than the India quake, and the Japan quake than the India quake.

$$\begin{array}{ll} \text{Japan in 2003,} & \text{R} = 8.3 \\ \text{Alaska in 2002,} & \text{R} = 7.9 \\ \text{India in 1993,} & \text{R} = 6.4 \end{array}$$

Exercises 63–69: Indicate True or False. Justify your answer.

63. The function $f(x) = \dfrac{1}{2^{-x}}$ is a decreasing function.

64. The domain of $g(x) = f(e^x)$ always coincides with the domain of f.

65. The domain of $g(x) = e^{f(x)}$ always coincides with the domain of f.

66. $\log_6(2 \cdot 3^x) = x$

67. The domain of $f(x) = \ln(\ln x)$ is $(1, \infty)$.

68. $10^{\log x + \log y} = x + y$, if $x > 0$ and $y > 0$.

69. $10^{\log x + \log y} = xy$, if $x > 0$ and $y > 0$.

CUMULATIVE REVIEW EXERCISES

Exercises 1–3: Simplify

1. $\left[\dfrac{-2^2(-ab^2)^3}{a^{-\frac{1}{3}}(-2b)^{-2}} \right]^2$

2. $\dfrac{(1+x^2)^2(2) - 2x(2)(1+x^2)(2x)}{(1+x^2)^4}$

3. $\dfrac{(e^x - e^{-x})^2 - (e^x + e^{-x})^2}{(e^x + e^{-x})^2} \cdot (-e^x - e^{-x})$

442   Chapter 9   Exponential and Logarithmic Functions

Exercise 4:   Find the partial fraction decomposition of $\dfrac{x^2 - 3x - 2}{x^3 + x^2 + x}$.

Then use a graphing calculator to check the decomposition, by graphing the original expression and the sum of the partial fractions, and checking that the two graphs coincide.

Exercise 5:   Express $\tan\left(\sin^{-1}\frac{3x}{4}\right)$ as an algebraic function of x, for $x > 0$.

Exercises 6–16:   Solve the equation or inequality.

6. $\dfrac{3x - 1}{x^2 + 4x + 3} + \dfrac{3}{x + 1} = \dfrac{4x}{x^2 - 2x - 3}$

7. $\dfrac{x - 2}{x^2} < \dfrac{1}{x + 1}$

8. $\dfrac{4^x}{16^{4x-1}} = \dfrac{1}{4}$

9. $2^{x-3} = e^{-2x}$

10. $3^{2x+2} + 17 \cdot 3^x - 2 = 0$

11. $4^{2\cos^2 x + 4} = 16^{2 - \frac{1}{2}\cos x}$, in $[0, 2\pi)$

12. $3^{4x} > 2$

13. $\ln(3x + 2) - \ln x = \ln(x + 2)$

14. $\log(x - 15) + \log x = 2$

15. $\ln 2 - \ln \sin x = \ln(6\sin x + 1)$, in $\left(-\frac{\pi}{2}, \frac{\pi}{2}\right)$

16. $\log_2(4 - x) \le 3$

Exercise 17:   Determine the amplitude, period, and phase shift, and sketch the graph of $f(x) = 4\cos(3x + \pi)$.

Exercise 18:   Find the domain and range of the function $f(x) = e^x - 1$.

Exercises 19–20:   Determine which, if any, of the following characteristics can be used to describe the function: increasing, decreasing, even, odd, one-to-one, invertible. In the event that the inverse function exists, find it. Justify your answers.

19. $h(x) = -\frac{1}{8}(x + 1)^3$

20. $k(x) = \log_3(x + 2)$

Exercise 21:   Evaluate each of the following.

(a) $\cot\frac{\pi}{3}$

(b) $\csc\frac{\pi}{6}$

(c) $\tan 0$

(d) $\dfrac{10^{-\log\frac{1}{2}} - 5\ln e^3 + e^{2\ln 3}}{37\ln 1 - \log 100}$

Exercise 22:   Express $\log_b\left(\frac{200}{9}\right)$ in terms of A, B, and C, where $A = \log_b 2$, $B = \log_b 3$, and $C = \log_b 5$.

Exercises 23–24:   Suppose that $x > 1$.

23. Express the following as a single logarithm:

$$2\ln(x - 1) - \frac{1}{2}\ln x + 3\ln(x + 1)$$

24. Express the following as a sum or difference of logarithms:

$$\log_{25}\left[\frac{x^{\frac{2}{3}}\sqrt{x-1}}{5(x+2)^4}\right]$$

Exercise 25: Determine the exact solution, and then approximate it to 4 decimal places.

$$\log_{\sqrt{2}}(x-1) = \pi$$

Exercise 26: (**Star Magnitudes**) Stars are classified into categories of brightness called magnitudes. The faintest stars (with light flux L_0) are assigned a magnitude of 6. Brighter stars are assigned magnitudes according to the formula $m = 6 - \frac{5}{2}\log\left(\frac{L}{L_0}\right)$, where L is the light flux from the star.

    (a) Find m if $L = 10^{0.4}L_0$.   (b) Solve the formula for L in terms of m and L_0.

Exercise 27: (**Radioactivity**) A certain radioactive substance loses $\frac{1}{3}$ of its original mass in 3 days. How long will it take to lose $\frac{1}{5}$ of its original mass?

Exercise 28: (**Sound Intensity**) The sound intensity level of a purring cat is 10 decibels while that of a power lawnmower is 100 decibels. How many times more intense is the sound of the mower than that of the purring cat?

Exercise 29: (**Compound Interest**) Suppose that you invest $10,000 at $6\frac{1}{2}\%$ compounded semi-annually, and $13,000 at 5% compounded quarterly. When will the value of the $10,000 investment exceed that of the $13,000 investment?

444

APPENDIX A

CONIC SECTIONS

Conic sections, or simply **conics**, are curves resulting from the intersection of a plane with a right circular cone of two napes. The conic obtained when the slicing plane is parallel to an edge of the cone is said to be a **parabola** [Figure A.1(a)]. As the plane takes on a different inclination, the curve of intersection may be an **ellipse** [Figure A.1(b)], or, in the event that the plane is parallel to the axis of the cone, a **hyperbola** [Figure A.1(c)]. The **circle** is a special case of the ellipse, and it is that curve obtained when the plane is parallel to the base of the cone [Figure A.1(b)].

Glancing at Figure A.1 and mentally moving the slicing plane so that it passes through the vertex of the cone you will arrive at what are called the **degenerate conics**: the parabola of (a) degenerates into a line; the ellipse (or circle) of (b) into a point; and the hyperbola of (c) into two lines intersecting at the vertex.

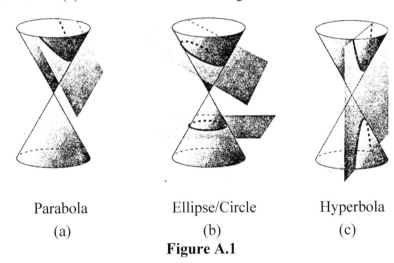

Parabola   Ellipse/Circle   Hyperbola
(a)   (b)   (c)

Figure A.1

The purpose of this section is to present the bare essentials of the study of conic sections: the link between the non-degenerate conics and equations of the form:

$$Ax^2 + By^2 + Cx + Dy + E = 0$$

where not both A and B are zero. We begin by asserting, without proof, that:

Geometrical properties of conic sections were extensively studied by the ancient Greeks (500 B.C.). A more analytical consideration emerged in the seventeenth century, and it paved the way for the development of the calculus. Nature itself underlines the importance of conic sections: elliptical and hyperbolic orbits of planets and comets; hyperbolic paths of alpha particles; parabolic paths of projectiles; and so on. We also use conical devices in numerous ways: parabolic- and hyperbolic-shaped radar receivers and transmitters, television disks, telescopes and other optical devices, construction arches and suspension devices, to name a few.

A-2   Appendix A

Please note that the graph of:

$$Ax^2 + By^2 + Cx + Dy + E = 0$$

could turn out to be a degenerate conic (lines or a single point), or even nothing at all [when no point (x, y) satisfies the equation].

THEOREM A.1

TYPE OF CONIC

If the graph of the equation

$$Ax^2 + By^2 + Cx + Dy + E = 0$$

is a non-degenerate conic, then that conic is:

A **PARABOLA**, if $AB = 0$ (with not both A and B zero).

A **CIRCLE**, if $A = B \neq 0$.

An **ELLIPSE** (not a circle), if $AB > 0$ and $A \neq B$.

A **HYPERBOLA**, if $AB < 0$.

CHECK YOUR UNDERSTANDING A.1

Assuming that the graph is a conic, is it the graph of a parabola, circle, ellipse, or hyperbola?

(a) $x^2 + 2x - 4y + 13 = 0$

(b) $4x^2 + 4y^2 - 4x + 16y + 5 = 0$

(c) $4x^2 - 9y^2 + 24x + 18y + 63 = 0$

(d) $4x^2 - 16x + 3y^2 - 6y + 7 = 0$

Answers: (a) parabola
(b) circle (c) hyperbola
(d) ellipse

STANDARD FORMS

Every equation of a conic, $Ax^2 + By^2 + Cx + Dy + E = 0$, can be written in one of the standard forms in the left column below. From the standard form of the equation it is easy to sketch the graph of the conic, as is illustrated by the examples in the accompanying right column.

Standard Form	Example
PARABOLA — **Axis parallel to *Y*-axis**	

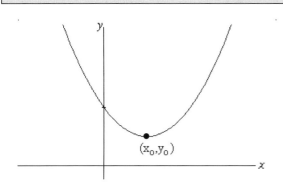

$y - y_0 = a(x - x_0)^2$, $a > 0$
(opens up)
Vertex: (x_0, y_0)

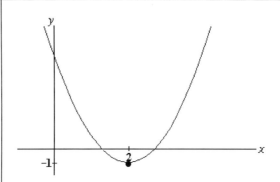

$y + 1 = 2(x - 2)^2$
(opens up, since $a = 2 > 0$)
Vertex: $(2, -1)$

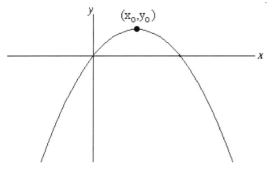

$y - y_0 = a(x - x_0)^2$, $a < 0$
(opens down)
Vertex: (x_0, y_0)

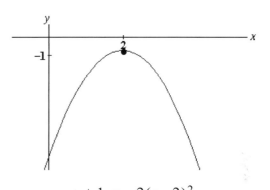

$y + 1 = -2(x - 2)^2$
(opens down, since $a = -2 < 0$)
Vertex: $(2, -1)$

PARABOLA
Axis Parallel to X-Axis

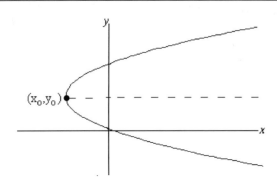

$x - x_0 = a(y - y_0)^2, \ a > 0$

(opens to the right)

Vertex: (x_0, y_0)

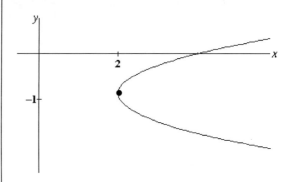

$x - 2 = 2(y + 1)^2$

(opens to the right, since $a = 2 > 0$)

Vertex: $(2, -1)$

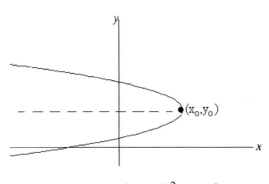

$x - x_0 = a(y - y_0)^2, \ a < 0$

(opens to the left)

Vertex: (x_0, y_0)

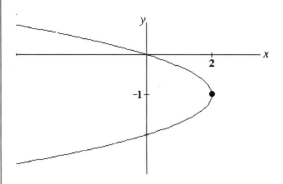

$x - 2 = -2(y + 1)^2$

(opens to the left, since $a = -2 < 0$)

Vertex: $(2, -1)$

Conic Sections   A-5

STANDARD FORM	EXAMPLE
CIRCLE	

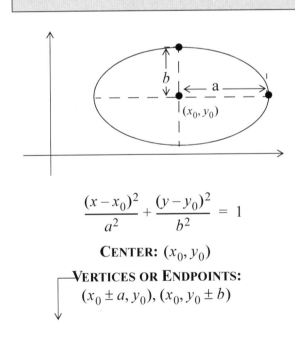

$(x - x_0)^2 + (y - y_0)^2 = r^2$	$(x - 2)^2 + (y + 1)^2 = 4$
CENTER: (x_0, y_0)	**CENTER:** $(2, -1)$
RADIUS: r	**RADIUS:** 2

ELLIPSE

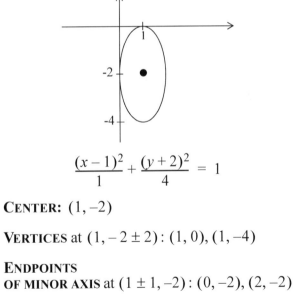

$\dfrac{(x - x_0)^2}{a^2} + \dfrac{(y - y_0)^2}{b^2} = 1$

CENTER: (x_0, y_0)

VERTICES OR ENDPOINTS:
$(x_0 \pm a, y_0), (x_0, y_0 \pm b)$

$\dfrac{(x - 1)^2}{1} + \dfrac{(y + 2)^2}{4} = 1$

CENTER: $(1, -2)$

VERTICES at $(1, -2 \pm 2)$: $(1, 0), (1, -4)$

ENDPOINTS OF MINOR AXIS at $(1 \pm 1, -2)$: $(0, -2), (2, -2)$

NOTE: The two dashed line segments intersecting at the center of the ellipse are called the **axes** of the ellipse. The longer axis is called the **major axis**, and the shorter one the **minor axis**. The two endpoints on the major axis are said to be the VERTICES of the ellipse, and the other two are called the ENDPOINTS of the minor axis. (If the axes are of the same length, $a = b$, then the ellipse is a circle.)

HYPERBOLA
TRANSVERSE AXIS PARALLEL TO THE X-AXIS
(bends about the x-axis)

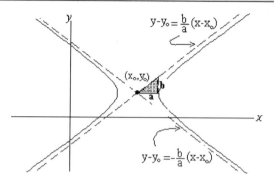

$$\frac{(x-x_0)^2}{a^2} - \frac{(y-y_0)^2}{b^2} = 1$$

CENTER: (x_0, y_0)

VERTICES: $(x_0 \pm a, y_0)$

OBLIQUE ASYMPTOTES:
$$y - y_0 = \pm \frac{b}{a}(x - x_0)$$

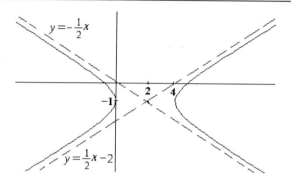

$$\frac{(x-2)^2}{4} - \frac{(y+1)^2}{1} = 1$$

CENTER: $(2, -1)$

VERTICES at $(2 \pm 2, -1)$: $(0, -1), (4, -1)$

OBLIQUE ASYMPTOTES: $y + 1 = \pm \frac{1}{2}(x - 2)$

HYPERBOLA
TRANSVERSE AXIS PARALLEL TO THE Y-AXIS
(bends about the y-axis)

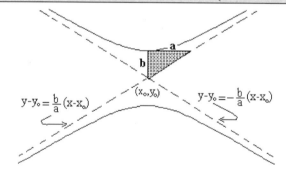

$$\frac{(y-y_0)^2}{b^2} - \frac{(x-x_0)^2}{a^2} = 1$$

CENTER: (x_0, y_0)

VERTICES: $(x_0, y_0 \pm b)$

OBLIQUE ASYMPTOTES:
$$y - y_0 = \pm \frac{b}{a}(x - x_0)$$

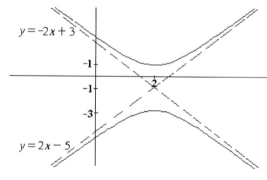

$$\frac{(y+1)^2}{4} - \frac{(x-2)^2}{1} = 1$$

CENTER: $(2, -1)$

VERTICES at $(2, -1 \pm 2)$: $(2, -3), (2, 1)$

OBLIQUE ASYMPTOTES: $y + 1 = \pm 2(x - 2)$

Conic Sections A-7

Answers:
(a) Ellipse; center: $(2, -3)$;
vertices: $(-1, -3)$, $(5, -3)$;
endpoints of minor axis:
$(2, -5), (2, -1)$
For graph, see page C-73.
(b) Parabola; vertex: $(3, -4)$
For graph, see page C-73.
(c) Hyperbola; center:
$(2, -3)$; vertices: $(-1, -3)$,
$(5, -3)$; asymptotes:

$$y + 3 = \pm \frac{2}{3}(x - 2)$$

For graph, see page C-74.

CHECK YOUR UNDERSTANDING A.2

Sketch the conic with given equation. If it is a parabola, find its vertex. If it is a circle, find its center and radius. If it is an ellipse, find its center, vertices, and endpoints of the minor axis. If it is a hyperbola, find its center, vertices, and the equations of its oblique asymptotes.

(a) $\dfrac{(x-2)^2}{9} + \dfrac{(y+3)^2}{4} = 1$ (b) $x - 3 = 4(y+4)^2$

(c) $\dfrac{(x-2)^2}{9} - \dfrac{(y+3)^2}{4} = 1$

We have noted that not every equation of the form $Ax^2 + By^2 + Cx + Dy + E = 0$ represents a conic section. These cases reveal their true nature when the equation is placed in standard form. For example, we can see that there can be no curve associated with the equation:

$$(x - 3)^2 + (y + 4)^2 = -3$$

as it has no solution (it wants to be a circle with center $(3, -4)$, but what would be its radius?). By the same token, the only point on the aspiring ellipse given by:

$$\frac{(x-2)^2}{3} + \frac{(y+4)^2}{4} = 0$$

is the point $(2, -4)$.

OBTAINING THE STANDARD-FORM EQUATION

As you will see in the following examples, the basic tool needed to place a given equation in standard-form is the completing-the-square method of page 53.

EXAMPLE A.1 Write the equation:

$$4x^2 + 4y^2 - 4x + 16y + 5 = 0$$

in standard form. Determine if it represents a conic section, and, if it does, identify it and sketch its graph.

SOLUTION: Since the coefficients of x^2 and y^2 are equal, if the equation represents a conic section, then it must be a circle. To gain more information, we need to rewrite the equation is standard-form. This is accomplished by completing the square in both the variables x and y:

$$4x^2 + 4y^2 - 4x + 16y + 5 = 0$$
$$x^2 + y^2 - x + 4y = -\frac{5}{4}$$
$$(x^2 - x \quad) + (y^2 + 4y \quad) = -\frac{5}{4}$$
$$\left(x^2 - x + \frac{1}{4}\right) + (y^2 + 4y + 4) = -\frac{5}{4} + \frac{1}{4} + 4$$
$$\left(x - \frac{1}{2}\right)^2 + (y+2)^2 = 3$$

We conclude that:
$$4x^2 + 4y^2 - 4x + 16y + 5 = 0$$
or: $\left(x - \frac{1}{2}\right)^2 + (y+2)^2 = 3$

is the equation of the circle with center at $(\frac{1}{2}, -2)$ and radius $\sqrt{3}$.

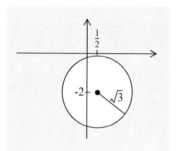

EXAMPLE A.2 Write the equation:
$$2x^2 + y^2 - 8x - 2y + 5 = 0$$
in standard form. Determine if it represents a conic section, and, if it does, identify it and sketch its graph.

SOLUTION: Since the coefficient of x^2 differs from that of y^2, but they are of the same sign, if the equation represents a conic section, then it must be an ellipse. We rewrite the equation in standard-form:
$$2x^2 + y^2 - 8x - 2y + 5 = 0$$
$$2(x^2 - 4x \quad) + y^2 - 2y = -5$$
$$2(x^2 - 4x + 4) + y^2 - 2y + 1 = -5 + 8 + 1$$
$$2(x-2)^2 + (y-1)^2 = 4$$
$$\frac{(x-2)^2}{2} + \frac{(y-1)^2}{4} = 1$$

We find that:
$$2x^2 + y^2 - 8x - 2y + 5 = 0$$
or: $\frac{(x-2)^2}{2} + \frac{(y-1)^2}{4} = 1$

is the equation of the ellipse with center at $(2, 1)$; vertices at:

$$(2, 1 \pm 2) = \begin{cases} (2, 3) \\ (2, -1) \end{cases}$$

and endpoints of its minor axis at:

$$(2 \pm \sqrt{2}, 1) = \begin{cases} (2 + \sqrt{2}, 1) \\ (2 - \sqrt{2}, 1) \end{cases}$$

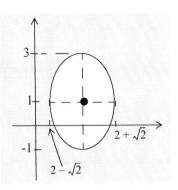

EXAMPLE A.3 Write the equation:
$$4x^2 - 9y^2 + 24x + 18y + 63 = 0$$
in standard form. Determine if it represents a conic section, and, if it does, identify it and sketch its graph.

SOLUTION: Since the **sign** of x^2 differs from that of y^2, if the equation represents a conic section, it is a hyperbola. Rewriting the equation in standard-form:

$$4x^2 - 9y^2 + 24x + 18y + 63 = 0$$
$$4(x^2 + 6x) - 9(y^2 - 2y) = -63$$
$$4(x^2 + 6x + 9) - 9(y^2 - 2y + 1) = -63 + 36 - 9$$
$$4(x + 3)^2 - 9(y - 1)^2 = -36$$
$$\frac{4(x+3)^2}{-36} - \frac{9(y-1)^2}{-36} = 1$$
$$\frac{(y-1)^2}{4} - \frac{(x+3)^2}{9} = 1$$

we see that $(-3, 1)$ is the center of the hyperbola
$$4x^2 - 9y^2 + 24x + 18y + 63 = 0$$
or: $\quad \dfrac{(y-1)^2}{4} - \dfrac{(x+3)^2}{9} = 1$

and that its axis is parallel to the y-axis (the positive sign is in front of the y-term). The vertices are $\sqrt{4} = 2$ units up and down from the center:

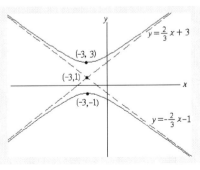

$$(-3, 1 \pm 2) = \begin{cases} (-3, 3) \\ (-3, -1) \end{cases}$$

and that the oblique asymptotes are:
$$y - 1 = \pm \frac{2}{3}(x + 3)$$

EXAMPLE A.4

Write the equation:
$$y^2 - 4x + 2y + 13 = 0$$
in standard form. Determine if it represents a conic section, and, if it does, identify it and sketch its graph.

SOLUTION: Since there is a y^2 but no x^2 in the equation, if the equation represents a conic section, it is a parabola with axis parallel to the x-axis. Rewriting the equation in standard-form:

$$y^2 - 4x + 2y + 13 = 0$$
$$4x - 13 = y^2 + 2y$$
$$4x - 13 + 1 = y^2 + 2y + 1$$
$$4x - 12 = (y + 1)^2$$
$$x - 3 = \frac{1}{4}(y + 1)^2$$

se now see that the vertex of the parabola:

$$y^2 - 4x + 2y + 13 = 0$$

or: $x - 3 = \frac{1}{4}(y + 1)^2$ (*)

is at $(3, -1)$, and that it opens to the right (since $a = \frac{1}{4} > 0$).

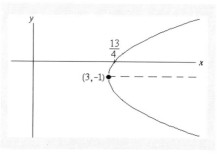

To get a better sense of its curvature, we determined an additional point on the parabola, its x-intercept, by setting $y = 0$ in (*) and then solving for x:

$$x - 3 = \frac{1}{4}(0 + 1)^2$$
$$x - 3 = \frac{1}{4}$$
$$x = \frac{13}{4}$$

Answers:
(a) Circle; center: $(-1, -1)$; radius: $\sqrt{5}$
For graph, see page C-74.
(b) Ellipse; center: $(1, -2)$ vertices: $(1, 1)$, $(1, -5)$; endpoints of minor axis:
$(1 - \sqrt{3}, -2), (1 + \sqrt{3}, -2)$
For graph, see page C-74.
(c) Parabola; vertex: $\left(\frac{1}{2}, \frac{1}{4}\right)$
For graph, see page C-75.
(d) Hyperbola; center: $(1, -2)$; vertices: $(0, -2)$, $(2, -2)$; asymptotes:
$$y + 2 = \pm 2(x - 1)$$
For graph, see page C-75.

CHECK YOUR UNDERSTANDING A.3

Sketch the conic with given equation. If it is a parabola, find its vertex. If it is a circle, find its center and radius. If it is an ellipse, find its center, vertices, and endpoints of the minor axis. If it is a hyperbola, find its center, vertices, and the equations of its oblique asymptotes.

(a) $x^2 + 2x - 3 + 2y = -y^2$ 　　(b) $3x^2 + y^2 = 6x - 4y + 2$

(c) $x^2 - x - 3y + 1 = 0$ 　　(d) $4x^2 - 4y - 4 = y^2 + 8x$

Conic Sections   A–11

EXERCISES

Exercises 1–10: Given that the equation represents a conic, identify it as a parabola, circle, ellipse, or hyperbola.

1. $3x^2 + 4y^2 - x - y - 20 = 0$

2. $-3x^2 + 4y^2 - x - y - 20 = 0$

3. $3x^2 - 4y^2 - x - y - 20 = 0$

4. $3x^2 + 4y^2 + x + y - 20 = 0$

5. $3x + 4y^2 + y - 20 = 0$

6. $-3x^2 - 4y^2 - x - y + 20 = 0$

7. $3x^2 + 4y - x - 20 = 0$

8. $3x^2 = 4y^2 + x + y + 20$

9. $3x^2 + 3y^2 - x - y - 20 = 0$

10. $3x^2 = 3y^2 + x + y + 20$

Exercises 11–20: Identify and sketch the given conic. If it is a parabola, find the vertex. If it is a circle, find the center and radius. If it is an ellipse, find the center, vertices, and endpoints of the minor axis. If it is a hyperbola, find the center, vertices and asymptotes.

11. $(x+3)^2 + (y-2)^2 = 25$

12. $y + \frac{1}{2} = 2\left(x - \frac{1}{3}\right)^2$

13. $\frac{(x-3)^2}{4} + \frac{(y-2)^2}{25} = 1$

14. $\frac{(x+1)^2}{9} - \frac{(y-3)^2}{25} = 1$

15. $x + 2 = -2(y-1)^2$

16. $16(y+3)^2 - (x-4)^2 = 16$

17. $4x^2 + y^2 = 1$

18. $x = -\frac{1}{2}(y-4)^2$

19. $(x-3)^2 - (y-2)^2 = 4$

20. $\frac{(x-5)^2}{4} + \frac{(y+1)^2}{4} = 4$

Exercises 21–28: Determine if the given equation is that of a parabola. If it is, find the vertex, the x- and y-intercepts, and sketch the graph.

21. $3y^2 + 12y - x + 13 = 0$

22. $4y^2 - 24y + x + 38 = 0$

23. $2y^2 + x - 20y + 42 = 0$

24. $4y^2 + 8y - x = 0$

25. $x^2 + y = -2x + 8$

26. $-2x^2 - y - 16x = 33$

27. $9x^2 - 2y = 6x - 1$

28. $y - 3\sqrt{2} = \sqrt{2}x^2 - 4x$

Exercises 29–37: Determine if the given equation is that of a circle. If it is, find the center and radius of the circle, and sketch the graph.

29. $x^2 + 2x + y^2 + 2y = -1$

30. $x^2 + 2x + y^2 + 2y = 1$

31. $x^2 + y^2 + 2x - 2y = 7$

32. $x^2 + 2x + y^2 + 2y = -3$

33. $y^2 - 16 = 6x - x^2 - 9$

34. $2x^2 + 3x + 2y^2 + y = 4$

35. $x^2 - 2x + y^2 - 6y + 11 = 0$

36. $x^2 + y^2 = 4x + 6y - 11$

37. $x^2 + y^2 = 4x + 6y - 8$

A–12   Appendix A

Exercises 38–47:   Determine if the given equation is that of an ellipse. If it is, find the center, vertices, and endpoints of the minor axis, and sketch the graph.

38. $2x^2 - 4x + 5y^2 - 2y = -12$ 39. $4x^2 + 9y^2 = 18y - 32x - 37$

40. $16x^2 + y^2 + 8y + 32x = -16$ 41. $y^2 + 6y = -3x^2$

42. $x^2 + 3y^2 - 4x + 6y = 2$ 43. $16x^2 + 9y^2 + 64x - 54y = -1$

44. $x^2 = -9y^2 + 36y$ 45. $x^2 + 4y^2 = 2x + 15$

46. $5x^2 + y^2 = 20x - 29 + 6y$ 47. $x^2 + 36y^2 - 4x + 72y + 31 = 0$

Exercises 48–55:   Determine if the given equation is that of a hyperbola. If it is, find the center, vertices, and asymptotes, and sketch the graph.

48. $3y^2 - 2x^2 = 6$ 49. $y^2 - x^2 + 2y - 8x = 31$

50. $y^2 + 6y = x^2$ 51. $x^2 - 9y^2 = 18(y + 1)$

52. $3y^2 - 6y - 4x^2 = 24x + 33$ 53. $y^2 - 16x^2 = 2y + 15$

54. $9x^2 + 108x - 4y^2 + 16y + 272 = 0$ 55. $16x^2 + 64x - 4y^2 + 8y = 4$

Exercises 56–63:   Sketch the given conic. If it is a parabola, find the vertex. If it is a circle, find the center and radius. If it is an ellipse, find the center, vertices, and endpoints of the minor axis. If it is a hyperbola, find the center, vertices and asymptotes.

56. $x^2 + 2x - y^2 - 4y = 3$ 57. $x^2 + y^2 - 6x + 8y = 0$

58. $-4x^2 - y - 24x - 31 = 0$ 59. $4x^2 - 8x + 9y^2 + 54y + 49 = 0$

60. $9x^2 - 36x + 9y^2 + 6y = -19$ 61. $\frac{1}{2}x^2 - y - 6x + 13 = 0$

62. $2x^2 - 8x - y^2 - 2y = -1$ 63. $3y^2 + 30y - \frac{16}{3}x^2 = -27$

Exercises 64–75:   Identify the conics represented by the two equations, and determine their points of intersection.

64. $x^2 - y^2 = 1$; $4y^2 - x^2 = 1$ 65. $x^2 + \dfrac{y^2}{4} = 1$; $\dfrac{x^2}{4} + y^2 = 1$

66. $x^2 + y^2 = 4$; $\dfrac{x^2}{9} + y^2 = 1$ 67. $(x+1)^2 + (y-1)^2 = 2$; $x^2 - (y-1)^2 = 3$

68. $y + 1 = 3(x - 2)^2$; $(x - 2)^2 - (y + 1)^2 = 1$

69. $2x^2 + y^2 = 9$; $x^2 - x + 2y^2 = 4$ 70. $\dfrac{x^2}{4} - \dfrac{y^2}{9} = 1$; $x^2 + y^2 = 4$

71. $2 - 2x = 3y^2$; $x^2 + \dfrac{y^2}{4} = 1$ 72. $x^2 - y^2 = 1$; $x + 1 = y^2$

73. $x^2 - 2y^2 = 8$; $x^2 + 2y^2 = 2$ 74. $x^2 + (y - 3)^2 = 50$; $(y - 3)^2 = 25x - 100$

75. $\dfrac{y^2}{2} + \dfrac{(x + 2)^2}{4} = 1$; $3y^2 - (x + 2)^2 = 1$

Conic Sections   A–13

Exercises GC76–GC82: Identify the conics represented by the two equations, and use a graphing calculator to determine the approximate coordinates of any points of intersection.

76. $14x^2 + 14y^2 = 165;\quad 16x^2 - 5x - 9(y-1)^2 = 3$

77. $\frac{1}{2}x^2 - \frac{2}{3}y^2 = 1;\quad \frac{(x-5)^2}{6} - \frac{(y+5)^2}{7} = 1$

78. $x^2 - 5x + y^2 - 4y = 6;\quad 7y^2 - 13(x-7)^2 = 1$

79. $x - 2 = -3(y+1)^2;\quad \frac{(x-2)^2}{3} - \frac{y^2}{5} = 1$

80. $-(x+1)^2 + (y-3)^2 = 6;\quad x = y^2 - 3$

81. $y^2 - 3x^2 + 4x = 2;\quad 0.03x^2 + 0.02x + 0.11y^2 = 1$

82. $0.1x - 0.01y = \frac{1}{11};\quad \frac{(x+0.2)^2}{0.33} - \frac{(y-0.4)^2}{0.41} = 1$

Exercise 83: In the calculus, you will be able to show that the area of an elliptical region is given by the formula $A = \pi ab$, where a and b are as on page A-5.

When a cylinder containing water is tilted, the water surface is elliptical. Find the area of the water surface, depicted in the adjacent figure.

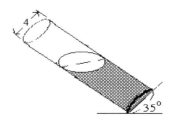

Exercise 84: Consider the equation

$$Ax^2 + Cy^2 + Dx + Ey + F = 0 \qquad (*)$$

where A and C are positive, and $A \neq C$. Let

$$r = \frac{D^2}{4A} + \frac{E^2}{4C} - F$$

Show that if

(a) $r > 0$, then (*) is the equation of an ellipse.

(b) $r = 0$, then only one point satisfies (*).

(c) $r < 0$, then no point satisfies (*).

Exercise 85: The vertices of a rectangle lie on the hyperbola

$$\frac{x^2}{9} - \frac{y^2}{4} = 1$$

and the x-coordinate of one of the vertices is 4. Find the coordinates of all of the vertices.

A–14   Appendix A

Exercise 86:   Find the area of the rectangle with each of its vertices on the hyperbola

$$\frac{x^2}{9} - \frac{y^2}{4} = 1$$

if its perimeter is 18 inches.

Exercises 87–90:   True or False. Justify your answer.

87. If $d < 0$, then the equation $ax^2 + bx + ay^2 + cy = d$ does not represent a circle.

88. If $d > 0$ and $a > 0$, then the equation $ax^2 + bx + ay^2 + cy = d$ represents a circle.

89. If both the equations

$$(1) \quad a_1x^2 + b_1x + a_1y^2 + c_1y = d_1$$
$$(2) \quad a_2x^2 + b_2x + a_2y^2 + c_2y = d_2$$

represent circles, and if $d_2 > d_1$, then the radius of circle (2) is larger than the radius of circle (1).

90. If both the equations

$$ax^2 + bx + ay^2 + cy = d_1$$
$$ax^2 + bx + ay^2 + cy = d_2$$

represent circles, then the circles will have the same center.

Solutions to Check Your Understanding Boxes   B–1

SOLUTIONS TO CHECK YOUR UNDERSTANDING BOXES

CHAPTER 1

Solution to CYU 1.1

(a)   As $a^0 = 1$ for $a \neq 0$, $\dfrac{\left(2 - \frac{1}{3}\right)^0 2^{-3}}{3^{-1}} = \dfrac{1 \cdot \frac{1}{2^3}}{\frac{1}{3}} = \dfrac{\frac{1}{8}}{\frac{1}{3}} = \dfrac{3}{8}.$

(b)   Since $\left(\dfrac{a}{b}\right)^{-1} = \dfrac{b}{a},$ $\left[\dfrac{-2^{-2} + 2^{-1}}{(-2)^2}\right]^{-1} = \dfrac{(-2)^2}{-2^{-2} + 2^{-1}} = \dfrac{2^2}{-\frac{1}{2^2} + \frac{1}{2}} = \dfrac{4}{\frac{1}{4}} = 16.$

Solution to CYU 1.2

(a) $\sqrt{9} - 16^{\frac{1}{4}} = 3 - 2 = 1$

(b) $(-27)^{\frac{2}{3}} - 4^{-\frac{3}{2}} = (-3)^2 - \dfrac{1}{4^{\frac{3}{2}}} = 9 - \dfrac{1}{2^3} = 9 - \dfrac{1}{8} = \dfrac{71}{8}$

Solution to CYU 1.3

(a) $\dfrac{\sqrt{a}}{(-a^{-2} a^{\frac{1}{3}})^{-2}} = a^{\frac{1}{2}} \left(-\dfrac{1}{a^2} \cdot a^{\frac{1}{3}}\right)^2 = a^{\frac{1}{2}} \cdot \dfrac{1}{a^4} \cdot a^{\frac{2}{3}} = \dfrac{a^{\frac{1}{2} + \frac{2}{3}}}{a^4} = \dfrac{1}{a^{4 - \frac{1}{2} - \frac{2}{3}}}$

$= \dfrac{1}{a^{\frac{24 - 3 - 4}{6}}} = \dfrac{1}{a^{\frac{17}{6}}}$

(b) $\left[\dfrac{(a-b)^2 (a-b)^5}{(2a - 2b)^3}\right]^3 = \left[\dfrac{(a-b)^7}{2^3 (a-b)^3}\right]^3 = \left[\dfrac{(a-b)^4}{8}\right]^3 = \dfrac{1}{8^3}(a-b)^{12}$

B–2   Solutions to Check Your Understanding Boxes

Solution to CYU 1.4

(a) $(2x^4 + 3x - 5)(x^2 - x) - 8(3x^3 - x^2 - 4)$

$$= 2x^6 - 2x^5 + 3x^3 - 3x^2 - 5x^2 + 5x - 24x^3 + 8x^2 + 32$$

$$= 2x^6 - 2x^5 - 21x^3 + 5x + 32$$

(b)

$$
\begin{array}{r}
2x - 3 \\
4x-1\overline{)8x^2-14x+10} \\
\underline{8x^2- 2x} \\
-12x+10 \\
\underline{-12x+ 3} \\
7
\end{array}
$$

(c)

$$
\begin{array}{r}
4x^2- 11x - 10 \\
x^2+2\overline{)4x^4-11x^3- 2x^2\qquad - 1} \\
\underline{4x^4\qquad + 8x^2} \\
-11x^3-10x^2\qquad - 1 \\
\underline{-11x^3\qquad -22x} \\
-10x^2+22x- 1 \\
\underline{-10x^2\qquad -20} \\
22x+19
\end{array}
$$

The quotient is $q(x) = 4x^2 - 11x - 10$, while the remainder is $r(x) = 22x + 19$.

Solution to CYU 1.5

(a) $25x^2 - 1 = (5x - 1)(5x + 1)$

(b) $3x^2 - 2 = (\sqrt{3}x - \sqrt{2})(\sqrt{3}x + \sqrt{2})$

(c) $2x^2 + 7x - 4 = (2x - 1)(x + 4)$

(d) $x^3 + x^2 - 16x - 16 = (x^3 + x^2) - (16x + 16) = x^2(x + 1) - 16(x + 1)$

$$= (x^2 - 16)(x + 1) = (x - 4)(x + 4)(x + 1)$$

Solution to CYU 1.6

Since $3(2^3) - 13(2^2) + 16(2) - 4 = 24 - 52 + 32 - 4 = 0$, 2 is a zero, which means that $x - 2$ is a factor. Dividing and factoring the resulting quadratic polynomial, we have:

$$3x^3 - 13x^2 + 16x - 4 = (x - 2)(3x^2 - 7x + 2) = (x - 2)(3x - 1)(x - 2)$$
$$= (x - 2)^2(3x - 1)$$

Solution to CYU 1.7

(a) $\quad 8x^3 + 27 \quad = (2x)^3 + 3^3 = (2x + 3)[(2x)^2 - 3(2x) + 3^2]$
$\qquad\qquad\qquad = (2x + 3)(4x^2 - 6x + 9)$

(b) $\quad x^6 - 1 \;= (x^3)^2 - 1^2 = (x^3 - 1)(x^3 + 1)$
$\qquad\qquad\quad = [(x - 1)(x^2 + x + 1)][(x + 1)(x^2 - x + 1)]$
$\qquad\qquad\quad = (x - 1)(x + 1)(x^2 + x + 1)(x^2 - x + 1)$

Solution to CYU 1.8

(a)
$$\frac{6x^2(5x + 2)^4 - 3x^3(5x + 2)^3}{6(5x + 2)^6} = \frac{3x^2(5x + 2)^3 \left[2(5x + 2) - x\right]}{6(5x + 2)^6}$$

$$= \frac{\cancel{3}x^2(5x \cancel{+} 2)^{\cancel{3}}(9x + 4)}{\underset{2}{\cancel{6}(5x + 2)^{\cancel{6}3}}} = \frac{x^2(9x + 4)}{2(5x + 2)^3}$$

(b) $2x(3x - 1)^{-1} + 4x^2(3x - 1)^{-3} = \dfrac{2x}{3x - 1} + \dfrac{4x^2}{(3x - 1)^3}$

$$= \frac{2x}{3x - 1} \cdot \frac{(3x - 1)^2}{(3x - 1)^2} + \frac{4x^2}{(3x - 1)^3}$$

$$= \frac{2x(3x - 1)^2 + 4x^2}{(3x - 1)^3}$$

$$= \frac{2x(9x^2 - 6x + 1) + 4x^2}{(3x - 1)^3} = \frac{18x^3 - 8x^2 + 2x}{(3x - 1)^3}$$

B–4 **Solutions to Check Your Understanding Boxes**

Solution to CYU 1.9

(a)
$$\frac{\dfrac{x}{2\sqrt{x+2}}-\sqrt{x+2}}{x+4}\cdot\frac{2\sqrt{x+2}}{2\sqrt{x+2}}=\frac{x-2(x+2)}{2\sqrt{x+2}\,(x+4)}=\frac{x-2x-4}{2\sqrt{x+2}\,(x+4)}$$

$$=\frac{-x-4}{2\sqrt{x+2}\,(x+4)}=-\frac{1}{2\sqrt{x+2}}$$

(b)
$$(2x+1)^{\frac{2}{3}}-3x(2x+1)^{-\frac{1}{3}}=(2x+1)^{\frac{2}{3}}-\frac{3x}{(2x+1)^{\frac{1}{3}}}$$

$$=(2x+1)^{\frac{2}{3}}\cdot\frac{(2x+1)^{\frac{1}{3}}}{(2x+1)^{\frac{1}{3}}}-\frac{3x}{(2x+1)^{\frac{1}{3}}}$$

$$=\frac{2x+1-3x}{(2x+1)^{\frac{1}{3}}}$$

$$=\frac{1-x}{(2x+1)^{\frac{1}{3}}}$$

Solution to CYU 1.10

To convert 12 ounces per minute to pounds per hour, we take the path
$$\frac{\text{oz}}{\text{min}}\to\frac{\text{oz}}{\text{min}}\cdot\frac{\text{min}}{\text{hr}}\cdot\frac{\text{lb}}{\text{oz}}\to\frac{\text{lb}}{\text{hr}}. \text{ Then}$$

$$\frac{12\text{ oz}}{\text{min}}=\frac{12\text{ oz}}{\text{min}}\cdot\frac{60\text{ min}}{1\text{ hr}}\cdot\frac{1\text{ lb}}{16\text{ oz}}=45\,\frac{\text{lb}}{\text{hr}}$$

Solution to CYU 1.11

To convert 0.3 gallons per square yard to quarts per square foot, we take the path
$$\frac{\text{gal.}}{\text{yd}^2}\to\frac{\text{gal.}}{\text{yd}^2}\cdot\frac{\text{qt}}{\text{gal.}}\cdot\frac{\text{yd}^2}{\text{ft}^2}\to\frac{\text{qt}}{\text{ft}^2}. \quad\text{Then}$$

$$\frac{0.3\text{gal.}}{\text{yd}^2}=\frac{0.3\text{gal.}}{\text{yd}^2}\cdot\frac{4\text{qt}}{1\text{gal.}}\cdot\frac{1^2\text{yd}^2}{3^2\text{ft}^2}=\frac{(0.3)(4)}{3^2}\frac{\text{qt}}{\text{ft}^2}=\frac{0.4}{3}\frac{\text{qt}}{\text{ft}^2}\approx0.13\frac{\text{qt}}{\text{ft}^2}.$$

Solutions to Check Your Understanding Boxes   B–5

Solution to CYU 1.12

To convert $4\dfrac{g}{cm^3}$ to $\dfrac{oz}{in.^3}$ take the path $\dfrac{g}{cm^3} \rightarrow \dfrac{g}{cm^3} \cdot \dfrac{cm^3}{m^3} \cdot \dfrac{m^3}{ft^3} \cdot \dfrac{ft^3}{in.^3} \cdot \dfrac{oz}{g} \rightarrow \dfrac{oz}{in.^3}.$
Then

$$4\dfrac{g}{cm^3} = 4\dfrac{g}{cm^3} \cdot \dfrac{100^3 cm^3}{1^3 m^3} \cdot \dfrac{1^3 m^3}{3.28^3 ft^3} \cdot \dfrac{1^3 ft^3}{12^3 in.^3} \cdot \dfrac{.0353 oz}{1g} = \dfrac{4(10)^6(.0353)}{(3.28)^3 12^3} \dfrac{oz}{in.^3} \approx 2.32\dfrac{oz}{in.^3}.$$

Solution to CYU 1.13

(a)   The slope of the line through the given points is $m = \dfrac{6-(-2)}{-1-3} = \dfrac{8}{-4} = -2.$
Since parallel lines have the same slope, the answer is -2.

(b)   With a slope of -2, one has to move 2 units down for every unit to the right to get to the line of (a). After moving 8 units to the right, that would be $8 \times 2 = 16$ units down. Thus we add 8 to the x-coordinate of $(-1,6)$ and -16 to the y-coordinate, arriving at the point $(-1+8, 6-16) = (7,-10)$.

Solution to CYU 1.14

(a)   The slope of the parallel line is the same as the slope of the line through the given points: $m = \dfrac{7-1}{2-(-6)} = \dfrac{6}{8} = \dfrac{3}{4}.$

(b)   The line perpendicular to the line in (a) has a slope which is the negative reciprocal of the slope in (a): $-\dfrac{1}{m} = -\dfrac{1}{\frac{3}{4}} = -\dfrac{4}{3}.$

Solution to CYU 1.15

First determine the slope of the given line:

$$x - 5y = 3 \implies 5y = x - 3 \implies y = \frac{1}{5}x - \frac{3}{5} \implies m = \frac{1}{5}$$

The line we seek, therefore, has an equation of the form $y = \frac{1}{5}x + b$. As the point $(-2, 4)$ lies on the line, $4 = \frac{1}{5}(-2) + b \implies b = 4 + \frac{2}{5} = \frac{22}{5}$. The equation is $y = \frac{1}{5}x + \frac{22}{5}$.

Solution to CYU 1.16

Begin by finding the slope of the given line: $x - 5y = 3 \implies y = \frac{1}{5}x - \frac{3}{5} \implies m = \frac{1}{5}$. The line we seek will then have slope $m = -\dfrac{1}{\frac{1}{5}} = -5$. A point-slope equation of the desired line is $y - 4 = -5(x + 2)$.

Solution to CYU 1.17

The form of the equation of the line of slope 2 is $y = 2x + b$. Because the line contains the point $(5, 4)$, $4 = 2(5) + b \implies b = -6$, and the equation is $y = 2x - 6$.

Solution to CYU 1.18

First multiply through by the least common denominator, namely 6, to eliminate working with fractions:

$$\overset{2}{\cancel{6}}\left(\frac{-2x}{\cancel{3}}\right) - \overset{3}{\cancel{6}}\left(\frac{x}{\cancel{2}}\right) + 6(1) = 6\left(\frac{2x + 1}{\cancel{6}}\right)$$

$$-4x - 3x + 6 = 2x + 1$$

$$-7x - 2x = 1 - 6$$

$$-9x = -5$$

$$x = \frac{5}{9}$$

Solutions to Check Your Understanding Boxes   B–7

Solution to CYU 1.19

Solving for x:

$$3x + 2(x - y) = y + 4x + 1$$
$$3x + 2x - 2y = y + 4x + 1$$
$$5x - 2y = y + 4x + 1$$
$$5x - 4x = 2y + y + 1$$
$$x = 3y + 1$$

Solving for y: $3y + 1 = x \implies 3y = x - 1 \implies y = \frac{1}{3}x - \frac{1}{3}$

Solution to CYU 1.20

First multiply both sides by 15 to obtain

$$\overset{3}{\cancel{15}}\left(\frac{3x}{\cancel{5}}\right) - \overset{5}{\cancel{15}}\left(\frac{2 - x}{\cancel{3}}\right) + 15(1) < 15\left(\frac{x - 1}{\cancel{15}}\right)$$
$$9x - 10 + 5x + 15 < x - 1$$
$$14x + 5 < x - 1$$
$$13x < -6$$
$$x < -\frac{6}{13}$$

Solution to CYU 1.21

Multiplying the second equation of $\left.\begin{array}{r}3x + 4y = -1 \\ x + 2y = 0\end{array}\right\}$ by -2

and adding the equations $\rightarrow \left.\begin{array}{r}3x + 4y = -1 \\ -2x - 4y = 0\end{array}\right\} \rightarrow x = -1 \implies$

$-1 + 2y = 0 \implies y = \frac{1}{2} \implies$ solution is: $\left(x = -1, y = \frac{1}{2}\right)$

Solution to CYU 1.22

$$\begin{array}{l}(1):\ \ x + y - 2z = 2\\(2):\ \ 2x + y + z = 0\\(3):\ -3x + 2y - z = -4\end{array}\Bigg\} \rightarrow$$

$$\begin{array}{l}-2\times(1):\ -2x - 2y + 4z = -4\\+\ (2):\ \ \underline{2x + y + z = 0}\\(4):\ \ \ \ \ \ \ \ \ \ \ -y + 5z = -4\end{array}\quad \text{and,}$$

$$\begin{array}{l}3\times(1):\ \ 3x + 3y - 6z = 6\\+\ (3):\ \underline{-3x + 2y - z = -4}\\(5):\ \ \ \ \ \ \ \ \ \ 5y - 7z = 2\end{array}\ \rightarrow\ \begin{array}{l}5\times(4):\ -5y + 25z = -20\\+\ (5):\ \ \underline{5y - 7z = 2}\\\ \ \ \ \ \ \ \ \ \ \ \ \ \ \ \ 18z = -18\end{array}$$

$\Rightarrow z = -1$. From (4): $-y - 5 = -4 \Rightarrow y = -1$, and from

(1): $x - 1 + 2 = 2 \Rightarrow x = 1$. The solution is $(x = 1, y = -1, z = -1)$.

Solution to CYU 1.23

$$\begin{array}{l}(1):\ x + y - 2z = -3\\(2): 2x + y = 4\\(3):\ \ \ \ \ \ y + z = 5\end{array}\Bigg\} \rightarrow \begin{array}{l}2\times(3):\ \ \ \ \ \ 2y + 2z = 10\\+\ (1):\underline{x + y - 2z = -3}\\(4):\underline{x + 3y = 7}\end{array} \rightarrow$$

$$\begin{array}{l}-2\times(4):\ -2x - 6y = -14\\+\ (2):\ \ \underline{2x + y = 4}\\\ \ \ \ \ \ \ \ \ \ \ -5y = -10\end{array}\ \Rightarrow\ y = 2.\ \text{From}\ (4):\ x + 6 = 7 \Rightarrow x = 1,$$

and from (3): $2 + z = 5 \Rightarrow z = 3$. The solution is: $(x = 1, y = 2, z = 3)$.

Solution to CYU 1.24

Let x be the ounces of X to be eaten, y the ounces of Y, and z the ounces of Z:

$$\begin{array}{l}(1):\ 5x + 3y + 3z = 24\\(2):\ 2x + y + 3z = 11\\(3):\ \ \ x + 5y + 2z = 15\end{array}\Bigg\} \rightarrow \begin{array}{l}5\times(3):\ 5x + 25y + 10z = 75\\-\ (1):\ \underline{5x + 3y + 3z = 24}\\(4):\ \ \ \ \ \ \ \ \ 22y + 7z = 51\end{array}\quad \text{and}$$

$$\begin{array}{l}2\times(3):\ 2x + 10y + 4z = 30\\-\ (2):\ \underline{2x + y + 3z = 11}\\(5):\ \ \ \ \ \ \ 9y + z = 19\end{array} \rightarrow \begin{array}{l}7\times(5):\ 63y + 7z = 133\\-\ (4):\ \underline{22y + 7z = 51}\\\ \ \ \ \ \ \ \ \ 41y = 82\end{array}\ \Rightarrow\ y = 2.$$

From (5): $18 + z = 19 \Rightarrow z = 1$. From (3): $x + 10 + 2 = 15 \Rightarrow x = 3$, and the solution is 3 oz of X, 2 oz of Y, and 1 oz of Z.

Solutions to Check Your Understanding Boxes   **B–9**

Solution to CYU 1.25

Moving all of the terms of the given quadratic equation $3x^2 + 3x = x^2 - 2x - 3$ to the left side of the equation, one obtains $2x^2 + 5x + 3 = 0$. Factoring, we find that $(2x + 3)(x + 1) = 0 \Rightarrow 2x + 3 = 0$ or $x + 1 = 0$. If $2x + 3 = 0$, then $x = -\frac{3}{2}$. If $x + 1 = 0$, then $x = -1$. Thus $x = -\frac{3}{2}$, or $x = -1$.

Solution to CYU 1.26

$(x^2 - 6x - 8)^2 = 64 \Rightarrow x^2 - 6x - 8 = \pm 8 \Rightarrow$ (i) $x^2 - 6x - 8 = 8$, or (ii) $x^2 - 6x - 8 = -8$. Solving (i): $x^2 - 6x - 16 = 0 \Rightarrow (x - 8)(x + 2) = 0 \Rightarrow x = 8, -2$. Solving (ii): $x^2 - 6x = 0 \Rightarrow x(x - 6) = 0 \Rightarrow x = 0, 6$. Hence the solutions are $-2, 0, 6, 8$.

Solution to CYU 1.27

(a)
$$x^2 - 5 = -2x^2 + 5x - 7$$
$$3x^2 - 5x = -2$$
$$x^2 - \frac{5}{3}x = -\frac{2}{3}$$
$$x^2 - \frac{5}{3}x + \left(\frac{5}{6}\right)^2 = -\frac{2}{3} + \left(\frac{5}{6}\right)^2$$
$$\left(x - \frac{5}{6}\right)^2 = -\frac{2}{3} + \frac{25}{36} = \frac{1}{36}$$
$$x - \frac{5}{6} = \pm\frac{1}{6}$$
$$x = \frac{5}{6} \pm \frac{1}{6}$$
$$x = \frac{4}{6} = \frac{2}{3} \text{ or } x = \frac{6}{6} = 1$$

(b) $3x^2 - 5x + 2 = 0 \Rightarrow (3x - 2)(x - 1) = 0 \Rightarrow x = \frac{2}{3}$ or $x = 1$

(c) $x = \dfrac{-(-5) \pm \sqrt{(-5)^2 - 4 \cdot 3 \cdot 2}}{2 \cdot 3} = \dfrac{5 \pm \sqrt{25 - 24}}{6} = \dfrac{5 \pm 1}{6} \Rightarrow x = \dfrac{4}{6}$ or $x = \dfrac{6}{6}$.

That is, $x = \frac{2}{3}$ or $x = 1$.

B–10   Solutions to Check Your Understanding Boxes

Solution to CYU 1.28

Let w be the width of the field, and l the length (see figure). Then $2w + l = 1200 \Rightarrow l = 1200 - 2w$, and the area $A = 160000 = wl = w(1200 - 2w)$. Solving for w:

l

160,000 ft² w

river

$$w(1200 - 2w) = 160000$$
$$-2w^2 + 1200w = 160000$$
$$2w^2 - 1200w + 160000 = 0$$
$$w^2 - 600w + 80000 = 0$$
$$(w - 400)(w - 200) = 0$$

Then $w = 400$ or $w = 200$. If $w = 400$, $l = 1200 - 2w = 1200 - 800 = 400$. If $w = 200$, $l = 1200 - 2w = 1200 - 400 = 800$. Thus there are two possibilities: the field is either a square of side 400 ft, or a rectangle 200 ft by 800 ft with the 800 ft parallel to the river.

CHAPTER 2

Solution to CYU 2.1

(a) $f(-2) = 3(-2) - 5 = -11$

(b) $f(t + 1) = 3(t + 1) - 5 = 3t - 2$

(c) $f(-2x + 1) = 3(-2x + 1) - 5 = -6x - 2$

(d) $f\left(-\frac{2}{x}\right) = 3\left(-\frac{2}{x}\right) - 5 = -\frac{6}{x} - 5 = \frac{-6 - 5x}{x}$

Solutions to Check Your Understanding Boxes   **B–11**

Solution to CYU 2.2

(a)   For $f(x) = \sqrt{x+3}$ to be defined, $x+3 \geq 0 \Rightarrow x \geq -3$, so that the domain is $D_f = [-3, \infty)$. The values of the square root are the nonnegative numbers, which means that the range is $R_f = [0, \infty)$.

(b)   The function $g(x) = \dfrac{1}{(x+1)(x-2)}$ is defined except at -1 and 2 where the denominator is 0. Thus $D_g = (-\infty, -1) \cup (-1, 2) \cup (2, \infty)$.

Solution to CYU 2.3

$x = -1$:   Since $-1 < 0$, $f(x) = -4x + 1 \Rightarrow f(-1) = -4(-1) + 1 = 5$

$x = 1$:   Since $0 \leq 1 < 5$, $f(x) = x^2$, and $f(1) = 1^2 = 1$

$x = 5$:   Since $5 \leq 5 < 10$, $f(x) = -2x$, and $f(5) = -2(5) = -10$

$x = 7$:   Since $5 \leq 7 < 10$, $f(7) = -2(7) = -14$.

The value $f(10)$ is not defined, since the domain of f is

$$\{x < 0\} \cup \{0 \leq x < 5\} \cup \{5 \leq x < 10\} = \{0 < x < 10\}$$

which does not include 10.

Solution to CYU 2.4

(a)   4 units, $|3 - 7| = |-4| = 4$

(b)   10 units, $|-3 - 7| = |-10| = 10$

(c)   10 units, $|-7 - 3| = |-10| = 10$

(d)   4 units, $|-7 - (-3)| = |-7 + 3| = |-4| = 4$

B–12   Solutions to Check Your Understanding Boxes

Solution to CYU 2.5

(a) $0.15(\$20{,}000) = \$3{,}000$

(b) $\$3{,}802.50 + 0.28(40{,}500 - 25{,}350) = \$8{,}044.50$

(c) $\$34{,}573.50 + 0.36(138{,}100 - 128{,}100) = \$38{,}173.50$

(d) $\$34{,}573.50 + 0.36(205{,}100 - 128{,}100) = \$62{,}293.50$

(e) $\$83{,}699.50 + 0.396(351{,}500 - 278{,}450) = \$112{,}627.30$

Solution to CYU 2.6

For $f(x) = x - 3$ and $g(x) = \dfrac{1}{x-3}$,

$$(f+g)(x) = x - 3 + \frac{1}{x-3} = \frac{x^2 - 6x + 10}{x-3}$$

$$(f-g)(x) = x - 3 - \frac{1}{x-3} = \frac{x^2 - 6x + 8}{x-3}$$

$$(fg)(x) = \frac{x-3}{x-3} = \begin{cases} 1 & \text{if } x \neq 3 \\ \text{undef} & \text{if } x = 3 \end{cases}$$

$$(5g)(x) = \frac{5}{x-3}$$

$$\left(\frac{f}{g}\right)(x) = \frac{x-3}{\frac{1}{x-3}} = (x-3)^2 = x^2 - 6x + 9$$

The domain of f is all real numbers and the domain of g is all numbers except 3. Thus the domain of each of the first four functions above is $(-\infty, 3) \cup (3, \infty)$. As the function g has no zeros, the domain of the function $\dfrac{f}{g}$ is also $(-\infty, 3) \cup (3, \infty)$.

Solution to CYU 2.7

(a) For $f(x) = x^2 + 2x - 2$ and $g(x) = 4x + 3$,

    (i) $(f \circ g)(-2) = f(g(-2)) = f(4(-2) + 3) = f(-5)$
$$= (-5)^2 + 2(-5) - 2 = 25 - 12 = 13$$

    (ii) $(f \circ g)(x) = f(g(x)) = f(4x + 3) = (4x + 3)^2 + 2(4x + 3) - 2$
$$= 16x^2 + 24x + 9 + 8x + 6 - 2 = 16x^2 + 32x + 13$$

(b) From $h(x) = \dfrac{x^2}{x^2 + 3} = (g \circ f)(x) = g(f(x))$, we see that one possibility is
$f(x) = x^2$ and $g(x) = \dfrac{x}{x + 3}$.

Solution to CYU 2.8

(a) Substituting for r in the formula $H(r)$:
$$H = 350 + 1000(0.1 - i - 0.02) = 350 + 100 - 1000i - 20 = 430 - 1000i$$

(b) When $i = 0.043$, $H = 430 - 1000(0.043) = 387$

Solution to CYU 2.9

(a) $(M \circ F)(i) = M(F(i))$ is the individual's father's mother, i.e. the paternal grandmother.

(b) $(M \circ M)(i) = M(M(i))$ is the individual's mother's mother, i.e. the maternal grandmother.

(c) $(F \circ M \circ M)(i) = (F \circ (M \circ M))(i)$ is the individual's maternal grandmother's father, i.e. great grandfather (twice maternal).

Solution to CYU 2.10

From the given graph:

(a) $f(-1) = 1$ (b) The x-intercepts are $\pm 2, 4$ (c) The y-intercept is 2

Solution to CYU 2.11

(a) $y = \sqrt{x}$ $y = -\sqrt{x}$ $y = -4\sqrt{x}$

 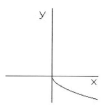 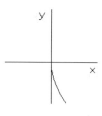

$y = -4\sqrt{x+2}$ $y = h(x) = -4\sqrt{x+2} + 4$

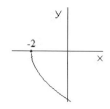 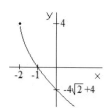

(b) The domain of h is $D_h = [-2, \infty)$, and the range is $R_h = (-\infty, 4]$.

(c) The x-intercept is -1, and the y-intercept is $-4\sqrt{2} + 4$.

Solution to CYU 2.12

(a) An even function's graph is symmetric wrt the y-axis (see figure).

(b) An odd function's graph is symmetric wrt the origin (see figure).

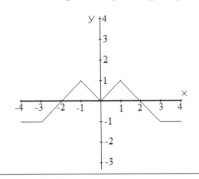
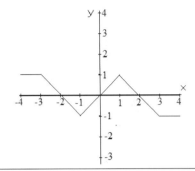

Solution to CYU 2.13

(a) The graph is not the graph of a function, by the Vertical Line Test. A circle is the graph of two functions: The top half is the graph of one function and the bottom half is the graph of another function.

(b) The graph is the graph of a function, as every vertical line that intersects the graph does so in only one point.

Solution to CYU 2.14

(a) (i) The solutions of $f(x) = 0$ are the x-intercepts of the graph of f: a, b, c
  (ii) $f < 0$ where the graph lies underneath the x−axis: $(a, b) \cup (c, \infty)$
  (iii) $f \geq 0$ where the graph lies above or on the x−axis: $(-\infty, a] \cup [b, c]$

(b) (i) $f = g$ at the points of intersection of the graphs of f and g: $-2, 3$
  (ii) $f < g$ where the graph of f lies underneath the graph of g: $(-2, 3)$
  (iii) $f \geq g$ where the graph of f lies above the graph of g or intersects the graph of g: $(-\infty, -2] \cup [3, \infty)$

Solution to CYU 2.15

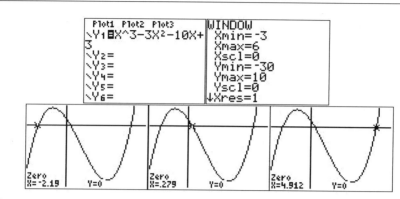

The solutions of the given equation are the zeros of
$f(x) = x^3 - 3x^2 - 10x + 3$ (the x-intercepts of the graph):
$$-2.19, 0.28, \text{ and } 4.91.$$

Solution to CYU 2.16

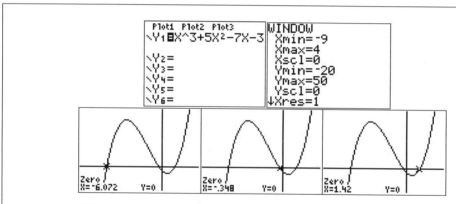

The solutions of the inequality are the intervals where
the graph of $f(x) = x^3 + 5x^2 - 7x - 3$ lies above the x-axis:
$$(-6.07, -0.35) \cup (1.42, \infty).$$

Solution to CYU 2.17

(a) A one-to-one function: It passes both the Vertical Line Test (for a function) and the Horizontal Line Test (for a 1-1 function).

(b) Not a function: The Vertical Line Test fails.

(c) A function, but not one-to-one: It passes the Vertical Line Test, but fails the Horizontal Line Test.

Solution to CYU 2.18

Begin by setting $f(a) = f(b)$:

$$\frac{a}{a+1} = \frac{b}{b+1}$$

$$a(b+1) = b(a+1)$$

$$ab + a = ba + b \quad (\text{but } ab = ba)$$

$$a = b$$

which shows that f is one-to-one.

Solution to CYU 2.19

To show that $f(x) = \dfrac{x^2 - 9}{-x + 2}$ is not one-to-one, for example, consider the numerator. As $x^2 - 9 = 0$ when $x^2 = 9$, that is, $x = \pm 3$, and the denominator is not zero there, $f(-3) = f(3) = 0$. This proves that f is not 1-1.

Solution to CYU 2.20

To find the inverse function, we set $y = f(x)$ and solve for x:

$$y = \frac{x}{x+1}$$

$$xy + y = x$$

$$xy - x = -y$$

$$x(y - 1) = -y$$

$$x = \frac{-y}{y-1} = \frac{y}{1-y}$$

Now swap x's and y's: $y = \dfrac{x}{1-x} = f^{-1}(x)$.

Verification: $(f \circ f^{-1})(x) = f\left(\dfrac{x}{1-x}\right) = \dfrac{\frac{x}{1-x}}{\frac{x}{1-x} + 1} \cdot \dfrac{1-x}{1-x} = \dfrac{x}{x+1-x} = \dfrac{x}{1} = x$

$(f^{-1} \circ f)(x) = f^{-1}\left(\dfrac{x}{x+1}\right) = \dfrac{\frac{x}{x+1}}{1 - \frac{x}{x+1}} \cdot \dfrac{x+1}{x+1} = \dfrac{x}{x+1-x} = \dfrac{x}{1} = x$

Solution to CYU 2.21

To show that the function $f(x) = \sqrt{x} - 2$ is one-to-one, set $f(a) = f(b)$:

$$\sqrt{a} - 2 = \sqrt{b} - 2$$
$$\sqrt{a} = \sqrt{b}$$
$$a = b$$

which proves that f is one-to-one. The domain and range of f are $D_f = [0, \infty)$ and $R_f = [-2, \infty)$, respectively. To find f^{-1}, first solve $y = f(x)$ for x:

$$y = \sqrt{x} - 2$$
$$\sqrt{x} = y + 2$$
$$x = (y + 2)^2$$

Now swap x's and y's: $y = (x + 2)^2$, so that $f^{-1}(x) = (x + 2)^2$, with $x \geq -2$. The domain and range of f^{-1} are $D_{f^{-1}} = R_f = [-2, \infty)$, and $R_{f^{-1}} = D_f = [0, \infty)$, respectively.

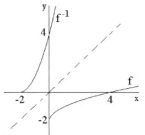

The graph of f^{-1} is obtained from the graph of f by reflecting the graph of f over the (dashed) line $y = x$ (see figure).

CHAPTER 3

Solution to CYU 3.1

With reference to the triangle at right, given $\cot\theta = 7$, first compute the hypotenuse $= \sqrt{7^2 + 1^2} = \sqrt{50} = 5\sqrt{2}$. Then

$$\sin\theta = \frac{\text{opp}}{\text{hyp}} = \frac{1}{5\sqrt{2}}, \quad \cos\theta = \frac{\text{adj}}{\text{hyp}} = \frac{7}{5\sqrt{2}}, \quad \tan\theta = \frac{\text{opp}}{\text{adj}} = \frac{1}{7}$$

$$\csc\theta = \frac{\text{hyp}}{\text{opp}} = 5\sqrt{2}, \quad \sec\theta = \frac{\text{hyp}}{\text{adj}} = \frac{5\sqrt{2}}{7}$$

Solution to CYU 3.2

(a) By definition, $\csc\theta = \dfrac{\text{hyp}}{\text{opp}}$, and this can be written as $\dfrac{1}{\frac{\text{opp}}{\text{hyp}}} = \dfrac{1}{\sin\theta}$.

(b)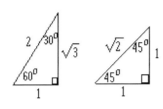

$\tan(90° - \theta) = \dfrac{b}{a} = \cot\theta$, and

$\cot(90° - \theta) = \dfrac{a}{b} = \tan\theta$

so that the complementary identities are:

$\tan(90° - \theta) = \cot\theta \quad \text{and} \quad \cot(90° - \theta) = \tan\theta.$

Solution to CYU 3.3

From the reference triangle for 30°:

$\sin\theta = \dfrac{\text{opp}}{\text{hyp}} = \dfrac{1}{2}, \quad \cos\theta = \dfrac{\text{adj}}{\text{hyp}} = \dfrac{\sqrt{3}}{2},$

$\tan\theta = \dfrac{\text{opp}}{\text{adj}} = \dfrac{1}{\sqrt{3}}, \quad \sec\theta = \dfrac{1}{\cos\theta} = \dfrac{2}{\sqrt{3}},$

$\text{and}\quad \cot\theta = \dfrac{1}{\tan\theta} = \sqrt{3}.$

From the reference triangle for 45°:

$\sin\theta = \dfrac{\text{opp}}{\text{hyp}} = \dfrac{1}{\sqrt{2}}, \quad \tan\theta = \dfrac{\text{opp}}{\text{adj}} = \dfrac{1}{1} = 1, \quad \csc\theta = \dfrac{1}{\sin\theta} = \sqrt{2}$

$\sec\theta = \dfrac{1}{\cos\theta} = \sqrt{2}, \quad \cot\theta = \dfrac{1}{\tan\theta} = \dfrac{1}{1} = 1.$

From the reference triangle for 60°:

$\sin\theta = \dfrac{\text{opp}}{\text{hyp}} = \dfrac{\sqrt{3}}{2}, \quad \cos\theta = \dfrac{\text{adj}}{\text{hyp}} = \dfrac{1}{2}, \quad \csc\theta = \dfrac{1}{\sin\theta} = \dfrac{2}{\sqrt{3}}$

$\sec\theta = \dfrac{1}{\cos\theta} = 2, \quad \cot\theta = \dfrac{1}{\tan\theta} = \dfrac{1}{\sqrt{3}}.$

Solution to CYU 3.4

(a) $\beta = 90° - 60° = 30°$

$\sin 60° = \dfrac{\text{opp}}{\text{hyp}} = \dfrac{a}{14} \Rightarrow a = 14\sin 60° = 14 \cdot \dfrac{\sqrt{3}}{2} = 7\sqrt{3}$

$\cos 60° = \dfrac{\text{adj}}{\text{hyp}} = \dfrac{b}{14} \Rightarrow b = 14\cos 60° = 14 \cdot \dfrac{1}{2} = 7.$

Thus $a = 7\sqrt{3}$, $b = 7$, and $\beta = 30°$.

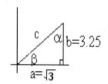

(b) Let $a = \sqrt{3}$, and $b = 3.25$. By the Pythagorean Theorem,

$c = \sqrt{(\sqrt{3})^2 + (3.25)^2} = \sqrt{13.5625} \approx 3.68.$

The value of $\tan \alpha = \dfrac{a}{b} = \dfrac{\sqrt{3}}{3.25} \approx 0.5329, \Rightarrow$
$\alpha = \tan^{-1} 0.5329 \approx 28°$, so that
$\beta = 90° - \alpha \approx 90° - 28° = 62°.$
Thus, $\alpha \approx 28°$ $\beta \approx 62°$ and $c \approx 3.68$.

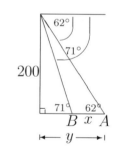

Solution to CYU 3.5

Let x ft be the width of the river (AB) and y ft be the distance from the base of the cliff to the far side of the river (see figure),

then (1) $\tan 62° = \dfrac{200}{y}$ and (2) $\tan 71° = \dfrac{200}{y - x}$.

From (2), $y - x = \dfrac{200}{\tan 71°} \Rightarrow x = y - \dfrac{200}{\tan 71°}.$

From (1), $y = \dfrac{200}{\tan 62°}$, and substituting into the previous equation gives

$x = \dfrac{200}{\tan 62°} - \dfrac{200}{\tan 71°} \approx 37.5$ ft.

Solution to CYU 3.6

Since $\sin 23° = \dfrac{34}{x}$, then $x = \dfrac{34}{\sin 23°} \approx 87.0$ miles.

Solution to CYU 3.7

From Theorem 3.9, $\dfrac{I_b}{I_a} = \dfrac{\sin \alpha}{\sin \beta} \Rightarrow I_b = \dfrac{I_a \sin \alpha}{\sin \beta} = \dfrac{1.0003 \sin 22.135°}{\sin 18.712°} \approx 1.1748$

Solution to CYU 3.8

(a) To convert θ to radian measure: $\theta = 120° = 120° \cdot \dfrac{\pi}{180°} = \dfrac{2\pi}{3}$

(b) To convert θ to degree measure: $\theta = \dfrac{\pi}{6} = \dfrac{\pi}{6} \cdot \dfrac{180°}{\pi} = 30°$

Solution to CYU 3.9

(a) Since $\theta = 999° - 720° = 279°$, θ is coterminal with $279°$. Therefore $\theta = 999°$ lies in the fourth quadrant.

(b) Since $-\dfrac{23\pi}{7} + 4\pi = \dfrac{5\pi}{7}$, θ is coterminal with $\dfrac{5\pi}{7}$. Consequently, $\theta = -\dfrac{23\pi}{7}$ lies in the second quadrant.

Solution to CYU 3.10

To use the formula $s = \theta r$, the angle θ must be in radian measure:
$$405° \cdot \frac{\pi}{180°} = \frac{81 \cdot 5}{90 \cdot 2} = \frac{5 \cdot 9 \cdot 9 \cdot \pi}{9 \cdot 5 \cdot 4} = \frac{9\pi}{4}$$
Then $s = \left(\frac{9\pi}{4}\right)\left(\frac{1}{2}\right) = \frac{9\pi}{8}$ ft.

Solution to CYU 3.11

(a) $\cos 420° = \cos(60° + 360°) = \cos 60° = \dfrac{\text{adj}}{\text{hyp}} = \dfrac{1}{2}$.

(b) $\csc\left(-\dfrac{11\pi}{6}\right) = \dfrac{1}{\sin\left(-\dfrac{11\pi}{6}\right)} = \dfrac{1}{\sin\left(-\dfrac{11\pi}{6} + 2\pi\right)} = \dfrac{1}{\sin\dfrac{\pi}{6}} = \dfrac{1}{\tfrac{1}{2}} = 2$.

Solution to CYU 3.12

From the given values for $0°$:
$$\csc 0° = \frac{1}{\sin 0°} = \frac{1}{0} \;-\; \text{is undefined, and}\;\; \sec 0° = \frac{1}{\cos 0°} = \frac{1}{1} = 1.$$

$90° = \dfrac{\pi}{2}$ radians, and corresponds to the point $(0,1)$ on the unit circle. Thus:
$$\sin 90° = 1, \;\; \cos 90° = 0, \;\; \tan 90° = \frac{\sin 90°}{\cos 90°} = \frac{1}{0} \;-\; \text{is undefined.}$$
$$\csc 90° = \frac{1}{\sin 90°} = \frac{1}{1} = 1, \;\; \sec 90° = \frac{1}{\cos 90°} = \frac{1}{0} \;-\; \text{is undefined.}$$
$$\cot 90° = \frac{\cos 90°}{\sin 90°} = \frac{0}{1} = 0$$

π radians is $180°$, and corresponds to the point $(-1, 0)$ on the unit circle. Thus:

$$\sin 180° = 0, \quad \cos 180° = -1, \quad \tan 180° = \frac{\sin 180°}{\cos 180°} = \frac{0}{-1} = 0.$$

$$\csc 180° = \frac{1}{\sin 180°} = \frac{1}{0} \ - \text{ is undefined}, \quad \sec 180° = \frac{1}{\cos 180°} = \frac{1}{-1} = -1.$$

$$\cot 180° = \frac{\cos 180°}{\sin 180°} = \frac{-1}{0} \ - \text{ is undefined}.$$

$270° = \frac{3\pi}{2}$ radians, and corresponds to the point $(0, -1)$ on the unit circle. Thus:

$$\sin 270° = -1, \quad \cos 270° = 0, \quad \tan 270° = \frac{\sin 270°}{\cos 270°} = \frac{-1}{0} \ - \text{ is undefined}.$$

$$\csc 270° = \frac{1}{\sin 270°} = \frac{1}{-1} = -1, \quad \sec 270° = \frac{1}{\cos 270°} = \frac{1}{0} \ - \text{ is undefined}.$$

$$\cot 270° = \frac{\cos 270°}{\sin 270°} = \frac{0}{-1} = 0$$

Solution to CYU 3.13

As $-\dfrac{29\pi}{7} + 4\pi = -\dfrac{\pi}{7}$, $\theta = -\dfrac{29\pi}{7}$ is coterminal with $-\dfrac{\pi}{7}$ and lies in the fourth quadrant. Therefore,

$$\sin \theta < 0, \qquad \cos \theta > 0, \qquad \tan \theta < 0$$

$$\csc \theta = \frac{1}{\sin \theta} < 0, \quad \sec \theta = \frac{1}{\cos \theta} > 0, \quad \cot \theta = \frac{1}{\tan \theta} < 0$$

Solution to CYU 3.14

(a) Since $-840° + 720° = -120°$, $-840°$ is coterminal with $-120°$. Thus $\sin(-840°) = \sin(-120°)$. As the reference angle of $-120°$ is $60°$, and $-120°$ lies in the third quadrant, $\sin(-840°) = \sin(-120°) = -\sin 60° = -\dfrac{\sqrt{3}}{2}$.

(b) As $\frac{11\pi}{4} - 2\pi = \frac{3\pi}{4}$, $\frac{11\pi}{4}$ is coterminal with $\frac{3\pi}{4}$ which lies in the second quadrant. This means that the sign of the cotangent is negative, and the reference angle is $\frac{\pi}{4}$. Thus $\cot \frac{11\pi}{4} = \cot \frac{3\pi}{4} = -\cot \frac{\pi}{4} = -1$.

(c) Since $-\frac{25\pi}{6} + 4\pi = -\frac{\pi}{6}$, $-\frac{25\pi}{6}$ is coterminal with $-\frac{\pi}{6}$ and lies in the fourth quadrant. The reference angle is $\frac{\pi}{6}$. The sign of the secant function is positive in the fourth quadrant. Thus

$$\sec\left(-\frac{25\pi}{6}\right) = \sec\left(-\frac{\pi}{6}\right) = \sec\left(\frac{\pi}{6}\right) = \frac{1}{\cos\left(\frac{\pi}{6}\right)} = \frac{1}{\frac{\sqrt{3}}{2}} = \frac{2}{\sqrt{3}}.$$

Solution to CYU 3.15

Since $\sec\theta = -\frac{7}{4}$ and $\sin\theta < 0$, θ lies in the third quadrant. Therefore, sine, cosine, secant and cosecant are negative, while tangent and cotangent are positive. With reference to the triangle at right, we compute the length of the remaining side: $\sqrt{7^2 - 4^2} = \sqrt{33}$. Then

$\sin\theta = -\sin\theta_r = -\frac{\text{opp}}{\text{hyp}} = -\frac{\sqrt{33}}{7}$, $\cos\theta = -\cos\theta_r = -\frac{\text{adj}}{\text{hyp}} = -\frac{4}{7}$

$\tan\theta = \tan\theta_r = \frac{\text{opp}}{\text{adj}} = \frac{\sqrt{33}}{4}$, $\csc\theta = \frac{1}{\sin\theta} = -\frac{7}{\sqrt{33}}$, $\cot\theta = \frac{1}{\tan\theta} = \frac{4}{\sqrt{33}}$

Solution to CYU 3.16

Given that the point $(-2, 1)$ lies on the terminal side of θ, $x = -2$, $y = 1$, and $r = \sqrt{x^2 + y^2} = \sqrt{4 + 1} = \sqrt{5}$ Then:

$\sin\theta = \frac{y}{r} = \frac{1}{\sqrt{5}}$, $\cos\theta = \frac{x}{r} = -\frac{2}{\sqrt{5}}$, $\tan\theta = \frac{y}{x} = \frac{1}{-2} = -\frac{1}{2}$

$\csc\theta = \frac{1}{\sin\theta} = \sqrt{5}$, $\sec\theta = \frac{1}{\cos\theta} = -\frac{\sqrt{5}}{2}$, $\cot\theta = \frac{1}{\tan\theta} = -2$

CHAPTER 4

Solution to CYU 4.1

(a) $\sin \frac{5\pi}{6} = \sin \frac{\pi}{6} = \frac{1}{2}$, and $\cos \frac{5\pi}{6} = -\cos \frac{\pi}{6} = -\frac{\sqrt{3}}{2}$

(b) $\sin\left(\frac{-11\pi}{3}\right) = \sin\left(\frac{-11\pi}{3} + 4\pi\right) = \sin \frac{\pi}{3} = \frac{\sqrt{3}}{2}$, and $\cos\left(\frac{-11\pi}{3}\right) = \cos \frac{\pi}{3} = \frac{1}{2}$

(c) As $361\pi = 360\pi + \pi$, is an integer multiple of 2π plus π, then
$\sin\left(\frac{\pi}{4} + 361\pi\right) = \sin\left(\frac{\pi}{4} + \pi\right) = \sin \frac{5\pi}{4} = -\frac{1}{\sqrt{2}}$, and
$\cos\left(\frac{\pi}{4} + 361\pi\right) = \cos \frac{5\pi}{4} = -\frac{1}{\sqrt{2}}$

Solution to CYU 4.2

The amplitude of $f(x) = -4\cos\left(x - \frac{\pi}{2}\right) + 3$ is $|-4| = 4$. To sketch the graph, reflect the basic curve $y = \cos x$ (figure 1) over the x-axis (figure 2), stretch it by a factor of 4 (figure 3) and then shift it right by $\frac{\pi}{2}$ (figure 4) and up 3 units (figure 5).

(1) $y = \cos x$ (2) $y = -\cos x$ (3) $y = -4\cos x$

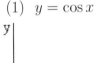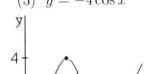

(4) $y = -4\cos\left(x - \frac{\pi}{2}\right)$ (5) $y = f(x) = -4\cos\left(x - \frac{\pi}{2}\right) + 3$

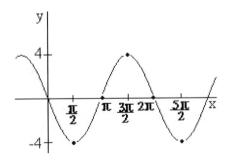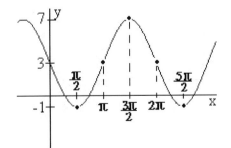

Solution to CYU 4.3

The amplitude of the graph of $f(x) = -\sin\left(\frac{1}{2}x + \frac{\pi}{2}\right) + 1$ is $|-1| = 1$. To sketch the graph, we determine the basic curve to be $y = \sin x$ (figure 1), and reflect it over the x-axis to obtain the graph of $y = -\sin x$ (figure 2). To determine the period and phase shift:

$$0 \leq \tfrac{1}{2}x + \tfrac{\pi}{2} \leq 2\pi$$

$$-\tfrac{\pi}{2} \leq \tfrac{1}{2}x \leq \tfrac{3\pi}{2}$$

$$-\pi \leq x \leq 3\pi$$

Thus the period is $3\pi - (-\pi) = 4\pi$, and the phase shift is $-\pi$. The graph of $y = -\sin\left(\frac{1}{2}x + \frac{\pi}{2}\right)$ appears in figure 3. We then shift the graph up 1 unit to obtain the final graph (figure 4).

(1) $y = \sin x$

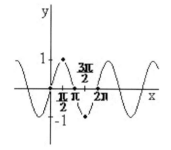

(2) $y = -\sin x$

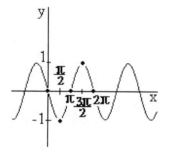

(3) $y = -\sin\left(\frac{1}{2}x + \frac{\pi}{2}\right)$

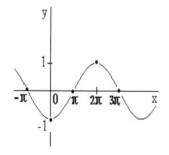

(4) $y = f(x) = -\sin\left(\frac{1}{2}x + \frac{\pi}{2}\right) + 1$

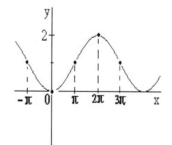

Solution to CYU 4.4

(a) To determine the graph of $f(x) = \cot x$ we note that $\cot x$ is the reciprocal of $\tan x$. Therefore it is positive where the tangent function is positive, negative where the tangent function is negative, and its graph has vertical asymptotes at the zeros of the tangent function $(n\pi)$. Consequently, the domain of $\cot x$ is all the real numbers except for the numbers of the form $n\pi$, where n is any integer. The graph of $\cot x$ appears below.

(b) The range of the cotangent function is all the real numbers, since the tangent function assumes all real numbers. As the tangent function has a period of π, so does the cotangent function:
$$\cot(x+\pi) = \frac{1}{\tan(x+\pi)} = \frac{1}{\tan x} = \cot x$$

(c) Since the tangent function is an odd function, so too is the cotangent function an odd function:
$$\cot(-x) = \frac{1}{\tan(-x)} = \frac{1}{-\tan x} = -\frac{1}{\tan x} = -\cot x$$

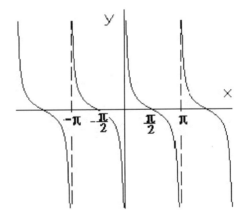

Solution to CYU 4.5

(a) To determine the graph of $f(x) = \sec x$, we note that $\sec x$ is the reciprocal of $\cos x$. Therefore $\sec x$ is undefined at the zeros of the cosine: $(2n+1)\frac{\pi}{2}$, and its graph has vertical asymptotes there. The domain is then the set of all real numbers excluding those of the form $(2n+1)\frac{\pi}{2}$, for any integer n. The graph of the secant function appears below.

(b) Since the range of the cosine function is $[-1, 1]$, its reciprocal has a range of $(-\infty, -1] \cup [1, \infty)$. The period of $f(x) = \sec x$ is the same as the period of the cosine function, 2π:
$$\sec(x + 2\pi) = \frac{1}{\cos(x+2\pi)} = \frac{1}{\cos x} = \sec x$$

(c) Since the cosine function is an even function, so too is the secant function an even function:
$$\sec(-x) = \frac{1}{\cos(-x)} = \frac{1}{\cos x} = \sec x$$

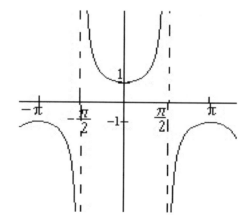

Solution to CYU 4.6

(a) (i) To establish the identity $1 + \tan^2 x = \sec^2 x$, begin with $\sin^2 x + \cos^2 x = 1$. Then divide both sides by $\cos^2 x$ to obtain

$$\frac{\sin^2 x}{\cos^2 x} + \frac{\cos^2 x}{\cos^2 x} = \frac{1}{\cos^2 x} \;\Rightarrow\; \left(\frac{\sin x}{\cos x}\right)^2 + 1 = \left(\frac{1}{\cos x}\right)^2 \;\Rightarrow\; \tan^2 x + 1 = \sec^2 x.$$

(ii) To establish the identity $1 + \cot^2 x = \csc^2 x$, begin with $\sin^2 x + \cos^2 x = 1$. Then divide both sides by $\sin^2 x$ to obtain

$$\frac{\sin^2 x}{\sin^2 x} + \frac{\cos^2 x}{\sin^2 x} = \frac{1}{\sin^2 x} \;\Rightarrow\; 1 + \left(\frac{\cos x}{\sin x}\right)^2 = \left(\frac{1}{\sin x}\right)^2 \;\Rightarrow\; 1 + \cot^2 x = \csc^2 x.$$

(b) (i) Begin with the definition of the tangent function, and then apply the addition identities for sine and cosine:

$$\tan(x + y) = \frac{\sin(x + y)}{\cos(x + y)} = \frac{\sin x \cos y + \cos x \sin y}{\cos x \cos y - \sin x \sin y}$$

$$= \frac{\sin x \cos y + \cos x \sin y}{\cos x \cos y - \sin x \sin y} \cdot \frac{\frac{1}{\cos x \cos y}}{\frac{1}{\cos x \cos y}}$$

$$= \frac{\dfrac{\sin x \cos y}{\cos x \cos y} + \dfrac{\cos x \sin y}{\cos x \cos y}}{\dfrac{\cos x \cos y}{\cos x \cos y} - \dfrac{\sin x \sin y}{\cos x \cos y}}$$

$$= \frac{\dfrac{\sin x}{\cos x} + \dfrac{\sin y}{\cos y}}{1 - \left(\dfrac{\sin x}{\cos x}\right)\left(\dfrac{\sin y}{\cos y}\right)}$$

$$= \frac{\tan x + \tan y}{1 - \tan x \tan y}$$

(ii) Apply (b) (i) to the function $\tan(x - y) = \tan[x + (-y)]$:

$$\tan[x + (-y)] = \frac{\tan x + \tan(-y)}{1 - \tan x \tan(-y)}$$

Since the tangent is an odd function, $\tan(-y) = -\tan y$, and we arrive at the other addition identity for the tangent:

$$\tan(x - y) = \tan[x + (-y)] = \frac{\tan x - \tan y}{1 + \tan x \tan y}$$

Solution to CYU 4.7

$$\cos \frac{\pi}{12} = \cos\left(\frac{\pi}{3} - \frac{\pi}{4}\right) = \cos \frac{\pi}{3} \cos \frac{\pi}{4} + \sin \frac{\pi}{3} \sin \frac{\pi}{4}$$

$$= \frac{1}{2} \cdot \frac{1}{\sqrt{2}} + \frac{\sqrt{3}}{2} \cdot \frac{1}{\sqrt{2}} = \frac{1+\sqrt{3}}{2\sqrt{2}} \cdot \frac{\sqrt{2}}{\sqrt{2}} = \frac{\sqrt{2}+\sqrt{6}}{4}$$

Solution to CYU 4.8

To determine
$\sin(\alpha - \beta) = \sin \alpha \cos \beta - \cos \alpha \sin \beta$,
we must evaluate $\sin \alpha$, $\sin \beta$, and $\cos \beta$. From the given information, we can draw the adjacent reference triangles, and calculate a and c:

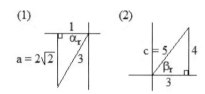

$a^2 + 1 = 3^2 \Rightarrow a = \sqrt{8} = 2\sqrt{2}$, and $3^2 + 4^2 = c^2 \Rightarrow c = \sqrt{25} = 5$

From triangle (1),
$$\sin \alpha = -\sin \alpha_r = -\frac{\text{opp}}{\text{hyp}} = -\frac{2\sqrt{2}}{3}$$

From triangle (2),
$$\sin \beta = \sin \beta_r = \frac{\text{opp}}{\text{hyp}} = \frac{4}{5} \text{ and } \cos \beta = \cos \beta_r = \frac{\text{adj}}{\text{hyp}} = \frac{3}{5}$$

Thus, $\sin(\alpha - \beta) = \left(-\frac{2\sqrt{2}}{3}\right)\left(\frac{3}{5}\right) - \left(-\frac{1}{3}\right)\left(\frac{4}{5}\right) = \frac{-6\sqrt{2}+4}{15}$.

Solution to CYU 4.9

From the addition identity for the tangent, see CYU 4.6 (b)(i):

$$\tan 2x = \tan(x + x) = \frac{\tan x + \tan x}{1 - \tan x \tan x} = \frac{2 \tan x}{1 - \tan^2 x}$$

Solution to CYU 4.10

Since the tangent is negative and the cosine is positive, the angle θ of radian measure x lies in the fourth quadrant. From $\tan x = -\frac{1}{2}$, then $\tan \theta_r = \frac{1}{2} = \frac{\text{opp}}{\text{adj}}$, resulting in the adjacent reference triangle.

The double angle identity for the cosine gives: $\cos 2x = 2\cos^2 x - 1$, and from the reference triangle,
$$\cos x = \cos \theta = \cos \theta_r = \frac{\text{adj}}{\text{hyp}} = \frac{2}{\sqrt{5}}$$

Thus, $\cos 2x = 2\left(\frac{2}{\sqrt{5}}\right)^2 - 1 = 2\left(\frac{4}{5}\right) - 1 = \frac{3}{5}$

Solution to CYU 4.11

$$\frac{\sin^2 x}{\cos x} - \sec x = \frac{\sin^2 x}{\cos x} - \frac{1}{\cos x} = \frac{\sin^2 x - 1}{\cos x} = \frac{-\cos^2 x}{\cos x} = -\cos x$$

Solution to CYU 4.12

$$\tan \frac{x}{2} = \frac{\sin \frac{x}{2}}{\cos \frac{x}{2}} = \frac{\pm\sqrt{\frac{1-\cos x}{2}}}{\pm\sqrt{\frac{1+\cos x}{2}}} = \pm\sqrt{\frac{\frac{1-\cos x}{2}}{\frac{1+\cos x}{2}}}$$

$$= \pm\sqrt{\frac{1-\cos x}{2} \cdot \frac{2}{1+\cos x}} = \pm\sqrt{\frac{1-\cos x}{1+\cos x}}$$

Solution to CYU 4.13

The angle θ of radian measure x lies in the fourth quadrant. From $\cot x = -\frac{24}{7}$, then
$$\cot \theta_r = \frac{24}{7} = \frac{1}{\tan \theta_r} \Rightarrow \tan \theta_r = \frac{7}{24} = \frac{\text{opp}}{\text{adj}},$$

resulting in the adjacent reference triangle.

Since $\frac{3\pi}{2} < x < 2\pi$, then $\frac{3\pi}{4} < \frac{x}{2} < \pi \Rightarrow \sec \frac{x}{2} < 0$. Thus
$$\sec \frac{x}{2} = \frac{1}{\cos \frac{x}{2}} = \frac{1}{-\sqrt{\frac{1+\cos x}{2}}} = -\frac{1}{\sqrt{\frac{1+\frac{24}{25}}{2}}}$$
$$= -\sqrt{\frac{2}{1+\frac{24}{25}}} = -\sqrt{\frac{2(25)}{25+24}} = -\frac{5\sqrt{2}}{7}$$

Solution to CYU 4.14

(a) Begin with the left side (LHS) of the identity $\dfrac{\csc x + \sec x}{1 + \tan x} = \csc x$.

$$\frac{\csc x + \sec x}{1 + \tan x} = \frac{\frac{1}{\sin x} + \frac{1}{\cos x}}{1 + \frac{\sin x}{\cos x}} \cdot \frac{\sin x \cos x}{\sin x \cos x}$$
$$= \frac{\cos x + \sin x}{\sin x \cos x + \sin^2 x}$$
$$= \frac{\cos x + \sin x}{(\sin x)(\cos x + \sin x)}$$
$$= \frac{1}{\sin x} = \csc x = \text{RHS}$$

(b) Applying the double angle identities to the left side of the given identity $\dfrac{1 + \cos 2x}{\sin 2x} = \cot x$, and choosing to express the $\cos 2x$ in terms of only cosines since the right side, $\cot x = \dfrac{\cos x}{\sin x}$:

$$\frac{1 + \cos 2x}{\sin 2x} = \frac{1 + (2\cos^2 x - 1)}{2 \sin x \cos x}$$
$$= \frac{2\cos^2 x}{2 \sin x \cos x} = \frac{\cos x}{\sin x} = \cot x = \text{RHS}$$

Solution to CYU 4.15

(a) From the figure, we see that
$$\sin^{-1}\left(\sin\tfrac{5\pi}{8}\right) = \sin^{-1}\left(\sin\tfrac{3\pi}{8}\right) = \tfrac{3\pi}{8}$$

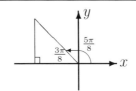

(b) Since the inverse sine of a number between -1 and 1 is the angle whose sine is that number, $\sin\left(\sin^{-1}\tfrac{5}{9}\right) = \tfrac{5}{9}$.

Solution to CYU 4.16

(a) The angle is in the 4th quadrant and the sine of the reference angle is $\tfrac{1}{2}$ $\Rightarrow$ the reference angle is $\tfrac{\pi}{6}$ (see figure). Thus $\sin^{-1}\left(-\tfrac{1}{2}\right) = -\tfrac{\pi}{6}$.

(b) Let $\theta = \sin^{-1}\left(-\tfrac{2}{3}\right)$. The angle θ is in the 4th quadrant, and the sine of the reference angle is $\tfrac{2}{3}$ $\Rightarrow$ the adjacent reference triangle (see figure). The cotangent is negative in the 4th quadrant $\Rightarrow$
$\cot\left[\sin^{-1}\left(-\tfrac{2}{3}\right)\right] = -\dfrac{\text{adj}}{\text{opp}} = -\dfrac{\sqrt{5}}{2}$.

Solution to CYU 4.17

(a) The inverse cosine is an angle between 0 and π, so $\cos^{-1}\left(\cos\tfrac{12\pi}{7}\right)$ cannot be $\tfrac{12\pi}{7}$. Since $\tfrac{12\pi}{7} = \pi + \tfrac{5\pi}{7}$, the angle $\tfrac{12\pi}{7}$ lies in the fourth quadrant and has a reference angle of $\tfrac{2\pi}{7}$. In the fourth quadrant, the cosine is positive $\Rightarrow \cos\tfrac{12\pi}{7} = \cos\tfrac{2\pi}{7}$. Therefore $\cos^{-1}\left(\cos\tfrac{12\pi}{7}\right) = \cos^{-1}\left(\cos\tfrac{2\pi}{7}\right) = \tfrac{2\pi}{7}$, as $0 < \tfrac{2\pi}{7} < \tfrac{\pi}{2}$.

(b) As $-1 \leq -\tfrac{1}{\pi} \leq 1$, $\cos\left[\cos^{-1}\left(-\tfrac{1}{\pi}\right)\right] = -\tfrac{1}{\pi}$.

Solution to CYU 4.18

(a) The angle is in the 2nd quadrant and the cosine of the reference angle is $\frac{1}{2}$ $\Rightarrow$ the reference angle is $\frac{\pi}{3}$ (see figure). Thus
$$\cos^{-1}\left(-\frac{1}{2}\right) = \frac{2\pi}{3}$$

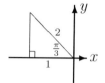

(b) Let $\theta = \cos^{-1}\left(-\frac{4}{5}\right)$. Then $\cos\theta = -\frac{4}{5}$. Applying the double angle identity for the cosine in terms only of cosines (since we have that value):
$$\cos\left[2\cos^{-1}\left(-\frac{4}{5}\right)\right] = \cos 2\theta = 2\cos^2\theta - 1 = 2\left(-\frac{4}{5}\right)^2 - 1$$
$$= 2\left(\frac{16}{25}\right) - 1 = \frac{32-25}{25} = \frac{7}{25}$$

Solution to CYU 4.19

(a) The angle $\frac{5\pi}{3}$ lies in the fourth quadrant and has a reference angle of $\frac{\pi}{3}$ (see figure). Then
$$\tan\frac{5\pi}{3} = -\tan\frac{\pi}{3} = -\sqrt{3}$$
Thus, $\tan^{-1}\left(\tan\frac{5\pi}{3}\right) = \tan^{-1}(-\sqrt{3}) = -\frac{\pi}{3}$.

(b) Since the domain of the inverse tangent function is all real numbers, $\tan(\tan^{-1} 500) = 500$.

Solution to CYU 4.20

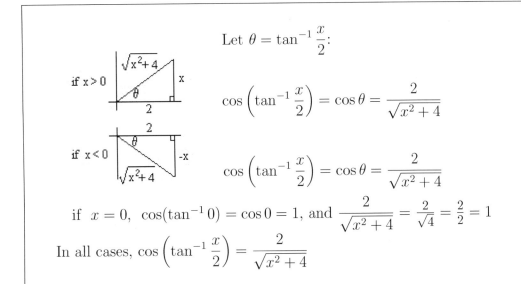

Let $\theta = \tan^{-1}\dfrac{x}{2}$:

$$\cos\left(\tan^{-1}\dfrac{x}{2}\right) = \cos\theta = \dfrac{2}{\sqrt{x^2+4}}$$

$$\cos\left(\tan^{-1}\dfrac{x}{2}\right) = \cos\theta = \dfrac{2}{\sqrt{x^2+4}}$$

if $x=0$, $\cos(\tan^{-1}0) = \cos 0 = 1$, and $\dfrac{2}{\sqrt{x^2+4}} = \dfrac{2}{\sqrt{4}} = \dfrac{2}{2} = 1$

In all cases, $\cos\left(\tan^{-1}\dfrac{x}{2}\right) = \dfrac{2}{\sqrt{x^2+4}}$

Solution to CYU 4.21

(a) $2\sin^2 x - 5\sin x - 3 = (2\sin x + 1)(\sin x - 3) = 0 \Rightarrow$
   (1) $\sin x = -\dfrac{1}{2}$, or (2) $\sin x = 3$
   Equation (2) has no solution, as $-1 \le \sin x \le 1$.
   From (1), θ lies in the third or fourth quadrants and
   $\sin\theta_r = \dfrac{1}{2} \Rightarrow \theta_r = \dfrac{\pi}{6}$.
   From the figure, and as $0 \le x < 2\pi$,
   $x = \pi + \dfrac{\pi}{6} = \dfrac{7\pi}{6}$, or $x = 2\pi - \dfrac{\pi}{6} = \dfrac{11\pi}{6}$.

(b) $2\cos x \sin x + \cos x = 2\sin x + 1 \Rightarrow 2\cos x \sin x + \cos x - 2\sin x - 1 = 0 \Rightarrow$
   $(\cos x - 1)(2\sin x + 1) = 0 \Rightarrow$ (1) $\cos x = 1$ or (2) $\sin x = -\dfrac{1}{2}$. From
   (1), $x = 0 + 2n\pi = 2n\pi$. From (2) and part (a) above, $x = \dfrac{7\pi}{6} + 2n\pi$ or
   $x = \dfrac{11\pi}{6} + 2n\pi$.

B–36   Solutions to Check Your Understanding Boxes

Solution to CYU 4.22

(a) Since $2\sin\left(\frac{x}{3}\right) = 1$, $\sin\left(\frac{x}{3}\right) = \frac{1}{2}$ $\Rightarrow$ the reference angle, θ_r is $\frac{\pi}{6}$, and θ lies in the 1st or 2nd quadrants, so that $\frac{x}{3} = \frac{\pi}{6} + 2n\pi$ or $\frac{x}{3} = \frac{5\pi}{6} + 2n\pi$ $\Rightarrow$ $x = \frac{\pi}{2} + 6n\pi$ or $x = \frac{5\pi}{2} + 6n\pi$. As $0 \le x < 2\pi$, then $x = \frac{\pi}{2}$ is the only solution.

(b) Substitute $Y = 4x - \pi$, in the equation $2\sin^2(4x - \pi) = 5\sin(4x - \pi) + 3$, to obtain the equation $2\sin^2 Y - 5\sin Y - 3 = 0$. This equation was solved in CYU 4.21(a) for $0 \le x < 2\pi$, with the solutions $\frac{7\pi}{6}$ and $\frac{11\pi}{6}$. Thus $Y = \frac{7\pi}{6} + 2n\pi$ or $Y = \frac{11\pi}{6} + 2n\pi$. Substituting back,

$$4x - \pi = \frac{7\pi}{6} + 2n\pi \qquad\qquad 4x - \pi = \frac{11\pi}{6} + 2n\pi$$

$$4x = \frac{13\pi}{6} + 2n\pi \qquad\qquad 4x = \frac{17\pi}{6} + 2n\pi$$

$$x = \frac{13\pi}{24} + \frac{n\pi}{2} \qquad\qquad x = \frac{17\pi}{24} + \frac{n\pi}{2}$$

Solution to CYU 4.23

$3\cos^2 x - 2\cos x - 1 = 0$ $\Rightarrow$ $(3\cos x + 1)(\cos x - 1) = 0$ $\Rightarrow$ $\cos x = -\frac{1}{3}$ or $\cos x = 1$. In $[0, 2\pi)$, if $\cos x = 1$, then $x = 0$. If $\cos x = -\frac{1}{3}$, then x lies in QII or QIII and the reference angle is $\cos^{-1}\frac{1}{3}$, so that $x = \pi - \cos^{-1}\frac{1}{3}$ or $x = \pi + \cos^{-1}\frac{1}{3}$.

Thus the solutions are 0, $\pi \pm \cos^{-1}\frac{1}{3}$.

$\left[\text{Note: } \pi - \cos^{-1}\frac{1}{3} \equiv \cos^{-1}\left(-\frac{1}{3}\right) \text{ and } \pi + \cos^{-1}\frac{1}{3} \equiv 2\pi - \cos^{-1}\left(-\frac{1}{3}\right).\right]$

CHAPTER 5

Solution to CYU 5.1

(a) The angle $\gamma = 180° - (\alpha + \beta) = 180° - (20° + 30°) = 130°$.

From the Law of Sines, $\dfrac{\sin 30°}{b} = \dfrac{\sin 130°}{4} \Rightarrow$

$b = \dfrac{4 \sin 30°}{\sin 130°} = \dfrac{2}{\sin 130°} \approx 2.6$. Also, $\dfrac{\sin 20°}{a} = \dfrac{\sin 130°}{4}$

$\Rightarrow a = \dfrac{4 \sin 20°}{\sin 130°} \approx 1.8$. Thus, $\gamma = 130°$, $b \approx 2.6$, $a \approx 1.8$.

(b) From the Law of Cosines, $c = \sqrt{a^2 + b^2 - 2ab\cos\gamma} =$

$\sqrt{9 + 16 - 2(3)(4)\cos 27°} = \sqrt{25 - 24\cos 27°} \approx 1.9$.

Applying the Law of Sines, $\dfrac{\sin\alpha}{3} = \dfrac{\sin 27°}{1.9} \Rightarrow$

$\alpha = \sin^{-1}\left(\dfrac{3\sin 27°}{1.9}\right) \approx 46° \Rightarrow$

$\beta = 180° - (\alpha + \gamma) \approx 180° - (46° + 27°) = 107°$.

The solution is $c \approx 1.9$, $\alpha \approx 46°$, $\beta \approx 107°$.

(c) By the Law of Cosines:

$c^2 = a^2 + b^2 - 2ab\cos\gamma$

$9^2 = 4^2 + 6^2 - 2 \cdot 4 \cdot 6 \cdot \cos\gamma$

$81 = 16 + 36 - 48\cos\gamma$

$81 = 52 - 48\cos\gamma$

$48\cos\gamma = 52 - 81 = -29$

$\cos\gamma = -\dfrac{29}{48}$

$\gamma = \cos^{-1}\left(-\dfrac{29}{48}\right) \approx 127°$

From the Law of Sines, we can determine either of the other two acute angles. For instance,

$\dfrac{\sin\beta}{b} = \dfrac{\sin\gamma}{c}$

$\sin\beta = \dfrac{b\sin\gamma}{c} = \dfrac{6\sin 127°}{9}$

$\beta = \sin^{-1}\left(\dfrac{6\sin 127°}{9}\right) \approx 32°$

As $\gamma \approx 127°$ and $\beta \approx 32°$, then

$\alpha = 180° - (\beta + \gamma) \approx 180° - (32° + 127°) = 180° - 159° = 21°$

Therefore $\alpha \approx 21°$, $\beta \approx 32°$, and $\gamma \approx 127°$.

Solution to CYU 5.2

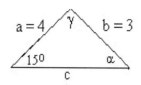

From the Law of Sines, we can determine α:
$$\frac{\sin\alpha}{4} = \frac{\sin 15°}{3} \Rightarrow \sin\alpha = \tfrac{4}{3}\sin 15° \text{ so that}$$
$\alpha \approx 20°$ or $\alpha' \approx 160°$ (both are possible).

In the triangle with $\alpha \approx 20°$, then
$$\gamma = 180° - (\alpha + \beta) \approx 180° - (20° + 15°) = 145°.$$
In the other triangle,
$$\gamma' \approx 180° - (160° + 15°) = 5°.$$
The remaining side of the triangles can be found from the Law of Sines:

In the first triangle: $\dfrac{\sin 145°}{c} = \dfrac{\sin 15°}{3} \Rightarrow c = \dfrac{3\sin 145°}{\sin 15°} \approx 6.6.$

In the second triangle, $\dfrac{\sin 5°}{c} = \dfrac{\sin 15°}{3} \Rightarrow c = \dfrac{3\sin 5°}{\sin 15°} \approx 1.0.$

Thus there are two triangles. In one
$$c \approx 6.6,\ \alpha \approx 20°,\ \text{and } \gamma \approx 145°.$$
In the second triangle,
$$c' \approx 1.0,\ \alpha' \approx 160°,\ \text{and } \gamma' \approx 5°.$$

Solution to CYU 5.3

From the Law of Cosines,
$$c = \sqrt{136^2 + 30^2 - 2(136)(30)\cos 150°}$$
$$= \sqrt{136^2 + 30^2 - 8160(-\cos 30°)}$$
$$= \sqrt{136^2 + 30^2 + 8160\left(\tfrac{\sqrt{3}}{2}\right)}$$
$$\approx 162.7 \text{ miles}$$

Solution to CYU 5.4

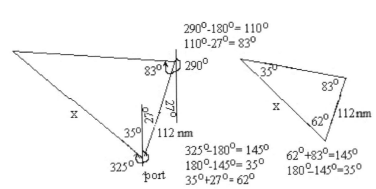

The determination of the angles is given above, and summarized in the triangle on the right. Applying the Law of Sines,
$$\frac{\sin 83°}{x} = \frac{\sin 35°}{112} \Rightarrow x = \frac{112 \sin 83°}{\sin 35°} \approx 193.8 \text{ nautical miles.}$$

CHAPTER 6

Solution to CYU 6.1

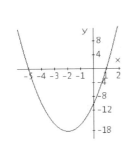

The vertex is at $x = -\frac{b}{2a} = \frac{-8}{2(2)} = -2$ and $y = 2(-2)^2 + 8(-2) - 10 = 8 - 26 = -18 \Rightarrow$ the vertex is $(-2, -18)$. The parabola opens up as $a = 2 > 0$. The y-intercept is -10, and next we find the x-intercepts:
$f(x) = 2(x^2 + 4x - 5) = 2(x+5)(x-1) = 0$ when $x = 1$ or $x = -5$.
From the graph, we find the range to be $[-18, \infty)$.

Solution to CYU 6.2

(a) As $b^2 - 4ac = 2^2 - 4(3)(1) = 4 - 12 < 0$, there are no x-intercepts.

(b) As $b^2 - 4ac = 8^2 - 4(-1)(-16) = 64 - 64 = 0$, there is exactly one x-intercept.

(c) As $b^2 - 4ac = (-4)^2 - 4(5)(-2) = 16 + 40 > 0$, there are two x-intercepts.

Solution to CYU 6.3

Now $C(x) = 2x^2 + 100x + 6400$. As before, $R(x) = -2x^2 + 500x$. The break-even points occur where $R(x) = C(x)$:

$$2x^2 + 100x + 6400 = -2x^2 + 500x$$
$$4x^2 - 400x + 6400 = 0$$
$$x^2 - 100x + 1600 = 0$$
$$(x - 80)(x - 20) = 0$$
$$x = 80, \text{ or } x = 20$$

We see that break-even occurs at production levels of 80 units and of 20 units per month.

Solution to CYU 6.4

(a) $(1): x^2 - 4x - y = 4$
$(2): x^2 + y = 2$

Then $(1)+(2)$: $2x^2 - 4x = 6 \Rightarrow x^2 - 2x - 3 = 0$
$\Rightarrow (x-3)(x+1) = 0 \Rightarrow x = -1$ or $x = 3$.
When $x = -1$, from (2), $y = 2 - x^2 = 2 - 1 = 1 \Rightarrow (x = -1, y = 1)$.
When $x = 3$, from (2), $y = 2 - 9 = -7 \Rightarrow (x = 3, y = -7)$.

(b)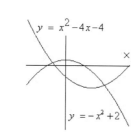
As $x^2 - 4x - y = 4 \Rightarrow y = x^2 - 4x - 4$ is the graph of a parabola opening upward, the vertex is at:
$x = -\frac{b}{2a} = \frac{4}{2} = 2 \Rightarrow y = 4 - 8 - 4 = -8 \Rightarrow (2, -8)$.
The graph of the equation $x^2 + y = 2 \Rightarrow y = -x^2 + 2$ is a parabola opening down with vertex at $(0,2)$. We also know the two points of intersection of the parabolas from part (a) above, and that helps us to sketch the adjacent graphs.

Solutions to Check Your Understanding Boxes   **B–41**

Solution to CYU 6.5

To solve the system, we let $x = \sin \alpha$ and $y = \cos \beta$:

$$\left.\begin{array}{r} 3\sin^2 \alpha - 2\sin \alpha + \cos \beta = \dfrac{9}{2} \\ 4\sin \alpha + 2\cos \beta = -5 \end{array}\right\} \quad \Rightarrow \quad \left.\begin{array}{rl} (1): & 3x^2 - 2x + y = \dfrac{9}{2} \\ (2): & 4x + 2y = -5 \end{array}\right\} \quad \Rightarrow$$

$$\left.\begin{array}{rl} -2 \times (1): & -6x^2 + 4x - 2y = -9 \\ (2): & 4x + 2y = -5 \end{array}\right\} \qquad \text{Adding the two equations:}$$

$$-6x^2 + 8x = -14$$
$$3x^2 - 4x = 7$$
$$3x^2 - 4x - 7 = 0$$
$$(3x - 7)(x + 1) = 0$$
$$x = \frac{7}{3}, \quad \text{or} \quad x = -1$$

As $x = \sin \alpha$, $-1 \le x \le 1$, the equation $\sin \alpha = \dfrac{7}{3}$ has no solution. When $x = -1$, from (2), $4(-1) + 2y = -5 \Rightarrow 2y = -1 \Rightarrow y = -\dfrac{1}{2}$. Substituting back $x = \sin \alpha$, $y = \cos \beta$:

$$\begin{array}{c|c} \sin \alpha = -1 & \cos \beta = -\dfrac{1}{2} \\[2mm] \alpha = \dfrac{3\pi}{2} & \beta = \dfrac{2\pi}{3}, \ \dfrac{4\pi}{3} \end{array}$$

The solution is $\left(\alpha = \dfrac{3\pi}{2}, \beta = \dfrac{2\pi}{3}\right)$ and $\left(\alpha = \dfrac{3\pi}{2}, \beta = \dfrac{4\pi}{3}\right)$.

Solution to CYU 6.6

(a) For $f(x) = -3(x + 2)^2 + 11$, $a = -3 < 0$ means that f has a maximum value. Since f is in standard form, we see that the maximum value is 11.

(b) For $f(x) = 3x^2 - 2x + 4$, $a = 3 > 0$ means that f has a minimum value. The minimum value occurs at $x = -\dfrac{b}{2a} = -\dfrac{-2}{2(3)} = \dfrac{1}{3}$, and that minimum value is $f\left(\dfrac{1}{3}\right) = 3\left(\dfrac{1}{3}\right)^2 - 2\left(\dfrac{1}{3}\right) + 4 = \dfrac{1}{3} - \dfrac{2}{3} + 4 = -\dfrac{1}{3} + 4 = \dfrac{11}{3}$.

B–42   Solutions to Check Your Understanding Boxes

Solution to CYU 6.7

The area to be maximized is $A = LW$, which expresses the area in terms of two variables. To eliminate one variable, consider the total cost, $2800, of fencing:

$$\$2800 = (6\$/\text{ft})(2L + W)\text{ft} + (8\$/\text{ft})W\text{ft}$$
$$1400 = 3(2L + W) + 4W$$
$$1400 = 6L + 3W + 4W = 6L + 7W$$
$$6L = 1400 - 7W$$
$$L = \frac{1400 - 7W}{6}$$

Substituting this for L in $A = LW$,

$$A = \left(\frac{1400 - 7W}{6}\right) W = -\tfrac{7}{6}W^2 + \tfrac{700}{3}W$$

The maximum of this quadratic function occurs at

$$W = -\frac{b}{2a} = -\frac{\frac{700}{3}}{2\left(-\frac{7}{6}\right)} = 100$$

Then $L = \dfrac{1400 - 7W}{6} = \dfrac{1400 - 700}{6} = \dfrac{700}{6} = \dfrac{350}{3}$, and the maximum area is

$$A = LW = \left(\tfrac{350}{3}\right)100 = \tfrac{35000}{3}\text{ft}^2 .$$

Solution to CYU 6.8

(a) To solve $(x^2 - 7)(x^4 - 81)(x^3 + 8)(x^3 - 7) = 0$, set each factor equal to zero:

$$x^2 = 7 \qquad x^4 = 81 \qquad x^3 = -8 \qquad x^3 = 7$$
$$x = \pm\sqrt{7} \qquad x = \pm 81^{\frac{1}{4}} \qquad x = (-8)^{\frac{1}{3}} \qquad x = 7^{\frac{1}{3}}$$
$$x = \pm 3 \qquad x = -2$$

The solution is $x = \pm\sqrt{7}, \ \pm 3, \ -2, \ 7^{\frac{1}{3}}$.

(b) Substitute $t = x^3$ in $x^6 + 2x^3 = 3$ to get the quadratic equation $t^2 + 2t = 3$:

$$t^2 + 2t = 3$$
$$t^2 + 2t - 3 = 0$$
$$(t - 1)(t + 3) = 0$$
$$t = 1, \text{ or } t = -3$$

Substituting back for x:

$$x^3 = 1 \qquad\qquad x^3 = -3$$
$$x = 1 \qquad\qquad x = -3^{\frac{1}{3}}$$

The solution is $x = 1$, or $x = -3^{\frac{1}{3}}$.

(c)
$$x^3 + x^2 = -9x - 9$$
$$x^3 + x^2 + 9x + 9 = 0$$
$$x^2(x + 1) + 9(x + 1) = 0$$
$$(x^2 + 9)(x + 1) = 0$$
$$x = -1$$

The solution is $x = -1$.

Solution to CYU 6.9

(a) We see that $x = 1$ is a zero of $2x^3 - 11x^2 + 4x + 5$, as $2(1)^3 - 11(1)^2 + 4(1) + 5 = 2 - 11 + 4 + 5 = 0$, which means that $x - 1$ is a factor. Dividing and factoring the resulting quadratic polynomial, we have:

$$2x^3 - 11x^2 + 4x + 5 = 0$$
$$(x - 1)(2x^2 - 9x - 5) = 0$$
$$(x - 1)(2x + 1)(x - 5) = 0$$
$$x = 1, \text{ or } x = -\frac{1}{2}, \text{ or } x = 5$$

Solution: $1, -\frac{1}{2}, 5$.

(b) The possible rational zeros of $p(x) = 2x^3 - x^2 - 41x - 20$ are:

$$\frac{c}{l} = \frac{\pm 1,\ \pm 2,\ \pm 4,\ \pm 5,\ \pm 10,\ \pm 20}{\pm 1,\ \pm 2} : \pm 1,\ \pm\tfrac{1}{2},\ \pm 2,\ \pm 4,\ \pm 5,\ \pm\tfrac{5}{2},\ \pm 10,\ \pm 20$$

Testing:

$$p(1) = 2 - 1 - 41 - 20 \neq 0$$
$$p(-1) = -2 - 1 + 41 - 20 \neq 0$$
$$p\left(\tfrac{1}{2}\right) = 2\left(\tfrac{1}{2}\right)^3 - \left(\tfrac{1}{2}\right)^2 - 41\left(\tfrac{1}{2}\right) - 20 = \left(\tfrac{1}{4}\right) - \left(\tfrac{1}{4}\right) - \left(\tfrac{41}{2}\right) - 20 \neq 0$$
$$p\left(-\tfrac{1}{2}\right) = 2\left(-\tfrac{1}{2}\right)^3 - \left(-\tfrac{1}{2}\right)^2 - 41\left(-\tfrac{1}{2}\right) - 20 = \left(-\tfrac{1}{4}\right) - \left(\tfrac{1}{4}\right) + \left(\tfrac{41}{2}\right) - 20$$
$$= \frac{-2 + 82 - 80}{4} = 0$$

As $x = -\tfrac{1}{2}$ is a zero, $x + \tfrac{1}{2}$ is a factor. Since the coefficients of $p(x)$ are integers, $2\left(x + \tfrac{1}{2}\right) = 2x + 1$ is also a factor. Dividing and factoring the resulting quadratic polynomial, we have:

$$2x^3 - x^2 - 41x - 20 = 0$$
$$(2x + 1)(x^2 - x - 20) = 0$$
$$(2x + 1)(x - 5)(x + 4) = 0$$
$$x = -\frac{1}{2}, \quad \text{or} \quad x = 5 \quad \text{or} \quad x = -4$$

Solution: $-\tfrac{1}{2},\ 5,\ -4$.

(c) $x^3 - \tfrac{1}{4}x^2 - \tfrac{1}{4}x + \tfrac{1}{16} = 0 \ \Rightarrow \ 16x^3 - 4x^2 - 4x + 1 = 0$. The possible rational zeros of $p(x) = 16x^3 - 4x^2 - 4x + 1$ are:

$$\frac{c}{l} = \frac{\pm 1}{\pm 1,\ \pm 2,\ \pm 4,\ \pm 8,\ \pm 16} : \pm 1,\ \pm\tfrac{1}{2},\ \pm\tfrac{1}{4},\ \pm\tfrac{1}{8},\ \pm\tfrac{1}{16}$$

Testing:

$$p(1) = 16 - 4 - 4 + 1 \neq 0$$
$$p(-1) = -16 - 4 + 4 + 1 \neq 0$$
$$p\left(\tfrac{1}{2}\right) = 16\left(\tfrac{1}{2}\right)^3 - 4\left(\tfrac{1}{2}\right)^2 - 4\left(\tfrac{1}{2}\right) + 1 = 2 - 1 - 2 + 1 = 0$$

As $x = \tfrac{1}{2}$ is a zero, $x - \tfrac{1}{2}$ is a factor. Since the coefficients of $p(x)$ are integers, $2\left(x - \tfrac{1}{2}\right) = 2x - 1$ is also a factor. Dividing and factoring the resulting quadratic polynomial, we have:

$$16x^3 - 4x^2 - 4x + 1 = 0$$
$$(2x - 1)(8x^2 + 2x - 1) = 0$$
$$(2x - 1)(4x - 1)(2x + 1) = 0$$
$$x = \frac{1}{2} \quad \text{or} \quad x = \frac{1}{4} \quad \text{or} \quad x = -\frac{1}{2}$$

Solution: $\pm\tfrac{1}{2},\ \tfrac{1}{4}$.

Solution to CYU 6.10

To solve $\csc^3 x + \csc^2 x - 6\csc x = 0$ for $0 \leq x \leq \pi$, first make the substitution $t = \csc x$:

$$t^3 + t^2 - 6t = 0$$
$$t(t^2 + t - 6) = 0$$
$$t(t+3)(t-2) = 0$$

Then $t = 0, -3,$ or 2. Substituting back for t: $\csc x = 0, -3,$ or 2.

As $\csc x = \dfrac{1}{\sin x}$:

$\sin x = \dfrac{1}{0}$ — undefined,

$\sin x = -\dfrac{1}{3}$ (no solution in the given interval, as $\sin x \geq 0$ there),

$\sin x = \dfrac{1}{2} \Rightarrow x$ lies in the first or second quadrants with a reference angle of $\dfrac{\pi}{6}$ (see figure).

Hence $x = \dfrac{\pi}{6}$ or $\dfrac{5\pi}{6}$.

Solution to CYU 6.11

(a) The equation $-2x^3 + x + 5 = 0$ has at most 3 solutions, since this polynomial equation is of degree 3.

(b) Since the coefficient of $x^3 - \dfrac{1}{2}x - \dfrac{5}{2}$ with the largest magnitude is $\dfrac{5}{2}$, $K = 1 + \dfrac{5}{2} = \dfrac{7}{2}$, and the interval is $\left[-\dfrac{7}{2}, \dfrac{7}{2}\right]$.

(c)
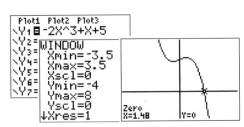

The solution is approximately 1.48.

Solution to CYU 6.12

The dimensions of the box are $12 - 2x$ by $12 - 2x$ by x, so that the volume is $V = x(12 - 2x)^2$. We chose a window of $[-1, 6]$ by $[-40, 150]$, since we must have $0 < x < 6$ and when $x = 1$, for example, $V = 100$. From the figure we see that the volume is a maximum for a square of side 2 inches.

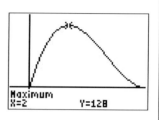

Solution to CYU 6.13

(a) $2x^2 + 3x - 2 \leq 0 \Rightarrow (2x - 1)(x + 2) \leq 0$. The x-intercepts of the graph of the parabola are $\frac{1}{2}$ and -2. From the graph of $f(x) = 2x^2 + 3x - 2$ in the figure, we see that the solution is $\left[-2, \frac{1}{2}\right]$ (where the graph lies on or below the x-axis).

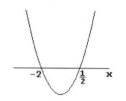

(b) The x-intercepts of the parabola which is the graph of $f(x) = 2x^2 + 7x + 2$ are gotten by the Pythagorean theorem:
$$x = \frac{-7 \pm \sqrt{7^2 - 4 \cdot 2 \cdot 2}}{4} = \frac{-7 \pm \sqrt{33}}{4}.$$
The parabola opens up, and from the figure we see that the solution of $2x^2 + 7x + 2 > 0$ is
$$\left(-\infty, \frac{-7 - \sqrt{33}}{4}\right) \cup \left(\frac{-7 + \sqrt{33}}{4}, \infty\right).$$

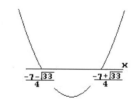

Solution to CYU 6.14

(a) Chart the sign of the polynomial on the left side of the inequality:

$$\overset{+\ \ \ \ \ -\ \ \ \ \ +\ \ \ \ \ -}{\underset{\substack{-2 \qquad\ 3 \qquad\ 5}}{\text{———}|\text{———}|\text{———}|\text{———}}}$$

$$\text{SIGN } (x-3)(x+2)(-x+5)$$

We want the intervals where the polynomial is negative: $(-2, 3) \cup (5, \infty)$.

(b) Chart the sign of the polynomial on the left side of the inequality:

$$\overset{-\ \ \ \ \ +\ \ \ \ \ +\ \ \ \ \ +}{\underset{\substack{-2 \qquad -1 \qquad\ 4}}{\text{———}|\text{———}|\text{———}|\text{———}}}$$

$$\text{SIGN } (x+1)^2(x+2)^3(x-4)^2$$

We want the intervals where the polynomial is non-negative: $[-2, \infty)$.

Solution to CYU 6.15

Begin by factoring: $x^2 + x - 2 = (x+2)(x-1)$. The polynomial $x^2 - x + 5$ has no factors as its discriminant $b^2 - 4ac = 1 - 20 < 0$, and when $x = 0$, $x^2 - x + 5 > 0$ indicating that this polynomial is always positive. The discriminant of the last polynomial, $-x^2 + x + 5$ is $b^2 - 4ac = 1 + 20 = 21$ implying that that polynomial has 2 zeros: $x = \dfrac{-1 \pm \sqrt{21}}{-2} = \dfrac{1 \pm \sqrt{21}}{2}$ and they are odd zeros (each corresponds to a linear factor of the polynomial). Charting the sign of the left side of the given inequality:

$$\overset{-\ \ \ \ \ +\ \ \ \ \ -\ \ \ \ \ +\ \ \ \ \ -}{\underset{\substack{-2 \qquad \frac{1-\sqrt{21}}{2} \qquad 1 \qquad \frac{1+\sqrt{21}}{2}}}{\text{———}|\text{———}|\text{———}|\text{———}|\text{———}}}$$

$$\text{SIGN } (x^2 + x - 2)(x^2 - x + 5)(-x^2 + x + 5)$$

We want the intervals where the polynomial is negative or zero:

$$(-\infty, -2] \cup \left[\frac{1 - \sqrt{21}}{2}, 1\right] \cup \left[\frac{1 + \sqrt{21}}{2}, \infty\right)$$

Solution to CYU 6.16

(a) To solve $(x+\pi)^2(\cos x + 1) \leq 0$, first find the zeros of the factors in $[0, 2\pi)$: $x + \pi = 0 \Rightarrow x = -\pi$ (not in the interval), and $\cos x + 1 = 0$ when $\cos x = -1 \Rightarrow x = \pi$. As $-1 \leq \cos x \leq 1$, for all x, $\cos x + 1 \geq 0$. This means that there is no sign change at π. When $x = \frac{\pi}{2}$, for example, $(x+\pi)^2(\cos x + 1) > 0$, as $\cos \frac{\pi}{2} = 0$, giving us the following sign chart:

```
        +    n    +
   +————+————+————+
   0         π         2π
   SIGN (x+π)² (cos x + 1)
```

The solution is π.

(b) To solve $(x+\pi)^2 \left(\cos x + \frac{\sqrt{3}}{2}\right) > 0$, first find the zeros of the factors in $[0, 2\pi)$: Again $x + \pi = 0 \Rightarrow x = -\pi$ is not in the interval. The factor $\cos x + \frac{\sqrt{3}}{2} = 0$ when $\cos x = -\frac{\sqrt{3}}{2}$. The reference angle is $\frac{\pi}{6}$ (see figure), and the angle lies in the 2nd or 3rd quadrants (as the cosine is negative) which means that

$$x = \pi - \frac{\pi}{6} = \frac{5\pi}{6} \quad \text{or} \quad \pi + \frac{\pi}{6} = \frac{7\pi}{6}$$

From the adjacent graph of $f(x) = \cos x + \frac{\sqrt{3}}{2}$, we see that there is a sign change at each of these zeros. Again choosing, for example $x = \frac{\pi}{2}$, gives the sign chart:

```
       +   c  _  c   +
   +———+——+——+———+
   0   5π  7π    2π
       6   6
   SIGN (x+π)² (cos x + √3/2)
```

The solution is $\left[0, \frac{5\pi}{6}\right) \cup \left(\frac{7\pi}{6}, 2\pi\right)$.

Solution to CYU 6.17

Rewriting the inequality as $3x^4 - 5x^2 - 4x - x^3 - 5 < 0$, that is, $3x^4 - x^3 - 5x^2 - 4x - 5 < 0$, and using the Confinement Theorem, we find that the x-intercepts of the graph of

$$f(x) = 3x^4 - x^3 - 5x^2 - 4x - 5$$
$$= 3\left(x^4 - \frac{1}{3}x^3 - \frac{5}{3}x^2 - \frac{4}{3}x - \frac{5}{3}\right)$$

are contained in the interval $(K = 1 + \frac{5}{3})$: $\left[-\frac{8}{3}, \frac{8}{3}\right]$.

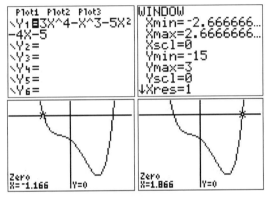

From the graph we see that the solution of the inequality (where the graph lies below the x-axis) is approximately: $(-1.17, 1.87)$.

Solution to CYU 6.18

(a) $f(x) = x^3 - x \Rightarrow g(x) = x^3$

(b) $f(x) = (2x^3)(x^2 - 5x + 1)(x - 1) = (2x^3)(x^2)(x) +$ lower power terms
$= 2x^6 +$ lower power terms

Therefore, $g(x) = 2x^6$.

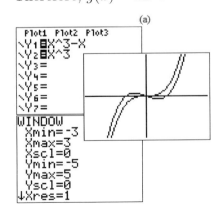
(a)

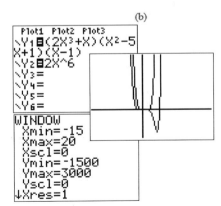
(b)

Solution to CYU 6.19

$$f(x) = -x^4 - 2x^3 + 3x^2 = -x^2(x^2 + 2x - 3) = -x^2(x+3)(x-1)$$

The y-intercept is 0, and the x-intercepts are $0, -3,$ and 1.
From the sign information of f:

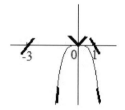

we see that the graph lies above the x-axis between -3 and 0 and between 0 and 1, and below the x-axis for $x < -3$ and $x > 1$. As $x \Rightarrow \pm\infty$, the graph resembles that of $g(x) = -x^4$.

We then sketched the graph (see figure).
[Note: Our analysis doesn't tell us how high those maxima really are.]

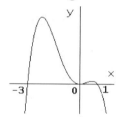

Solution to CYU 6.20

To graph $f(x) = x^5 - 6x^4 + 8x^3 - 2$, we first graph
$h(x) = x^5 - 6x^4 + 8x^3 = x^3(x^2 - 6x + 8) = x^3(x-4)(x-2)$.

The y-intercept is 0, and the x-intercepts are $0, 4,$ and 2.

From the sign information of h:

we see that the graph lies above the x-axis between 0 and 2 and after 4, and below the x-axis for $x < 0$ and $2 < x < 4$. As $x \Rightarrow \pm\infty$, the graph resembles that of $g(x) = x^5$.

We sketched the graph of h, and then lowered that graph by 2 units to obtain the graph of f. [Note: Our analysis doesn't tell us how high the graph rises or how low it falls.]

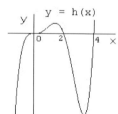 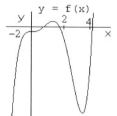

Solutions to Check Your Understanding Boxes   B-51

CHAPTER 7

7.1 (a) Since a rational function will only be undefined where the denominator takes on a value

of zero, the domain of $f(x) = \dfrac{x^2+4}{x^2} - 4 = \dfrac{x^2+4-4x^2}{x^2} = \dfrac{-3x^2+4}{x^2}$ consists of all real

numbers except 0: $D_f = (-\infty, 0) \cup (0, \infty)$.

(b) Since $g(x) = \dfrac{x^2-4}{x^2+x-6} = \dfrac{x^2-4}{(x+3)(x-2)}$, $D_f = (-\infty, -3) \cup (-3, 2) \cup (2, \infty)$

Note: While $\dfrac{x^2-4}{(x+3)(x-2)} = \dfrac{(x+2)(x-2)}{(x+3)(x-2)}$

$g(x) = \dfrac{x^2-4}{x^2+x-6} = \dfrac{(x+2)(x-2)}{(x+3)(x-2)}$ is **NOT equal to** $h(x) = \dfrac{x+2}{x+3}$ $\boxed{h(2) = \frac{4}{5} \text{ while } g(2) \text{ is undefined}}$

7.2 (a)

$$\frac{x-2}{x^2-4} - \frac{5}{4} = \frac{1}{x-3}$$

$$\frac{x-2}{(x+2)(x-2)} - \frac{5}{4} = \frac{1}{x-3}$$

$$\frac{1}{x+2} - \frac{5}{4} = \frac{1}{x-3}$$

clear denominators: $4(x-3) - 5(x+2)(x-3) = 4(x+2)$

$$4x - 12 - 5(x^2 - x - 6) = 4x + 8$$

regroup: $-5x^2 + 5x + 10 = 0$

$$x^2 - x - 2 = 0$$

$$(x-2)(x+1) = 0$$

$\cancel{x = 2}$ or $\boxed{x = -1}$

not a solution since the denominator of $\dfrac{x-2}{x^2-4}$ is 0 when $x = 2$

(b)

$$3\left(\frac{x}{x^2+1}\right) + \left(\frac{x^2+1}{x}\right) = 4$$

let $y = \dfrac{x}{x^2+1}$:

$$3y + \frac{1}{y} = 4$$

$$3y^2 + 1 = 4y$$

$$3y^2 - 4y + 1 = 0$$

$$(3y - 1)(y - 1) = 0$$

$y = \dfrac{1}{3}$

$$\frac{x}{x^2+1} = \frac{1}{3}$$

$$x^2 + 1 = 3x$$

$$x^2 - 3x + 1 = 0$$

$$x = \frac{3 \pm \sqrt{9-4}}{2} = \boxed{\frac{3 \pm \sqrt{5}}{2}}$$

both are solutions

$y = 1$

$$\frac{x}{x^2+1} = 1$$

$$x^2 + 1 = x$$

$\boxed{x^2 - x + 1 = 0 \atop \text{no solution since the discriminant is negative.}}$

7.3

$$3\sin x + 1 = \frac{1}{2\sin x - 1}$$
$$(3\sin x + 1)(2\sin x - 1) = 1$$
$$6\sin^2 x - \sin x - 2 = 0$$
$$(3\sin x - 2)(2\sin x + 1) = 0$$

$\sin x = \frac{2}{3}$ $\quad$ $\sin x = -\frac{1}{2}$

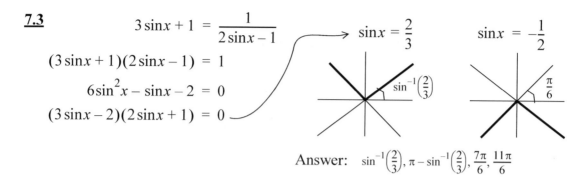

Answer: $\sin^{-1}\left(\frac{2}{3}\right), \pi - \sin^{-1}\left(\frac{2}{3}\right), \frac{7\pi}{6}, \frac{11\pi}{6}$

7.4 From Step 1 of Example 7.4 we can express the amount of material, the total surface area, M (in square inches) in terms of the radius r and height h of the aluminum can:

$$M = 2\pi r^2 + 2\pi r h \quad (*)$$

From the given information that the volume of the can is 160 cubic inches, we have:
$$\pi r^2 h = 160$$
$$h = \frac{160}{\pi r^2} \quad (**)$$

Substituting in (*):
$$M = 2\pi r^2 + 2\pi r h = 2\pi r^2 + 2\pi r \frac{160}{\pi r^2} = 2\pi r^2 + \frac{320}{r}$$

Turning to a graphing calculator:

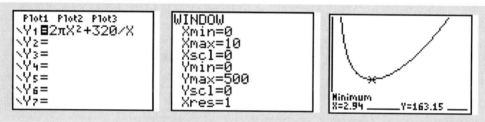

Conclusion: Radius of can for minimal material: $r \approx 2.94$ in. From (**) $h \approx \dfrac{160}{\pi(2.94)^2} \approx 5.89$ in.

7.5 (a) $\dfrac{x+2}{x^2 - 2x - 3} \geq 0$

$\dfrac{x+2}{(x-3)(x+1)} \geq 0$:

$\begin{array}{c} - \quad c \quad + \quad c \quad - \quad c \quad + \\ \hline \bullet \quad \circ \quad\quad \circ \\ -2 \quad -1 \quad\quad 3 \end{array}$

Solution: $[-2, -1) \cup (3, \infty)$

(b) $x < \dfrac{1}{3x+2}$

$x - \dfrac{1}{3x+2} < 0$

$\dfrac{3x^2 + 2x - 1}{3x+2} < 0$

$\dfrac{(3x-1)(x+1)}{3x+2} < 0$:

$\begin{array}{c} - \quad c \quad + \quad c \quad - \quad c \quad + \\ \hline \bullet \quad \circ \quad\quad \bullet \\ -1 \quad -\frac{2}{3} \quad\quad \frac{1}{3} \end{array}$

Sollution: $(-\infty, -1) \cup (-\frac{2}{3}, \frac{1}{3})$

7.6

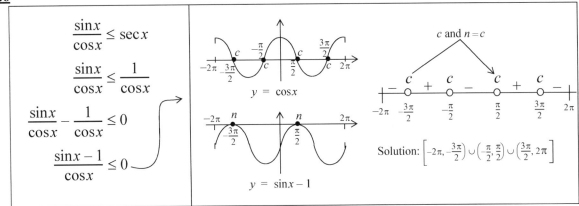

$$\frac{\sin x}{\cos x} \le \sec x$$

$$\frac{\sin x}{\cos x} \le \frac{1}{\cos x}$$

$$\frac{\sin x}{\cos x} - \frac{1}{\cos x} \le 0$$

$$\frac{\sin x - 1}{\cos x} \le 0$$

Solution: $\left[-2\pi, -\frac{3\pi}{2}\right) \cup \left(-\frac{\pi}{2}, \frac{\pi}{2}\right) \cup \left(\frac{3\pi}{2}, 2\pi\right]$

7.7 (a)

$$\frac{x}{x^2 - 3} = \frac{x - 1}{3x^4 + 1}$$

multiply both sides of the equation by $(x^2 - 3)(3x^4 + 1)$: $x(3x^4 + 1) = (x^2 - 3)(x - 1)$

regroup: $3x^5 - x^3 + x^2 + 4x - 3 = 0$

Divide by 3 to get a leading coefficient of 1: $x^5 - \frac{x^3}{3} + \frac{x^2}{3} + \frac{4x}{3} - 1 = 0$

Applying the Confinement Theorem (page 266) to the equation: $x^5 - \frac{x^3}{3} + \frac{x^2}{3} + \frac{4x}{3} - 1 = 0$ we conclude that all of the zeros of the polynomial are contained in the interval $[-K, K]$, where $K = 1 + \text{largest}\left\{\left|-\frac{1}{3}\right|, \left|\frac{1}{3}\right|, \left|\frac{4}{3}\right|, |-1|\right\} = 1 + \frac{4}{3} = \frac{7}{3}$. Accordingly, by setting the x-range of our window to be $-5 \le x \le 5$, we know that we will capture all solutions of the given equation; bringing us to:

From the above we conclude that the given equation has one solution: $x \approx 0.64$.

(b) To solve the inequality $\frac{x}{x^2 - 3} < \frac{x - 1}{3x^4 + 1}$ we first bring all terms to the left of the inequality sign: $\frac{x}{x^2 - 3} - \frac{x - 1}{3x^4 + 1} < 0$ (**DON'T** multiply both sides of the inequality by $(x^2 - 3)(3x^4 + 1)$ — why not?). We know that the sign of the function

$$f(x) = \frac{x}{x^2 - 3} - \frac{x - 1}{3x^4 + 1} = \frac{x(3x^4 + 1) - (x - 1)(x^2 - 3)}{(x^2 - 3)(3x^4 + 1)} = \frac{3x^5 - x^3 + x^2 + 4x - 3}{(x + \sqrt{3})(x - \sqrt{3})(3x^4 + 1)}$$

can only change about a zero of the numerator or of the denominator. The zeros of the denominator are easily seen to be at $x = \pm\sqrt{3}$. We know that $f(x) = \dfrac{x}{x^2-3} - \dfrac{x-1}{3x^4+1}$ has vertical asymptotes at $x = \pm\sqrt{3}$. We also observed, in (a) that the only zero of the numerator occurs at $x \approx 0.64$. From below, we conclude that $(-\infty, -\sqrt{3}) \cup (0.64, \sqrt{3})$ is the approximate solution set of the given inequality.

7.8 (a) Since the degree of the numerator of $f(x) = \dfrac{3x^4 - 4x^2 + 1}{x^4 - 5x^2 + 4}$ equals that of the denominator, the graph has a horizontal asymptote. Specifically, as $x \to \pm\infty$, the graph of the function approaches the horizontal line $y = \dfrac{3x^4}{x^4} = 3$.

(b) Since the degree of the numerator of $f(x) = \dfrac{4x^3 - 2x^2 - 4}{3x^4 + 1}$ is less than that of the denominator, the graph has a horizontal asymptote. Specifically, it will approach the x-axis ($y = 0$) since, $\dfrac{4x^3}{3x^4} = \dfrac{4}{3x}$ approaches zero as $x \to \pm\infty$.

(c) Since the degree of the numerator of $f(x) = \dfrac{6x^3 - 9x^2 - 4x}{3x^2 - 1}$ is one more than that of the denominator, the graph has an oblique asymptote. As $x \to \pm\infty$, the graph of f will get arbitrarily close to the line $y = 2x - 3$, since $\dfrac{6x^3 - 9x^2 - 4x}{3x^2 - 1} = 2x - 3 + \dfrac{-2x-3}{3x^2-1}$, and $\dfrac{-2x-3}{3x^2-1} \to 0$ as $x \to \pm\infty$

$$\begin{array}{r}2x-3\\3x^2-1\overline{\smash{\big)}6x^3-9x^2-4x}\\6x^3-2x\\\hline-9x^2-2x\\-9x^2+3\\\hline-2x-3\end{array}$$

7.9 Identical to that about the vertical asymptote at $x = 2$, since the signs on either side of the vertical asymptote at $x = 2$ coincide with those of the corresponding side of the vertical asymptote at $x = -1$.

7.10 (a) **Step 1. Factor:** Already in factored form

Step 2. y-intercept: $y = f(0) = \dfrac{0-1}{0+1} = -1$.

x-intercepts: $f(x) = 0: \frac{x-1}{x+1} = 0 \Rightarrow x = 1$.

Vertical Asymptotes: The line $x = -1$

SIGN $f(x)$: From the sign information: $\underset{-1}{\overset{+ \quad c \quad - \quad c \quad +}{\circ \quad \bullet}}$, we conclude that the graph goes from below the x-axis to above the x-axis, as you move, from left to right, across the x-intercept at $x = 1$. Since the function is positive just to the left of the vertical asymptote at $x = -1$, the graph must tend to $+\infty$ as x approaches -1 from the left. Since the function is negative just to the right of -1, the graph tends to $-\infty$ as x approaches -1 from the right.

Step 3. As $x \to \pm\infty$: The graph resembles that of $g(x) = \frac{x}{x} = 1$. This tells us that the line $y = 1$ is a horizontal asymptote for the graph.

Indeed, from $\frac{x-1}{x+1} = 1 - \frac{2}{x+1}$ we conclude that the graph will approach the horizontal asymptote $y = 1$ from below as $x \to +\infty$ (since $\frac{2}{x+1}$ is positive), and from above as $x \to -\infty$ (since $\frac{2}{x+1}$ will be negative).

Step 4. Sketch the anticipated graph: The above information is reflected on the left below; leading us to the anticipated graph at the right.

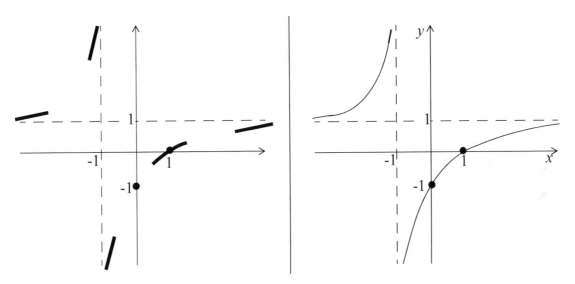

(b) **Step 1. Factor:** $\frac{x^7}{x^4 - 16} = \frac{x^7}{(x^2+4)(x^2-4)} = \frac{x^7}{(x^2+4)(x+2)(x-2)}$

Step 2. y-intercept: $y = f(0) = \frac{0}{-16} = 0$.

x-intercepts: $f(x) = 0: \frac{x^7}{x^4 - 16} = 0 \Rightarrow x = 0$.

Vertical Asymptotes: The lines $x = \pm 2$

SIGN $f(x)$: From the sign information: $\underset{-2 \quad 0 \quad 2}{\overset{- \quad c \quad + \quad c \quad - \quad c \quad +}{\circ \quad \bullet \quad \circ}}$, we conclude that the graph goes from above the x-axis to below the x-axis, as you move, from left to right, across the x-intercept at $x = 0$. Since the function is negative just to the left of the vertical asymptote at

$x = -2$ and the vertical asymptote at $x = 2$ the graph must tend to $-\infty$ as x approaches either of those two asymptotes from the left. Since the function is positive just to the right of those two asymptotes, the graph tends to $+\infty$ as they are approached from the right.

Step 3. As $x \to \pm\infty$: The graph resembles that of $g(x) = \dfrac{x^7}{x^4} = x^3$.

Step 4. Sketch the anticipated graph: The above information is reflected on the left below; leading us to the anticipated graph to the right.

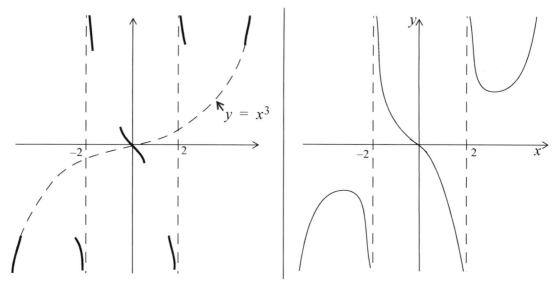

7.11 Taking advantage of the double angle identity $\sin 2x = 2\sin x \cos x$ (page 187), we can rewrite the given function in the form $f(x) = \dfrac{1}{2}\dfrac{\sin 2x}{(x+1)(x-1)}$. There are two odd zeros in the denominator, at $x = -1$ and $x = 1$. Since $\sin 2x$ has zeros when $2x$ is a multiply of π, or when x is a multiple of $\dfrac{\pi}{2}$, over the interval $[-\pi, \pi]$ the numerator will be zero at $0, \pm\dfrac{\pi}{2}, \pm\pi$. Moreover $\sin 2x$ will change sign when traversing any of those zeros. Bringing us to:

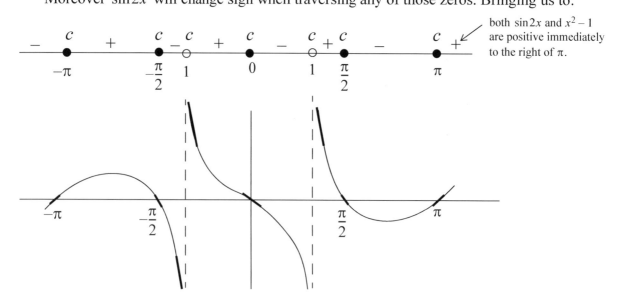

7.12 To arrive at the general form of the partial fraction decomposition of

$$\frac{x-4}{(x-3)(x^2+x+5)(2x+1)^2(x^2+5)^2}, \text{ we turn to Figure 7.5:}$$

$(x-3)$ gives rise to $\dfrac{A}{x-3}$ Case 1 in Figure

(x^2+x+5) gives rise to $\dfrac{Bx+C}{x^2+x+5}$ Case 3 in Figure (note that x^2+x+5 is irreducible)

$(2x+1)^2$ gives rise to $\dfrac{D}{2x+1}+\dfrac{E}{(2x+1)^2}$ Case 2 in Figure

$(x^2+5)^2$ gives rise to $\dfrac{Fx+G}{x^2+5}+\dfrac{Hx+I}{(x^2+5)^2}$ Case 4 in Figure (note that x^2+5 is irreducible)

Bringing us to:
$$\frac{x-4}{(x-3)(x^2+x+5)(2x+1)^2(x^2+5)^2}$$
$$= \frac{A}{x-3}+\frac{Bx+C}{x^2+x+5}+\frac{D}{2x+1}+\frac{E}{(2x+1)^2}+\frac{Fx+G}{x^2+5}+\frac{Hx+I}{(x^2+5)^2}$$

7.13 The general decomposition: $\dfrac{-x^2-4x-3}{x^3+x^2+x} = \dfrac{-x^2-4x-3}{x(x^2+x+1)} = \dfrac{A}{x}+\dfrac{Bx+C}{x^2+x+1}$

Clearing denominators: $-x^2-4x-3 = A(x^2+x+1)+(Bx+C)x$

Setting $x=0$: $\boxed{-3=A}$.

Equating the coefficients of x^2: $-1=A+B \Rightarrow -1=-3+B \Rightarrow \boxed{B=2}$

Equating the coefficients of x: $-4=A+C \Rightarrow -4=-3+C \Rightarrow \boxed{C=-1}$

Answer: $\dfrac{-x^2-4x-3}{x^3+x^2+x} = \dfrac{-3}{x}+\dfrac{2x-1}{x^2+x+1}$

7.14 Since the degree of the numerator of $\dfrac{x^4-x^3+x^2+2x-10}{x^2-x-2}$ is not less than that of the denominator, we first perform the necessary division process:

$$
\begin{array}{r}
x^2+3 \\
x^2-x-2\,\overline{\smash{\big)}\,x^4-x^3+x^2+2x-10} \\
\underline{x^4-x^3-2x^2} \\
\text{subtract:} \quad 3x^2+2x-10 \\
\underline{3x^2-3x-6} \\
5x-4
\end{array}
$$

Bringing us to: $\dfrac{x^4-x^3+x^2+2x-10}{x^2-x-2} = x^2+3+\dfrac{5x-4}{x^2-x-2}$

Then: $\dfrac{5x-4}{x^2-x-2} = \dfrac{5x-4}{(x-2)(x+1)} = \dfrac{A}{x-2}+\dfrac{B}{x+1}$

Clearing denominators: $5x-4 = A(x+1)+B(x-2)$

Setting $x=-1$: $-9=B(-3) \Rightarrow \boxed{B=3}$. Setting $x=2$: $6=A(3) \Rightarrow \boxed{A=2}$.

Answer: $\dfrac{x^4-x^3+x^2+2x-10}{x^2-x-2} = x^2+3+\dfrac{2}{x-2}+\dfrac{3}{x+1}$

Chapter 8

8.1 The graph of $f(x) = |x^4 - x^3 - 6x^2|$ can be obtained by reflecting the negative portion of the graph of $f(x) = x^4 - x^3 - 6x^2$ of Example 6.22, page 288, about the x-axis:

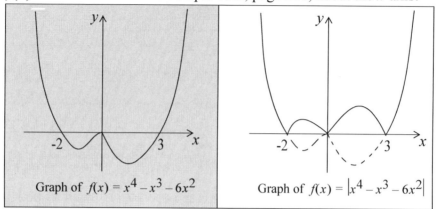

8.2 For $a = -5$ and $b = 3$:

(i) $|a| \geq 0 : |-5| = 5 \geq 0$ (ii) $|a| = |-a| : |-5| = 5 = |-(-5)|$

(iii) $|ab| = |a||b|$ and, if $b \neq 0$, $\left|\dfrac{a}{b}\right| = \dfrac{|a|}{|b|}$: $|(-5)(3)| = |-15| = 15 = |-5||3|$

and: $\left|\dfrac{-5}{3}\right| = \left|-\dfrac{5}{3}\right| = \dfrac{5}{3} = \dfrac{|-5|}{|3|}$

(iv) $|a|^2 = a^2 : |-5|^2 = 5^2 = 25 = (-5)^2$

(v) $\sqrt{a^2} = |a| : \sqrt{(-5)^2} = \sqrt{25} = 5 = |-5|$

(vi) $|a + b| \leq |a| + |b| : |(-5) + 3| = |-2| = 2 \leq 5 + 3 = |-5| + |3|$

8.3 (a)
$$|3x - 5| = 7$$
$3x - 5 = 7$ OR $3x - 5 = -7$
$3x = 12$ $3x = -2$
$x = 4$ $x = -\dfrac{2}{3}$

(b)
$$|x^2 - x - 4| = 2$$
$x^2 - x - 4 = 2$ OR $x^2 - x - 4 = -2$
$x^2 - x - 6 = 0$ $x^2 - x - 2 = 0$
$(x - 3)(x + 2) = 0$ $(x - 2)(x + 1) = 0$
$x = 3, x = -2$ $x = 2, x = -1$

(c)
$$|4x + 3| = |2 - 5x|$$
$4x + 3 = 2 - 5x$ OR $4x + 3 = -(2 - 5x)$
$9x = -1$ $4x + 3 = -2 + 5x$
$x = -\dfrac{1}{9}$ $x = 5$

8.4 $|2\tan x - \cot x| = 1$ if $2\tan x - \cot x = 1$ or $2\tan x - \cot x = -1$. In Example 4.25, page 214, we showed that $\pi - \tan^{-1}\left(\frac{1}{2}\right)$, $2\pi - \tan^{-1}\left(\frac{1}{2}\right)$, $\frac{\pi}{4}$ and $\frac{5\pi}{4}$ are the solutions of $2\tan x - \cot x = 1$ in the interval $[0, 2\pi)$. Below, we observe that $\tan^{-1}\left(\frac{1}{2}\right)$, $\pi + \tan^{-1}\left(\frac{1}{2}\right)$ (the tangent is positive in the first and third quadrant), and $\pi - \frac{\pi}{4}$, $2\pi - \frac{\pi}{4}$ are solutions of $2\tan x - \cot x = -1$ (the tangent is negative in the second and fourth quadrant):

$$2\tan x - \cot x + 1 = 0$$

$$2\tan x - \frac{1}{\tan x} + 1 = 0$$

$$2\tan^2 x - 1 + \tan x = 0$$

$$(2\tan x - 1)(\tan x + 1) = 0$$

$$\tan x = \frac{1}{2} \quad \text{or} \quad \tan x = -1$$

Conclusion: $\tan^{-1}\left(\frac{1}{2}\right)$, $\pi - \tan^{-1}\left(\frac{1}{2}\right)$, $\pi + \tan^{-1}\left(\frac{1}{2}\right)$, $2\pi - \tan^{-1}\left(\frac{1}{2}\right)$, $\frac{\pi}{4}, \frac{3\pi}{4}, \frac{5\pi}{4}, \frac{7\pi}{4}$ are the solutions of $|2\tan x - \cot x| = 1$ in the interval $[0, 2\pi)$.

8.5 (a)
$$|5x - 4| \le 2$$
$$-2 \le 5x - 4 \le 2$$
$$2 \le 5x \le 6$$
$$\frac{2}{5} \le x \le \frac{6}{5}$$
Solution Set: $\left[\frac{2}{5}, \frac{6}{5}\right]$

(b)
$$|2x - 3| > 5$$
$$2x - 3 < -5 \quad \text{OR} \quad 2x - 3 > 5$$
$$2x < -2 \qquad\qquad 2x > 8$$
$$x < -1 \qquad\qquad x > 4$$
Solution Set: $(-\infty, -1) \cup (4, \infty)$

8.6 (a)
$$\sqrt{x-2} + \sqrt{x+3} = 5$$
$$\sqrt{x-2} = 5 - \sqrt{x+3}$$
$$(\sqrt{x-2})^2 = (5 - \sqrt{x+3})^2$$
$$x - 2 = 25 - 10\sqrt{x+3} + x + 3$$
$$10\sqrt{x+3} = 30$$
$$\sqrt{x+3} = 3$$
$$x + 3 = 9$$
$$x = 6$$
Check: $\sqrt{6-2} + \sqrt{6+3} = 2 + 3 = 5$ Yes

(b)
$$(x^2 + 2x - 1)^{\frac{1}{5}} = (x+7)^{\frac{1}{5}}$$
$$\left[(x^2 + 2x - 1)^{\frac{1}{5}}\right]^5 = \left[(x+7)^{\frac{1}{5}}\right]^5$$
$$x^2 + 2x - 1 = x + 7$$
$$x^2 + x - 8 = 0$$
$$x = \frac{-1 \pm \sqrt{1 - 4(-8)}}{2} = \frac{-1 \pm \sqrt{33}}{2}$$

(Raising both sides of an equation to an odd power cannot introduce extraneous roots.)

B-60 Solutions to Check Your Understanding Boxes

8.7 (a)

$$\sqrt{4\sin x - 1} = 2\sin x$$
$$4\sin x - 1 = 4\sin^2 x$$
$$4\sin^2 x - 4\sin x + 1 = 0$$
$$(2\sin x - 1)^2 = 0$$
$$2\sin x - 1 = 0$$
$$\sin x = \frac{1}{2}$$

$$x = \pi - \frac{\pi}{6} = \frac{5\pi}{6} \qquad x = \frac{\pi}{6}$$

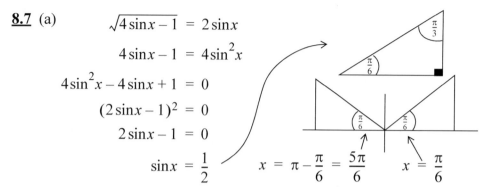

From the above, we see that $\frac{\pi}{6}$ and $\frac{5\pi}{6}$ are the only possible solutions of $\sqrt{4\sin x - 1} = 2\sin x$ in the interval $[0, 2\pi)$. Given that we squared both sides of the equation in our solution process, we must challenge each candidate:

$\sqrt{4\sin\frac{\pi}{6} - 1} \stackrel{?}{=} 2\sin\frac{\pi}{6}$	$\sqrt{4\sin\frac{5\pi}{6} - 1} \stackrel{?}{=} 2\sin\frac{5\pi}{6}$
$\sqrt{4\cdot\frac{1}{2} - 1} \stackrel{?}{=} 2\cdot\frac{1}{2}$	$\sqrt{4\cdot\frac{1}{2} - 1} \stackrel{?}{=} 2\cdot\frac{1}{2}$
$\sqrt{1} \stackrel{?}{=} 1$ **Yes!**	$\sqrt{1} \stackrel{?}{=} 1$ **Yes!**

Conclusion: $\frac{\pi}{6}$ and $\frac{5\pi}{6}$ are the solutions of $\sqrt{4\sin x - 1} = 2\sin x$ in $[0, 2\pi)$

(b)

$$\cot x = \csc x + \frac{1}{\sqrt{3}}$$
$$\cot^2 x = \csc^2 x + \frac{2}{\sqrt{3}}\csc x + \frac{1}{3}$$

CYU 4.6 (a-ii) page 185:
$$\csc^2 x - 1 = \csc^2 x + \frac{2}{\sqrt{3}}\csc x + \frac{1}{3}$$
$$-1 = \frac{2}{\sqrt{3}}\csc x + \frac{1}{3}$$
$$\csc x = -\frac{2}{\sqrt{3}}$$

$$\sin x = -\frac{\sqrt{3}}{2}$$

$$x = \pi + \frac{\pi}{3} = \frac{4\pi}{3} \qquad x = 2\pi - \frac{\pi}{3} = \frac{5\pi}{3}$$

Mandatory Check:

$\cot\frac{4\pi}{3} \stackrel{?}{=} \csc\frac{4\pi}{3} + \frac{1}{\sqrt{3}}$	$\cot\frac{5\pi}{3} \stackrel{?}{=} \csc\frac{5\pi}{3} + \frac{1}{\sqrt{3}}$
$\frac{\cos\frac{4\pi}{3}}{\sin\frac{4\pi}{3}} \stackrel{?}{=} \frac{1}{\sin\frac{4\pi}{3}} + \frac{1}{\sqrt{3}}$	$\frac{\cos\frac{5\pi}{3}}{\sin\frac{5\pi}{3}} \stackrel{?}{=} \frac{1}{\sin\frac{5\pi}{3}} + \frac{1}{\sqrt{3}}$
$\frac{-\frac{1}{2}}{-\frac{\sqrt{3}}{2}} \stackrel{?}{=} \frac{1}{-\frac{\sqrt{3}}{2}} + \frac{1}{\sqrt{3}}$	$\frac{\frac{1}{2}}{-\frac{\sqrt{3}}{2}} \stackrel{?}{=} \frac{1}{-\frac{\sqrt{3}}{2}} + \frac{1}{\sqrt{3}}$
$\frac{1}{\sqrt{3}} \stackrel{?}{=} -\frac{1}{\sqrt{3}}$ **No!**	$-\frac{1}{\sqrt{3}} \stackrel{?}{=} -\frac{1}{\sqrt{3}}$ **Yes!**

Conclusion $\frac{5\pi}{3}$ is the only solution of $\cot x = \csc x + \frac{1}{\sqrt{3}}$ in $[0, 2\pi)$

8.8 (a) $(x+1)^{\frac{3}{4}} = 125 \Rightarrow x+1 = 125^{\frac{4}{3}} = \left(125^{\frac{1}{3}}\right)^4 = 5^4 = 625 \Rightarrow x = 624$

(b) $x^{\frac{2}{3}} = \frac{1}{4} \Rightarrow x = \pm\left(\frac{1}{4}\right)^{\frac{3}{2}} = \pm\left[\left(\frac{1}{4}\right)^{\frac{1}{2}}\right]^3 = \pm\left(\frac{1}{2}\right)^3 = \pm\frac{1}{8}$

8.9 (a) $x^{\frac{4}{5}}(x-1)^{\frac{1}{3}}(-x+3)^3 < 0$

Solution Set: $(-\infty, 0) \cup (0, 1) \cup (3, \infty)$

(b) $x^{\frac{3}{5}}(x-1)^{\frac{2}{3}} \geq 0$

Solution Set: $[0, \infty)$

8.10 $(-x+3)(x+2)^{\frac{5}{2}}(x+1)^{\frac{2}{3}} \leq 0$

Domain: $x+2 \geq 0$
$x \geq -2$

exponent: 1

Solution Set: $\{-2, -1\} \cup [3, \infty)$

8.11 In the SEE THE PROBLEM box of Example 8.19, replace the "x miles (at 4 mi/hr)" with "x miles (at 3 mi/hr)," and the "y miles (at 5 mi/hr)" with "y miles (at 4 mi/hr)." Then:

$$\text{Time for man to reach the dock} = \frac{x \text{ mi}}{3\frac{\text{mi}}{\text{hr}}} + \frac{y \text{ mi}}{4\frac{\text{mi}}{\text{hr}}} = \left(\frac{x}{3} + \frac{y}{4}\right) \text{ hr} \quad (*)$$

As was determined in the solution of Example 8.19: $y = 10 - \sqrt{x^2 - 9}$. Substituting in (*):

$$T = \frac{x}{3} + \frac{10 - \sqrt{x^2 - 9}}{4}$$

Turning to a graphing calculator:

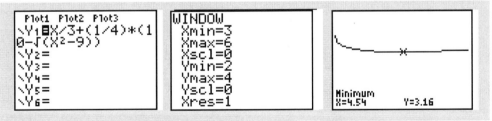

From the above, we see that the minimum time is approximately 3.16 hours, or approximately 3 hours and 10 minutes.

CHAPTER 9

Solution to CYU 9.1

(a) As the base of $f(x) = \left(\frac{8}{3}\right)^x$ is $b = \frac{8}{3} > 1$, the graph (see Figure below) follows that of Figure 9.4(a). The x-axis is a horizontal asymptote of the graph. The y-intercept is 1.

(b) As the base of $f(x) = \left(\frac{3}{8}\right)^x$ is $b = \frac{3}{8} < 1$, the graph (see Figure below) follows that of Figure 9.4(b). The x-axis is a horizontal asymptote of the graph. The y-intercept is 1.

(a) 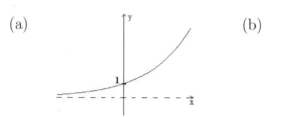 (b)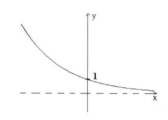

Solution to CYU 9.2

To solve $3^{2x^2-1} = \left(\frac{1}{27}\right)^x$ we first decide on a common base, 3.

$$3^{2x^2-1} = (3^{-3})^x$$
$$3^{2x^2-1} = 3^{-3x}$$
$$2x^2 - 1 = -3x$$
$$2x^2 + 3x - 1 = 0$$
$$x = \frac{-3 \pm \sqrt{3^2 - 4(2)(-1)}}{2(2)}$$
$$= \frac{-3 \pm \sqrt{17}}{4}$$

Solutions to Check Your Understanding Boxes   **B–63**

Solution to CYU 9.3

To solve $9^{\tan x} = \dfrac{3^{\cot x}}{3}$ for $0 \leq x < 2\pi$:

$$9^{\tan x} = \frac{3^{\cot x}}{3}$$

$$3^{2\tan x} = 3^{\cot x - 1}$$

$$2\tan x = \cot x - 1$$

$$= \frac{1}{\tan x} - 1$$

$$2\tan^2 x = 1 - \tan x$$

$$2\tan^2 x + \tan x - 1 = 0, \quad \text{substitute } t = \tan x$$

$$2t^2 + t - 1 = 0$$

$$(2t - 1)(t + 1) = 0$$

$$t = \frac{1}{2} \quad \text{or} \quad t = -1$$

Substituting back $t = \tan x$:

$\tan x = \frac{1}{2}$	$\tan x = -1$
(reference angle is $\tan^{-1}\frac{1}{2}$)	(reference angle is $\frac{\pi}{4}$)
$\tan x > 0 \;\Rightarrow\; $ QI or QIII	$\tan x < 0 \;\Rightarrow\; $ QII or QIV
$x = \tan^{-1}\frac{1}{2}$ or $x = \pi + \tan^{-1}\frac{1}{2}$	$x = \pi - \frac{\pi}{4} = \frac{3\pi}{4}$ or $x = 2\pi - \frac{\pi}{4} = \frac{7\pi}{4}$

B–64   Solutions to Check Your Understanding Boxes

Solution to CYU 9.4

To solve $3^{2x-1} + 3^{x-1} = \frac{2}{3}$, first multiply by 3:

$$3 \cdot 3^{2x-1} + 3 \cdot 3^{x-1} = 2$$
$$3^{2x} + 3^x = 2$$
$$(3^x)^2 + 3^x = 2$$
$$(3^x)^2 + 3^x - 2 = 0, \quad \text{substitute } t = 3^x$$
$$t^2 + t - 2 = 0$$
$$(t+2)(t-1) = 0$$
$$t = -2 \quad \text{or} \quad t = 1$$

Substituting back $t = 3^x$:

$$3^x = -2 \qquad\qquad 3^x = 1$$
$$\text{no solution} \qquad\qquad 3^x = 3^0$$
$$(3^x > 0 \text{ for all } x) \qquad\qquad \text{x} = 0$$

Solution to CYU 9.5

(a)   First determine a common base, 2.

$$4^{x+1} > 8 \cdot 2^x$$
$$(2^2)^{x+1} > 2^3 \cdot 2^x$$
$$2^{2x+2} > 2^{3+x} \qquad \text{then, since the base} > 1,$$
$$2x + 2 > 3 + x$$
$$2x - x > 3 - 2$$
$$x > 1$$

and the solution is $(1, \infty)$.

(b)   As the base is < 1, the inequality between the exponents will be reversed:

$$\left(\frac{2}{3}\right)^{5x-2} > \left(\frac{2}{3}\right)^{2x}$$
$$5x - 2 < 2x$$
$$3x < 2$$
$$x < \frac{2}{3}$$

and the solution is $\left(-\infty, \frac{2}{3}\right)$.

Solution to CYU 9.6

(a) (i) Let $\log_2 16 = x$ and change to exponential form: $2^x = 16 \Rightarrow 2^x = 2^4 \Rightarrow x = 4$

  (ii) Let $\log_{\frac{1}{3}} 9 = x$ and change to exponential form: $\left(\frac{1}{3}\right)^x = 9 \Rightarrow 3^{-x} = 3^2 \Rightarrow$
  $-x = 2 \Rightarrow x = -2$

  (iii) As $\log_b 1 = 0$ for all bases b, $\log_5 1 = 0$

(b) The exponential form of $\log_3 \left(\frac{1}{27}\right) = -3$ is $3^{-3} = \frac{1}{27}$.

(c) The logarithmic form of $\left(\frac{1}{2}\right)^4 = \frac{1}{16}$ is $\log_{\frac{1}{2}} \left(\frac{1}{16}\right) = 4$.

(d) (i) $\dfrac{\ln 37.3}{\ln 1.8 + \ln 13.2} \approx 1.14$ (ii) $\dfrac{\log 50}{5 - \log 12} \approx 0.43$

Solution to CYU 9.7

(a) $\log_3 9 = x \Rightarrow 3^x = 9 \Rightarrow 3^x = 3^2 \Rightarrow x = 2$, and $\log_2 2^{\sqrt{5}} = \sqrt{5}$, as
 $\log_b b^x = x$. Then $15 \log_3 9 + \log_2 2^{\sqrt{5}} = 15(2) + \sqrt{5} = 30 + \sqrt{5}$

(b) Since $3^{\log_3 5} = 5$, because $b^{\log_b x} = x$, and because $\ln e^x = x$, then
 $\ln e^{3^{\log_3 5}} = 5$

(c) From $\ln e^x = x$, then $\ln e = \ln e^1 = 1$, and $\ln \sqrt{e} = \ln e^{\frac{1}{2}} = \frac{1}{2}$. Also,
 $\log 100 = \log_{10} 10^2 = 2$. Putting it all together: $\dfrac{\ln e - \ln \sqrt{e}}{3 - \log 100} = \dfrac{1 - \frac{1}{2}}{3 - 2} = \dfrac{1}{2}$

Solution to CYU 9.8

The function $\dfrac{\ln(x-3)}{x-5}$ is defined when the logarithm is defined ($x > 3$) and the denominator is nonzero ($x \ne 5$). Therefore the domain is $D_f = (3, 5) \cup (5, \infty)$.

B–66   **Solutions to Check Your Understanding Boxes**

Solution to CYU 9.9

(a)   To solve $\ln(x^2 - 15) = \ln(-2x)$ use Theorem 9.4, then

$$x^2 - 15 = -2x$$
$$x^2 + 2x - 15 = 0$$
$$(x + 5)(x - 3) = 0$$
$$x = -5 \quad \text{or} \quad x = 3$$

When $x = 3$, then $-2x = -2(3) = -6$ which is not positive (and logarithms are only defined on positive quantities), so that 3 is not a solution of the given equation. When $x = -5$, then $-2x$ and $x^2 - 15$ are both positive. The solution is therefore -5.

(b)   To solve $\log(\cos 2x + 2\cos x) = \log 3$ use Theorem 9.4, then

$$\cos 2x + 2\cos x = 3$$
$$2\cos^2 x - 1 + 2\cos x = 3$$
$$\cos^2 x + \cos x - 2 = 0$$
$$(\cos x + 2)(\cos x - 1) = 0$$
$$\cos x = -2 \quad \text{or} \quad \cos x = 1$$

As $-1 \leq \cos x \leq 1$, only $\cos x = 1$ leads to a possible solution, namely $x = 0$. Indeed when $x = 0, \cos 2x + 2\cos x = 1 + 2 > 0$. The solution is 0.

Solution to CYU 9.10

As the base of the logarithmic functions is less than 1, the inequality between the arguments of the logarithms will be reversed.

$$\log_{\frac{1}{2}}(3x - 1) > \log_{\frac{1}{2}}(2x + 5)$$
$$3x - 1 < 2x + 5$$
$$x < 6$$

For both logarithms to be defined, $3x - 1 > 0 \Rightarrow x > \frac{1}{3}$ and $2x + 5 > 0 \Rightarrow x > -\frac{5}{2}$. Thus, $\frac{1}{3} < x < 6$, and the solution is $\left(\frac{1}{3}, 6\right)$.

Solution to CYU 9.11

(a) $2\log(x+1) - 3\log x + \log 12 = \log(x+1)^2 - \log x^3 + \log 12$

$$= \log \frac{(x+1)^2}{x^3} + \log 12$$

$$= \log \frac{12(x+1)^2}{x^3}$$

(b) We can express 45 as $45 = 9 \cdot 5 = 3^2 \cdot 5$. Then

$$\ln 45 = \ln(3^2 \cdot 5)$$
$$= \ln 3^2 + \ln 5$$
$$= 2\ln 3 + \ln 5$$
$$= 2A + B$$

(c)

$$\frac{\ln \sin x - \ln \cos x}{\ln \left(e^{\sin^2 x} e^{\cos^2 x} \right)} = \frac{\ln \dfrac{\sin x}{\cos x}}{\ln e^{\sin^2 x + \cos^2 x}}$$

$$= \frac{\ln \tan x}{\ln e^1}$$

$$= \frac{\ln \tan x}{1}$$

$$= \ln \tan x$$

Solution to CYU 9.12

(a) $\log_5 32 = \dfrac{\ln 32}{\ln 5} \approx 2.15$

(b) $\log_{\frac{1}{2}} 17 = \dfrac{\ln 17}{\ln \frac{1}{2}} \approx -4.09$

B–68   Solutions to Check Your Understanding Boxes

Solution to CYU 9.13

Apply the natural logarithm to both sides, and use the properties in Theorem 9.6:

(a)
$$(3x - 1)\ln 2 = (x + 2)\ln 15$$
$$3x \ln 2 - \ln 2 = x \ln 15 + 2\ln 15$$
$$(3\ln 2)x - (\ln 15)x = \ln 2 + 2\ln 15$$
$$x = \frac{\ln 2 + 2\ln 15}{3\ln 2 - \ln 15}$$
$$= \frac{\ln 2 \cdot 15^2}{\ln 2^3 - \ln 15} = \frac{\ln 450}{\ln \frac{8}{15}}$$

(b)
$$7 \cdot 3^{x+1} = 4 \cdot 2^{x+2}$$
$$\ln 7 + (x + 1)\ln 3 = \ln 4 + (x + 2)\ln 2$$
$$\ln 7 + x \ln 3 + \ln 3 = \ln 4 + x \ln 2 + 2\ln 2$$
$$x(\ln 3 - \ln 2) = \ln 4 + 2\ln 2 - \ln 7 - \ln 3$$
$$x = \frac{\ln 4 + 2\ln 2 - \ln 7 - \ln 3}{\ln 3 - \ln 2}$$
$$= \frac{\ln 4 + \ln 4 - \ln 21}{\ln \frac{3}{2}}$$
$$= \frac{\ln 16 - \ln 21}{\ln \frac{3}{2}} = \frac{\ln \frac{16}{21}}{\ln \frac{3}{2}}$$

Solution to CYU 9.14

If $\sin(e^{2x+1}) = 1$, then $e^{2x+1} = \frac{\pi}{2} + 2n\pi$, where n is any nonnegative integer (as exponential function values are always positive).

$$e^{2x+1} = \frac{\pi}{2} + 2n\pi, \qquad \text{then, apply the natural logarithm}$$
$$2x + 1 = \ln\left(\frac{\pi}{2} + 2n\pi\right)$$
$$2x = -1 + \ln\left(\frac{\pi}{2} + 2n\pi\right)$$
$$x = \frac{1}{2}\left[-1 + \ln\left(\frac{\pi}{2} + 2n\pi\right)\right]$$

Solution to CYU 9.15

(a) Apply the natural logarithm to both sides, and use the properties in Theorem 9.6:

$$2 \cdot 3^{x+2} \geq 5^{x-1}$$
$$\ln 2 + \ln 3^{x+2} \geq \ln 5^{x-1}$$
$$\ln 2 + (x+2)\ln 3 \geq (x-1)\ln 5$$
$$\ln 2 + x\ln 3 + 2\ln 3 \geq x\ln 5 - \ln 5$$
$$x\ln 3 - x\ln 5 \geq -\ln 2 - 2\ln 3 - \ln 5$$
$$x(\ln 3 - \ln 5) \geq -(\ln 2 + \ln 9 + \ln 5)$$
$$x\ln \tfrac{3}{5} \geq -\ln 90$$
$$x \leq -\frac{\ln 90}{\ln \tfrac{3}{5}} \qquad \left(\text{as } \ln \tfrac{3}{5} \text{ is negative}\right)$$
$$\leq \frac{\ln 90}{\ln \tfrac{5}{3}} \qquad \left(\text{as } \ln \tfrac{3}{5} = -\ln \tfrac{5}{3}\right)$$

Thus the answer is $\left(-\infty, \dfrac{\ln 90}{\ln \tfrac{5}{3}}\right]$.

(b) Apply the natural logarithm to both sides, and use the properties in Theorem 9.6:

$$3^{x-1} < \frac{1}{5^{x+1}}$$
$$\ln 3^{x-1} < \ln 1 - \ln 5^{x+1}$$
$$(x-1)\ln 3 < \ln 1 - (x+1)\ln 5$$
$$x\ln 3 - \ln 3 < 0 - x\ln 5 - \ln 5$$
$$x(\ln 3 + \ln 5) < \ln 3 - \ln 5$$
$$x\ln 15 < \ln \tfrac{3}{5}$$

As $\ln 15 > 0$, $x < \dfrac{\ln \tfrac{3}{5}}{\ln 15}$. The solution set is therefore $\left(-\infty, \dfrac{\ln \tfrac{3}{5}}{\ln 15}\right)$.

B-70   Solutions to Check Your Understanding Boxes

Solution to CYU 9.16

(a) $\log_{\frac{1}{3}}\left(\frac{1}{x}\right) = 3 \Rightarrow \left(\frac{1}{3}\right)^3 = \frac{1}{x} \Rightarrow \frac{1}{3^3} = \frac{1}{x} \Rightarrow 3^3 = x \Rightarrow x = 27.$

As $\frac{1}{x} = \frac{1}{27} > 0$, the logarithm is defined, and the solution is 27.

(b)
$$\log_8(x - 6) = 2 - \log_8(x + 6)$$
$$\log_8(x - 6) + \log_8(x + 6) = 2$$
$$\log_8(x - 6)(x + 6) = 2$$
$$\log_8(x^2 - 36) = 2$$

Changing to exponential form: $8^2 = x^2 - 36 \Rightarrow x^2 = 100 \Rightarrow x = \pm 10.$ We must check that each original logarithm is defined, that is, that both $x - 6$ and $x + 6$ are positive: If $x = -10$ then $x - 6 = -16 < 0 \Rightarrow x = -10$ is not a solution. If $x = 10$ then $x - 6 = 4$ and $x + 6 = 16$ are both positive. The solution is 10.

Solution to CYU 9.17

(a) For the logarithms to be defined, $\tan x$ must be positive, which restricts the solution interval to $\left(0, \frac{\pi}{2}\right)$.

$$\ln \tan x + \ln(2 \tan x + 1) = \ln 3, \qquad \text{now apply theorem 9.6(i)}$$
$$\ln[(\tan x)(2 \tan x + 1)] = \ln 3$$
$$(\tan x)(2 \tan x + 1) = 3$$
$$2 \tan^2 + \tan x - 3 = 0$$
$$(2 \tan x + 3)(\tan x - 1) = 0$$
$$\tan x = -\frac{3}{2} \quad \text{or} \quad \tan x = 1$$

The first equation has no solution in $\left(0, \frac{\pi}{2}\right)$. The second equation has only one solution there, namely, $x = \frac{\pi}{4}$.

(b) Since $\cos(\ln x) = -\frac{1}{2} < 0$, the angle whose radian measure is $\ln x$ lies in the second or third quadrants. The reference angle has a cosine of $\frac{1}{2}$ which means that it is $\frac{\pi}{3}$. Therefore, $\ln x = \pi - \frac{\pi}{3} + 2n\pi = \frac{2\pi}{3} + 2n\pi$ or $\pi + \frac{\pi}{3} + 2n\pi = \frac{4\pi}{3} + 2n\pi$, where n is any integer. Solving for x, $x = e^{\ln x} = e^{\frac{2\pi}{3} + 2n\pi}$ or $x = e^{\frac{4\pi}{3} + 2n\pi}$.

Solution to CYU 9.18

(a) Since the function $\left(\frac{2}{3}\right)^x$ is a decreasing function, when we apply it to both sides of the inequality, we must reverse the inequality sign:

$$\log_{\frac{2}{3}}(1 - 2x) > 3$$

$$\left(\frac{2}{3}\right)^{\log_{\frac{2}{3}}(1 - 2x)} < \left(\frac{2}{3}\right)^3$$

$$1 - 2x < \frac{8}{27}$$

$$-2x < \frac{8}{27} - 1 = -\frac{19}{27}$$

$$x > \frac{19}{54}$$

Since we must have $1 - 2x > 0$, for the logarithm to be defined, then $-2x > -1$ and $x < \frac{1}{2}$. The solution set is then $\left(\frac{19}{54}, \frac{1}{2}\right)$.

(b) Since the base is greater than 1, the logarithmic function is an increasing function. Therefore

$$\log_3(3x - 6) \leq \log_3(x + 2)$$

$$3x - 6 \leq x + 2$$

$$2x \leq 8$$

$$x \leq 4$$

Since we must have both $3x - 6 > 0$ and $x + 2 > 0$, for the logarithms to be defined, then $x > 2$ and $x > -2$, so that $x > 2$.
The solution set is then $(2, 4]$.

Solution to CYU 9.19

In the inequality, $-\log(x - 450) < -(x - 475)^2 - 5$, the logarithm is only defined when $x - 450 > 0$. Thus we chose a window of $[449, 500] \times [-5, 5]$, and graphed each side of the inequality separately. We then determined the x-coordinates of the points of intersection.

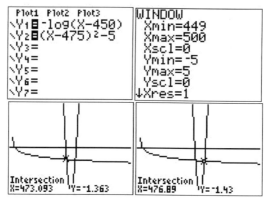

The solution is $(450, 473.09) \cup (476.89, \infty)$.

Solution to CYU 9.20

Let P(t) be the population of the town at time t. Then $P(t) = P_0 e^{kt}$. Since the initial population is 500, $P_0 = 500$, so that $P(t) = 500 e^{kt}$. In 9 years ($t = 9$) the population, $P(9) = 500 + \frac{15}{100}(500) = 575$. Thus

$$575 = 500(e^k)^9$$
$$(e^k)^9 = \frac{575}{500} = \frac{23}{20}$$
$$e^k = \left(\frac{23}{20}\right)^{\frac{1}{9}}, \text{ so that}$$
$$P(t) = 500\left(\frac{23}{20}\right)^{\frac{t}{9}}$$

Next, we seek t such that $P(t) = 3(500) = 1500$:

$$P(t) = 1500$$
$$1500 = 500\left(\frac{23}{20}\right)^{\frac{t}{9}}$$
$$3 = \left(\frac{23}{20}\right)^{\frac{t}{9}}$$
$$\ln 3 = \frac{t}{9} \ln\left(\frac{23}{20}\right)$$
$$t = \frac{9 \ln 3}{\ln\left(\frac{23}{20}\right)}$$
$$\approx 70.75 \text{ years } (= 70 \text{ yrs } 9 \text{ mos})$$

Solution to CYU 9.21

(a) Take 2004 as $t = 0$, then $A_0 = 35$ so that the formula becomes $A(t) = 35e^{-\frac{1}{4}t}$

The year 2007 corresponds to $t = 3$, and

$$A(3) = 35e^{-\frac{1}{4}(3)} = 35e^{-\frac{3}{4}} \approx 16.5 \text{ grams.}$$

(b) Since $A(t) = 35e^{-\frac{1}{4}t}$, we will find t such that $A(t) = \frac{1}{2}(35)$. So we solve for t in the equation

$$\frac{1}{2}(35) = 35e^{-\frac{1}{4}t}$$

$$\frac{1}{2} = e^{-\frac{1}{4}t}$$

$$\ln\left(\frac{1}{2}\right) = \ln\left(e^{-\frac{1}{4}t}\right)$$

$$-\ln 2 = -\frac{1}{4}t$$

$$t = 4\ln 2 \approx 2.77 \text{ years}$$

Solution to CYU 9.22

Let r be the annual interest rate. Since there are $n = 10 \cdot 365 = 3650$ time periods (i.e. days) in ten years, we apply Theorem 9.11 with $n = 3650$ and $A = 2A_0$, and

obtain $2A_0 = A_0\left(1 + \dfrac{r}{365}\right)^{3650}$. Solving for r, we find

$$2 = \left(1 + \frac{r}{365}\right)^{3650}$$

$$2^{\frac{1}{3650}} = 1 + \frac{r}{365}$$

$$\frac{r}{365} = 2^{\frac{1}{3650}} - 1$$

$$r = 365\left(2^{\frac{1}{3650}} - 1\right) \approx .0693$$

Thus, the annual interest rate is about 6.93%.

B–74   **Solutions to Check Your Understanding Boxes**

Solution to CYU 9.23

From Theorem 9.13, the amount after t years at 4% is given by the formula, $A_c(t) = A_0 e^{0.04t}$. We want to find A_0 such that $A_c(5) = 10,000$. Putting that into the formula: $10000 = A_0 e^{(0.04)(5)}$. Solving for A_0, we find

$$A_0 = \frac{10000}{e^{0.2}} \approx \$8,187.31$$

Solution to CYU 9.24

Let I_T and I_C denote the intensities of the earthquakes in Turkey and California, respectively.

$$7.4 = \log \frac{I_T}{I_0} \qquad \qquad 5.0 = \log \frac{I_C}{I_0}$$

$$10^{7.4} = \frac{I_T}{I_0} \qquad \qquad 10^{5.0} = \frac{I_C}{I_0}$$

$$I_T = 10^{7.4} I_0 \qquad \qquad I_C = 10^{5.0} I_0$$

Forming the quotient, $\dfrac{I_T}{I_C} = \dfrac{10^{7.4} I_0}{10^{5.0} I_0} = 10^{2.4} \approx 251.2$, we find that the intensity of the earthquake in Turkey was about 251 times that of the California earthquake.

APPENDIX A

Solution to CYU A.1

From Theorem A.1:

(a)   Since $B = 0$ and $A = 1 \neq 0$ (there is an x^2 term but no y^2 term), if this is the graph of a conic, it is a parabola.

(b)   As $A = B = 4 \neq 0$, if this is the graph of a conic, it is a circle.

(c)   As $A = 4$ and $B = -9$, so that $AB < 0$, if this is the graph of a conic, it is a hyperbola.

(d)   As $A = 4$ and $B = 3$ so that $A \neq B$ and $AB > 0$, if this is the graph of a conic, it is an ellipse.

Solution to CYU A.2

(a) The equation $\dfrac{(x-2)^2}{9} + \dfrac{(y+3)^2}{4} = 1$ is in the standard form for the equation of an ellipse: $\dfrac{(x-x_0)^2}{a^2} + \dfrac{(y-y_0)^2}{b^2} = 1$ with $(x_0, y_0) = (2, -3)$, $a = \sqrt{9} = 3$, and $b = \sqrt{4} = 2$. It follows that the center of the ellipse is at the point $(2, -3)$. The vertices are at $(2 \pm 3, -3)$, i.e. $(-1, -3)$ and $(5, -3)$. The endpoints of the minor axis are $(2, -3 \pm 2)$, i.e. $(2, -5)$ and $(2, -1)$. The graph appears in Figure 1 below.

(b) The equation $x - 3 = 4(y + 4)^2$ is in the standard form for the equation of a parabola with its axis parallel to the x-axis and opening to the right: $(x - x_0) = a(y - y_0)^2$, $a > 0$, with $(x_0, y_0) = (3, -4)$, $a = 4$. The x-intercept of the graph is $3 + 4(0 + 4)^2 = 67$. The graph appears in Figure 2 below.

(c) The equation $\dfrac{(x-2)^2}{9} - \dfrac{(y+3)^2}{4} = 1$ is in the standard form for the equation of a hyperbola with transverse axis parallel to the x-axis: $\dfrac{(x-x_0)^2}{a^2} - \dfrac{(y-y_0)^2}{b^2} = 1$, with $(x_0, y_0) = (2, -3)$, $a = \sqrt{9} = 3$, and $b = \sqrt{4} = 2$. It follows that the center of the hyperbola is at the point $(2, -3)$. The vertices are $(2 \pm 3, -3)$, i.e. $(-1, -3)$ and $(5, -3)$. The oblique asymptotes are the lines $y + 3 = \pm\dfrac{2}{3}(x - 2)$. The graph appears in Figure 3 below.

Figure 1

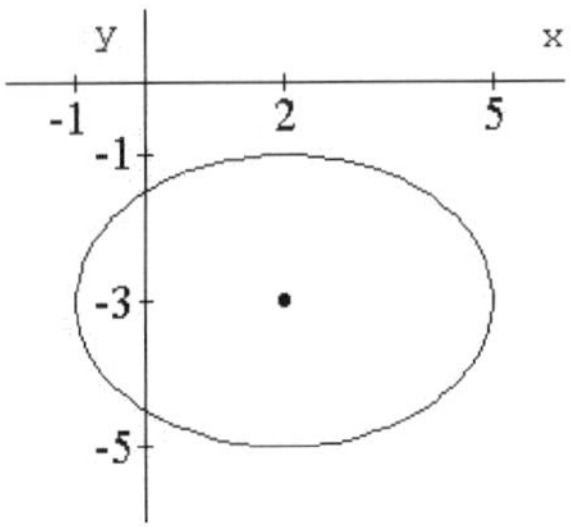

Figure 2

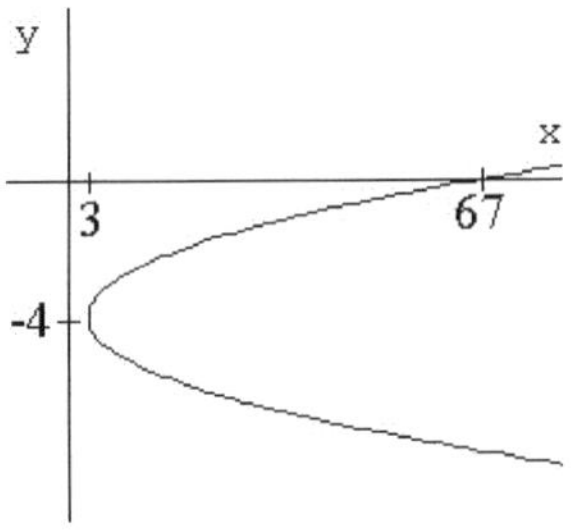

Figure 3

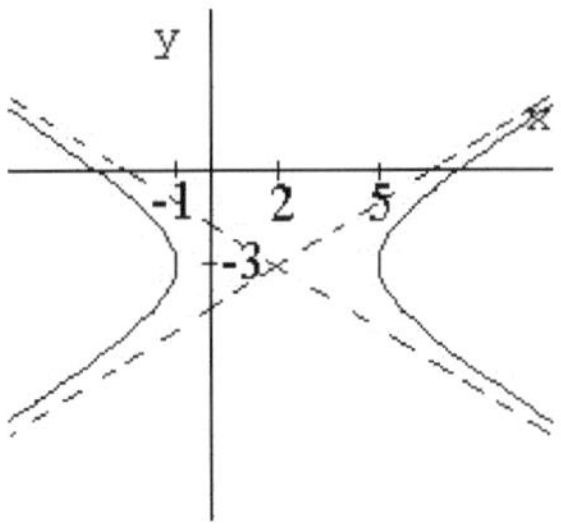

B–76   Solutions to Check Your Understanding Boxes

Solution to CYU A.3

(a) First put the equation $x^2 + 2x - 3 + 2y = -y^2$ in standard form, by completing the square:

$$x^2 + 2x + y^2 + 2y = 3$$
$$x^2 + 2x + 1 + y^2 + 2y + 1 = 3 + 1 + 1$$
$$(x + 1)^2 + (y + 1)^2 = 5$$

The center of the circle is at $(-1, -1)$, and the radius is $\sqrt{5}$.
See Figure 1 (below) for the graph.

(b) First put the equation $3x^2 + y^2 - 6x + 4y - 2 = 0$ in standard form, by completing the square:

$$3x^2 + y^2 - 6x + 4y - 2 = 0$$
$$3x^2 - 6x + y^2 + 4y = 2$$
$$3(x^2 - 2x) + (y^2 + 4y) = 2$$
$$3(x^2 - 2x + 1) + (y^2 + 4y + 4) = 2 + 3 \cdot 1 + 4 = 9$$
$$3(x - 1)^2 + (y + 2)^2 = 9$$
$$\frac{(x - 1)^2}{3} + \frac{(y + 2)^2}{9} = 1$$

The center of the ellipse is at $(1, -2)$.
The vertices are $(1, -2 + 3) = (1, 1)$ and $(1, -2 - 3) = (1, -5)$.
The endpoints of the minor axis are $(1 - \sqrt{3}, -2)$ and $(1 + \sqrt{3}, -2)$.
See Figure 2 (below) for the graph.

(c) First put the equation $x^2 - x - 3y + 1 = 0$ in standard form, by completing the square:

$$x^2 - x - 3y + 1 = 0$$
$$3y = x^2 - x + 1$$
$$3y = x^2 - x + \left(\tfrac{1}{2}\right)^2 - \left(\tfrac{1}{2}\right)^2 + 1$$
$$3y = \left(x - \tfrac{1}{2}\right)^2 + \tfrac{3}{4}$$
$$y = \tfrac{1}{3}\left(x - \tfrac{1}{2}\right)^2 + \tfrac{1}{4}$$

The vertex of the parabola is at $\left(\tfrac{1}{2}, \tfrac{1}{4}\right)$, and the parabola opens up.

The y-intercept is at $\tfrac{1}{3}\left(0 - \tfrac{1}{2}\right)^2 + \tfrac{1}{4} = \tfrac{1}{12} + \tfrac{1}{4} = \tfrac{4}{12} = \tfrac{1}{3}$.

See Figure 3 (below) for the graph.

(d) First put the equation $4x^2 - y^2 - 8x - 4y - 4 = 0$ in standard form, by completing the square:

$$4(x^2 - 2x) - (y^2 + 4y) = 4$$
$$4(x^2 - 2x + 1) - (y^2 + 4y + 4) = 4 + 4 \cdot 1 - 4 = 4$$
$$4(x-1)^2 - (y+2)^2 = 4$$
$$(x-1)^2 - \frac{(y+2)^2}{4} = 1$$

Thus the center of the hyperbola is at $(1, -2)$.
The vertices are at $(1+1, -2) = (2, -2)$ and $(1-1, -2) = (0, -2)$.
The asymptotes are $y - (-2) = \pm \frac{\sqrt{4}}{\sqrt{1}}(x-1)$ i.e. $y + 2 = \pm 2(x-1)$.
See Figure 4 for the graph.

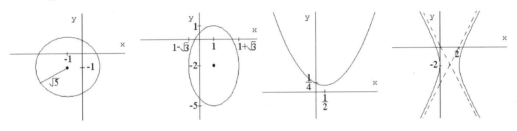

Figure 1 Figure 2 Figure 3 Figure 4

B–78

Answers to Odd-Numbered Exercises   C–1

ANSWERS TO ODD-NUMBERED EXERCISES

CHAPTER 1

SECTION 1.1, PAGE 6

#1: $\frac{5}{6}$ #3: $\frac{1}{128}$ #5: 72 #7: -3 #9: $-\frac{1}{9}$ #11: $\frac{13}{4}$

#13: undefined   #15: $11\sqrt{2}$ #17: a^8 #19: $\frac{1}{a^8}$ #21: a^3b^3

#23: $-\frac{1}{ab}$ #25: $-a$ #27: $\frac{b^8}{1-ab}$ #29: $\frac{1}{a^8b^{12}}$ #31: $\frac{1}{a^{\frac{5}{2}}}$ #33: $\frac{1}{a^{\frac{2}{3}}}$

#35: $4(a-b)^{\frac{4}{3}}$ #37: $\frac{1}{8a^2b^{\frac{5}{2}}c^{\frac{7}{4}}}$ #39: $b\sqrt{c}$ #41: $\frac{c^{\frac{7}{4}}}{a^{\frac{59}{8}}b^{\frac{11}{2}}}$ #43: -4

#45: 0 #47: $\frac{\sqrt{5}}{5}$ #49: $\frac{5+\sqrt{7}}{2}$ #51: $\frac{2\sqrt{3}-\sqrt{5}}{7}$ #53: $\frac{2\sqrt{3}-21\sqrt{2}}{6}$

#55: $-\frac{\sqrt{2}+3\sqrt{7}}{5}$ #57: $\sqrt{2}-1$ #59: 3×10^8 #61: 1.07×10^{-37}

#63: 2.997925×10^{10} cm/sec   #65: 9.1×10^{-28} gms   #67: 1.987×10^{30} kg

SECTION 1.2, PAGE 19

#1: $-x^2-2x-2$ #3: $-2x^5-x^4-2x^3+x-1$ #5: $-4x^4-5x^3+13x^2+5x-6$

#7: $-3x^4+60x^2-5x-13$ #9: q: $6x-5$, r: -2 #11: q: $3x+9$, r: 20

#13: q: $-3x^2+8x-5$, r: -4 #15: q: $-x^3-x^2-x+1$, r: 1 #17: $(x+6)(x^3-x-1)-3$

#19: $(x-2)(x^2-x-11)-20$ #21: $(x^2-3x-1)(x+3)+10x+3$ #23: $(x^3-2)(x^2+1)-3x$

#25: $5x-2+\frac{11x-4}{x^2-2}$ #27: $4x^2+3+\frac{-2x+5}{4x^3-6x+3}$ #29: $(3x-4)(2x+5)$

#31: $4(x-2)^2$ #33: $\frac{1}{5}(x-5)(x+5)$ #35: $(3x-a)(3x+a)(9x^2+a^2)$

#37: $(x+1)(x-1)^2$ #39: $(x^4+3)(x^2+1)$ #41: $(x-2)(x+2)(2x+1)$

#43: $(x-2)(3x-1)(2x+3)$ #45: $(x-1)(x+1)(x+3)(x-2)$ #47: $(x+1)^2(x-2)^2$

#49: $(2x-y)(4x^2+2xy+y^2)$ #51: $(a+3b)(a^2-3ab+9b^2)$ #53: $(x+1)(x-1)(x^2+x+1)$

C-2   Answers to Odd-Numbered Exercises

#55: $(7a+b)(19a^2+17ab+7b^2)$ #57: $(3x-2)(3x+2)(9x^2+6x+4)(9x^2-6x+4)$

#59: $(3x+5)^2$ #61: $x^2(2x-1)(2x+1)$ #63: $(\sqrt{2}a-\sqrt{3}b)(\sqrt{2}a+\sqrt{3}b)$

#65: $3(x-2)(x+2)(x^2+4)$ #67: $(3x^2+4)(\sqrt{2}x-1)(\sqrt{2}x+1)$ #69: $\dfrac{x^2+14x-6}{6x^2-2x}$

#71: $\dfrac{11x^2-18x-17}{12x^3-36x+24}$ #73: $-x-1$ #75: $\dfrac{18x-12}{(3x+1)^4}$ #77: $\dfrac{(3x+2)(5x-4)}{3x-2}$

#79: $\dfrac{50}{(x+3)^3}$ #81: $\dfrac{x(3x-2)}{8(3x+1)^4}$ #83: $\dfrac{x-3}{24x-8}$ #85: $\dfrac{9-2x}{4(x-4)^3}$

#87: $\dfrac{2}{\sqrt{x}(x+4)^{\frac{3}{2}}}$ #89: $\dfrac{-4x^2+28x-40}{(2x-5)^{\frac{5}{2}}}$ #91: $\dfrac{3(x-2)}{2\sqrt{x}}$ #93: $\dfrac{2-3x}{2\sqrt{1-x}}$

#95: $-\dfrac{4}{x^{\frac{2}{3}}(x^{\frac{1}{3}}-2)^2(x^{\frac{1}{3}}+2)^2}$ #97: $4x^3-8x^2-20x+24$

SECTION 1.3, PAGE 27

#1: ≈ 0.000606 mi   #3: ≈ 1893.939 ml   #5: ≈ 22.663 kg   #7: ≈ 80.67 ft/sec

#9: ≈ 1.42 mph   #11: ≈ 0.305 in.^{3} #13: ≈ 86.07 ft^2 #15: ≈ 6771.3713 kg/m^3

#17: ≈ 5.06 lb/yd^3

#19: (a) 2.9979×10^5 km/sec   (b) $\approx 186,233.18$ mi/sec   (c) $\approx 6.7044 \times 10^8$ mph

#21: 20 quires   #23: ≈ 172.573 mi

#25: (a) ≈ 14.695 lb/in.^{2} (b) ≈ 1.0319 kg/cm^2 (c) 7.268×10^{-6} atmospheres

#27: ≈ 1.474 mm   #29: ≈ 3.157 l/sec   #31: 4 in./sec   #33: ≈ 2.473 ft^2

#35: ≈ 3.13 min

SECTION 1.4, PAGE 37

#1: -9 #3: 0 #5: -1 #7: $\dfrac{2b}{a}$ #9:

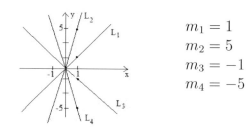

$m_1 = 1$
$m_2 = 5$
$m_3 = -1$
$m_4 = -5$

#11: for ex: $(0,-1)$ #13: for ex: $(5,8)$ #15: $y = -2x + 4$ #17: $y = -\frac{1}{2}x + \frac{3}{2}$

#19: $y + 2 = -(x - 3)$ #21: $y - 8 = -(x + 1)$ #23: $y = 1$ #25: $y = -2$

#27: $y = 2$ #29: $x = 1$ #31: $x = 0$ #33: $x = 3$

#35: $y = \frac{9}{2}x - \frac{13}{2}$ #37: $y = \frac{7}{3}x$ #39: $y = \frac{3}{2}x + 3$

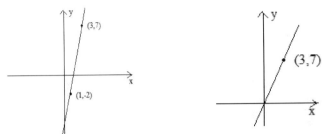

#41: $x = 2$ #43: $y = -\frac{1}{3}x + \frac{13}{3}$ #45: $\frac{3}{2}$ #47: $\frac{1}{2}$

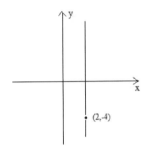 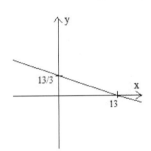

#49: $-\frac{19}{5}$ #51: $x = -y + 2$; $y = -x + 2$ #53: $x = \frac{21}{5}y$, $y = \frac{5}{21}x$

#55: $x = \frac{5y+3}{y+2}$, $y = \frac{3-2x}{x-5}$ #57: $a = \frac{4b}{1+4b}$ and $b = \frac{a}{4-4a}$ #59: 1

#61: $x \geq -\frac{1}{2}$ #63: $x \leq \frac{1}{2}$ #65: all real numbers #67: $x \geq \frac{12}{5}$

#69: $x \geq \frac{9}{2}$ #71: $x < \frac{35}{6}$ #73: $-2 \leq x < 2$ #75: $\frac{7}{10} \leq x < \frac{19}{10}$

#77: no solution #79: $x \leq -4$ #81: $1 < x < 2$

SECTION 1.5, PAGE 48

#1: $(x = -2, y = 5)$ #3: $(x = 3, y = 7)$ #5: $(x = 11, y = 12, z = 13)$

#7: $(x = 0, y = 1, z = 2)$

#9: (a) $\approx (0.67, 2)$ (b) $\left(x = \frac{2}{3}, y = 2\right)$ (c) same: $\frac{2}{3} \approx 0.67$

C–4   Answers to Odd-Numbered Exercises

#11: (a) $(-0.048, -1.143)$ (b) $\left(x = -\frac{1}{21},\ y = -\frac{8}{7}\right)$

(c) same: $-\frac{1}{21} \approx -0.048$ and $-\frac{8}{7} \approx -1.143$

#13: $(10, 10)$ #15: (a) $c = 6,\ d = 8$ (b) $c = 3,\ d = 4$ #17: $\frac{9}{2},\ -\frac{11}{2}$

#19: $\frac{13}{3}$ mph   #21: $f(x) = 4x^2 - 3x + 1$

#23: $\frac{25}{11}$ oz of 40% sol., $\frac{50}{11}$ oz of 50% sol., $\frac{35}{11}$ oz of 10% sol.   #25: 110 spec., 50 blue

#27: 30 1-pt, 1 2-pt, 27 3-pt   #29: \$3,000   #31: $(x = -4, y = 2, z = -3)$

#33: $(x \approx -5.3,\ y \approx 5.7,\ z \approx -3.8)$

#35: 50 TI-83s, 30 TI-85s, 20 TI-89s, 10 TI-92s

SECTION 1.6, PAGE 59

#1: $5,\ -3$ #3: $-4,\ 1$ #5: $-\frac{1}{2},\ \frac{5}{2}$ #7: $\pm\frac{2}{7}$ #9: $-\frac{2}{3},\ -\frac{8}{3}$

#11: $\frac{5}{3},\ 1$ #13: $-5,\ 2,\ 0,\ -3$ #15: $-4 \pm 3\sqrt{2}$ #17: $\frac{-2 \pm \sqrt{5}}{2}$

#19: (a) $\frac{1 \pm \sqrt{22}}{3}$ (b) $-3\left(x - \frac{1 + \sqrt{22}}{3}\right)\left(x - \frac{1 - \sqrt{22}}{3}\right)$ #21: no solution

#23: (a) $-2 \pm \sqrt{7}$ (b) $-(x + 2 - \sqrt{7})(x + 2 + \sqrt{7})$

#25: (a) $\frac{7 \pm \sqrt{17}}{4}$ (b) $2\left(x - \frac{7 - \sqrt{17}}{4}\right)\left(x - \frac{7 + \sqrt{17}}{4}\right)$ #27: $\frac{5}{4},\ -\frac{3}{2}$

#29: $\frac{5 \pm 2\sqrt{7}}{2}$ #31: $\frac{3}{2}$ #33: $-2 \pm \sqrt{3}$ #35: $0,\ -1$ #37: $\frac{1}{2}$ in.

#39: 12 in. by 36 in.   #41: (a) 175 ft   (b) ≈ 35.8 mph   #43: $\frac{5}{2}$ sec

#45: ≈ 2.4 sec   #47: (a) $\frac{\pi}{8}w^2 + 3w$ (b) ≈ 3.95 ft

#49: One possible answer: $6x^2 - 7x + 2 = 0$

CHAPTER REVIEW, PAGE 66

#1: $-\frac{7}{10}$ #3: $9a^6b^2$ #5: $\frac{9b^4}{a^6}$ #7: $8a - b$ #9: $a^{\frac{20}{3}}b^2$

Answers to Odd-Numbered Exercises   **C–5**

#11: $ab^{\frac{3}{2}}$ #13: $-4a^3b^2$ #15: $-\dfrac{27\sqrt{a}}{b^{\frac{11}{4}}c^{\frac{3}{2}}}$ #17: $(x-4)(3x^2-4x+5)-6$

#19: $(x^4-3)(x^3+2)-4x+5$ #21: $(3x+1)(x+2)$ #23: $2x^2(x+3)(x^2-3x+9)$

#25: $(x-y+2)(x+y+8)$ #27: $\dfrac{3}{2}$ #29: $-\dfrac{3c}{3c+1}$ #31: $\dfrac{14x-5}{9x^2-4}$

#33: $\dfrac{4-17x}{32(3x-1)^3}$ #35: $\dfrac{x-2}{x^2+4}$ #37: $-\dfrac{5x+5}{3x^2+2x-5}$ #39: $\dfrac{10x^{\frac{1}{3}}-4}{3x^{\frac{2}{3}}}$

#41: $\dfrac{2x-1}{2(1-x)^{\frac{3}{2}}}$ #43: $\approx 1.31\ \$/\text{m}$ #45: ≈ 25.41 threads/in.   #47:   for ex. $(4,12)$

#49:   for ex. $(5,10)$ #51: $y=-\dfrac{5}{3}x-1$ #53: $y=2x+2$ #55: $y-2=-3(x-4)$

#57: $y-5=2(x+2)$ #59: $y=-\dfrac{4}{13}x+\dfrac{11}{13}$ #61: $y=1$ #63: $x=4$

#65: $y=\dfrac{5}{2}x-5$ #67: $m_1=m_2=5$ #69:   no solution   #71: $\dfrac{16}{9}$

#73: $x\le\dfrac{20}{7}$ #75: $x<-\dfrac{1}{3}$ #77: $(x=5,\ y=-2)$ #79: $\left(x=-3,\ y=4,\ z=\dfrac{1}{2}\right)$

#81: $(x=0,\ y=1,\ z=2)$ #83: $\dfrac{7}{2},\ \dfrac{3}{2}$ #85: $\dfrac{4}{3},\ 2$ #87: $-2\pm2\sqrt{2}$

#89: $\dfrac{2}{3},\ \dfrac{4}{3}$ #91:   436 ft $\times$ 364 ft   #93:   11 in.   #95: ≈ 5.59 gal.

#97: ≈ 7.14 lbs of A, ≈ 42.86 lbs of B

CUMULATIVE REVIEW, PAGE 70

#1: $\dfrac{2a^2-1}{(a^2+1)^{\frac{5}{2}}}$ #2: $\dfrac{(2-a)(a+b)}{2}$ #3: $\dfrac{2x+1}{3x-4}$ #4: $\dfrac{8}{3}$ #5: $\dfrac{1}{x+1}$ #6: $\dfrac{4b^{\frac{16}{3}}}{9a^{\frac{4}{3}}}$

#7: $\dfrac{9x-2}{(3x+2)^4}$ #8: $\dfrac{9x^2-10}{(x^2-2)^{\frac{3}{2}}}$ #9:   F: (for ex: take $a=b=c=1$, LHS$=2$, RHS$=\frac{1}{2}$)

#10:   F: $[-5^2=-(5^2)=-25]$ #11: $(3x-2)(4x+1)$ #12: $(x+3)(x-3)$

#13: $(x-1)(x^2+x+1)$ #14: $(x+5)(x^2-5x+25)$ #15: $x(x-1)(2x+3)$

#16: $(x+1)(3x-1)(2x+1)$ #17: $(3x+1)(x-\sqrt{2})(x+\sqrt{2})$ #18: $(x^2+2)(\sqrt{2}+x)(\sqrt{2}-x)$

#19: $\approx 0.63\ l/\text{sec}$ #20: $\approx 2.76\times10^6$ kg/m^3 #21:   8   #22: $y=-\dfrac{19}{2}x+\dfrac{15}{2}$

C–6   **Answers to Odd-Numbered Exercises**

#23: $y = \frac{3}{2}x + \frac{5}{2}$ #24: -14 #25: $x < \frac{11}{6}$ #26: $\frac{1 \pm \sqrt{13}}{2}$

#27: $\frac{5 \pm \sqrt{2}}{3}$ #28: $(x = -7,\ y = 3,\ z = -4)$ #29: $\frac{4 \pm \sqrt{6}}{2}$ #30:   75 ft by 150 ft

#31:   64 oz   #32: $\frac{3}{4}$ sec

CHAPTER 2

SECTION 2.1, PAGE 84

#1: $f(0) = -9,\ f(4) = 11,\ f(-2) = 5$ #3: $f(0)$ undefined, $f(4) = 6,\ f(-2) = 3$

#5: $f(0)$ and $f(-2)$ undefined, $f(4) = 0$

#7:   (a) $18x - 5$ (b) $21x - 17$ (c) $3x^2 - 5$ (d) $3x + 3h - 5$ (e)   3

#9:   (a) $\dfrac{1 - 12x}{6x}$ (b) $\dfrac{9 - 14x}{7x - 4}$ (c) $\dfrac{1 - 2x^2}{x^2}$ (d) $\dfrac{1 - 2x - 2h}{x + h}$ (e) $-\dfrac{1}{x(x + h)}$

#11:   (graph: 0 to 3)   #13:   (graph: -4 to -2)   #15:   (graph at -1)   #17: $[5, 9]$ (graph: 5 to 9)

#19: $(-5, -3)$ (graph: -5 to -3)   #21: $(1, 9]$ #23: $(-\infty, 9)$ #25: $(2, 4]$

#27: $\left(-\infty, \frac{3}{5}\right) \cup \left(\frac{3}{5}, \infty\right)$ #29: $(-\infty, 1) \cup (1, \infty)$ #31: $[0, 1) \cup (1, \infty)$ #33: $\left(\frac{5}{3}, 3\right]$

#35: $D_f = (-1, 3]$ #37: $(-\infty, -1) \cup [0, 4] \cup (6, \infty)$ #39: $D_g = R_g = (-\infty, \infty)$

#41: $D_g = [1, \infty),\ R_g = [0, \infty)$ #43: $D_f = [0, \infty),\ R_f = [-1, \infty)$

#45: $f(-1) = 2,\ f(0) = 1,\ f(5) = -14$ #47: $f(-1) = f(0) = 0,\ f(5)$ undefined

#49:   (graph: -2 to 5) $; 7$ #51:   (graph: 2 to 5) $; 3$ #53:   (graph: -6 to 0) $; 6$

#55: $4x + 4,\ D = (-\infty, \infty)$ #57: $3x^2 + 2x - 5,\ D = (-\infty, \infty)$ #59: $2x^2 - 1 + \dfrac{1}{x} + \sqrt{x + 1}$

#61: $\left(2x - \dfrac{1}{x}\right)\sqrt{x + 1}$ #63: $\dfrac{1}{x\sqrt{x + 1}}$ #65: $\dfrac{1}{2x^3 - x} - \sqrt{x + 1}$ #67: $-\dfrac{1}{9}$

#69: $\dfrac{2x - 7}{x - 2}$ #71: $-\dfrac{1}{3x}$ #73: $-\dfrac{3}{4}$ #75:   1   #77: $\dfrac{1}{(x - 1)^2}$ #79:   0

#81: -1 #83: For ex: $f(x) = x^2 + 1$, $g(x) = x^{-3}$

#85: For ex: $f(x) = \dfrac{2x}{3x-1}$, $g(x) = \sqrt{x}$ #87: (a) 3,000 (b) 33,000 (c) out of range

#89: (a) $67,500 (b) $115,000 (c) $16.42   #91: (a) $68.75 (b) $89.75 (c) $90

#93: (a) $C(x) = \begin{cases} 7.65, & \text{if } 0 \le x < 2 \\ 7.95, & \text{if } 2 \le x < 10 \\ 8.55, & \text{if } 10 \le x \le 15 \end{cases}$ (b) $C(x) = \begin{cases} 8.05, & \text{if } 0 \le x \le 2 \\ 8.65, & \text{if } 2 < x \le 10 \\ 9.30, & \text{if } x > 10 \end{cases}$

SECTION 2.2, PAGE 98

#1: $D_f = [-\infty, \infty)$
$R_f = [0, \infty)$
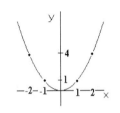

#3: $D_g = R_g = (-\infty, 0) \cup (0, \infty)$
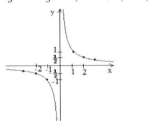

#5: (a) (i) -3, (ii) -2, (iii) 1.5,   (b) (i) -8, (ii) $-6, 5$, (iii) $-4, -2.2, 6.6$

#7: $D_f = (-\infty, 0] \cup [2, \infty)$; $R_f = (-\infty, 0] \cup [1, \infty)$; x-int: -1; y-int: -2

#9: 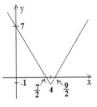   #11: 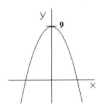   #13: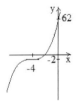

#15:   #17: 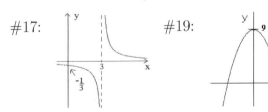   #19:   #21:

#23: odd   #25: neither   #27: even

#29:   (a) 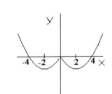   (b) 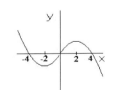   (c) For ex: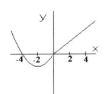

#37: a function #39: not a function #41: not a function

#43: (a) 0, 2 (b) 0 (c) 0, 2 (d) (0, 2) (e) $(-\infty, 0) \cup (2, \infty)$

#45: (a) $-1, 3$ (b) $\approx \frac{3}{2}$ (c) $-1, 3$ (d) $(-\infty, -1) \cup (3, \infty)$ (e) $(-1, 3)$

#47: (a) $-2, 0, 2$ (b) $(-2, 0) \cup (2, \infty)$ (c) $(-\infty, -2) \cup (0, 2)$ #49: $-1.09, -0.13, 1.21$

#51: $\approx \pm 1.41$ #53: $(-2.17, 0.39) \cup (1.78, \infty)$ #55: $(-\infty, -1) \cup (1, \infty)$

#57: (a) $(0.33, \infty)$ (b) $(-\infty, 0.33)$

#59: (a) $(-\infty, -5.45) \cup (-0.55, 1)$ (b) $(-5.45, -0.55) \cup (1, \infty)$

#61: between 124 and 200 units, inclusive #63: between 132 and 200 units, inclusive

#65: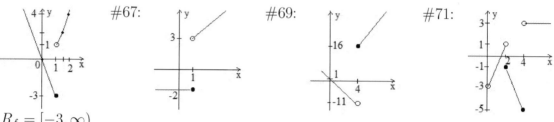

$R_f = [-3, \infty)$

#67:

#69:

#71:

#73: For example:

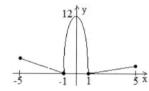

#75: A function of x can have at most one y-intercept.

#77: $f(x) = \begin{cases} -1 & \text{if } -1 \le x < 0 \\ 1 & \text{if } 0 \le x < 1 \\ 2 & \text{if } 1 \le x < 2 \\ 3 & \text{if } 2 \le x < 3 \end{cases}$

#79: $f(x) = \begin{cases} 2x + 2 & \text{if } -1 \le x < 0 \\ 2x & \text{if } 0 \le x < 1 \\ 2x - 2 & \text{if } 1 \le x < 2 \end{cases}$

SECTION 2.3, PAGE 112

#1: not 1-1 #3: 1-1 #17: For ex: $f(-1) = f(1) = -1$ #19: For ex: $f(-4) = f(4) = 0$

#21: For ex: $f(-2) = f\left(-\frac{2}{5}\right) = 0$ #23: For ex: $f(0) = f(2) = -4$

#25: For ex: $g(x) = 3$; $g\left(\frac{3 \pm \sqrt{3}}{2}\right) = 3$ #27: For ex: $h(x) = 5$; $h\left(-\frac{5}{2}\right) = h(-1) = 5$

#29: $f^{-1}(x) = \frac{1}{2}(x + 3)$; $D_{f^{-1}} = R_{f^{-1}} = (-\infty, \infty)$

#31: $f^{-1}(x) = \frac{x^2}{4} - 3$; $D_{f^{-1}} = R_f = [0, \infty)$; $R_{f^{-1}} = D_f = [-3, \infty)$

#33: $f^{-1}(x) = (8-x)^{\frac{1}{3}}$ #35: $f^{-1}(x) = \dfrac{1}{x}$ #37: $f^{-1}(x) = \dfrac{x+1}{x}$

#39: $f^{-1}(x) = -\dfrac{4x}{4+3x}$ #41: $f^{-1}(x) = \dfrac{x-2}{x+3}$

#43: (a) $C = 10°$ (b) $F = 50°$ (c) inverse functions

#45: 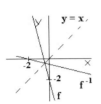 #47: 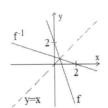 #49: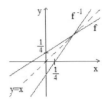

#51: 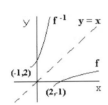 #53: 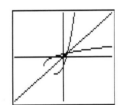 #55: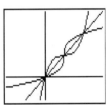

#57: For ex: 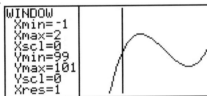 #59: For ex: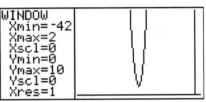

CHAPTER REVIEW, PAGE 119

#1: $(4, 6)$ #3: $(-\infty, \infty)$ #5: $(-\infty, \infty)$ #7: $\left[-\frac{4}{3}, 0\right) \cup (0, \infty)$

#9: $(-\infty, \infty)$ #11: $(0, \infty)$ #13: $(-\infty, 5)$; $f(1) = -5$, $f(3) = -9$, $f(4) = 31$

#15: 13 #17: $18x^2 + 9x - 1$ #19: $2x^2 + 4xh + 2h^2 - 3x - 3h - 1$

#21: $f(3) = \frac{1}{4}$ #23: $f\left(\dfrac{1}{x}\right) = \dfrac{x}{1+x}$ #25: $f(x^2) = \dfrac{1}{x^2+1}$

#27: $-\dfrac{1}{(x+h+1)(x+1)}$ #29: $\dfrac{3x^2 + 17x + 20}{x}$ #31: $8 - 14x$

#33: $\dfrac{9x - 4x^2}{x+4}$ #35: $\dfrac{4-7x}{8-7x}$ #37: $\dfrac{x}{5x+16}$

#39: For ex: $f(x) = 3x^3 + 2x - 1$, $g(x) = x^{10}$ #41: For ex: $f(x) = 2x + 7$, $g(x) = x^{\frac{1}{4}}$

C–10   Answers to Odd-Numbered Exercises

#43: $D_g = (-\infty, \infty)$
      y–int: 3
      $R_g = [-2, \infty)$
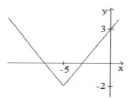

#45: $D_K = R_K = (-\infty, \infty)$
      y–int: 1
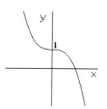

#47: $D_g = [5, \infty)$
      y–int: none
      $R_g = (-\infty, 3]$
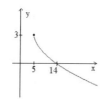

#49: odd   #51: even

#53: (a) ± 1
      (b) $(-\infty, -1) \cup (1, \infty)$
      (c) $(-1, 1)$

#55: (a) $-7, 1$
      (b) $(-7, 1)$
      (c) $(-\infty, -7) \cup (1, \infty)$

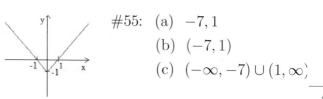

#57: (a) 5
      (b) $(5, \infty)$
      (c) $\emptyset$

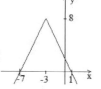

#59: (a) $-3, 2$ (b) 3
      (c) $-3, 2$ (d) $(-\infty, -3)$
      (e) $(-3, 2) \cup (2, \infty)$

#61: (a) -1,   (b) $(-\infty, -1)$ (c) $(-1, \infty)$

#63: (a) $-4, -1, 2$ (b) $(-4, -1) \cup (2, \infty)$ (c) $(-\infty, -4) \cup (-1, 2)$

#65: $\approx -1.39, 1.90$ #67: $(-\infty, -0.62) \cup (1.62, \infty)$ #69: not 1-1   #71: 1-1

#73: (a) -1, (b) $g(0) = g(2) = -1$ #75: 1-1; $f^{-1}(x) = \frac{3}{2}x - 6$

#77: 1-1, $f^{-1}(x) = \dfrac{4}{x}$ #79: not 1-1   #81: $f^{-1}(x) = \dfrac{3 - 4x}{x}$

#83: $f^{-1}(x) = \dfrac{-x - 3}{3x + 5}$ #85: $f^{-1}(x) = \frac{1}{2}(x^2 - 1)$ #87:

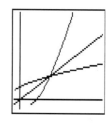

CUMULATIVE REVIEW, PAGE 122

#1: $\dfrac{5x^2 + 6x - 21}{2x^2 - 5x - 3}$ #2: $\dfrac{3}{x^{\frac{7}{2}}}$ #3: $\dfrac{-15x + 32}{(3x + 4)^3}$ #4: $\dfrac{5 \pm \sqrt{97}}{6}$ #5: $\dfrac{4}{5}, -\dfrac{2}{5}$

#6: ≈ 5.1 m^3/hr

#7: $f(3) = 2$, $f(t) = \dfrac{t + 3}{t}$, $f(2t) = \dfrac{2t + 3}{2t}$, $f(x - 3) = \dfrac{x}{x - 3}$, $f\left(\dfrac{1}{x}\right) = 1 + 3x$

#8: $f(3) = 7$, $f(t) = \sqrt{t - 2} + 2t$, $f(2t) = \sqrt{2t - 2} + 4t$, $f(x - 3) = \sqrt{x - 5} + 2x - 6$,

$f\left(\dfrac{1}{x}\right) = \sqrt{\dfrac{1 - 2x}{x}} + \dfrac{2}{x}$

#9: $2x^2 - x - 15$ #10: $\dfrac{-2x^2 + x + 16}{x - 3}$ #11: 23   #12: $\dfrac{5x - 13}{x - 3}$ #13: $\dfrac{x - 3}{10 - 3x}$

#14: $\dfrac{1}{2x + 2}$ #15: $D_g = [6, \infty)$, $R_g = [0, \infty)$ #16: $D_g = (-\infty, \infty)$, $R_g = (-\infty, 5]$

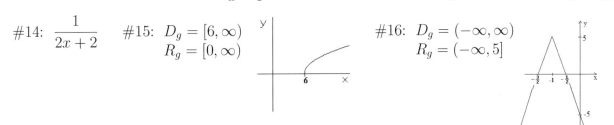

#17: (a) $C(x) = \begin{cases} 8.50, & \text{if } 0 \leq x < 10 \\ 8.00, & \text{if } 10 \leq x < 20 \\ 7.50, & \text{if } x \geq 20 \end{cases}$

(b) $C(9) = \$76.50$, $C(10) = \$80.00$, (get 10th drive for $\$3.50$)

(c) $C(19) = \$152$, $C(20) = \$150$, (cheaper to buy 20 than 19)

#18: (a)   (b)   #19: $f^{-1}(x) = \dfrac{1 + 2x}{3x}$

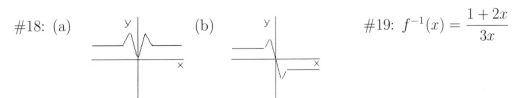

#20: For ex: $f(0) = f(1) = 0$ #21: $f^{-1}(x) = x^2 + 4$, $x \geq 0$

#22: (a) $-5, 2$ (b) -1 (c) $-5, 2$ (d) $(-5, 2) \cup (2, \infty)$ (e) $(-\infty, -5)$

#23: (a) $-2, 1$ (b) $(-2, 1)$ (c) $(-\infty, -2) \cup (1, \infty)$

#24: (a) $-0.94, 0.94$ (b) $(-0.94, 0.94)$ (c) $(-\infty, -0.94) \cup (0.94, \infty)$

C–12   Answers to Odd-Numbered Exercises

#25: $(-\infty, -1.78) \cup (0.42, 5.35)$

#26: (a) between 221 units and 13,579 units, inclusive   #27: $\frac{80}{7}$ ounces

   (b) between 1,081 units and 9,719 units, inclusive

CHAPTER 3

SECTION 3.1, PAGE 132

#1: $x = \frac{3}{2}, \ y = \frac{15}{4}$ #3: $\alpha = \beta = 30^\circ, \ x = \frac{10\sqrt{3} - 15}{3}$

#5: $\sin\theta = \frac{2}{3}, \quad \cos\theta = \frac{\sqrt{5}}{3}, \quad \tan\theta = \frac{2}{\sqrt{5}}, \quad \csc\theta = \frac{3}{2}, \quad \sec\theta = \frac{3}{\sqrt{5}}, \quad \cot\theta = \frac{\sqrt{5}}{2}$

#7: $\sin\theta = \frac{\sqrt{15}}{4}, \quad \tan\theta = \sqrt{15}, \quad \sec\theta = 4, \quad \csc\theta = \frac{4}{\sqrt{15}}, \quad \cot\theta = \frac{1}{\sqrt{15}}$

#9: $\sin\theta = \frac{2}{\sqrt{5}}, \ \cos\theta = \frac{1}{\sqrt{5}}, \ \tan\theta = 2, \ \csc\theta = \frac{\sqrt{5}}{2}, \ \sec\theta = \sqrt{5}, \ \cot\theta = \frac{1}{2}$

#11: $\sin\theta = \frac{4}{5}, \ \cos\theta = \frac{3}{5}, \ \tan\theta = \frac{4}{3}, \ \csc\theta = \frac{5}{4}, \ \sec\theta = \frac{5}{3}, \ \cot\theta = \frac{3}{4}$

#13: $\sin\theta = \frac{2\sqrt{2}}{3}, \ \cos\theta = \frac{1}{3}, \ \tan\theta = 2\sqrt{2}, \ \csc\theta = \frac{3}{2\sqrt{2}}, \ \sec\theta = 3, \ \cot\theta = \frac{1}{2\sqrt{2}}$

#15: $\sin\theta = \frac{5}{13}, \ \cos\theta = \frac{12}{13}, \ \tan\theta = \frac{5}{12}, \ \sec\theta = \frac{13}{12}, \ \cot\theta = \frac{12}{5}$

#17: $\frac{2}{3}$ #21: $\sin(90^\circ - 30^\circ) = \sin 60^\circ = \cos 30^\circ = \frac{\sqrt{3}}{2}$

#23: $\cos^2 45^\circ + \sin^2 45^\circ = \left(\frac{1}{\sqrt{2}}\right)^2 + \left(\frac{1}{\sqrt{2}}\right)^2 = \frac{1}{2} + \frac{1}{2} = 1$

#25: $\cos(60^\circ - 30^\circ) = \cos 30^\circ = \frac{\sqrt{3}}{2}; \quad \cos 60^\circ - \cos 30^\circ = \frac{1 - \sqrt{3}}{2}$

#27: $\tan(60^\circ - 30^\circ) = \tan 30^\circ = \frac{1}{\sqrt{3}}; \quad \tan 60^\circ - \tan 30^\circ = \frac{2}{\sqrt{3}}$

#29: $\sec(60^\circ - 30^\circ) = \sec 30^\circ = \frac{2}{\sqrt{3}}; \quad \sec 60^\circ - \sec 30^\circ = 2 - \frac{2}{\sqrt{3}}$

#31: $\sin 2(30^\circ) = \sin 60^\circ = \frac{\sqrt{3}}{2}; \quad 2\sin 30^\circ = 1$

#33: $\tan 2(30^\circ) = \tan 60^\circ = \sqrt{3}; \quad 2\tan 30^\circ = \frac{2}{\sqrt{3}}$

#35: $\sec 2(30^\circ) = \sec 60^\circ = 2; \quad 2\sec 30^\circ = \frac{4}{\sqrt{3}}$ #37: $b = 3, \ c = 3\sqrt{2}, \ \alpha = 45^\circ$

#39: $\beta = 30^\circ, \ b = 6, \ a = 6\sqrt{3}$ #41: $b = 5, \ \alpha = 60^\circ, \ \beta = 30^\circ$

#43: $c = \sqrt{74}, \ \alpha \approx 54^\circ, \ \beta \approx 36^\circ$ #45: $\beta = 37^\circ, \ a \approx 2.7, \ c \approx 3.3$

#47: $\alpha = 69^\circ, \ a \approx 9.9, \ c \approx 10.6$ #49: $\approx 102.5^\circ$

#53: False: The larger the acute angle, the smaller is its cosine.

Answers to Odd-Numbered Exercises   C–13

#55: False: $\tan\theta = \dfrac{5}{\sqrt{36-25}} = \dfrac{5}{\sqrt{11}} > \dfrac{5}{6}$.   #57: False: $\cos\theta = \dfrac{\sqrt{36-25}}{6} = \dfrac{\sqrt{11}}{6} < \dfrac{5}{6}$.

#59: True: As $\cos\theta < 1$, then $\sin\theta < \dfrac{\sin\theta}{\cos\theta} = \tan\theta$

SECTION 3.2, PAGE 141

#1: $12\sqrt{3}$ ft to top; 24 ft long   #3: ≈ 62.9 ft   #5: ≈ 112.4 m   #7: ≈ 249.9 m

#9: ≈ 54.9 ft   #11: ≈ 90.3 m   #13: ≈ 157.1 ft   #15: ≈ 36 min

#17: $\approx 2.997\times 10^8$ m/sec   #19: ≈ 1.8738 #21: (a) ≈ 1.5513 (b) $\approx 1.933\times 10^8$ m/sec

SECTION 3.3, PAGE 149

#1: $\dfrac{3\pi}{2}$ #3: $-\dfrac{\pi}{6}$ #5: $-\dfrac{2\pi}{3}$ #7: π #9: $-\dfrac{5\pi}{12}$ #11: $\dfrac{11\pi}{4}$

#13: $\dfrac{5\pi}{4}$ #15: $-\dfrac{7\pi}{4}$ #17: $300°$ #19: $330°$ #21: $-135°$ #23: $150°$

#25: $225°$ #27: $\left(\dfrac{1314}{\pi}\right)° \approx 418°$ #29: $-\left(\dfrac{4032}{\pi}\right)° \approx -1283°$

#31:   #33:   #35:   #37:

#39:   #41:   #43   #45:

#47: QII   #49: QIII   #51: QIV   #53: QI   #55: QII   #57: QIII   #59: QII

#61: $-88°$ (or $272°$)   #63: $65°$ (or $-295°$)   #65: $-40°$ (or $320°$)

#67: $-\dfrac{\pi}{4}$ $\left(\text{ or }\dfrac{7\pi}{4}\right)$ #69: $\dfrac{\pi}{3}$ $\left(\text{ or }-\dfrac{5\pi}{3}\right)$ #71: $450°$, $\dfrac{5\pi}{2}$ #73: $180°$, π

#75: π cm   #77: ≈ 553.3 mi   #79: $\approx 6,086.7$ ft   #81: Pluto, by $\approx 12,463$ ft

#83: 400 grads   #85: $\dfrac{9\pi}{10}$ #87: $A_\theta = \frac{1}{2}\theta r^2$ #89: 350 ft/min.

C–14   **Answers to Odd-Numbered Exercises**

#91: (a) greater on $33\frac{1}{3}$ (b) larger closest to rim

   (c) For ex: $d_2 = 27, d_1 = 20$, as $d_1 = \frac{20}{27}\,d_2$ #93: 7926π miles/day

SECTION 3.4, PAGE 162

#1:   (a) $\sin 30°$ (b) $\frac{1}{2}$ #3:   (a) $\sec 60°$ (b) 2   #5: $(-1, 0)$ #7: $(0, -1)$

#9: $(0, 1)$ #11: $\sin 5\pi = 0$, $\cos 5\pi = -1$, $\tan 5\pi = 0$

   $\csc 5\pi$ undefined , $\sec 5\pi = -1$, $\cot 5\pi$ undefined

#13: $\sin\left(-\frac{7\pi}{2}\right) = 1$, $\cos\left(-\frac{7\pi}{2}\right) = 0$, $\tan\left(-\frac{7\pi}{2}\right)$ undefined

   $\csc\left(-\frac{7\pi}{2}\right) = 1$, $\sec\left(-\frac{7\pi}{2}\right)$ undefined , $\cot\left(-\frac{7\pi}{2}\right) = 0$

#15: $\sin\left(-\frac{\pi}{2}\right) = -1$, $\cos\left(-\frac{\pi}{2}\right) = 0$, $\tan\left(-\frac{\pi}{2}\right)$ undefined

   $\csc\left(-\frac{\pi}{2}\right) = -1$, $\sec\left(-\frac{\pi}{2}\right)$ undefined , $\cot\left(-\frac{\pi}{2}\right) = 0$

#17: positive   #19: negative   #21: negative   #23: positive   #25: positive

#27:   positive   #29:   QII   #31:   QI   #33:   QII   #35:   60°

#37: $\frac{\pi}{8}$ #39: 30°   #41: $\frac{\pi}{3}$ #43: $\sin 35°$ #45: $-\sec 15°$ #47: $\sec 35°$

#49: $-\cos\frac{\pi}{4}$ #51: $\csc\frac{\pi}{5}$ #53: 1   #55: -2 #57: 0   #59: $\frac{1}{\sqrt{2}}$ #61: 0

#63: $\frac{1}{\sqrt{3}}$ #65: undefined   #67: $\sqrt{3}$ #69: $-\frac{1}{\sqrt{2}}$ #71: $-\frac{1}{\sqrt{3}}$

#73: -1 #75: $-\frac{\sqrt{3}}{2}$ #77: impossible (implies $\sin\theta = 4$, but $-1 \le \sin\theta \le 1$)

#79: $\cos\theta = -\frac{7}{25}$, $\tan\theta = \frac{24}{7}$, $\cot\theta = \frac{7}{24}$, $\sec\theta = -\frac{25}{7}$, $\csc\theta = -\frac{25}{24}$

#81: $\sin\theta = -\frac{12}{13}$, $\cos\theta = \frac{5}{13}$, $\tan\theta = -\frac{12}{5}$, $\csc\theta = -\frac{13}{12}$, $\cot\theta = -\frac{5}{12}$

#83: In QII: $\sin\theta = \frac{\sqrt{3}}{2}$, $\cos\theta = -\frac{1}{2}$, $\tan\theta = -\sqrt{3}$, $\csc\theta = \frac{2}{\sqrt{3}}$, $\cot\theta = -\frac{1}{\sqrt{3}}$

   In QIII: $\sin\theta = -\frac{\sqrt{3}}{2}$, $\cos\theta = -\frac{1}{2}$, $\tan\theta = \sqrt{3}$, $\csc\theta = -\frac{2}{\sqrt{3}}$, $\cot\theta = \frac{1}{\sqrt{3}}$

#85: In QII: $\sin\theta = \frac{1}{\sqrt{5}}$, $\cos\theta = -\frac{2}{\sqrt{5}}$, $\csc\theta = \sqrt{5}$, $\sec\theta = -\frac{\sqrt{5}}{2}$, $\cot\theta = -2$

   In QIV: $\sin\theta = -\frac{1}{\sqrt{5}}$, $\cos\theta = \frac{2}{\sqrt{5}}$, $\csc\theta = -\sqrt{5}$, $\sec\theta = \frac{\sqrt{5}}{2}$, $\cot\theta = -2$

#87: In QI: $\sin\theta = \frac{4}{5}$, $\tan\theta = \frac{4}{3}$, $\cot\theta = \frac{3}{4}$, $\csc\theta = \frac{5}{4}$, $\sec\theta = \frac{5}{3}$

In QIV: $\sin\theta = -\frac{4}{5}$, $\tan\theta = -\frac{4}{3}$, $\cot\theta = -\frac{3}{4}$, $\csc\theta = -\frac{5}{4}$, $\sec\theta = \frac{5}{3}$

#89: $\sin\theta = \frac{4}{5}$, $\cos\theta = -\frac{3}{5}$, $\tan\theta = -\frac{4}{3}$, $\csc\theta = \frac{5}{4}$, $\sec\theta = -\frac{5}{3}$, $\cot\theta = -\frac{3}{4}$

#91: $\sin\theta = \frac{2}{\sqrt{5}}$, $\cos\theta = \frac{1}{\sqrt{5}}$, $\tan\theta = 2$, $\csc\theta = \frac{\sqrt{5}}{2}$, $\sec\theta = \sqrt{5}$, $\cot\theta = \frac{1}{2}$

#93: $\sin\theta = -1$, $\cos\theta = 0$, $\tan\theta$ undefined, $\csc\theta = -1$, $\sec\theta$ undefined, $\cot\theta = 0$

#95: $\sin\theta = -\frac{5}{13}$, $\cos\theta = -\frac{12}{13}$, $\tan\theta = \frac{5}{12}$, $\csc\theta = -\frac{13}{5}$, $\sec\theta = -\frac{13}{12}$, $\cot\theta = \frac{12}{5}$

#97: $\sin\theta = -\frac{5}{13}$, $\cos\theta = \frac{12}{13}$, $\tan\theta = -\frac{5}{12}$, $\csc\theta = -\frac{13}{5}$, $\sec\theta = \frac{13}{12}$, $\cot\theta = -\frac{12}{5}$

#99: $\sin\theta = -\frac{2}{\sqrt{13}}$, $\cos\theta = -\frac{3}{\sqrt{13}}$, $\tan\theta = \frac{2}{3}$, $\csc\theta = -\frac{\sqrt{13}}{2}$, $\sec\theta = -\frac{\sqrt{13}}{3}$, $\cot\theta = \frac{3}{2}$

#101: For ex: $\alpha = 127.5°$, $\beta = 82.5°$ #103: For ex: $\alpha = 30°$, $\beta = 7.5°$

CHAPTER REVIEW, PAGE 168

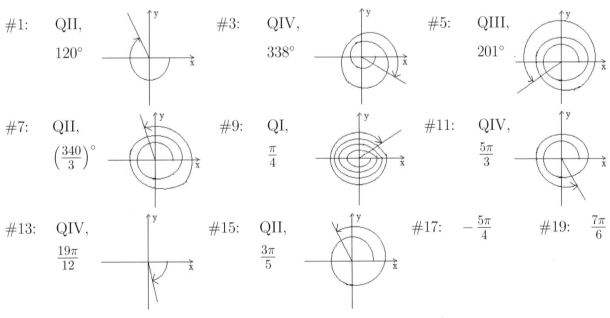

#1: QII, 120°

#3: QIV, 338°

#5: QIII, 201°

#7: QII, $\left(\frac{340}{3}\right)°$

#9: QI, $\frac{\pi}{4}$

#11: QIV, $\frac{5\pi}{3}$

#13: QIV, $\frac{19\pi}{12}$

#15: QII, $\frac{3\pi}{5}$

#17: $-\frac{5\pi}{4}$ #19: $\frac{7\pi}{6}$

#21: $-30°$ #23: $420°$ #25: -1 #27: -1 #29: $\frac{1}{\sqrt{2}}$ #31: 1 #33: 2

#35: $\beta = 45°$, $b = \sqrt{2}$, $= 2$ #37: $\alpha = 60°$, $a = \frac{3\sqrt{3}}{2}$, $b = \frac{3}{2}$

C–16   Answers to Odd-Numbered Exercises

#39: $\beta = 58°$, $b \approx 2.3$, $c \approx 2.7$

#41: $\cos\theta = -\dfrac{\sqrt{15}}{4}$, $\tan\theta = \dfrac{1}{\sqrt{15}}$, $\csc\theta = -4$, $\sec\theta = -\dfrac{4}{\sqrt{15}}$, $\cot\theta = \sqrt{15}$

#43: In QII: $\tan\theta = -\dfrac{3}{4}$, $\sin\theta = \dfrac{3}{5}$, $\cos\theta = -\dfrac{4}{5}$, $\csc\theta = \dfrac{5}{3}$, $\sec\theta = -\dfrac{5}{4}$

In QIV: $\tan\theta = -\dfrac{3}{4}$, $\sin\theta = -\dfrac{3}{5}$, $\cos\theta = \dfrac{4}{5}$, $\csc\theta = -\dfrac{5}{3}$, $\sec\theta = \dfrac{5}{4}$

#45: ≈ 447.8 m   #47: northernmost ranger is ≈ 23.3 mi away, other is ≈ 18.9 mi away

#49: ≈ 93.6 ft   #51: ≈ 8.6 in

CUMULATIVE REVIEW, PAGE 169

#1: $4a^{\frac{5}{2}}b^{\frac{9}{2}}$ #2: $\dfrac{1}{\sqrt{x}(x-3)}$ #3: $\dfrac{3 \pm \sqrt{11}}{2}$ #4: $-\dfrac{1}{3}, 4$ #5: $\approx 6.71 \times 10^8$ mph

#6: (a) $\left(-\infty, -\dfrac{4}{3}\right) \cup \left(-\dfrac{4}{3}, \infty\right)$,   (b) $\dfrac{-x^2 + 1}{3x^2 + 1}$,   (c) 0,   (e) $-\dfrac{4x}{3x+1}$

#7: (a) domain: $[-4, \infty)$ (c)   (d) range: $(-\infty, 0]$
    (b) x-int: -4
        y-int: -2

#8: $\dfrac{3}{2}$ #9: 1   #10: -5 #11: 8

#12: $\cos 225° = -\dfrac{1}{\sqrt{2}}$, $\sin 225° = -\dfrac{1}{\sqrt{2}}$, $\tan 225° = 1$,

$\sec 225° = -\sqrt{2}$, $\csc 225° = -\sqrt{2}$, $\cot 225° = 1$

#13: $\cos(-240°) = -\dfrac{1}{2}$, $\sin(-240°) = \dfrac{\sqrt{3}}{2}$, $\tan(-240°) = -\sqrt{3}$,

$\sec(-240°) = -2$, $\csc(-240°) = \dfrac{2}{\sqrt{3}}$, $\cot(-240°) = -\dfrac{1}{\sqrt{3}}$

#14: $\cos 690° = \dfrac{\sqrt{3}}{2}$, $\sin 690° = -\dfrac{1}{2}$, $\tan 690° = -\dfrac{1}{\sqrt{3}}$,

$\sec 690° = \dfrac{2}{\sqrt{3}}$, $\csc 690° = -2$, $\cot 690° = -\sqrt{3}$

#15: $\cos \dfrac{3\pi}{2} = 0$, $\sin \dfrac{3\pi}{2} = -1$, $\tan \dfrac{3\pi}{2}$ undefined,

$\sec \dfrac{3\pi}{2}$ undefined, $\csc \dfrac{3\pi}{2} = -1$, $\cot \dfrac{3\pi}{2} = 0$

#16: $\cos\left(-\frac{7\pi}{4}\right) = \frac{1}{\sqrt{2}}$, $\sin\left(-\frac{7\pi}{4}\right) = \frac{1}{\sqrt{2}}$, $\tan\left(-\frac{7\pi}{4}\right) = 1$,

$\sec\left(-\frac{7\pi}{4}\right) = \sqrt{2}$, $\csc\left(-\frac{7\pi}{4}\right) = \sqrt{2}$, $\cot\left(-\frac{7\pi}{4}\right) = 1$

#17: $\cos\frac{5\pi}{6} = -\frac{\sqrt{3}}{2}$, $\sin\frac{5\pi}{6} = \frac{1}{2}$, $\tan\frac{5\pi}{6} = -\frac{1}{\sqrt{3}}$,

$\sec\frac{5\pi}{6} = -\frac{2}{\sqrt{3}}$, $\csc\frac{5\pi}{6} = 2$, $\cot\frac{5\pi}{6} = -\sqrt{3}$

#18: $\cos\theta = -\frac{24}{25}$, $\tan\theta = -\frac{7}{24}$, $\sec\theta = -\frac{25}{24}$, $\csc\theta = \frac{25}{7}$, $\cot\theta = -\frac{24}{7}$

#19: $\sin\theta = -\frac{3}{5}$, $\tan\theta = \frac{3}{4}$, $\sec\theta = -\frac{5}{4}$, $\csc\theta = -\frac{5}{3}$, $\cot\theta = \frac{4}{3}$

#20: (a) ≈ 660.5 mi   (b) S32°E

CHAPTER 4

SECTION 4.1, PAGE 179

#1: $\sin\frac{7\pi}{6} = -\frac{1}{2}$, $\cos\frac{7\pi}{6} = -\frac{\sqrt{3}}{2}$, $\tan\frac{7\pi}{6} = \frac{1}{\sqrt{3}}$,

$\csc\frac{7\pi}{6} = -2$, $\sec\frac{7\pi}{6} = -\frac{2}{\sqrt{3}}$, $\cot\frac{7\pi}{6} = \sqrt{3}$

#3: $\sin\left(-\frac{5\pi}{4}\right) = \frac{1}{\sqrt{2}}$, $\cos\left(-\frac{5\pi}{4}\right) = -\frac{1}{\sqrt{2}}$, $\tan\left(-\frac{5\pi}{4}\right) = -1$,

$\csc\left(-\frac{5\pi}{4}\right) = \sqrt{2}$, $\sec\left(-\frac{5\pi}{4}\right) = -\sqrt{2}$, $\cot\left(-\frac{5\pi}{4}\right) = -1$

#5: amplitude: 4

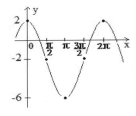

#7: amplitude: 2

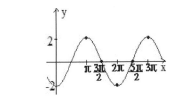

#9: amplitude: 3

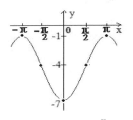

#11: amplitude: 2

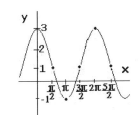

#13: amplitude: 2

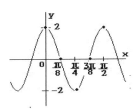

#15: amplitude: $\frac{5}{2}$
period: $\frac{\pi}{4}$

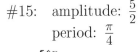
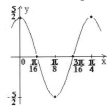

C–18   Answers to Odd-Numbered Exercises

#17: amplitude: 3
      period: $\frac{2\pi}{3}$

#19: amplitude: 1
      period: 3π

#21: amplitude: $\frac{1}{3}$
      period: 2π
      phase shift: $\frac{3\pi}{4}$

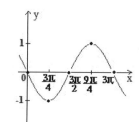
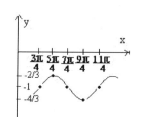
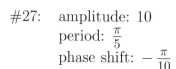

#23: amplitude: 1
      period: 4π
      phase shift: 2π

#25: amplitude: 3
      period: $\frac{\pi}{2}$
      phase shift: none

#27: amplitude: 10
      period: $\frac{\pi}{5}$
      phase shift: $-\frac{\pi}{10}$

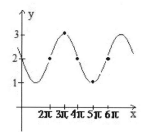
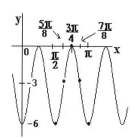
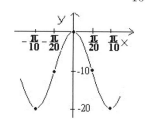

#29: va: $x = \pm\frac{\pi}{4}$, $x = \pm\frac{3\pi}{4}$, $x = \pm\frac{5\pi}{4}$, $x = \pm\frac{7\pi}{4}$

      x-ints: 0, $\pm\frac{\pi}{2}$, $\pm\pi$, $\pm\frac{3\pi}{2}$, $\pm 2\pi$; y-int: 0

#31: va: $x = \pm\frac{\pi}{2}$, $x = \pm\frac{3\pi}{2}$; x-ints: $-\frac{7\pi}{4}$, $-\frac{3\pi}{4}$, $\frac{\pi}{4}$, $\frac{5\pi}{4}$; y-int: -1

#33: va: $x = \pm\frac{\pi}{2}$, $x = \pm\frac{3\pi}{2}$; x-ints: 0, $\pm\pi$, $\pm 2\pi$; y-int: 0

#35: va: $x = \pm\frac{\pi}{4}$, $x = \pm\frac{3\pi}{4}$, $x = \pm\frac{5\pi}{4}$, $x = \pm\frac{7\pi}{4}$; x-ints: none; y-int: 1

#37: va: $x = \pm\frac{\pi}{2}$, $x = \pm\frac{3\pi}{2}$; x-ints: 0, $\pm 2\pi$; y-int: 0

#39: va: $x = 0$, $x = \pm\pi$, $x = \pm 2\pi$; x-ints: $-\frac{\pi}{2}$, $\frac{3\pi}{2}$; y-int: none

#41: period: $\frac{\pi}{2}$

#43: period: π

#45: period: π

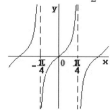
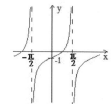
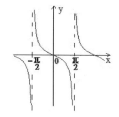

#47: period: 2π #49: period: 2π #51: period: 2π

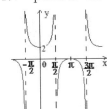
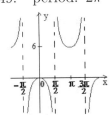
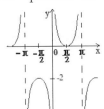

#53: For ex. take $a = \frac{\pi}{2}$, then $f(a) = 2$ and $g(a) = 0$

#55: For ex. take $a = 0$, then $f(a) = 1$ and $g(a) = -1$

#57: For ex. take $a = 0$, then $f(a) = -1$ and $g(a) = -\frac{\pi}{4}$

#63: $f(x) = 2\cos 2x$ #65: $f(x) = -3\cos\frac{\pi}{4}(x-2)$ #67: $f(x) = -5\sin x$

#69: $\tan^2 x - \sec^2 x = -1$ #71: For ex: $f(x) = \tan\frac{1}{3}x$ #73: For ex: $f(x) = -\sec\frac{1}{2}x$

#75: $\sin x = x - \frac{x^3}{3!} + \frac{x^5}{5!} - \frac{x^7}{7!} + \cdots$ #77: For ex: $f(x) = \begin{cases} 1, & 9n < x \leq 9n+1 \\ 2, & 9n+1 < x \leq 9n+9 \end{cases}$

#79: For ex: $f(x) = \begin{cases} 3, & 9n \leq x < 9n+1 \\ 0, & 9n+1 \leq x < 9n+9 \end{cases}$ #81: (a) $k \approx -0.13$ (b) ≈ 18 sec

#83: True: The values of both the sine and the cosine are at most 1, so their product is at most 1.

#85: True: As $\sin x < 1$ and $\cos x < 1$, then $\sin x + \cos x < 2$.

#87: False: f is undefined when the cotangent is undefined, i.e. when $\sin x = 0$, for example, when $x = 0$, but $\tan^2 0 = 0$ is defined then.

#89: False: For ex: for some x_0, $\cot x_0 = 4 \Rightarrow \tan x_0 = \frac{1}{4} \Rightarrow \tan x_0 + 1 = \frac{5}{4}$, while, $\cot x_0 - 1 = 3$.

#91: True: Because $tan x$ has infinitely many x-intercepts.

#93: False: The graph has vertical asymptotes at those of the tangent and at those of the cotangent.

SECTION 4.2, PAGE 193

#1: For ex., take $x = y = \frac{\pi}{2}$, then $\sin(x+y) = 0$ and $\sin x + \sin y = 2$

#3: For ex., take $x = y = \frac{\pi}{4}$, then $\tan(x+y) = $ undefined and $\tan x + \tan y = 2$

#5: $\sin^2 x \cos^2 y - \cos^2 x \sin^2 y$ #11: $\sin(45° + 30°) = \frac{\sqrt{6}+\sqrt{2}}{4}$

#13: $\tan\left(\frac{\pi}{3} + \frac{\pi}{4}\right) = -2 - \sqrt{3}$

C–20   Answers to Odd-Numbered Exercises

#15: $\sin(\alpha - \beta) = \frac{63}{65},\quad \cos(\alpha - \beta) = -\frac{16}{65},\quad \tan(\alpha - \beta) = -\frac{63}{16}$

#17: $\sin(\alpha - \beta) = \frac{18\sqrt{13}}{65},\quad \cos(\alpha - \beta) = -\frac{\sqrt{13}}{65},\quad \tan(\alpha - \beta) = -18$

#19: $\sin 2x = -\frac{24}{25},\quad \cos 2x = \frac{7}{25},\quad \tan 2x = -\frac{24}{7}$

#21: $\sin 2x = \frac{15}{17},\quad \cos 2x = -\frac{8}{17},\quad \tan 2x = -\frac{15}{8}$ #23: $\cos 79°$ #25: $\sin 150°$

#27: $\frac{1}{2}\sin 174°$ #29: -1 #31: 2 #33: $\sec x + \tan x$ #35: $\frac{1}{8}\tan^2 x$

#37: $\frac{3}{8} - \frac{1}{2}\cos 2x + \frac{1}{8}\cos 4x$ #39: $\dfrac{2}{1 + \cos 4x}$ #41: $8\cos^4 x - 8\cos^2 x + 1$

#43: $\frac{3}{8}$ #45: $\sin\left[\frac{1}{2}\left(\frac{\pi}{4}\right)\right] = \frac{1}{2}\sqrt{2 - \sqrt{2}}$

#47: $\sin x = -\frac{\sqrt{14}}{4},\ \cos x = -\frac{\sqrt{2}}{4},\ \tan x = \sqrt{7},\ \csc x = -\frac{4}{\sqrt{14}},\ \sec x = -2\sqrt{2},\ \cot x = \frac{1}{\sqrt{7}}$

#49: $\sin\left(\frac{x}{2}\right) = \frac{2}{\sqrt{5}},\ \cos\left(\frac{x}{2}\right) = \frac{1}{\sqrt{5}},\ \tan\left(\frac{x}{2}\right) = 2,\ \csc\left(\frac{x}{2}\right) = \frac{\sqrt{5}}{2},\ \sec\left(\frac{x}{2}\right) = \sqrt{5},\ \cot\left(\frac{x}{2}\right) = \frac{1}{2}$

#51: $\sin\left(\frac{x}{2}\right) = -\frac{1}{\sqrt{10}},\quad \cos\left(\frac{x}{2}\right) = \frac{3}{\sqrt{10}},\quad \tan\left(\frac{x}{2}\right) = -\frac{1}{3},$

$\csc\left(\frac{x}{2}\right) = -\sqrt{10},\quad \sec\left(\frac{x}{2}\right) = \frac{\sqrt{10}}{3},\quad \cot\left(\frac{x}{2}\right) = -3$

#53: $\frac{1}{2}[\cos 9x + \cos x]$ #55: $\frac{1}{2}[\sin(7x + 1) + \sin(x - 3)]$ #59: $-2\sin 4x \sin 2x$

#61: $2\cos\left(\frac{15x}{2}\right)\cos\left(\frac{3x}{2}\right)$ #63: $a\tan\theta$ #65: $a\cos\theta$ #67: $f(x) = \sqrt{2}\cos\left(x + \frac{\pi}{4}\right)$

#69: $f(x) = \frac{2}{\sqrt{3}}\sin\left(x + \frac{\pi}{3}\right)$ #111: For ex: $x = 0 \Rightarrow \text{RHS} = 1$, LHS is undefined

#115: Not identity   #117: Identity

SECTION 4.3, PAGE 207

#1: $-\frac{\pi}{2}$ #3: $\frac{1}{2}$ #5: $\frac{3\pi}{10}$ #7: $\frac{5}{13}$ #9: $\frac{7}{25}$ #11: does not exist   #13: $\frac{5}{9}$

#15: $\frac{5\pi}{8}$ #17: $\frac{\sqrt{5}}{3}$ #19: $-\frac{7}{8}$ #21: $-\frac{\pi}{6}$ #23: $-\frac{\pi}{9}$ #25: $\frac{5}{4}$ #27: $\frac{3}{5}$

#29: $\frac{2}{\sqrt{13}}$ #31: $-\frac{2}{11}$ #35: $\dfrac{\sqrt{1 - x^2}}{x}$ #37: $\dfrac{2x}{\sqrt{1 - 4x^2}}$

#39: $\dfrac{2x^2-1}{2x\sqrt{1-x^2}}$ #41: $\dfrac{1}{\sqrt{1+x^2}}$

#43: $y = 4\cos^{-1}\left(x+\dfrac{\pi}{4}\right)$; $-1-\dfrac{\pi}{4} \le x \le 1-\dfrac{\pi}{4}$ #45: $\sqrt{3}$ #47: $-2+\dfrac{3}{\sqrt{10}}$

#49: $-1-\dfrac{\pi}{6}+2n\pi$, $-1+\dfrac{7\pi}{6}+2n\pi$ #51: -200.167 #53: $0, \pm 1.557$ #55: 0.695

#57: 0.707 #59: $y = \csc^{-1}x$ #61: (a) $\dfrac{\pi}{3}$ (b) $\dfrac{4}{3}$ (c) $\dfrac{2\pi}{5}$

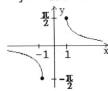

#63: $D = (-\infty, -1] \cup [1, \infty)$, $R = [0, \dfrac{\pi}{2}) \cup (\dfrac{\pi}{2}, \pi]$, x in $(-\infty, -1] \cup [1, \infty)$, $x = \sec y$

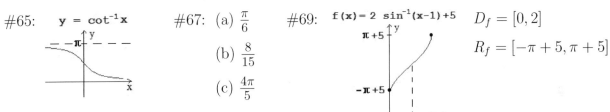

#65: $y = \cot^{-1}x$ #67: (a) $\dfrac{\pi}{6}$ (b) $\dfrac{8}{15}$ (c) $\dfrac{4\pi}{5}$ #69: $f(x) = 2\sin^{-1}(x-1)+5$ $D_f = [0, 2]$ $R_f = [-\pi+5, \pi+5]$

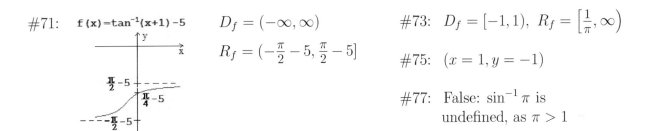

#71: $f(x) = \tan^{-1}(x+1)-5$ $D_f = (-\infty, \infty)$ $R_f = (-\dfrac{\pi}{2}-5, \dfrac{\pi}{2}-5]$ #73: $D_f = [-1, 1)$, $R_f = \left[\dfrac{1}{\pi}, \infty\right)$

#75: $(x=1, y=-1)$

#77: False: $\sin^{-1}\pi$ is undefined, as $\pi > 1$

#79: True: Defined, as $-1 \le \sin x \le 1$ #81: True: Defined, as domain of sine is $(-\infty, \infty)$

#83: False: For ex., $\dfrac{\sin^{-1}\frac{1}{2}}{\cos^{-1}\frac{1}{2}} = \dfrac{\pi/6}{\pi/3} = \dfrac{1}{2}$, while $\tan^{-1}\dfrac{1}{2} \approx 0.46$

SECTION 4.4, PAGE 216

#1: $\dfrac{\pi}{3}, \dfrac{2\pi}{3}$ #3: $0, \pi, \dfrac{\pi}{3}, \dfrac{5\pi}{3}$ #5: $\dfrac{\pi}{6}, \dfrac{5\pi}{6}, \dfrac{7\pi}{6}, \dfrac{11\pi}{6}$ #7: $\dfrac{\pi}{2}, \dfrac{3\pi}{2}$

#9: $\dfrac{\pi}{2}, \pi, \dfrac{3\pi}{2}$ #11: $\dfrac{3\pi}{4}+2n\pi, \dfrac{7\pi}{4}+2n\pi$ #13: $\dfrac{7\pi}{6}+2n\pi, \dfrac{11\pi}{6}+2n\pi, \dfrac{\pi}{2}+2n\pi$

#15: $\dfrac{3\pi}{4}+2n\pi, \dfrac{7\pi}{4}+2n\pi, \dfrac{2\pi}{3}+2n\pi, \dfrac{5\pi}{3}+2n\pi$

#17: $\frac{\pi}{3} + 2n\pi$, $\frac{2\pi}{3} + 2n\pi$, $\frac{4\pi}{3} + 2n\pi$, $\frac{5\pi}{3} + 2n\pi$ #19: $\frac{2\pi}{3}$, $\frac{5\pi}{3}$, $\frac{5\pi}{6}$, $\frac{11\pi}{6}$

#21: π, $\frac{\pi}{2}$, $\frac{3\pi}{2}$ #23: $\frac{\pi}{6}$, $\frac{\pi}{2}$, $\frac{5\pi}{6}$, $\frac{7\pi}{6}$, $\frac{3\pi}{2}$, $\frac{11\pi}{6}$, $\frac{7\pi}{12}$, $\frac{11\pi}{12}$, $\frac{19\pi}{12}$, $\frac{23\pi}{12}$

#25: $\frac{\pi}{3} + 2n\pi$, $\frac{5\pi}{3} + 2n\pi$ #27: $\frac{n\pi}{3}$ #29: $\cos^{-1}\frac{3}{5}$ #31: $\pi - \sin^{-1}\frac{3}{4}$

#33: $-\tan^{-1}\frac{3}{2} + n\pi$ #35: $\pi - \tan^{-1}8 \approx 1.695$, $2\pi - \tan^{-1}8 \approx 4.837$

#37: $\pi + \sin^{-1}\frac{5}{6} \approx 4.127$, $2\pi - \sin^{-1}\frac{5}{6} \approx 5.298$, $\sin^{-1}\frac{3}{4} \approx 0.848$, $\pi - \sin^{-1}\frac{3}{4} \approx 2.294$

#39: $\frac{\pi}{6}$, $\frac{5\pi}{6}$, $\pi + \sin^{-1}\frac{2}{3} \approx 3.871$, $2\pi - \sin^{-1}\frac{2}{3} \approx 5.553$

#41: $\frac{\pi}{2}$, $\frac{3\pi}{2}$, $\sin^{-1}\frac{3}{8} \approx 0.384$, $\pi - \sin^{-1}\frac{3}{8} \approx 2.757$

#43: $\sin^{-1}\frac{3}{4} \approx 0.848$, $\pi - \sin^{-1}\frac{3}{4} \approx 2.294$

#45: $\frac{3\pi}{4}$, $\frac{7\pi}{4}$, $\tan^{-1}\frac{1}{3} \approx 0.322$, $\pi + \tan^{-1}\frac{1}{3} \approx 3.463$

#47: $\left(x = \frac{\pi}{6}, y = 0\right)$, $\left(x = \frac{5\pi}{6}, y = 0\right)$, $(x = \pi, y = -1)$ #49: ≈ 0.28, 1.57

#51: ≈ 0.76 #53: $\left(0, \frac{\pi}{2}\right) \cup (2.38, \pi]$ #55: $[0, 0.52) \cup (0.9, 2.06)$

CHAPTER REVIEW, PAGE 221

#1: period: 2
y-int: π
amplitude: 1
phase shift: none

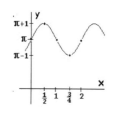

#3: period: π
y-int: 2
amplitude: 4
phase shift: $\frac{\pi}{4}$

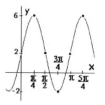

#5: period: π
y-int: none
vertical asymps:
$x = n\pi$

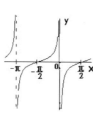

#7: period: 2π
y-int: none
vertical asymps:
$x = n\pi$

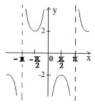

#9: period: $\frac{\pi}{2}$
y-int: none
vertical asymps:
$x = \frac{n\pi}{2}$

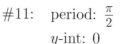

#11: period: $\frac{\pi}{2}$
y-int: 0
vertical asymps:
$x = \frac{(2n+1)\pi}{8}$

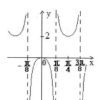

#13: For ex: $f\left(\frac{\pi}{2}\right) = 3$, $g\left(\frac{\pi}{2}\right) = -1$ #15: For ex: $f(0) = 0$, $g(0)$ undefined

#17: $\dfrac{1}{1+\cos x}$ #31: $\sin(\alpha+\beta) = -\dfrac{33}{65}$, $\cos(\alpha-\beta) = \dfrac{16}{65}$, $\sin 2\alpha = -\dfrac{24}{25}$, $\cos 2\beta = \dfrac{119}{169}$

#33: $\sin\theta = \dfrac{4}{\sqrt{17}}$, $\cos\theta = \dfrac{1}{\sqrt{17}}$, $\tan\theta = 4$, $\sec\theta = \sqrt{17}$, $\csc\theta = \dfrac{\sqrt{17}}{4}$, $\cot\theta = \dfrac{1}{4}$

#35: $-\sqrt{17 + 12\sqrt{2}}$ #37: $5\sin x - 20\sin^3 x + 16\sin^5 x$ #39: 0   #41: $\dfrac{\pi}{6}$

#43: $\dfrac{4\sqrt{3} - 3}{10}$ #45: $\dfrac{\sqrt{1+x^2}}{x}$ #47: $\dfrac{\sqrt{1-4x^2}}{2x}$ #49: $\dfrac{\pi}{4}, \dfrac{5\pi}{4}$

#51: $\dfrac{\pi}{4} + n\pi$, $\dfrac{3\pi}{4} + n\pi$ #53: $(2n+1)\dfrac{\pi}{3}$

#55: $\pi - \cos^{-1}\dfrac{3}{4} \approx 2.42$, $\pi + \cos^{-1}\dfrac{3}{4} \approx 3.86$, $\cos^{-1}\dfrac{2}{3} \approx 0.84$, $2\pi - \cos^{-1}\dfrac{2}{3} \approx 5.44$

#57: $\approx 0.32, 1.78, 2.01$ #59: no solution   #61: ≈ -399.16 #63: $\tan^{-1}\left(\dfrac{H-h}{d}\right)$

#65: True   #67: False: For ex. let $x = \dfrac{1}{\sqrt{2}}$ #69: True

CUMULATIVE REVIEW, PAGE 223

#1: $\dfrac{40(x-2)^3}{x^2}$ #2: $\dfrac{-x^3 - 2x^2 - 3x - 1}{x^{\frac{3}{2}}(x+1)}$ #3: $6x + 3h - 4$

#4: $D_f = [-2, \infty)$
$R_f = [0, \infty)$

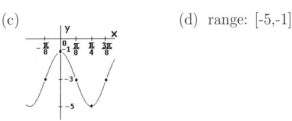

#5: $\sin\theta = -\dfrac{7}{25}$, $\cos\theta = -\dfrac{24}{25}$, $\tan\theta = \dfrac{7}{24}$, $\csc\theta = -\dfrac{25}{7}$, $\sec\theta = -\dfrac{25}{24}$

#6: $\dfrac{\pi}{4} + n\pi$, $-\tan^{-1}\dfrac{1}{2} + n\pi$ #7: $0, \pi, \sin^{-1}\dfrac{3}{4}, \pi - \sin^{-1}\dfrac{3}{4}$

#9: $6x\sqrt{1-9x^2}$ #10: ≈ 355.4 ft

#11: (a) x-ints: none   (c)   (d) range: $[-5,-1]$
  y-int: -1
 (b) amplitude: 2
   period: $\dfrac{\pi}{2}$
   phase shift: $-\dfrac{\pi}{8}$

C-24   **Answers to Odd-Numbered Exercises**

#12: $\dfrac{63}{65}$ #13: $\dfrac{16}{65}$ #14: $\dfrac{63}{16}$ #15: $-\dfrac{24}{25}$ #16: $-\dfrac{3}{\sqrt{10}}$

#17: $\sqrt{2-\sqrt{3}}$ #18: $-\dfrac{1}{\sqrt{2}}$ #19: $-\dfrac{\pi}{6}$ #20: $-\dfrac{\pi}{4}$ #21: $-\dfrac{\pi}{2}$

#22: $-\dfrac{120}{169}$ #23: $\dfrac{6\pi}{7}$ #24: $\dfrac{3}{2\sqrt{2}}$ #25: $\dfrac{-2\sqrt{2}-\sqrt{3}}{6}$

CHAPTER 5

SECTION 5.1, PAGE 232

#1: $a \approx 11.4$, $c \approx 4.5$, $\alpha = 129°$ #3: $a \approx 23.3$, $b \approx 19.4$, $\gamma = 25°$

#5: $b \approx 9.4$, $c \approx 7.4$, $\alpha = 125°$ #7: $\alpha \approx 18°$, $\beta = 110°$, $\gamma \approx 52°$

#9: $a \approx 11.7$, $\beta \approx 78°$, $\gamma \approx 52°$ #11: $\alpha \approx 22°$, $\beta = 120°$, $\gamma \approx 38°$

#13: $a \approx 3.7$, $\beta \approx 107°$, $\gamma \approx 43°$ #15: No such triangle   #17: $a \approx 2.6$, $c \approx 5.4$, $\alpha = 29°$

#19: $\alpha \approx 57°$, $\beta = 17°$, $\gamma \approx 106°$ #21: $b \approx 1.5$, $c \approx 1.1$, $\gamma = 37°$

#23: $b \approx 17.9$, $\alpha \approx 50°$, $\gamma \approx 30°$ #25: $b = c \approx 6.5$, $\beta = \gamma = 40°$

#27: $b \approx 13.4$, $\alpha \approx 12°$, $\beta \approx 138°$ #29: No such triangle

#31:   No such triangle   #33: $b \approx 7.6$, $c \approx 11.7$, $\beta \approx 40°$, $\gamma \approx 90°$

#35: 2 triangles; $c \approx 7.8$, $\beta \approx 35°$, $\gamma \approx 118°$, and $c' \approx 1.2$, $\beta' \approx 145°$, $\gamma' \approx 8°$

#37: $c \approx 14.0$, $\beta \approx 15°$, $\gamma \approx 134°$

#39: 2 triangles; $a \approx 5.8$, $\alpha \approx 52°$, $\gamma \approx 73°$ and $a' \approx 2.3$, $\alpha' \approx 18°$, $\gamma' \approx 107°$

#41: 2 triangles; $a \approx 6.4$, $\alpha \approx 39°$, $\beta \approx 79°$, and $a' \approx 3.0$, $\alpha' \approx 17°$, $\beta' \approx 101°$

#43: 2 triangles; $c \approx 5.4$, $\beta \approx 82°$, $\gamma \approx 74°$, and $c' \approx 4.8$, $\beta' \approx 98°$, $\gamma' \approx 58°$

#51: (a) yes, 4.4, (b) ≈ 12.3 sq. units   #53: $3 < a < 6$ #55: $c = 3$, $c' = 8$

#57: No, from the previous exercise, α must be $12°$ not $10°$.

SECTION 5.2, PAGE 240

#1: ≈ 7.7 ft   #3: ≈ 81.6 m   #5: ≈ 13.2 mi   #7: ≈ 96.7 mi

#9: $150°, 30°, 150°$; 6 in^2 #11: position: ≈ 34.8 ft up hill from post; cable: ≈ 29.0 ft

#13: (a) $\approx 50°$ (b) ≈ 7 min

CHAPTER REVIEW, PAGE 243

#1: $c \approx 6.2$, $\alpha \approx 24°$, $\beta \approx 126°$ #3: $b \approx 8.5$, $\alpha \approx 40°$, $\gamma \approx 25°$

#5: No such triangle

#7: 2 triangles; $a \approx 18.8$, $\alpha \approx 113°$, $\beta \approx 47°$, and $a' \approx 9.3$, $\alpha' \approx 27°$, $\beta' \approx 133°$

#9: $c \approx 14.8$, $\beta \approx 28°$, $\gamma \approx 82°$ #11: No such triangle

#13: $\alpha \approx 56°$, $\beta = 94°$, $\gamma \approx 30°$

#15: 2 triangles; $c \approx 2.6$, $\alpha \approx 80°$, $\gamma \approx 59°$, and $c' \approx 1.9$, $\alpha' \approx 100°$, $\gamma' \approx 39°$

#17: 4.7 mi #19: 60.3 ft #21: 5.5°

CUMULATIVE REVIEW, PAGE 244

#1: $1024(2-a)^7 b^{\frac{29}{2}}$ #2: $-\dfrac{3x^2 - 4x - 1}{2x^{\frac{3}{2}}}$ #3: $y = -4x + \dfrac{3}{2}$ #4: $\dfrac{1}{2}$ #5: 1

#6: $\sqrt{x-1}$ #7: $f^{-1}(x) = \dfrac{x}{1-x}$ #8:

#10: $\dfrac{\pi}{12}, \dfrac{5\pi}{12}, \dfrac{7\pi}{12}, \dfrac{11\pi}{12}, \dfrac{13\pi}{12}, \dfrac{17\pi}{12}, \dfrac{19\pi}{12}, \dfrac{23\pi}{12}$

#11: $\gamma = 77°$, $a \approx 14.7$, $b \approx 23.8$ #12: $c = 124$, $\alpha \approx 30°$, $\beta \approx 122°$

#13: two triangles: $\beta \approx 35°$, $\gamma \approx 121°$, $c \approx 25.3$ and $\beta' \approx 145°$, $\gamma' \approx 11°$, $c' \approx 5.6$

#14: two triangles: $\alpha \approx 50°$, $\beta \approx 92°$, $b \approx 66.6$ and $\alpha' \approx 130°$, $\beta' \approx 12°$, $b' \approx 13.9$

#15: $\approx 46°$ #16: #17: $\dfrac{7}{8}$ #18: $-\dfrac{1}{\sqrt{3}}$ #19: $\dfrac{1}{\sqrt{2}}$

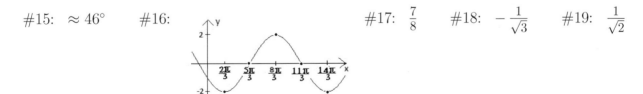

#20: -1 #21: 2 #22: $\dfrac{3x - 1}{\sqrt{6x - 9x^2}}$

CHAPTER 6

SECTION 6.1, PAGE 256

#1: $(-2, 3)$ #3: $(2, 2)$ #5: $(-1, -3)$

#7:   vertex: $(4, -4)$
      x-ints: 2, 6
      y-int: 12
      $R_f = [-4, \infty)$

#9:   vertex: $(-5, 16)$
      x-ints: $-7, -3$
      y-int: -84
      $R_f = (-\infty, 16]$

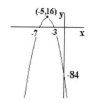

#11: vertex: $(1, 1)$
      x-ints: $\dfrac{3 \pm \sqrt{3}}{3}$
      y-int: -2
      $R_f = (-\infty, 1]$

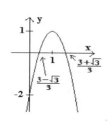

#13: vertex: $(1, 3)$
      x-ints: none
      y-int: 5
      $R_f = [3, \infty)$

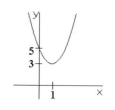

#15: $y = \dfrac{4}{9}x^2 - \dfrac{16}{9}x - \dfrac{11}{9}$ #17: $y = -\dfrac{1}{4}(x+3)^2 + 1$ #19:   2

#21: 50 units, 9325 units   #23: 20 units, 1313 units

#25: $\left(x = -\dfrac{1}{2}, y = -8\right), (x = 1, y = -5)$ #27: $(x = \pm 1, y = -1)$

#29: $(x = \pm 2, y = -3), (x = \pm 3, y = 2)$ #31: $(6, \sqrt{5}), (2, -\sqrt{5})$

#33: 35 ft $\times$ 65 ft   #35: $\left(\alpha = \dfrac{3\pi}{2}, \beta = \pi\right)$

#37: $\left(\alpha = \dfrac{\pi}{4}, \beta = \dfrac{3\pi}{2}\right), \left(\alpha = \dfrac{3\pi}{4}, \beta = \dfrac{3\pi}{2}\right), \left(\alpha = \dfrac{5\pi}{4}, \beta = \dfrac{3\pi}{2}\right), \left(\alpha = \dfrac{7\pi}{4}, \beta = \dfrac{3\pi}{2}\right)$

#39: $\left(\alpha = \dfrac{\pi}{2}, \beta = \dfrac{\pi}{2}\right), \left(\alpha = \dfrac{\pi}{2}, \beta = \dfrac{3\pi}{2}\right)$ #41:   min: 2   #43:   min: 2

#45: max: $\dfrac{9}{4}$ #47: $40\sqrt{2}$ ft/sec   #49: 112 ft/sec   #51: 10,000 ft^2 #53: 250 units

#55: $31,730   #57: 55 trees   #59: $25/month   #61: For ex: $x^2 - 3x + 4$

#63: (a) 0, 1, or 2 (b) 0, 1, or 2 (c) 0, 1, 2, 3, or 4   #65: For ex: $a = 0, b = \dfrac{2}{3}$

#67: For ex: $a = 0, b = 1$

#69:  #71: 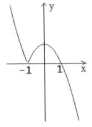 #73: 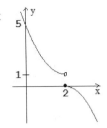 #75: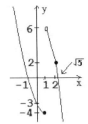

appears 1-1   appears not 1-1

#77: (a) 6 miles

(b) $h(t) = \begin{cases} t^2 + 2, & 0 \le t \le 1 \\ t + 2, & 1 < t < 4 \\ 6, & t \ge 4 \end{cases}$

(c)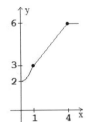

(d) $h(t) = \begin{cases} t^2 + 2, & 0 \le t \le 1 \\ 0.1t^2 + 2.9, & 1 < t < 4 \\ 4.5, & t \ge 4 \end{cases}$

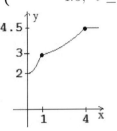

SECTION 6.2, PAGE 271

#1: -2 #3: ± 1 #5: $0, \pm 2$ #7: $0, \dfrac{4 \pm \sqrt{10}}{2}$ #9: 1

#11: $4, -1, 0, 3$ #13: $\pm \dfrac{2}{\sqrt{3}}$ #15: $\pm 2\sqrt{2}$ #17: $\pm \dfrac{1}{\sqrt{3}}, \pm \sqrt{5}$

#19: $-1, 3, 4$ #21: $1, \dfrac{3 \pm \sqrt{15}}{2}$ #23: $\dfrac{1}{2}, \dfrac{1}{3}, \dfrac{2}{3}$ #25: $\dfrac{1}{3}, \pm \dfrac{2}{3}$

#27: $-\dfrac{1}{2}, \dfrac{3}{5}, -\dfrac{1}{3}$ #29: $2, -3, 5, -4$ #31: $\dfrac{\pi}{2}, \dfrac{3\pi}{2}, \dfrac{3\pi}{4}, \dfrac{7\pi}{4}, \dfrac{\pi}{3}, \dfrac{4\pi}{3}$

#33: $0, \pi, \dfrac{\pi}{2}$ #35: $\left[-\dfrac{7}{3}, \dfrac{7}{3}\right]$; 6   #37: $[-5, 5]$; 4   #39: $\left[-\dfrac{9}{5}, \dfrac{9}{5}\right]$; 4

#41: $[-4, 4]$; 7   #43: $\approx -0.35, 0.94, 11.41$ #45: $\approx -1.46, 4.79$

#47: (a) at 4.19 hrs and 8.48 hrs   (b) 104.3°   (c) 96.9°

#49: (a) 7.4 parts/million; after 3.9 hrs   (b) when given, and after 6.8 hrs

#51: $\approx$ 4.1 in. × 20.4 in. × 3.4 in. high   #53: ≈ 0.746 units

#55: (a) min: 60 ft, max $\approx$ 138.7 ft   (b) $\dfrac{25}{14}$ sec; ≈ 74 ft   (c) down

#57: (a) 1, 0   (b) 3   (c) 1, 0.89   (d) -0.43

#59: (a) 1, -0.26 (b) 1.15   (c) 2, -1.15 and 0.54 (d) -1.35 and 0.49

C-28   **Answers to Odd-Numbered Exercises**

#61: For ex: $x(x - 0.001)(x - 50)$

#63: (a) $x^6 - 30x - 4 = 0$ (b) $x = 2$ (c) $[-31, 31]$ (d) none

#65: True: Since $p(c) = 0 \Rightarrow p(c)q(c) = 0$ #67: True: p((c-1)+1)=p(c)=0

#69: False: since it is true for some $p(x)$ and c (for ex: $P(x) = x$ and $c = 0$), false for others (for ex: $p(x) = x - 1$ and $c = 1$)

SECTION 6.3, PAGE 284

#1: $\emptyset$ #3: $(-\infty, -2] \cup \left[\frac{2}{3}, \infty\right)$ #5: $(-\infty, -6) \cup (5, \infty)$ #7: $(-1 - \sqrt{10}, -1 + \sqrt{10})$

#9: $(-\sqrt{7}, \sqrt{7})$ #11: $(-\infty, -1) \cup \left(\frac{1}{3}, \infty\right)$ #13: $\left(-\infty, \frac{1}{5}\right) \cup \left(\frac{2}{3}, 3\right)$ #15: $[-2, 5]$

#17: $(-\infty, -3] \cup \left[\frac{1}{2}, \infty\right)$ #19: $(-\infty, -2] \cup [1, \infty) \cup \{0\}$ #21: $(-\infty, -3) \cup (4, \infty)$

#23: $[3, \infty) \cup \{2\}$ #25: $(1, \infty)$ #27: $\left(\frac{1}{4}, \frac{1}{3}\right) \cup \left(\frac{1}{3}, \frac{1}{2}\right)$

#29: $\left(-\infty, -\frac{1}{\sqrt{2}}\right] \cup \left[\frac{1}{\sqrt{2}}, \infty\right) \cup \{0\}$ #31: $(-\infty, -1) \cup (1, \infty)$

#33: $\left[\frac{-1 - \sqrt{3}}{2}, \frac{-1 + \sqrt{3}}{2}\right] \cup [1, \infty)$ #35: $(-4, 0) \cup (4, \infty)$

#37: $(-\infty, -1) \cup (-1, 0) \cup (1, \infty)$ #39: $\left[0, \frac{\pi}{2}\right] \cup [\pi, 2\pi)$ #41: $\left[\frac{\pi}{6}, 2\pi\right)$

#43: $\left[0, \frac{\pi}{4}\right] \cup \left[\frac{3\pi}{4}, \frac{5\pi}{4}\right] \cup \left[\frac{7\pi}{4}, 2\pi\right)$ #45: $\left[0, \frac{\pi}{3}\right] \cup \left[\frac{\pi}{2}, \frac{3\pi}{2}\right] \cup \left[\frac{5\pi}{3}, 2\pi\right)$

#47: $\approx (-\infty, -1.27) \cup (2.24, \infty)$ #49: $\approx (-0.48, 0.25) \cup (1.22, \infty)$

#51: Between $5\sqrt{2}$ feet and $5\sqrt{3}$ feet, inclusive

#53: For the first half second, and between $\frac{3}{2}$ seconds and 3 seconds after it was thrown

#55: (a) $(0, 4)$ (b) $\frac{32}{3}x - \frac{2}{3}x^3$ (c) between 1 in. and 3.41 in.

#57: (a) $V = \frac{5}{12}x(425 - 10x^2)$ (b) $(0, \sqrt{42.5}$ (c) $(0, 0.86) \cup (6.05, 6.52)$

#59: (a) $0 \le t < 3.42$, and $9.56 < t \le 10$ (b) $3.86 < t < 9.36$

SECTION 6.4, PAGE 293

#1: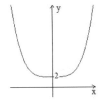

#3: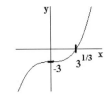

#5: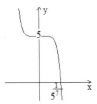

#7: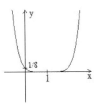

#9: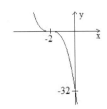

#11:

#13: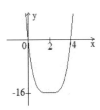

#15: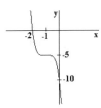

#17: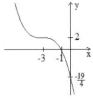

#19: $g(x) = x^4$ #21: $g(x) = x^7$ #23: $g(x) = 24x^4$

#25: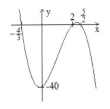

#27: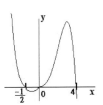

#29: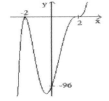

#31:

#33: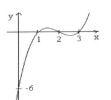

#35: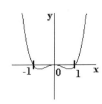

#37: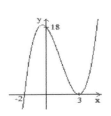

#39: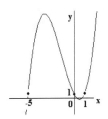

#41: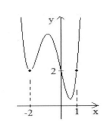

C–30   Answers to Odd-Numbered Exercises

CHAPTER REVIEW, PAGE 297

#1: vertex: $\left(\frac{3}{4}, -\frac{23}{8}\right)$; maximum: $-\frac{23}{8}$

#3: vertex: $(3, -4)$
x-ints: $1, 5$
y-int: 5

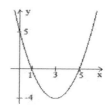

#5: $(x = 2, y = 4)$, $(x = -1, y = -5)$

#7: $\left(\alpha = \frac{\pi}{3}, \beta = \frac{7\pi}{6}\right)$, $\left(\alpha = \frac{2\pi}{3}, \beta = \frac{7\pi}{6}\right)$, $\left(\alpha = \frac{4\pi}{3}, \beta = \frac{7\pi}{6}\right)$, $\left(\alpha = \frac{5\pi}{3}, \beta = \frac{7\pi}{6}\right)$,
$\left(\alpha = \frac{\pi}{3}, \beta = \frac{11\pi}{6}\right)$, $\left(\alpha = \frac{2\pi}{3}, \beta = \frac{11\pi}{6}\right)$, $\left(\alpha = \frac{4\pi}{3}, \beta = \frac{11\pi}{6}\right)$, $\left(\alpha = \frac{5\pi}{3}, \beta = \frac{11\pi}{6}\right)$

#9: $2, \frac{-5 \pm \sqrt{33}}{4}$ #11: $\pm 2\sqrt{2}, -2, 4$ #13: $\pm 1, 1 \pm \sqrt{3}$ #15: $\frac{1}{2}, \frac{3}{4}, -\frac{3}{2}$

#17: ± 1 #19: $-2, 1$ #21: $\left(-\infty, -\frac{5}{3}\right) \cup \left(-\frac{4}{5}, \infty\right)$ #23: $\left[-\frac{1}{2}, \frac{1}{4}\right]$

#25: $\left(-\infty, 1 - \sqrt{2}\right] \cup \left[1 + \sqrt{2}, \infty\right)$ #27: $(-\infty, \infty)$ #29: $(-\infty, -2) \cup (1, 2)$

#31: $\left(-\infty, -\frac{5}{2}\right) \cup \left(-\frac{5}{2}, \frac{1}{3}\right) \cup (1, \infty)$ #33: $\left[\frac{3 - \sqrt{13}}{2}, 1\right] \cup \left[\frac{3 + \sqrt{13}}{2}, \infty\right)$

#35: $\left[0, \frac{\pi}{6}\right) \cup \left(\frac{\pi}{2}, \frac{5\pi}{6}\right)$ #37: $-2.46, \quad 0.17, \quad 2.46$ #39: $(-0.89, \infty)$

#41: $\approx (-1.38, \infty)$ #43: $\left[-\frac{7}{2}, \frac{7}{2}\right]; \quad 7$

#45: x-int: none
y-int: -54

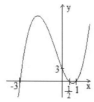

#47: x-int: $\frac{1}{2}$
y-int: -7

#49: $6x^3$
#51: $-128x^6$

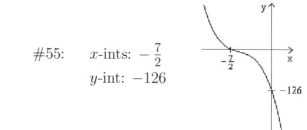

#53: x-ints: $-3, \frac{1}{2}, 1$
y-int: 3

#55: x-ints: $-\frac{7}{2}$
y-int: -126

#57: x-ints: $\pm 1, 2$
y-int: -2

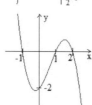

#59: y-int: 8

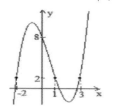

#61: $R(x) = -x^2 + 450x$; 197 units   #63: 144 ft   #65: 100 ft; 3 sec   #67: $\frac{1}{2}$

#69: $\frac{300}{7}$ ft by 200 ft, with the 200 ft parallel to the river

#71: 25,000 ft^2 #73: $50   #75: ≈ $17,980

#77: (a) 39 to 185 units   (b) 73 to 168 units   (c) less than 39 or more than 185 units

#79: at most 1 in., or between 4.42 in. and 5 in.

#81: between $\frac{15}{\pi^{\frac{1}{3}}} \approx 10.24$ ft and 20 ft

CUMULATIVE REVIEW, PAGE 301

#1: $\dfrac{2a+1}{1-a^2}$ #2: $16\sqrt{2}(a+b)^{\frac{3}{2}}$ #3: ≈ 124.9 lb^2/in.   #4: $\dfrac{7\pi}{6}, \dfrac{11\pi}{6}, \dfrac{3\pi}{2}$

#5: $D_f = (-\infty, \infty)$ #6: $\left(x = \dfrac{1}{3}, y = \dfrac{2}{9}\right), \left(x = -2, y = -\dfrac{11}{3}\right)$
   $R_f = [-2, 4]$

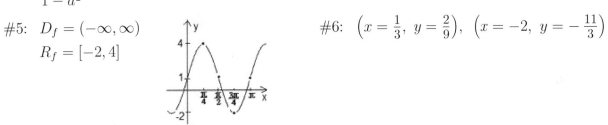

#7: $\pm\dfrac{1}{2}$ #8: 0, 2, −2, 4   #9: $\pm 1, -\dfrac{3}{2}$ #10: $0, -\dfrac{2}{5}, \dfrac{3}{5}$ #11: $\dfrac{1}{2}, -\dfrac{2}{3}$

#12: (a) $-\dfrac{1}{2}$ (b) π (c) $\dfrac{\pi}{4}$ (d) $\dfrac{4}{5}$ (e) $\dfrac{1}{8}$ #13: $(-\infty, -2] \cup [2, \infty)$

#14: $\left(0, \dfrac{\pi}{2}\right] \cup \left[2, \dfrac{3\pi}{2}\right]$ #15: $\left[\dfrac{3-\sqrt{33}}{4}, \dfrac{3+\sqrt{33}}{4}\right]$ #16: $(-\infty, -2) \cup (-\sqrt{3}, 1) \cup (\sqrt{3}, \infty)$

#17: $\left(-1, -\dfrac{2}{3}\right) \cup \left(-\dfrac{2}{3}, 2\right)$ #18: $\left[\dfrac{\pi}{3}, \dfrac{5\pi}{3}\right]$ #19:   −101.30,   0.21,   1.09

#20: $(-1.88, 0.35) \cup (1.53, \infty)$ #21:   Vertex: $(-2, 2)$
                                                    x-ints: $-2 \pm \sqrt{2}$
                                                    y-int: -2

#22: −13 is the minimum value   #23: x-int: 2
                                     y-int: −4

C–32   Answers to Odd-Numbered Exercises

#24: x-ints: $\frac{1}{2}$, −3, 4   #25: x-ints: $-\frac{3}{4}$, 0, $\frac{1}{2}$
       y-int: 12   y-int: 0

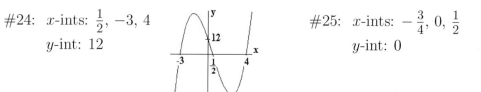

#26: $14.50   #27: $\frac{256}{3} \approx 85.3$ ft/sec   #28: ≈ 349.9 in^3

CHAPTER 7

SECTION 7.1, PAGE 315

#1: $(-\infty, -1) \cup (-1, 2) \cup (2, \infty)$ #3: $(-\infty, -\frac{3}{2}) \cup (-\frac{3}{2}, \frac{4}{3}) \cup (\frac{4}{3}, \infty)$

#5: $(-\infty, 0) \cup (0, \frac{3}{5}) \cup (\frac{3}{5}, \frac{5}{2}) \cup (\frac{5}{2}, \infty)$ #7: $(-\infty, -\frac{1}{3}) \cup (-\frac{1}{3}, \infty)$

#9: $-\frac{5}{3}$ #11: $\frac{1 \pm \sqrt{15}}{2}$ #13: 5   #15: 25   #17: $-\frac{1}{3}$ #19: ± 1 #21: $5 \pm \sqrt{2}$

#23: $\frac{7\pi}{6}, \frac{11\pi}{6}$ #25: $\frac{\pi}{3}, \frac{2\pi}{3}, \frac{4\pi}{3}, \frac{5\pi}{3}$ #27: $\tan^{-1} \frac{2}{3}, \pi + \tan^{-1} \frac{2}{3}, \frac{3\pi}{4}, \frac{7\pi}{4}$

#29: ≈ 11.81 in. wide by 15.75 in. high   #31: ≈ 4.05 in. by 4.05 in. by 12.17 in. high

#33: 56.57 ft by 141.42 ft, with the 56.57 ft side parallel to the inside fences

#35: $(-3, \frac{1}{2})$ #37: $(-\infty, -2] \cup (\frac{2}{3}, 2]$ #39: $(-\infty, \frac{2}{3}) \cup (1, \infty)$

#41: $(-1, -\frac{2}{3}) \cup (\frac{1}{3}, \infty)$ #43: $[1, \frac{3}{2}) \cup [\frac{5}{2}, \infty)$ #45: $(-5, -\frac{3}{2}] \cup (-\frac{1}{3}, \infty)$

#47: $(\frac{\pi}{2}, \pi) \cup (\frac{3\pi}{2}, 2\pi)$ #49: $(-\frac{3\pi}{2}, -\frac{\pi}{2}) \cup [0, \frac{\pi}{2}) \cup (\frac{3\pi}{2}, 2\pi] \cup \{\pi, -2\pi\}$

#51: -1.42, -0.83, -0.21, 1.06; $(-\infty, -1.42) \cup (-1, -0.83) \cup (-0.21, 1) \cup (1.06, \infty)$

#53: -2.93, -1.31, 1.50; $(-4, -3) \cup (-2.93, -1.31) \cup (1.50, \infty)$ #55: 0.68; $(0.68, 1)$

#57: 4.34; $(-\infty, -1) \cup (0, 4.34)$

#59: (a) 772 units   (b) between 614 and 1500 units   (c) between 649 and 1343 units

#61: $P - \frac{8 - 2t}{8t^3 + t^2 - 4t}$ #63. $-\frac{1}{25}$ #65: False: Not so if $k(x) \equiv p(x)$

#67: True: $\dfrac{p((c-1)+1)}{q((c-1)+1)} = \dfrac{p(c)}{q(c)} = 0$

#69: False: For ex.: 1 is a solution of $\dfrac{x-1}{x} = 0$, and 0 is not a solution of $\dfrac{x}{x} = 0$, as $\dfrac{x}{x}$ is undefined when $x = 0$.

#75: $\dfrac{625}{3}$ in.^{3}

SECTION 7.2, PAGE 329

#1: $g(x) = -3$ #3: $g(x) = \tfrac{4}{3}x^5$ #5: $g(x) = 0$ #7: $g(x) = -2x$

#9: va: $x = \tfrac{1}{4}$, ha: $y = \tfrac{3}{8}$ #11: va: $x = 1$, oa: $y = -3x - 3$

#13: va: $x = 2$, $x = -2$, $x = 1$, ha: $y = 1$

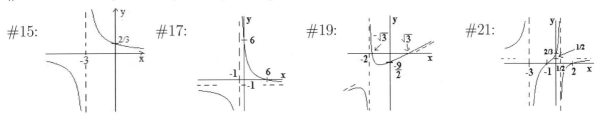

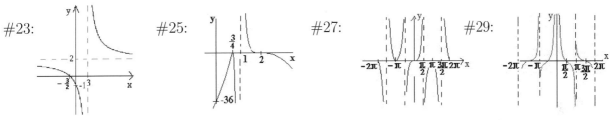

#31: $x = -1.72$, $x = 1.53$

#33: $x = 1.21$

#39: $D_f = (-\infty, 0) \cup (0, \infty)$ #41: $D_f = (-\infty, 0) \cup (0, \infty)$ #43: False: undefined there
no vertical asymptotes   no vertical asymptotes

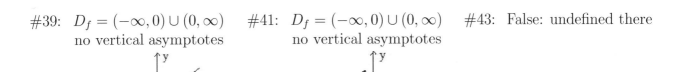

C–34   Answers to Odd-Numbered Exercises

#45: True: For ex.: $f(x) = \dfrac{x-1}{x^2+1}$ (see figure) has $x = 0$ as a horiz. asymptote

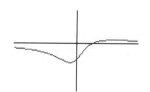

SECTION 7.3, PAGE 338

#1: $\dfrac{A}{x-3} + \dfrac{B}{2x-5}$

#3: $\dfrac{A}{x+1} + \dfrac{B}{(x+1)^2} + \dfrac{C}{(x+1)^3} + \dfrac{D}{x-2} + \dfrac{E}{(x-2)^2}$

#5: $\dfrac{Ax+B}{x^2+1} + \dfrac{C}{x+4} + \dfrac{D}{x-1} + \dfrac{Ex+F}{x^2+2x+3}$

#7: $\dfrac{A}{x-1} + \dfrac{B}{(x-1)^2} + \dfrac{C}{x+1} + \dfrac{D}{(x+1)^2} + \dfrac{Ex+F}{x^2+1} + \dfrac{Gx+H}{(x^2+1)^2}$

#9: $\dfrac{A}{x-1} + \dfrac{B}{(x-1)^2} + \dfrac{C}{(x-1)^3} + \dfrac{D}{(x-1)^4} + \dfrac{Ex+F}{x^2+1}$

#11: $3 + \dfrac{A}{x-3} + \dfrac{B}{x+1} + \dfrac{C}{(x+1)^2}$

#13: $\dfrac{4}{x-3} + \dfrac{2}{x+2}$

#15: $\dfrac{2}{x+1} - \dfrac{1}{x-1} + \dfrac{5}{x-2}$

#17: $\dfrac{8}{x-5} + \dfrac{2}{(x-5)^2} - \dfrac{1}{x+1}$

#19: $-\dfrac{2}{x} + \dfrac{3x-1}{x^2+x+1}$

#21: $-\dfrac{1}{x^2+3} + \dfrac{3}{x+1} + \dfrac{1}{(x+1)^2}$

#23: $\dfrac{4x-3}{x^2+x+1} + \dfrac{2x}{(x^2+x+1)^2}$

#25: $\dfrac{6x+1}{x^2+1} + \dfrac{-2x+1}{(x^2+1)^2}$

#27: $5x - 13 + \dfrac{21}{x+1} - \dfrac{7}{(x+1)^2}$

#29: $-x + 5 - \dfrac{9}{3x-1} + \dfrac{13x+1}{x^2+1}$

CHAPTER REVIEW, PAGE 342

#1: $\left(-\infty, -\tfrac{1}{2}\right) \cup \left(-\tfrac{1}{2}, \infty\right)$ #3: $(-\infty, -1) \cup (-1, 0) \cup (0, \infty)$ #5: $\pm\sqrt{7}$

#7: $\tfrac{2}{3}, -1$ #9: $-4 \pm \sqrt{15}$ #11: $\pi, \tfrac{2\pi}{3}, \tfrac{4\pi}{3}$ #13: ≈ -2.70 and -1.56

#15: ≈ 25.35 ft by 118.34 ft, with 118.34 ft parallel to river

#17: (a) 119 units through 300 units,   (b) 96.6 units,   (c) 131 units through 300 units

#19: $(-\infty, 4) \cup \left(\tfrac{31}{7}, \infty\right)$ #21: $\left[-\tfrac{2}{5}, 0\right) \cup (1, \infty)$ #23: $(-\infty, -9) \cup (-1, 0)$

#25: $(0, 2)$ #27: $(-\infty, -1) \cup (1.07, \infty)$ #29: 12   #31: $g(x) = 0$

#33: va: $x = -\tfrac{4}{5}$; ha: $y = \tfrac{3}{5}$ #35: va: $x = 0, x = -1, x = 1$; oa: $y = 4x + 2$

#37: 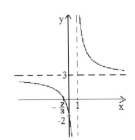   #39: 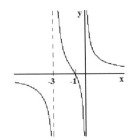   #41: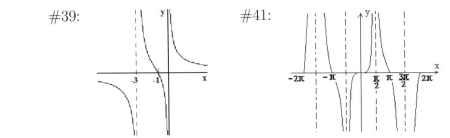

#43: $\dfrac{A}{x} + \dfrac{B}{2x-3} + \dfrac{C}{5x-2}$ #45: $\dfrac{Ax+B}{3x^2+1} + \dfrac{C}{3x-2} + \dfrac{D}{(3x-2)^2} + \dfrac{E}{(3x-2)^3}$

#47: $\dfrac{A}{x+1} + \dfrac{B}{(x+1)^2} + \dfrac{C}{(x+1)^3} + \dfrac{D}{(x+1)^4} + \dfrac{Ex+F}{x^2+x+3}$

#49: $\dfrac{6}{3x-1} - \dfrac{5}{x+3} + \dfrac{2}{(x+3)^2}$ #51: $\dfrac{3}{4(x-5)} - \dfrac{1}{2(x+3)}$

CUMULATIVE REVIEW, PAGE 345

#1: $\dfrac{1}{4(x-a)}$ #2: $\dfrac{-x^2-8x-6}{x(4x+3)^2}$ #3: $\dfrac{1-\pi}{2}$ #4: $\sin^2(\sin^2 x)$ #5: $\dfrac{1}{4}$

#6: $\alpha = \gamma \approx 69°,\ \beta \approx 42°$ #7: x-ints: $2, 4$ y-int: -80

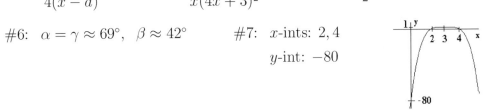

#8: x-ints: $-4 \pm 2\sqrt{5}$
y-int: -2
vertex: $(-4, -10)$

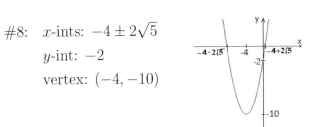

#9: (a) -1 (b) 1 (c) $\dfrac{1}{\sqrt{3}}$

#10: (a) $\pm\dfrac{120}{169}$ (b) $\dfrac{119}{169}$ #11: $\left(x = -\dfrac{5}{2}, y = -\dfrac{7}{4}\right),\ (x = 6, y = 11)$ #12: $\dfrac{10}{3}$

#13: -7 #14: $-\dfrac{1}{2}$ #15: $\dfrac{n\pi}{2}$ #16: $1,\ 4$ #17: $\dfrac{\pi}{3},\ \dfrac{2\pi}{3},\ \dfrac{4\pi}{3},\ \dfrac{5\pi}{3}$

#18: $\left[\dfrac{3-\sqrt{11}}{2}, \dfrac{3+\sqrt{11}}{2}\right]$ #19: $(-\infty, -1) \cup (1, \infty)$ #20: $(-2, -1)$

#21: $[-2, -1) \cup \left[\dfrac{1}{2}, \infty\right)$ #22: $(-15, 4)$

#23: x-ints: -2
y-int: 2
ha: $y = -1$
va: $x = 1$
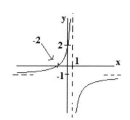

#24: x-int: $-\frac{1}{4}$
y-int: $-\frac{1}{6}$
ha: $y = 0$
va: $x = -3$, $x = 2$
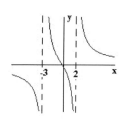

#25: $D_f = \left(\frac{1}{2}, \frac{5}{2}\right] \cup (4, \infty)$

#26: $\dfrac{3}{x-2} - \dfrac{1}{(x-2)^2} + \dfrac{2}{x+3}$

#27: $4x - 1 + \dfrac{4x-3}{x^2+1} + \dfrac{6}{x}$

#28: 3364.71 in.^{2}

#29: 8.255 in. by 16.510 in. by 11.006 in.

CHAPTER 8

SECTION 8.1, PAGE 354

#1: 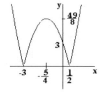 #3: 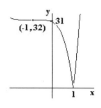 #5: 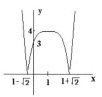 #7: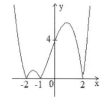

#9: 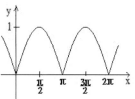 #11: $y = 4x - 1$ $y = |4x - 1|$ $y = 4|x| - 1$
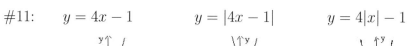

#13: $y = 3x^2 - 12$ $y = |3x^2 - 12|$ $y = 3|x|^2 - 12$ #15: $-2.94, -0.44, 0.38$
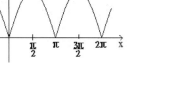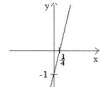

#17: $-0.84, 1.68$

#19: F #21: F

#23: F #25: T

#27: T #29: 3, 7

#31: $-\dfrac{11}{3}, -1$ #33: $\pm 2, \pm\sqrt{2}$ #35: 2, 0 #37: $\emptyset$ #39: $-1, \dfrac{5}{7}$

#41: $0, \dfrac{\pi}{2}, \dfrac{3\pi}{2}, \pi$ #43: $\dfrac{\pi}{4}, \dfrac{3\pi}{4}, \dfrac{5\pi}{4}, \dfrac{7\pi}{4}$ #45: 1 #47: $(-\infty, 5) \cup (5, \infty)$

Answers to Odd-Numbered Exercises   **C–37**

#49: $(0, \infty)$ #51: $\left(-\infty, -\frac{1}{2}\right]$ #53: $(-2, 8)$ #55: $\left[\frac{3}{2}, \frac{7}{2}\right]$ #57: $(-\infty, 6) \cup (6, \infty)$

#59: $(-\infty, -2] \cup [3, \infty)$ #61: $\left[2, \frac{8}{3}\right]$ #63: $\left(-\infty, \frac{5}{4}\right] \cup \left[\frac{7}{4}, \infty\right)$

#65: $(-\infty, -4) \cup \left(-4, -\frac{1}{3}\right) \cup (7, \infty)$ #67: $(1, \infty)$ #69: $\{x \mid x \neq n\pi, n \text{ any integer}\}$

#71:   (a)   between 6.84 sec and 7.08 sec,   (b) between 2.19 sec and 3.14 sec

#73: 3.9 units   #75: 3.9 units   #77: $\begin{cases} 7x - 4 & \text{if } x \geq \frac{4}{7} \\ -7x + 4 & \text{if } x < \frac{4}{7} \end{cases}$

#79: $\begin{cases} 2x - 1 & \text{if } x \geq 0 \\ -2x - 1 & \text{if } x < 0 \end{cases}$

#81:

#83:   min at $x \approx \pm 2.15$,   max at $x \approx 0$

#85:   1.83, 1.98   #87: $-2.5,\ 0.5$ #89: $\approx (0.80, \infty)$

#95: $\left(\frac{1}{2^{\frac{1}{3}}}, \infty\right)$ #97: $\left(-2, \frac{6}{5}\right)$

SECTION 8.2, PAGE 367

#1: 36   #3: 35   #5: $\dfrac{1 + \sqrt{17}}{8}$ #7: 3   #9: $\emptyset$ #11: 5   #13: 3, $\dfrac{3 \pm \sqrt{5}}{2}$

#15: $\dfrac{3\pi}{2}, \dfrac{\pi}{6}$ #17: $\dfrac{\pi}{2}$ #19: π #21: 0, $\dfrac{3\pi}{2}$ #23: $2\pi - \cos^{-1} \dfrac{4}{5}$ #25: $\dfrac{\pi}{8}, \dfrac{5\pi}{8}$

#27: 0   #29: $\emptyset$ #31: 11   #33: $\dfrac{17}{3}$ #35: $-\dfrac{1}{2}, -1$ #37: 8   #39: -2

#41:   no solution: square roots are nonnegative   #43:   0

#45:   no solution: left side is positive, right side is negative or 0   #47: $(-\infty, 1] \cup \{3\}$

#49: $[3, \infty) \cup \{0\}$ #51: $(-\infty, -1) \cup (-1, 0)$ #53: $[-3, 1)$ #55: $(-\infty, \infty)$ #57: $\emptyset$

#59: ≈ 19.83 ft   #61: ≈ 0.75 km from Q #63: between 4,167 and 10,000

CHAPTER REVIEW, PAGE 372

#1: $\dfrac{1}{3}, -3$ #3: $-\dfrac{2}{3}, -8$ #5:   5, $-\dfrac{7}{9}$ #7: $\left[-\dfrac{3}{4}, -\dfrac{1}{4}\right]$

#9: $\left(-\infty, -\frac{5}{2}\right] \cup \left[\frac{1}{2}, \infty\right)$ #11: $\left(-\frac{2}{3}, 0\right) \cup \left(0, \frac{2}{3}\right)$ #13: $\frac{\pi}{6}, \frac{5\pi}{6}, \frac{3\pi}{2}$ #15: $\frac{31}{9}$

#17: $\emptyset$ #19: -64 #21: $\frac{34}{3}$ #23: $\emptyset$ #25: $\frac{5\pi}{6}$ #27: $\emptyset$ #29: $\pm 5\sqrt{5}$

#31: 1

#33: $[-2, 0) \cup (2, \infty)$

#35: $(1, \infty)$

#37:

#39: ≈ 1.06 km from B

CUMULATIVE REVIEW, PAGE 373

#1: $\dfrac{3}{x^{\frac{2}{3}}}$

#2: (a) $2x^2 + 4xh + 2h^2 - 5x - 5h + 1$ (b) $9x - 8$ (c) $18x^2 - 39x + 19$ (d) $\dfrac{2 - 5x + x^2}{x^2}$

#3: $\frac{\pi}{3}, \frac{\pi}{2}, \frac{5\pi}{3}, \frac{3\pi}{2}$ #4: $3 - 14\sin^2 x + 8\sin^4 x$ #5: $\dfrac{-x+1}{x^2+2} + \dfrac{2}{x-1} + \dfrac{3}{x+1}$

#6: $\left(-\infty, \frac{3}{5}\right] \cup (5, \infty)$ #7: $-5, -\frac{4}{5}$ #8: $-3, 2, \dfrac{-1 \pm \sqrt{17}}{2}$ #9: $\frac{3\pi}{2}, \frac{\pi}{2}$

#10: $\left(-\infty, \frac{3}{5}\right] \cup \left[\frac{9}{5}, \infty\right)$ #11: $(1 - \sqrt{6}, -1) \cup (3, 1 + \sqrt{6})$ #12: -29 #13: $-\frac{1}{5}$

#14: 58   #15: $\frac{\pi}{6}$ #16: $\pm 4\sqrt{2}$ #17: 5   #18: $[0, 9]$

#19: $(-\infty, -4] \cup \{0\}$ #20: $[3, 6]$ #21: x-ints: $-\frac{3}{2}, \frac{4}{5}$
y-int: 12

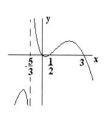

#22: x-int: 1
y-int: 7

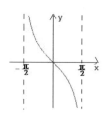

#23: x-ints: $0, \frac{1}{2}, 3$
y-int: 0
vertical asymp:
$x = -\frac{5}{3}$

#24: x-int: 0
y-int: 0
vertical asymp:
$x = \pm \frac{\pi}{2}$

#25: 34.8 sq. units   #26: 3.7 units

CHAPTER 9

SECTION 9.1, PAGE 383

#1: y-int: 3
horiz. asymp: $y = 0$
$R_f = (0, \infty)$

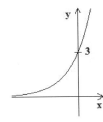

#3: y-int: -1
horiz. asymp: $y = 1$
$R_f = (-\infty, 1)$

#5: y-int: $-\frac{7}{4}$
horiz. asymp:
$y = -2$
$R_f = (-2, \infty)$

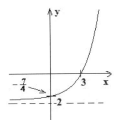

#7: y-int: $\frac{10}{3}$
horiz. asymp: $y = 3$
$R_f = (3, \infty)$

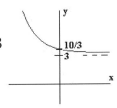

#9: $-\frac{3}{5}$ #11: -6 #13: $\frac{1}{2}, -3$ #15: $-\frac{5}{4}$ #17: $\frac{5}{7}$

#19: $\frac{3\pi}{4}, \frac{7\pi}{4}, \tan^{-1}\frac{2}{3}, \pi + \tan^{-1}\frac{2}{3}$ #21: $\frac{2\pi}{3}, \frac{4\pi}{3}, 0$ #23: -1

#25: ± 1 #27: left side positive, right side negative   #29: left side > 1, right side ≤ 1

#31: $(0, \infty)$ #33: $\left(-\infty, -\frac{1}{2}\right]$ #35: $\left(-3, \frac{1}{2}\right)$ #37: $\left(-\infty, -\frac{1}{2}\right) \cup (5, \infty)$

#39: $\left[\frac{1}{2}, \frac{5}{2}\right]$ #41: $(-\infty, -1) \cup \left(-\frac{1}{2}, \infty\right)$ #43: ≈ 1.10 #45: $0, \approx 0.68$

#47: $N(x) = 300(0.5749)^{\frac{1}{10}x}$, ≈ 4777.05 gm in 1900, ≈ 1.18 gm in 2050

#51: local max ≈ 1.04, no x-int, y-int $\frac{3}{4}$ #53: local min ≈ -0.37, x-int $=$ y-int $= 0$

#55: $(x = 0, y = 0)$ #57: False: For ex.: $f(0) = 1, g(0) = 2$

#59: True: Only part lies below   #61: True: $y = b^x + c$ lies above $y = b^x - c$

#63: True: Domain=Range $= (-\infty, \infty)$ #65: True: Since b must be > 1, $\frac{1}{b} < 1$

#67: True: At $x = 1$ #69: True: $\left(\frac{a}{b}\right)^x = 1$ only for $x = 0$

#71: False: For ex.: $4^x < 2^x$ for $x < 0$

SECTION 9.2, PAGE 397

#1: -3 #3: -1 #5: $\frac{1}{2}$ #7: $64^{\frac{1}{3}} = 4$ #9: $\left(\frac{1}{2}\right)^0 = 1$

#11: $\log_5 125 = 3$ #13: $\log_{\frac{1}{2}} \frac{1}{4} = 2$ #15: ≈ 5.18 #17: 2.74 #19: 4

#21: -2 #23: 13 #25: 2 #27: $\frac{3}{5}$ #29: $(-\infty, -3) \cup (3, 4)$

#31: $(-\infty, 0)$ #33: $\left(-\frac{7}{2}, -\frac{1}{3}\right) \cup (0, \infty)$ #35: $(0, 1) \cup (1, 2) \cup (2, \infty)$

#37: domain: $(0, \infty)$ #39: domain: $(0, \infty)$ #41: domain: $(1, \infty)$

 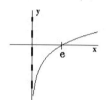 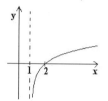

#43: domain: $(3, \infty)$ #45: domain: $(-1, \infty)$ #47: -1 #49: -4

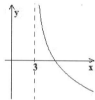 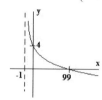

#51: $-\frac{3}{2}$ #53: $\emptyset$

#55: $\frac{\pi}{4}, \frac{5\pi}{4}$ #57: $\frac{7\pi}{6}, \frac{11\pi}{6}$

#59: $\left(\frac{10}{3}, 4\right)$ #61: $(1, \infty)$

#63: $\ln 1000 x^4$ #65: $\log_3 \frac{\sqrt{14}\,(2x-5)}{(x+4)^4}$ #67: $4\log_2 x + 4\log_2 y - 8\log_2 z$

#69: $\frac{1}{2}\log x - \frac{1}{2}\log y + 2\log z$ #71: $2A + B$ #73: $2A - 2B + C$ #75: $\frac{47}{8}$

#77: 1 #79: $\ln \sin 2x$ #81: $\frac{1}{\ln \tan x}$ #83: $\frac{\ln \frac{3}{5}}{\ln 2}$ #85: $-\frac{\ln 3x^{16}}{\ln 2}$ #87: $\frac{2}{3}$

#89: 25 #91: 0 #93: 1 #95: ≈ 0.82 #97: ≈ 0.88 #99: $\frac{1}{2}$ #101: $\frac{1}{3}$

#103: False: $\ln x$ undefined when $x < 0$ #105: False: $\ln x$ only defined for $x > 0$

#107: True: As $|x|$ is even #109: False: $f(0)$ is undefined, $g(0) = \ln 10$

#111: False: For ex.: Let $x = y = 10$ #113: False: For ex.: Let $x = y = b$

#115: True: By Change of Base Formula #117: True: By Theorem 9.6(iii)

#119: True: By Theorem 9.6(ii)

Answers to Odd-Numbered Exercises **C–41**

SECTION 9.3, PAGE 408

#1: $\dfrac{5\ln 3 + \ln 12}{\ln 27}$ #3: $\dfrac{\ln 25}{\ln 10}$ #5: $\dfrac{\ln 108}{\ln \frac{16}{243}}$ #7: $-\dfrac{\ln 24}{\ln \frac{4}{3}}$ #9: $\dfrac{\ln 36}{\ln \frac{20}{3}}$

#11: -2; $\dfrac{\ln \frac{5}{6}}{\ln 2}$ #13: $\dfrac{-1 + \ln\left(\frac{2\pi}{3} + 2n\pi\right)}{4}$, $\dfrac{-1 + \ln\left(\frac{4\pi}{3} + 2n\pi\right)}{4}$, integer $n \geq 0$

#15: $\dfrac{2 - \ln\left(\frac{5\pi}{6} + 2n\pi\right)}{3}$, $\dfrac{2 - \ln\left(\frac{7\pi}{6} + 2n\pi\right)}{3}$, integer $n \geq 0$ #17: $-\dfrac{1}{2}\ln 3$

#19: x-int: -1
y-int: 12
horiz. asymp:
$\quad y = -6$

#21: $\left(-\infty, \dfrac{\ln 160}{\ln 8}\right)$ #23: $\left(\dfrac{\ln 25}{\ln 10}, \infty\right)$

#25: $\left(-\infty, \dfrac{\ln \frac{243}{20}}{\ln \frac{3}{4}}\right]$ #27: e^8 #29: 9 #31: 28 #33: 4 #35: $\dfrac{3\sqrt{6}}{2}$ #37: 8

#39: $\dfrac{3}{2}$ #41: 7 #43: $\dfrac{1}{4}, -\dfrac{1}{2}$ #45: $\dfrac{7}{4}$ #47: $\emptyset$ #49: $e^{-\frac{1}{3}} - 3$, $e - 3$

#51: $\sin^{-1}\dfrac{3}{4}$ #53: $\pm\dfrac{\pi}{3}$ #55: $1 + e^{\frac{\pi}{3} + n\pi}$, any integer n

#57: $e^{\sin^{-1}\frac{1}{4} + 2n\pi}$, $e^{-\sin^{-1}\frac{1}{4} + (2n+1)\pi}$, any integer n #59: $2 + e^{-\frac{3}{4}}$

#61: x-int: 14
y-int: -3
vertical asymp:
$\quad x = -2$

#63: $(0, e^4)$ #65: $[103, \infty)$
#67: $(3, \infty)$ #69: $\emptyset$
#71: $(0, 1) \cup (3, \infty)$

#73:   For ex: $[50, 55] \times [-700, -500]$ #75:   For ex: $[5, 7] \times [49, 51]$

#77: ≈ 342.67, ≈ 357.35 #79: $\approx (25, 42.48)$ #81: $\log 2 \approx 0.3$

#83: $\ln 7.3 \approx 2.0$ #85: $1 + (\sqrt{2})^\pi \approx 4.0$ #87: $\dfrac{1}{3}(\ln\ln 12 - \ln\ln 3 + 2) \approx 0.9$

#89: 1.85; $(1.85, \infty)$ #91: 0.29; $(0.29, \infty)$ #93: 0.55; $(-\infty, 0.55)$

#95: 4.56; $(-1, 4.56)$ #97: -4.08; $(-5, -4.08)$ #99: $\dfrac{\ln 26}{\ln 32}$

#101: $(x = -5, y = -13), (x = 2, y = 1)$ #103: $e^{-4} - 3$ #105: $e^3 - 2$

SECTION 9.4, PAGE 420

#1: 0.863   #3: (a) 6.93 min   (b) 24,365 bacteria   (c) 736 bacteria

#5: (a) 329.4 million   (b) 233.8 million   (c) 2014   (d) 2050

#7: (a) 148 years   (b) 216.14 grams   #9: 121 years   #11: 600 years   #13: ≈ 2.9 min

SECTION 9.5, PAGE 430

#1: $2,427.10   #3: $19,055.60; $19,121.80   #5: (a) 1.61 years;   (b) 1.59 years

#7: $3,101.20   #9: $95,398.45   #11: ≈ 18.45 decibels

#13: SF quake about 50 times more intense   #15: 29.3 days

#17: (a) (i) 8.2   (ii) 3.5   (iii) 7.4   (b) (i) 2.51×10^{-7} (ii) 0.005012   (iii) 1.26×10^{-10}

#19: $8,193.96   #21: $131.69

#23: (a) 1495   (b) As $t \to \infty$, $\left(\frac{1}{5}\right)^t \to 0$ so that $P(t) \to 1500$; As time advances, the population approaches its maximum of 1500 pheasants   (c) ≈ 1.8 quarters

CHAPTER REVIEW, PAGE 439

#1: $D_f = (-\infty, \infty)$
$R_f = (-1, \infty)$
x-int: -2
y-int: 3
horizontal asymp:
$y = -1$

#3: $D_f = (3, \infty)$
$R_f = (-\infty, \infty)$
x-int: 7
y-int: none
vertical asymp:
$x = 3$

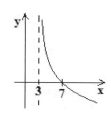

#5: $\log_3 81 = 4$ #7: $2^{-2} = \frac{1}{4}$ #9: 2   #11: -3 #13: $\log_4 \frac{75}{x}$

#15: $\frac{3}{2} \ln x + \frac{1}{4} \ln y$ #17: -2.1333 #19: 507.58   #21: $\pm \frac{\sqrt{37}}{2}$ #23: $\frac{7}{5}$

#25: $\frac{\pi}{3}, \frac{2\pi}{3}, \frac{4\pi}{3}, \frac{5\pi}{3}$ #27: $\frac{\ln 1200}{\ln 320}$ #29: $\frac{1 + \ln\left(\frac{\pi}{4} + n\pi\right)}{6}$, integer $n \geq 0$ #31: 3

#33: $\frac{\pi}{6} + n\pi, \frac{5\pi}{6} + n\pi$ #35: 5   #37: 63, $-\frac{3}{4}$ #39: $\frac{\pi}{6} + 2n\pi$ #41: $\left(\frac{2}{3}, \infty\right)$

#43: $\left(-\infty, \frac{\ln 6}{\ln 3}\right]$ #45: $\left[\frac{\ln 2}{\ln \frac{81}{4}}, \infty\right)$ #47: $\left(\frac{9}{2}, \infty\right)$ #49: $[-1, \infty)$ #51: $-10.52, -3.51, 2.10$

Answers to Odd-Numbered Exercises   **C–43**

#53: $297.42, 302.59$ #55: $(-\infty, 2.26) \cup (3.91, \infty)$ #57: $\approx 379,420$ atoms

#59: ≈ 5.66 days #61: (a) ≈ 9.98 yrs (b) ≈ 9.79 yrs (c) ≈ 9.75 yrs (d) ≈ 9.73 yrs

#63: False: $f(x) = 2^x$ #65: True: e^x is defined for all x

#67: True: $\ln \ln x$ is defined for $\ln x > 0$ #69: True: LHS $= 10^{\log xy}$

CUMULATIVE REVIEW, PAGE 441

#1: $256 a^{\frac{20}{3}} b^{16}$ #2: $\dfrac{2(1 - 3x^2)}{(1 + x^2)^3}$ #3: $\dfrac{4e^x}{e^{2x} + 1}$ #4: $-\dfrac{2}{x} + \dfrac{3x - 1}{x^2 + x + 1}$

#5: $\dfrac{3x}{\sqrt{16 - 9x^2}}$ #6: 12 #7: $(-\infty, -2) \cup (-1, 0) \cup (0, \infty)$

#8: $\dfrac{3}{7}$ #9: $\dfrac{\ln 8}{2 + \ln 2}$ #10: -2 #11: $\dfrac{2\pi}{3}, \dfrac{4\pi}{3}, \dfrac{\pi}{2}, \dfrac{3\pi}{2}$ #12: $\left(\dfrac{\ln 2}{\ln 81}, \infty \right)$

#13: 2 #14: 20 #15: $\dfrac{\pi}{6}$ #16: $[-4, 4)$

#17: amplitude: 4
period: $\dfrac{2\pi}{3}$
phase shift: $-\dfrac{\pi}{3}$

#18: $D_f = (-\infty, \infty)$,
$R_f = (-1, \infty)$

#19: decreasing, one-to-one, invertible, $h^{-1}(x) = -1 - 2x^{\frac{1}{3}}$

#20: increasing, one-to-one, invertible, $k^{-1}(x) = 3^x - 2$

#21: (a) $\dfrac{1}{\sqrt{3}}$ (b) 2 (c) 0 (d) 2 #22: $3A + 2C - 2B$ #23: $\ln \dfrac{(x - 1)^2 (x + 1)^3}{\sqrt{x}}$

#24: $\dfrac{2}{3} \log_{25} x + \dfrac{1}{2} \log_{25}(x - 1) - \dfrac{1}{2} - 4 \log_{25}(x + 2)$ #25: $1 + 2^{\frac{\pi}{2}} \approx 3.9707$

#26: (a) 5 (b) $L = L_0 \cdot 10^{\frac{2}{5}(6 - m)}$ #27: ≈ 1.7 days #28: a billion times #29: 18.4 yrs

APPENDIX A, PAGE A-11

#1: ellipse   #3: hyperbola   #5: parabola   #7: parabola   #9: circle

C–44   Answers to Odd-Numbered Exercises

#11: circle
      center: $(-3, 2)$
      radius: 5

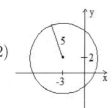

#13: ellipse
      center: $(3, 2)$
      vertices:
      $(3, 7), (3, -3)$
      endpts minor axis:
      $(1, 2), (5, 2)$

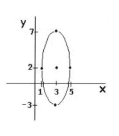

#15: parabola
      vertex: $(-2, 1)$

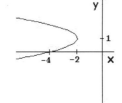

#17: ellipse
      center: $(0, 0)$
      vertices: $(0, \pm 1)$
      endpts minor axis:
      $\left(\pm \frac{1}{2}, 0\right)$

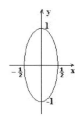

#19: hyperbola
      center: $(3, 2)$
      vertices: $(5, 2), (1, 2)$
      asymptotes:
      $y - 2 = \pm(x - 3)$

#21: parabola
      vertex: $(1, -2)$
      x-int: 13
      y-int: none

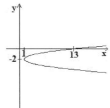

#23: parabola
      vertex: $(8, 5)$
      x-int: -42
      y-ints: 3, 7

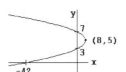

#25: parabola
      vertex: $(-1, 9)$
      x-ints: $-4, 2$
      y-int: 8

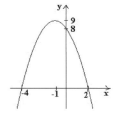

#27: parabola
      vertex: $\left(\frac{1}{3}, 0\right)$
      x-int: $\frac{1}{3}$
      y-int: $\frac{1}{2}$

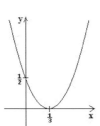

#29: circle
      center: $(-1, -1)$
      radius: 1

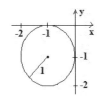

#31: circle
      center: $(-1, 1)$
      radius: 3

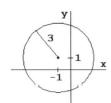

#33: circle
      center: $(3, 0)$
      radius: 4

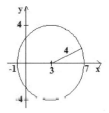

#35: degenerate

#37: circle
　　center: $(2, 3)$
　　radius: $\sqrt{5}$

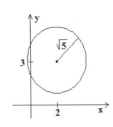

#39: ellipse
　　center: $(-4, 1)$
　　vertices:
　　$(-7, 1)$, $(-1, 1)$
　　endpts minor axis:
　　$(-4, 3)$, $(-4, -1)$

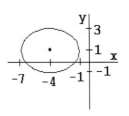

#41: ellipse
　　center: $(0, -3)$
　　vertices:
　　$(0, 0)$, $(0, -6)$
　　endpts minor axis:
　　$(\pm\sqrt{3}, -3)$

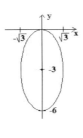

#43: ellipse
　　center: $(-2, 3)$
　　vertices:
　　$(-2, -1)$, $(-2, 7)$
　　endpts minor axis:
　　$(-5, 3)$, $(1, 3)$

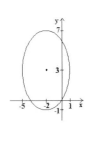

#45: ellipse
　　center: $(1, 0)$
　　vertices:
　　$(5, 0)$, $(-3, 0)$
　　endpts minor axis:
　　$(1, \pm 2)$

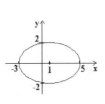

#47: ellipse
　　center: $(2, -1)$
　　vertices:
　　$(-1, -1)$, $(5, -1)$
　　endpts minor axis:
　　$\left(2, -\frac{3}{2}\right)$, $\left(2, -\frac{1}{2}\right)$

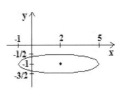

#49: hyperbola
　　center: $(-4, -1)$
　　vertices:
　　$(-4, -5), (-4, 3)$
　　asymptotes:
　　$y + 1 = \pm(x + 4)$

#51: hyperbola
　　center: $(0, -1)$
　　vertices:
　　$(\pm 3, -1)$
　　asymptotes:
　　$y + 1 = \pm\frac{1}{3}x$

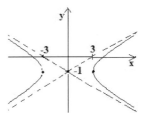

#53: hyperbola
　　center: $(0, 1)$
　　vertices:
　　$(0, -3), (0, 5)$
　　asymptotes:
　　$y - 1 = \pm 4x$

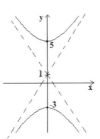

#55: hyperbola
　　center: $(-2, 1)$
　　vertices:
　　$(-4, 1)$, $(0, 1)$
　　asymptotes:
　　$y - 1 = \pm 2(x + 2)$

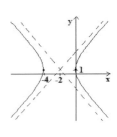

#57: circle

center:

(3, −4)

radius: 5

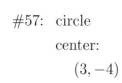

#59: ellipse

center: $(1, -3)$

vertices:

$(-2, -3), (4, -3)$

endpoints minor axis:

$(1, -5), (1, -1)$

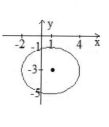

#61: parabola

vertex: $(6, -5)$

y-int: 13

x-ints: $6 \pm \sqrt{10}$

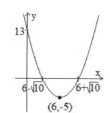

#63: hyperbola

center: $(0, -5)$

vertices:

$(0, -1), (0, -9)$

asymptotes:

$y + 5 = \pm \frac{4}{3} x$

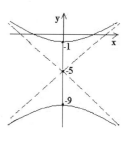

#65: two ellipses; $\left(-\frac{2}{\sqrt{5}}, \pm \frac{2}{\sqrt{5}}\right), \left(\frac{2}{\sqrt{5}}, \pm \frac{2}{\sqrt{5}}\right)$

#67: circle and hyperbola; $(-2, 0), (-2, 2)$ #69: two ellipses; $(2, \pm 1)$

#71: parabola and ellipse; $\left(-\frac{5}{6}, \pm \frac{\sqrt{11}}{3}\right)$ and $(1, 0)$

#73: hyperbola and ellipse; no points of intersection

#75: circle and parabola; $(5, 8), (5, -2)$

#77: two hyperbolas; $(7.95, -6.77), (48.84, 42.28)$

#79: parabola and hyperbola; $(0.26, -0.24), (-0.26, -1.87)$

#81: hyperbola and ellipse; $(-1, \pm 3), (2.17, \pm 2.72)$

#83: ≈ 15.34 #85: $\left(4, \pm \frac{2\sqrt{7}}{3}\right), \left(-4, \pm \frac{2\sqrt{7}}{3}\right)$

#87: False: For ex.: $x^2 - 4x + y^2 = -3 \Rightarrow (x-2)^2 + y^2 = 1$

#89: False: For ex.: Consider $x^2 - 6x + y^2 = 0$ and $x^2 + y^2 = 1$

INDEX

A

Absolute Maximum; Absolute Minimum, 252
Absolute Value, 77, 347
  Equation, 350
  Function, 347
  Inequality, 352
  Properties, 348
Algebraic Expression, 8
Alternate Interior Angle, 134
Ambiguous Case of the Law of Sines, 229
Amplitude, 175
Angle, 125
  Acute, 125
  Alternate Interior, 134
  Central, 125
  Coterminal, 146
  Degree Measure, 125
  Obtuse, 125
  Of Depression, 134
  Of Elevation, 134
  Of Refraction, 140
  Oriented, 145
  Quadrant, 155
  Radian Measure, 144
  Reference, 157
  Right, 126
  Vertex, 125
  Vertical, 134
Approximate Solutions of Equations, 96
Approximate Solutions of Inequalities, 96
Arc Length, 147
Arithmetic of Functions, 79
Asymptote
  Horizontal, 320
  Oblique, 320
  Vertical, 321

B

Bearing, 136, 237
Bel and Decibel, 427

C

Cancellation Law, 4
Cartesian Plane, 88
Center of a Circle, A-5
Center of a Hyperbola, A-6

Center of an Ellipse, A-5
Change of Base Formula, 395
Circle, A-1, A-5
Coefficient of a Polynomial, 263
Common Logarithm, 387
Complementary Identities, 129
Completing the Square, 53
Composition of Functions, 80
Compound Interest, 422, 425
  Compounded Continuously, 426
  Future Value, 423, 425
Confinement Theorem, 267, 313
Conic Sections, A-1
Constant Coefficient of a Polynomial, 263
Cosecant Function, 127, 154, 177
Cosine Function, 127, 154, 171
Cotangent Function, 127, 154
Coterminal Angle, 146

D

Degree Measure, 125
Dependent System of Equations, 45
Difference of Two Squares, 11
Discriminant, 55
Distance, 77, 109
Division Algorithm, 10
Domain of a Function, 75
Doubling Time, 414

E

Earthquake Magnitude, 429
Ellipse, A-1, A-5
Empty Set, 74
Equations
  Absolute Value, 350
  Exponential, 378, 401
  Involving Roots and Powers, 361
  Linear, 34
  Logarithmic, 391, 402
  Polynomial, 263
  Quadratic, 52, 248
  Radical, 358
  Rational, 304
  Linear System, 41
  Nonlinear System, 250
  Trigonometric, 211
Equilateral Triangle, 130
Even Function, 92

Exponent, 1
  Rational, 3
  Rules, 4
Exponential Equation, 378, 401
Exponential Function, 376
   Graph, 378
Exponential Growth and Decay, 411
  Doubling Time, 41
  Half-Life, 416
Exponential Inequality, 381, 401
Extraneous solution, 304

F

Factoring, 11
  Difference of Two squares, 11
  Sum and Difference of Two Cubes, 14
  Zeros and Factors, 12, 52
Function, 73
  Absolute Value, 347
  Composition, 80
  Domain, 75, 88
  Even 92
  Exponential, 376
  Graph, 88
  Inverse, 107
  Inverse Trigonometric, 198
  Linear, 245
  Logarithmic, 387
  Odd, 92
  One-To-One, 104
  Periodic, 172
  Piecewise-Defined, 76
  Polynomial, 263, 245, 263
  Quadratic, 245
  Range, 75, 88
  Rational, 303
  Sum, Difference, Product, and Quotient, 79
  Trigonometric, 127, 171
    Cosecant,127, 154, 177
    Cosine, 127, 154, 173
    Cotangent,127, 154
    Inverse Cosine Function, 201
    Inverse Sine Function, 198
    Inverse Tangent Function, 204
    Secant,127, 154
    Sine,127, 154, 172
    Tangent,127, 154, 176
  Future Value, 423, 425

G

Graph of a Function, 88
  Absolute Value Function, 347
  Cosecant Function, 177
  Cosine Function, 173
  Exponential Function, 378
  Polynomial Functions, 287
  Intercept, 90
  Inverse Function, 109
  Logarithmic Function, 389
  Polynomial Function, 287
  Rational Function, 320
  Sine Function, 172
  Tangent Function, 176

H

Half-Life, 418
Horizontal Asymptote, 320
Horizontal Line, 30
Horizontal Line Test, 104
Hyperbola, A-1, A-6
Hypotenuse, 126

I

Identities
  Addition, 184
  Basic, 128, 184
  Complementary, 129
  Double Angle, 187
  Fundamental, 128
  Half-Angle, 189
  Pythagorean, 128, 184
  Verifying, 191
Inconsistent System of Equations, 45
Independent System of Equations, 45
Index of Refraction, 139
Inequalities
  Absolute Value, 352
  Exponential, 381, 401
  Involving Rational Exponents, 362
  Linear, 36
  Logarithmic, 405
  Polynomial, 277
  Quadratic, 276
  Rational, 310
Intensity Level of Sound, 428

Intersection of Sets, 75
Interval
  Closed, 75
  Open, 75
Interval Notation, 73
Inverse Function, 107
Isosceles Triangle, 129

L

Law of Cosines, 225
Law of Sines, 225
  Ambiguous Case, 229
Leading Coefficient of a Polynomial, 263
Line, 29
  Horizontal, 33
  Parallel, 30
  Perpendicular, 30, 33
  Point-Slope Equation, 32
  Slope, 29
  Slope-Intercept Equation, 31
  Vertical, 30, 33
Linear Equation, 34
  System, 41
Linear Function, 245
Linear Inequality, 36
Local Maximum; Local Minimum, 252
Logarithmic Equation, 391, 402
Logarithmic Function, 387
Logarithmic Inequality, 392, 405
Logarithmic Properties, 393

M

Metric System, 25

N

Natural Logarithm, 387
Number of Solutions of a Polynomial Equation, 266

O

Oblique Asymptote, 320
Oblique Asymptotes of a Hyperbola, A-6
Oblique Triangle, 126
  Applications, 235
  Solving, 225
Odd Function, 92
One-To-One Function, 104

Optimization Problem, 252
  Solution Process, 253
Oriented Angle, 145

P

Parabola, 245, A-1, A-3, A-4
Partial Fraction Decomposition, 331, 334
  General Form, 332
Periodic Function 172
Perfect Square, 53
Phase Shift, 175
Piecewise-Defined Function 76
Point-Slope Equation of a Line, 32
Polynomial, 8
  Coefficients, 263
  Constant Coefficient, 263
  Degree, 263
  Equation, 263
  Function, 245
    Graphing, 287
  Inequality, 277
  Leading Coefficient, 263
Pythagorean Identity, 128, 184
Pythagorean Theorem, 127

Q

Quadrant, 146
Quadrant Angle, 155
Quadratic Equation, 52, 254
Quadratic Formula, 55
Quadratic Function, 248
Quadratic Inequality, 276

R

Radian Measure, 144
Radical Equation, 358
Radical Expression, 17
Radical Inequality, 358
Range of a Function, 75
Rational Equation, 304
Rational Expression, 15
Rational Function, 303
  Graphing, 320
Rational Inequality, 310
Rational Zeros Theorem, 265
Reference Angle, 157

Refraction, 139
Richter Scale, 429
Right Triangle, 126
   Applications, 134
   Hypotenuse, legs, 126
   Solving, 131
Roots, 2

S

Scientific Notation, 7
Secant Function, 127, 154
Set, 74
   Empty, 74
   Intersection, 75
   Union, 75
Similar Triangles, 126
Simple Harmonic Motion, 182
Sine Function, 127, 154, 172
Slope of a Line, 29
Slope-Intercept Equation of a Line, 31
Snell's Law, 139
Sound Intensity, 427
Systems of Equations
   Dependent, Inconsistent, Independent, 45
   Linear, 41
   Nonlinear, 250

T

Tangent Function, 127, 154, 176
Triangle, 126
   Equilateral, 130
   Isosceles, 129
   Oblique, 126
   Reference, 129
   Right, 126
   Similar, 126
Triangle Inequality, 348
Trigonometric Equation, 211
Trigonometric Function, 127, 154, 171
Trigonometric Identities, 184
   Double Angle, 187
   Half Angle, 189
   Pythagorean, 184
   Sum and Difference, 184
   Verifying, 191

U

Understood Domain of a Function, 75
Union of sets, 75
Unit Conversion, 23

V

Vertex of a Parabola, 246, A-3, A-4
Vertical Angles, 134
Vertical Asymptote, 321
Vertical Line, 30
Vertical Line Test, 93
Vertices
   Of a Parabola, 246, A-3, A-4
   Of a Hyperbola, A-6
   Of an Ellipse, A-5

X

x-intercept, 90

Y

y-intercept, 90

Z

Zero of a Polynomial, 12, 52

Made in the USA
Middletown, DE
28 August 2020